T5-AOU-626

Modeling Methods *for* Environmental Engineers

Modeling Methods *for* Environmental Engineers

Isam Mohammed Abdel-Magid, Ph.D.
Sultan Qaboos University
Sultanate of Oman

Abdel-Wahid Hago Mohammed, Ph.D.
Sultan Qaboos University
Sultanate of Oman

Donald R. Rowe, Ph.D.
D.R. Rowe Engineering Services
Bowling Green, Kentucky, USA

LEWIS PUBLISHERS
Boca Raton New York London Tokyo

Acquiring Editor:	Joel Stein
Project Editor:	Andrea Demby
Marketing Manager:	Greg Daurelle
Direct Marketing Manager:	Becky McEldowney
Cover design:	Denise Craig
PrePress:	Greg Cuciak
Manufacturing:	Sheri Schwartz

Library of Congress Cataloging-in-Publication Data

'Abd al-Majīd, 'Isām Muhammad.
Modeling methods for environmental engineers / Isam Mohammed Abdel -Magid, Abdel-Wahid Hago Mohammed, Donald R. Rowe.
p. cm.
Includes bibliographical references and index.
ISBN 1-56670-172-4 (alk. paper)
1. Pollution control equipment--Design and construction--Data processing. 2. Pollution control equipment--Design and construction--Computer programs. I. Mohammed, Abdel-Wahid Hago. II. Rowe, Donald R. III. Title.
TD192.A25 1996
628--dc20 96-4980
CIP

Direct all inquiries to CRC Press, Inc., 2000 Corporate Blvd. N.W., Boca Raton, Florida 33431.

Lewis Publishers is an imprint of CRC Press.

International Standard Book Number 1-56670-172-4
Library of Congress Card Number 96-4980
Printed in the United States of America 1 2 3 4 5 6 7 8 9 0
Printed on acid-free paper

Preface

Environmental engineering is that branch of engineering dealing with activities in the areas of air, water, wastewater, solid waste, radiation, and hazardous materials. Many textbooks have been published in all of these areas; however, most of them lack the fundamental utilization and application of computer programming. Computer programming is now of paramount importance not only for the students, but also for the practicing engineers. This book presents computer programs in the air, water, and wastewater fields that can be used to evaluate and assist in the design of environmental control systems. This book will be of value to environmental engineers, sanitary engineers, civil engineers, town planners, industrial engineers, agricultural engineers, treatment plant designers, consultation firms, and researchers. The BASIC programming language was selected for its simplicity and used throughout this book.

The broad objectives of this book are to facilitate improving teaching, learning, and research in the application of computer programming based on well-established models, equations, formulas, and procedures used in the environmental engineering field.

The more specific objectives are as follows:

1. To help the reader in the design of treatment and disposal facilities
2. To introduce the basic concepts of computer design
3. To aid in the teaching of environmental engineering unit operations and processes design
4. To improve the understanding by students and researchers in the environmental engineering field

This book contains many illustrative figures, schematic diagrams, solved practical examples, design problems, and tutorial homework problems. A wide range and spectrum of environmental engineering subjects are covered. Many computer programs are included and they cover both the environmental design principles and the structural design parameters using different materials. This book is written in a clear, concise manner in order to enhance understandability. SI units are used throughout this book.

The format of the book is divided into four sections. The first section deals with introductory material that contains functions and layout of the computer, programming and modeling concepts, and the basic outlines for the BASIC language.

The second section covers water sources and resources and is divided into three chapters. Chapter Two deals with characteristics of water and wastewater, supported by mathematical equations in order to help programming, for water resources and sources, and groundwater flow. Emphasis is placed on water abstraction from wells with the consequent drawdown relationships. Chapter Three discusses water treatment unit operations/processes and water storage and distribution. This chapter outlines the design concepts adopted in treatment plant unit operations (e.g., screens, sedimentation and flotation, and aeration) and treatment unit processes (e.g., coagulation and flocculation, filtration, desalinization, and water stabilization). Each treatment process is briefly introduced and emphasis is given to the design of the units using appropriate equations, assumptions, and design tools. Chapter Four deals with water storage and distribution. This chapter highlights design of water storage tanks (communal, dwelling) and the methods used for both urban and rural water distribution.

The third section of this book deals with wastewater engineering, measurement, collection, treatment units, and disposal. This section is divided in two main chapters, with Chapter Five focused on wastewater collection systems for rural areas and urban fringes. Sewerage system design is introduced within the context of this chapter. Chapter Five also discusses selected methods for wastewater disposal for rural inhabitants. Chapter Six is introduced to aid design of wastewater treatment and disposal units. More stress is placed on indepth design considerations and parameters for selected systems universally adopted for both rural and urban regions. This chapter emphasizes preliminary, primary, tertiary, and sludge treatment methods and the disposal of treated wastewater to rivers, lakes, or oceans.

The fourth section of this book deals with air pollution control technology. The first section presents a brief overview of the air pollution field followed by fundamental concepts that are necessary for calculations dealing with air pollution control. Also included are computer programs based on mathematical equations and models relating to these fundamental concepts. The most commonly used air pollution control devices for gaseous and particulate pollutants are also presented (e.g., absorption, adsorption and combustion for gaseous contaminants and settling chambers, cyclones, electrostatic precipitators, venturi scrubbers, and baghouse filters for controlling particulate emissions). Each control device or technique is accompanied by a computer program that can aid in the design or evaluation of air pollution control equipment. This section concludes with mathematical models that can be used for determination of effective stack heights (plume rise), as well as dispersion models that can help estimate the concentrations of air pollutants dispersed into the atmosphere. Computer programs for both stack height calculations and dispersion models are included. All computer programs in the book are included on diskettes.

Isam Mohammed Abdel-Magid
Abdel-Wahid Hago Mohammed
Donald R. Rowe

Authors

Isam Mohammed Abdel-Magid, Ph.D., received his B.Sc. degree in Civil Engineering from Khartoum University, Sudan; Diploma in Hydrology from Padova University, Italy; M.Sc. in Sanitary Engineering from Delft University of Technology, The Netherlands, and Ph.D. in Public Health Engineering from Strathclyde University, U. K. in 1977, 1978, 1979, and 1982 respectively.

Dr. Abdel-Magid has authored over 28 publications, 10 scientific textbooks, and numerous technical reports and lecture notes dealing with water supply and resources; wastewater treatment, disposal, reclamation, and reuse; industrial wastes; and slow sand filtration, both in English and Arabic. He has participated in several workshops, symposia, seminars, and conferences. He has edited and co-edited many conference proceedings and college bulletins.

The Sudan Engineering Society awarded him the prize for the Best Project in Civil Engineering in 1977. In 1986, the Khartoum University Press (KUP) awarded him an Honourly Scarf & Badge for Enrichment of Knowledge.

Dr. Abdel-Magid has taught at Khartoum University, Sudan; United Arab Emirates University, United Arab Emirates; Sultan Qaboos University, Oman; and Omdurman Islamic University, Sudan.

Abdel-Wahid Hago Mohammed, Ph.D., received his B.Sc. degree in Civil Engineering from Khartoum University, Sudan in 1976. He received his Ph.D. degree from Glasgow University, Scotland in 1982. Dr. Hago has authored over 13 publications, numerous reports, lecture notes, and over 95 translated articles. He headed the Design Office of Newtech Industrial and Engineering Group, Sudan for three years, where he contributed to the design of buildings and factories in different parts of that country. In 1986, he acted as a Senior Analyst for Associated Consultants and Partners, Khartoum, Sudan for the design of the ElObied town water distribution system.

Dr. Hago has participated in several conferences and was a member of the organizing committee of the First National Conference on Technology of Buildings organized by the Sudan Engineering Society in 1984. Dr. Hago is a member of the Sudan Engineering Society. He has taught both undergraduate and postgraduate courses at Khartoum University and Sultan Qaboos University. He has supervised and co-supervised many M.Sc. and Ph.D. students.

Donald R. Rowe, Ph.D., is President of D. R. Rowe Engineering Services, Inc., Bowling Green, Kentucky.

He received his M.S. (1962) and Ph.D. (1965) from the University of Texas at Austin, and his B.S. degree (1948) in Civil Engineering from the University of Saskatchewan.

From 1964 to 1969, he was Associate Professor, Department of Civil Engineering, Tulane University; from 1969 to 1981, he was Associate Professor, Department of Engineering Technology, Western Kentucky University; and from 1981 to 1988, he was Professor, Department of Civil Engineering, King Saud University, Saudi Arabia. From 1971 to 1993, he was Vice-President of the Larox Research Corporation, and from 1990 to the present, he has been President of his own company.

Dr. Rowe has authored or co-authored over 70 reports and publications in the areas of air pollution, solid waste, water treatment and wastewater reclamation and reuse. He also co-authored *Environmental Engineering* (McGraw-Hill Publishers) and *The Handbook of Wastewater Reclamation and Reuse* (CRC Press/Lewis Publishers).

Dr. Rowe holds, with co-inventors, over 10 patents on catalytic conversion removal processes of air contaminants in the United States, Canada, Great Britain, and Japan.

In 1980, he was recipient of a Fulbright-Hays Award to Ege University in Izmir, Turkey. Dr. Rowe also carried out research on wastewater reclamation and reuse while at King Saud University.

From 1984 to the present, he has served on the Peer Review Evaluation Committee for research projects funded by King Abdulaziz City for Science and Technology (KACST), Riyadh, Saudi Arabia.

Acknowledgments

The authors wish to gratefully acknowledge the support, patience, encouragement, and understanding of their families while the authors were engaged in the enjoyable task of writing this book.

CRC Press/Lewis Publishers editors are acknowledged for their forebearance and patience during all stages of the manuscript preparation, review, editing, and publishing. Particular thanks and appreciation go to Andrea Demby, Jan Boyle, and Victoria Held for their real contributions and outstanding efforts to bring this book to existence.

The authors are indebted to a number of institutions, organizations, and individuals for permission to reproduce and draw published material, and these are acknowledged in the references. Thanks are also extended to the reviewers of the manuscript for their helpful comments and numerous constructive suggestions.

The authors extend their thanks to all who contributed in different ways towards the production of this book. In the air pollution control section, special acknowledgment goes to Asa Raymond for typing, reviewing, and developing the computer programs. Also, a special thanks to Gary Whittle for his significant contribution in locating pertinent literature in the air pollution control field. Special thanks are also extended to Majid Soud Rashid Al-Khanjri and Taiseer Marhoon Al-Riyami for redrawing and tracing Figures 7.16, 7.17, 7.32, and 7.33.

The authors trust that the book will serve the needs of the profession and provide help and stimulation for candidates developing their own computer models. The authors invite and welcome any comments, suggestions, improvements, and constructive criticism from users of the text. Although diligent efforts have been made by the authors, reviewers, editors, and publisher to produce an errorfree book, some errors may have escaped the scrutiny of all. The authors would appreciate if any omissions or errors found be brought to their attention so that corrections can be made which will enhance the usefulness of this publication.

Dedications

To all my former teachers and professors
I. M. Abdel-Magid

To all knowledge seekers
A. W. Hago

To Amjad and Nasra
D. R. Rowe

Abbreviations and Symbols

Symbol	Description	Unit
A	Area	m^2
AI	Aggressiveness index	
Alk	Total alkalinity	mg/L $CaCO_3$
Ast	Area of steel needed	m^2
A_i	Area of heat exchanger number i	m^2
A_n	Area of individual unit	m^2
A_t	Total area	m^2
a	Constant, coefficient, factor, fraction	
a′	Effective sulfide flux coefficient	m/hr
a_a	Percentage open area for clogged screen	%
a_c	Cell constant	
a_o	Percentage open area for a clean screen	%
B	Width, thickness	m
B_c	Width of cyclone	m
BOD	Biochemical oxygen demand	mg/L
BOD_e	Effective biochemical oxygen demand	mg/L
BOD_{load}	BOD load	kg/m^3
BOD_s	BOD of standard sewage	$gBOD_5/c/d$
BOD_t	BOD that has been exerted in time interval t	mg/L
BOD_5	5-d BOD of wastewater	mg/L
b′	Constant	
b	Slope of the straight line of t/V vs. V	s/m^6

Symbol	Description	Unit
C′	Solubility of oxygen at barometric pressure P and given temperature	mg/L
C	Concentration	mg/L
CA	Cell age	d
CON	Conductance	ohm^{-1}
C_B	Breakthrough capacity	fractional
C_D	Newton's drag coefficient	dimensionless
C_e	Concentration in effluent	mg/L, g/m^3
C_g	Gas concentration in gas phase	g/m^3
C_f	Friction coefficient, coefficient of DeChezy	$m^{0.5}/s$
C_m	Concentration of the pollutant in mixture of river water and wastewater discharge, or concentration of pollutant in the river downstream of point of discharge	mg/L
C_0	Concentration at time t = 0 (initial)	mg/L
C_s	Saturation concentration	mg/L
C_r	Concentration of same pollutant in river upstream discharge point	mg/L
C_v	Volumetric concentration of particles (volume of particles divided by the total volume of the suspension)	dimensionless
C_w	Concentration of pollutant in wastewater	kg/m^3
c	Coefficient, constant	
ch	Chlorinity	g/kg
D	Diameter	m
D	Flexural rigidity	KNm^2
Diff	Molecular diffusion coefficient	m^3/s
D_a	Actual diameter of tank	m
d_g	Geometric mean diameter of particle	m
D_m	Mean hydraulic depth	ft
DO_c	Critical oxygen deficit	mg/L
D_r	Dielectric constant	
DO_0	Initial oxygen deficit at the point of waste discharge, at time t = 0	mg/L
DO_t	Oxygen deficit at time t	mg/L
DR	Dilution rate	dimensionless

Symbol	Description	Unit
DWF	Dry weather flow	L/d
d	Depth, diameter	m
dP	Differential change in pressure	Pa
du/dy	Velocity gradient (rate of angular deformation, rate of shear)	s^{-1}
dV	Differential change in volume	m^3
d_{max}	Maximum depth	m
d_p	Distance between blades or the scroll pitch	m
d′	Diameter of truncated cone (hopper floor)	m
E_b	Bulk modulus	$N*m^{-2}$
E_v	Rate of evaporation	L/d
E′	Activation energy	calories
E	Coefficient of eddy diffusion, or turbulent mixing	m^2/s
E	Young's modulus of the material, sorption efficiency	
Eff	Efficiency	%
EC	Electrical conductivity	ohm/m
E_e	Electrochemical equivalent	g/Coulomb
EM	Electrophoretic mobility	m/s/V/m
e_e	Expanded bed porosity	dimensionless
e	Porosity	dimensionless
Far	Faraday's constant	A*h
F/M	Food-to-microorganisms ratio	d^{-1}
Fr	Froude number	dimensionless
F(d)	Frequency of particle occurrence of diameter d	
F_τ	Shear force	N
F_1,F_2	Recirculation factor	dimensionless
f	Factor, function, coefficient	
f_r	Contaminated gas flow rate	mL/s
f_{cu}	Concrete strength	N/m^2
f_{st}	Design steel stress	N/m2
f_Y	Yield strength of steel	N/m^2
G	Velocity gradient	s^{-1}

Symbol	Description	Unit
G_1	Gas flow	kg*mol/h
g	Gravitational acceleration	m/s^2
H	Total energy (total head, energy head)	m
Hard	Hardness	mg equivalent $CaCO_3$/L
HCa	Calcium hardness	mg/L $CaCO_3$
HE_i	Heat exchanged in effect number i	J
H_c	Height of inlet duct	m
H_0	Initial resistance for a clean screen	m
H_s	Resistance for a clogged screen	m of water
$[H^+]$	Concentration of hydrogen ions	mol/L
h	Height, depth, thickness	m
h_f	Friction head	m
h_l	Head loss	m
h_t	Height of proposed tank	m
h'	Depth of truncated cone	m
I	Mean rainfall intensity for a duration equal to the time of concentration	mm/h
I_r	Average infiltration into the sewer owing to poor joints or pervious material	L/d
i	Current	amp
i	Current density	A/cm^2
j	Slope, gradient	m/m
K	Constant, factor, coefficient	
K_b	Bunsen absorption coefficient	g/J
K_D	Distribution coefficient	
K_s	Half-velocity constant	mg/L
k	Coefficient, constant	
k_N	Efficiency coefficient of step N of a cascade	%
k_H	Henry's constant	g/m^3*Pa (g/J)
k_l	Loss coefficient	
k_n	Kinetic coefficient	dimensionless

Symbol	Description	Unit
k_p	Removal rate constant for waste stabilization pond	d^{-1}
k_t	Decay constant for the particular decay reaction	d^{-1}
k''	Reaeration constant	d^{-1}
k'	First-order reaction rate constant	d^{-1}
k_1	Bacterial–die away rate	d^{-1}
LI	Langelier index	
L_b	BOD_5^{20} of benthal deposit	g/kg volatile matter
L_c	Internal length of first compartment of septic tank, length of vertical cylinder	m
L_e	Effluent BOD	mg/L
L_i	Influent BOD	mg/L
L_m	Maximum daily benthal oxygen demand	g/m^2
L_0	BOD remaining at time $t = 0$ (total or ultimate first stage BOD initially present)	mg/L
L_t	Amount of first stage BOD remaining in the sample at time t	mg/L
L_1	Liquid flow	kg*mol/h
l	Length, depth	m
l_e	Expanded bed depth	m
l_e	Equivalent pipe length	m
M	Moment	N*m
MCRT	Mean cell residence time	d
MLVSS	Mixed liquor volatile suspended solids	mg/L
MTZ	Mass transfer zone	m
MW	Molecular weight of the gas	g.mol
m	Mass	kg
m.c.	Moisture content	%, fractional
N	Number	dimensionless
N_e	Effluent bacterial number, number of bacteria	/100 mL
N_i	Influent bacterial number, number of bacteria	/100mL
N_0	Number of viable microorganisms of one type at time $t = 0$	/100mL
n	Roughness factor, Manning and Kutter factor	$m^{1/6}$

Symbol	Description	Unit
n	Number	dimensionless
OL	Organic loading rate of the sewage effluent	g/L
P	Pressure	mmHg, N/m^2 (Pa)
PE	Population equivalent	dimensionless
PF	Packing factor	fractional
POP	Number of people served by the sewer	dimensionless
POW	Power	J/s, W
ppb	Parts per billion	
pphm	Parts per hundred million	
ppm	Part per million	
P_g	Partial pressure of the respective gas in the gas phase	Pa, atm
P_{osm}	Osmotic pressure	atm
pAlk	Negative logarithm of total alkalinity	eq. $CaCO_3$/L
pCa^{++}	Negative logarithm of calcium ion	moles/L
pf	Peaking factor	dimensionless
pH	Negative logarithm of the hydrogen ion concentration	
pH_a	Actual (measured) pH of the water	
pH_s	Saturation pH	
pOH	Negative logarithm of the hydroxyl ion concentration	
p_w	Pressure of saturated water vapor	mmHg, N/m^2
Q	Rate of flow, fire demand	m^3/s
Q_a	Average flow	m^3/s
Q_I	Total inflow volume during a specified period	m^3/s
Q_{max}	Maximum flow rate	m^3/s
Q_{min}	Minimum wastewater flow	m^3/d
Q_o	Total outflow volume during a specified period	m^3/s
Q_p	Peak flow	m^3/s
Q_R	Recirculated flow	m^3/s
Q_w	Waste sludge flow	m^3/d
q	Flow that can be applied per unit area	L/m^2*d

Symbol	Description	Unit
q^+	Charge on the colloid	Coulomb
R	Universal gas constant	J/K*mole, L*atm/mole*K, J/kg*K
Re	Reynolds number	dimensionless
RES	Resistance of a conductor	Ω
RES_s	Specific resistance of the suspension	Ω*cm
RI	Ryzner index	
R_m	Resistance of filter medium	m^{-1}
R_{mr}	Rate of corrosion	
R_s	Specific resistance	Ω^{-1}
R_T	Total removal	%
R_u	Recycling ratio	dimensionless
r	Radius	m
r_D	Rate of deoxygenation	
r_H	Hydraulic radius	m
r_r	Rate of reaeration	
r_s	Specific resistance of sludge cake	m/kg
r_s'	Constant	
S	Salinity correction term	
SA	Sludge age	d
SDI	Sludge density index	g/mL
SLR	Sludge loading rate	d^{-1}
SRT	Solids retention time	m
SS	Suspended solids	mg/L
SVI	Sludge volume index	mL/g
STP	Standard temperature and pressure	
S_c	Specific capacity of well	m^3/d*m
S_e	Effective residual drag	
S_u	Sulfide concentration	mg/L*h
S_{udis}	Dissolved sulfide concentration in wastewater	mg/L
sa	Salinity	g/kg
s.g.	Specific gravity of settling particle	dimensionless
s_c	Space around catalyst	mL

Symbol	Description	Unit
s*	Growth-limiting substrate concentration in solution	mg/L
T	Temperature	degrees C, F, K
TDS	Total dissolved solids	mg/L
TOR	Net torque input	dyne/cm
TR	Transmissibility of aquifer	$m^3/d*m$
T_c	Temperature correction factor	dimensionless
T_w	Average trade-waste discharge	L/d
t	Time, detention time	d
t_c	Critical time	d
t_w	Maximum shear at pipe wall (wall shear stress)	N/m^2
$t_{1/2}$	Half life of the particular nuclide	s, d
t_{25}	Time required for water surface to fall 25 mm	s
U	Velocity at free plate surface	ms^{-1}
u_a	Average velocity	m/s
u_i	Heat transfer coefficient for heat exchanger number i	J
u	Mean wind speed	m/s
V	Volume	m^3
VOL	Volumetric organic loading rate	kg/m^3*s
VS	Concentration of volatile solids	kg/m^3
V_g	Volumetric gas production rate (specific yield)	m^3 gas/m^3 digester/d
V_s	Settled volume of sludge in a 1000-mL graduated cylinder in 30 minutes	mL/L, %
v	Velocity	m/s
v′	Hindered settling velocity	m/s
v_a	Approach, actual velocity	m/s
v_c	Centerline velocity	m/s
v_f	Filtration velocity, gas velocity of filter surface	m/s
v_H	Velocity of horizontal water movement	m/s
v_r	Rotational velocity of the bowl	rad/s
v_s	Settling rate, displacement velocity, loading rate, overflow rate	m/s

Symbol	Description	Unit
v_{sc}	Self-cleansing velocity	m/s
v_{so}	Design settling velocity	m/s
v_{sco}	Scour velocity	m/s
v_w	The difference in the rotational velocity between the bowl and the conveyor	rad/s
v_z	Updraft velocity	m/s
W	Organic loading of the trickling filter	g BOD/d
WAE	Weight of adsorbate	kg
WAT	Weight of adsorbent	kg
Wt	Weight	kg
W	Loading rate	kg/d
W_c	Working capacity of adsorbent	fractional
w	Lateral deflection of tank, drift velocity	
w_p	Wetted perimeter	m
X/M	Mass X of element or contaminant adsorbed from solution per unit mass of adsorbent M	
X_A	Mole fraction of pollutant in liquid phase into column	
X_B	Mole fraction of pollutant in liquid phase out of column	
X_T	Total removal	%
x_g	Mole fraction of gas	dimensionless
x_c	Critical distance	m
x	Distance, drawdown	m
Y	Distance from fixed plate	m
Y_A	Mole fraction of pollutant in gas phase out of column	
Y_B	Mole fraction of pollutant in gas phase into column	
Y_t	Ultimate gas yield	m^3 gas/kg VS added
y	Distance, depth, width	m
Z	Depth, elevation, lever arm	m
ZP	Zeta potential	V

Symbol	Description	Unit
Z_c	Length of cone section	m
α	Constant	
β	Beta function for a centrifuge, constant	
σ	Surface tension	N/m
σ_g	Standard geometric deviation of data	
σ_y, σ_z	Dispersion coefficient	m
σ_{ye}	Adjusted lateral dispersion coefficient	m
σ_{ze}	Adjusted vertical dispersion coefficient	m
Φ	Angle	°
φ	Particle shape factor	dimensionless
φ_{sw}	Hydrogen sulfide flux to pipe wall	g/m^2*h
γ	Specific weight	N/m^3
η	Fractional collection efficiency	%, fractional
τ	Shear stress	$N*m^{-2}$
Σ	Parameter related to the characteristics of centrifuge	
ΣF	Sum of all forces acting on fluid contained between two cross sections	N
ρ	Density	kg/m^3
δ	Thickness of laminar layer	m
Ψ	Rate of deposition	m^2/s
Ψ	Interial impaction parameter	
ν	Kinematic viscosity	m^2*s^{-1}
ζ	Eddy viscosity	$N*s/m^2$
μ	Dynamic (absolute) viscosity	$N*s/m^2$
ξ	Ionic strength	
q	Temperature coefficient	dimensionless
ε	Roughness of the pipe wall	m
λ_o	Filtration coefficient	m^{-1}
ω	Impeller rotations	number of rotations/min
χ	Constant, ground level concentration	
μ_s	Growth rate of microorganisms	d^{-1}
$(\mu_s)_{max}$	Maximum specific growth rate of microorganisms	d^{-1}

Symbol	Description	Unit
υ	Specific volume of the fluid	m^3/kg
ΔH	Heat of sorption	
Δh	Distance-dependent phase rise	m
ΔT	Change in temperature	°C
ΔS	Change in storage volume during specified interval	
ΔP	Pressure difference	Pa
ν	Poission's ratio	dimensionless

Contents

Contents

Section I
Introduction

Chapter 1

Programming Concepts

Contents

This chapter presents an introduction to the basic functions and layout of a computer. Computer programming languages are briefly outlined and summarized, with emphasis being placed on the BASIC language.

The management of water resources, wastewater collection, treatment, and disposal, as well as air pollution control, involves complex processes. This is due to many interacting parameters, some of which are difficult to present in a straightforward mathematical model, equation, or formula. Mathematical modeling techniques can be used to aid in predicting the quality and sequence of relationships that can help management in its effort to solve potential environmental problems. A model can be regarded as an assembly of concepts in the form of one or more mathematical equations that approximate the behavior of a natural system or phenomena.[1] Models can be divided into simulation, optimization, and computer-aided design.[2] Simulation models address the formulation of a mathematical model that simulates a specific situation, with the formulation of a mathematical relationship and solution through a structured and valid process. Optimization models use mathematical techniques to achieve a reasonable solution from a range of possibilities. Computer-aided design models help in the preparation of design drawings and computation of quantities.

Modeling methods rely on computers for their validation, testing, and speedy and efficient use. The basic concepts of mathematical modeling and formulation of a well written operative computer program form the central theme of this book. The aim is to train individuals to write, develop, validate, and use their own environmental engineering control programs.

1.1 Components of a Computer

A computer is a machine that can perform complex, long, and repetitive sequences of operations at very high speeds.[3] It captures data in the form of instructions or electronic signals, processes it, and then supplies the results in the form of information or electronic signals to control some other device or process.[4] A typical personal computer (PC) includes five hardware components: a processor (central processing unit, CPU), disk drives, full-sized keyboard, video display device, and an output device (such as a printer), in addition to other peripherals.

The main components of a personal computer system including both hardware and software are (see Figures 1.1, 1.2, and 1.3)

1. **The central processing unit (CPU)** — This component represents the brain or the control center of the computer. It carries out the sequence of operations specified by a program, and it controls all computer activities. The functions of the CPU can be divided as follows:[5,6]
   a. The control unit controls electronic signals passing through the computer, and it monitors the operation of the whole system.
   b. The arithmetic/logical unit executes arithmetic and logical operations.
   c. The primary memory is volatile; that is, when the computer is turned off the information is lost (also known as main store, random access memory or RAM, read-write memory). It is used by the processor as a temporary information storage device. The information can be an input to the system or it may be generated from internal calculations. The RAM also is used to store and retrieve computer programs. The basic measurement unit of computer memory is denoted as a byte (corresponds to eight bits, a bit being the smallest unit of information handled by a computer and either 1 or 0 in the binary number system). The immutable memory is called read-only memory (ROM) and contains the basic input/output system (BIOS). The BIOS controls the major I/O devices in the system. ROM is not volatile memory, it is built into the computer, and this information is retained even when the computer is off.
2. **Input devices** — There are several programmable input devices such as the keyboard, light pen, joystick, mouse, and digitizer. These input devices function much the same as human sensory organs. The keyboard is the primary input device of the computer for entering commands and data. It transfers characters to the computer store by the depression of keys. A typical keyboard has 83 keys that can generate 128 characters in ASCII (the American Standard Code for Information Interchange), as well as special symbols and shapes. In total, the keyboard can generate 256 characters, symbols and shapes. The light pen is a pointing device that can be programmed to sense a location on the screen and then activate some function.[7] The joystick or the mouse is a positioning device that locates the cursor and activates a function through the use of its

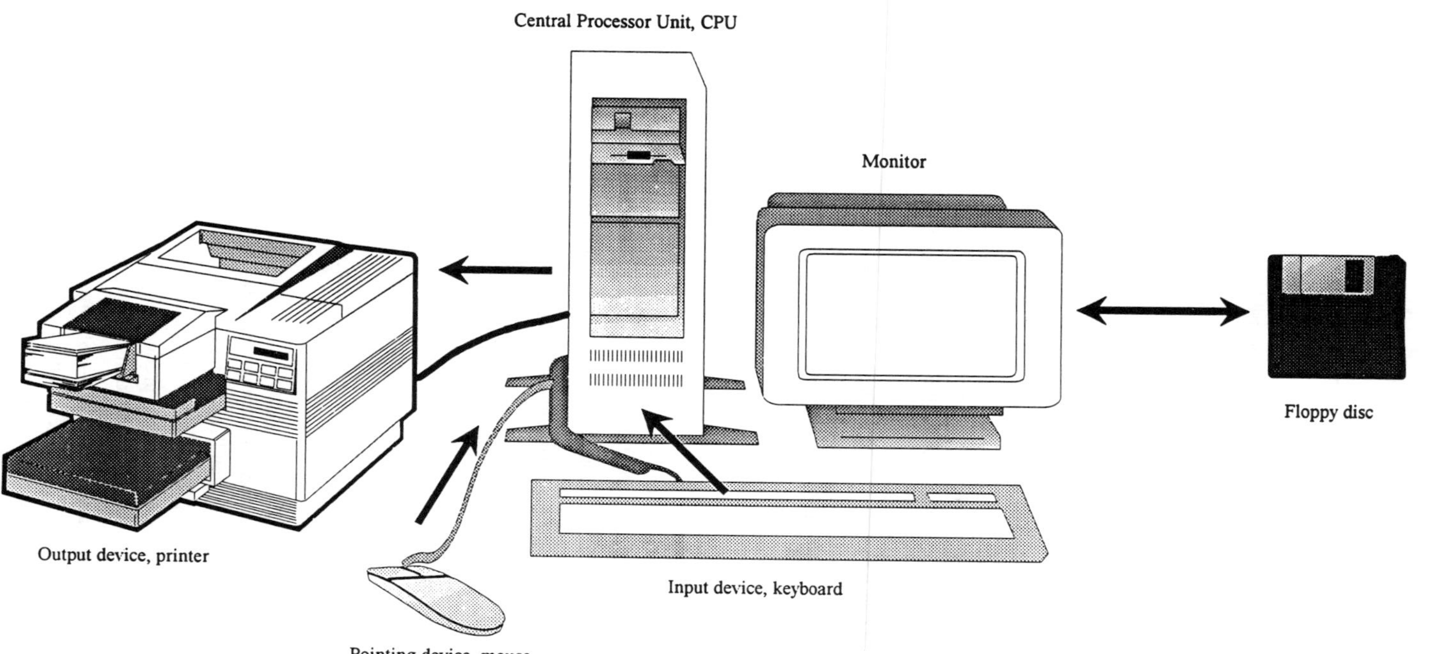

Figure 1.1
The components of the computer.

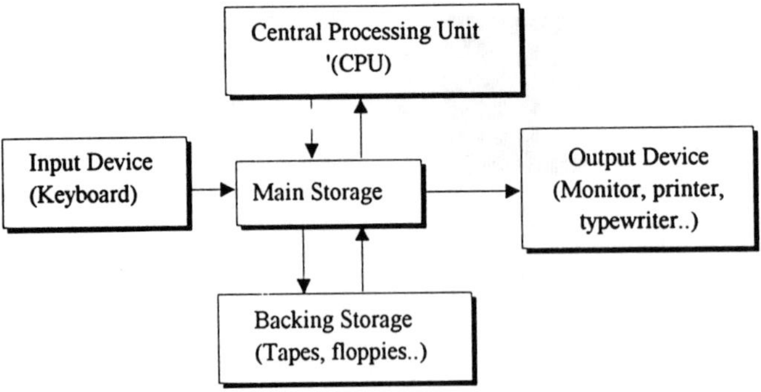

Figure 1.2
Components of a computer.

buttons. The mouse is a small hand-held device that sends signals to the computer. A digitizer inputs data and is also used for pointing, picking, and giving commands and for tracing an irregular sketch.

3. **Output device such as a monitor (visual display, display screen)** — This unit permits the display of input information as entered and allows the display of the output produced by the computer. Many types of monitors are found on the market, such as monochrome or color video monitors. The monitor is driven by a graphics board inside the computer. There are many manufacturers of graphic boards: Hercules, IBM color graphics adapter (CGA), IBM enhanced graphics adapter (EGA), or IBM video graphics adapter (VGA).[8]

4. **Printers and plotters (output device)** — A printer allows the generation of a permanent record or hard copy of the output. Printers can be broadly classified according to:
   - Mechanisms of character formation:
     - Dot-matrix printers, which form individual characters by using a series of dots one line at a time.
     - Solid font mechanisms, which produce fully formed characters due to formation of sharp images. Solid font mechanisms are to be preferred to dot-matrix printers when letter-quality output is needed.
   - Method of transfer or print to page:
     - Impact printers, which form characters by striking the type against an inked ribbon that presses the image onto the paper.
     - Nonimpact printers, which do not involve physical contact but instead transfer the image to the paper using ink spray, heat, xerography, or laser. Nonimpact printers are usually quieter and faster than impact printers.

   A plotter allows the sequential drawing of objects rather than using the raster-scan principle adopted for printers. The plotter produces professional quality drawings on different paper sizes and allows the production of final, multiple-colored drawings.

5. **Data storage media (software, filestore)** — Storage or programs and data can be made on disks, magnetic tapes, drums, cassettes, and cartridges. A disk is an auxiliary storage device. Generally, external memory media, such as magnetic disks, are used to store information. They may be divided into:

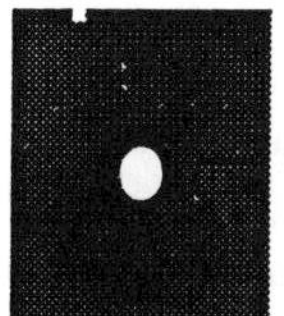
5.25" Floppy

CD-ROM disc

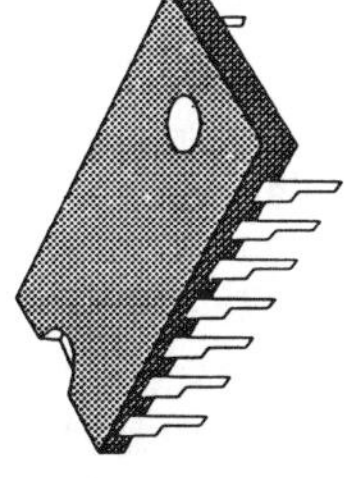
Computer chip

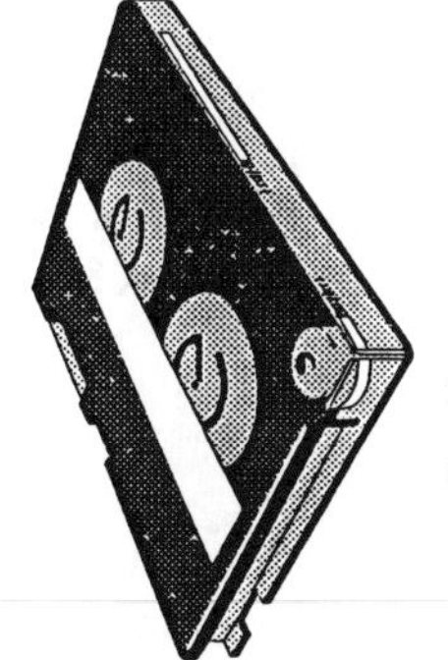
Cartridge tape

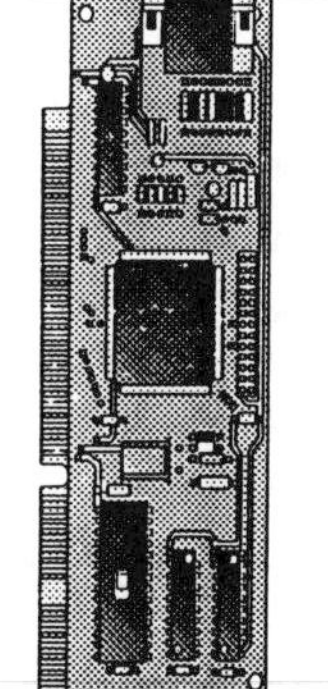
Circuit board

Figure 1.3
Computer items.

- **Floppy disks (diskettes)** — These portable media are made of flexible plastics that can be magnetized. Data is stored on the disk as magnetized spots or concentric tracks on its magnetic coating by the head of the disk drive.
- **Hard disks (Winchester disks)** — These are made of inflexible metal and are installed and fixed in the computer. They are more reliable, with larger storage capacities and faster access times than the floppies.
- **Magnetic tapes** — These are secondary permanent storage devices usually used as a mass storage medium for archive or backup.

Figure 1.2 shows the main components of a computer. The inside of the computer contains the main board, power supply, adapter cards, disk drives, processor, RAM, ROM, math co-processor, and support chips. For effective, trouble-free operation, routine corrective and preventive maintenance should be performed regularly for the computer. Computer failures can be caused by excessive heat, dust buildup, noise interference, power-line problems, corrosion, and the presence of magnetic flux.[9]

1.2 Programming and Languages

A program is an orderly collection of instructions to the computer to perform tasks. Programs collectively available to the computer are termed software. Software is needed to direct the computer to perform required tasks. Software can be divided into:

Operating system. This is a group of programs that efficiently control the operation of the computer. The operating system permanently resides in the computer memory, continuously supervising and coordinating other programs running in the machine and monitoring activities. Many operating systems have been, and will continue to be, developed, such as Apple-DOS, TRS-DOS, Microsoft Disk Operating System (MS-DOS), control program/microprocessors (CP/M) from Digital Research, XENIX of Microsoft Corporation, and UNIX, developed by Bell Laboratories.

Utility programs. These programs are designed by manufacturers or software companies to perform specific tasks. Examples of utility programs include: text editors, debugging aids, disk formatters, copiers, sorters, and data file mergers.

Language processor. This is a program that translates a user source program to machine language understood by the computer. The machine language is a collection of very detailed, cryptic instructions that control the internal circuitry of the computer.[7] Programs written for the computer can be divided into:

- Binary code, which represents a sequence of instructions and operands in binary, listed as they appear in the memory of the computer.
- Octal or hexadecimal code, which signifies translation to octal or hexadecimal representation.
- Symbolic code, which uses symbols for instructions. Each instruction can be translated into binary code by a certain program termed assembler.
- High-level programming language (HIL) addresses procedures used in problem solving. The instruction set of the language is more compatible with human languages and human thought processes.[7] This results in simplicity, uniformity, and machine independence. There are various programming languages such as: Beginner's All-Purpose Symbolic Instruction Code (BASIC), Formula Translation (FORTRAN),

Pascal (developed by Nicklaus Wirth and named after Blaise Pascal), COBOL (Common Business-Oriented Language), C (a general purpose, portable programming language featuring concise expressions and a design that permits well-structured programs; originally developed at Bell Laboratories as a language for writing systems software)[5], etc. High-level programs need to be translated to machine language through compilation or interpretation programming processes carried out by the computer.

- Application programs are designed to solve certain settings and may be used in various applications. Examples of these programs include word processing packages, mathematical and graphical packages (e.g., QuatroPro or Excel spreadsheets), data base management programs, AutoCAD from Autodesk, Inc., and others.

Writing programs has many benefits over packaged programs. These advantages include:

- Solving specific problems
- Speed, performance and interpretation of calculations, compared to manual interpretation of data
- Enhancing graphic work
- Formulating cost-effective design and operation of products
- Computer-aided design and manufacturing
- Performing preliminary testing
- Controlling manufacturing and assembly operations
- Control and automation of processes
- Effective business and management operations
- Communicating technical results
- Educational and training programs

Once a program has been written, it needs to be verified for its accuracy through the use of specific data. This results in a cycle of checking and correcting the program to obtain required accuracy and specifications. This process is called debugging, and it is carried out by running the program on a computer with suitable test data to eliminate all compile-time, run-time, and logical errors. Figure 1.4 shows the steps to be followed in order to obtain an error-free program.

High level programming languages are machine independent. The features considered essential to the success of a program include simplicity of the algorithm or technique, program clarity, and efficiency in executing the program. Some of the most commonly used high-level languages are described briefly below.

1. FORTRAN (or FORmula TRANslation) uses a notation that facilitates writing of mathematical formulae. The language is useful especially when dealing with large amounts of data such as in the case of business-oriented work. The language is excellent for sophisticated computations and is fast and precise with relatively powerful input/output capabilities. FORTRAN programs execute quite rapidly because the language is almost a compiled one. In program writing, a number of commands are used throughout the different stages of the program to its completion. Each instruction is written as a single line and, in some systems, aligned in specific columns to constitute the source program. The compiler translates the program for the computer to yield the object program.[5,10]

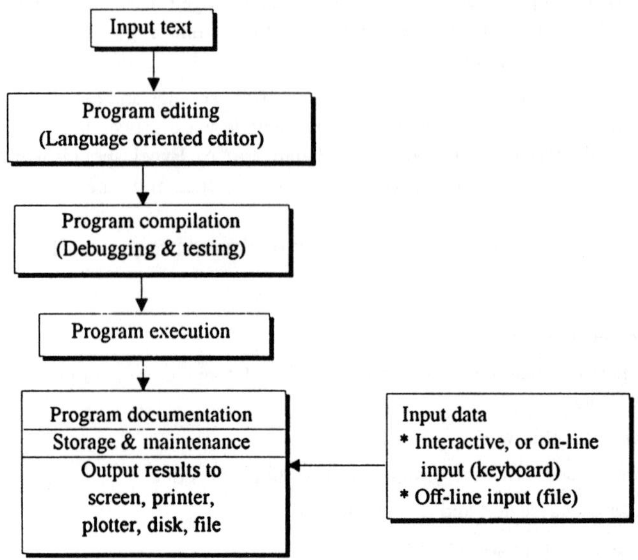

Figure 1.4
Steps of programming.

2. BASIC (or Beginner's All-Purpose Symbolic Instruction Code) is widely used as it is one of the easiest programming languages to learn. It is also used in most educational settings. BASIC is interactive by nature and simple to use. Each program instruction in BASIC is usually entered as a separate line, referred to as a statement. BASIC progresses from one line to the next, unless programmed to do otherwise. System commands direct the computer to perform specific tasks in the program.

3. Pascal is a structured language used to perform large-scale programming. The language uses a restricted set of design structures and other organizing principles. The language rules are made up of syntax and symantics. The former rules define how the words of the language can be put together to form sentences; the latter ascribes meaning and significance to these combinations of words.[3] The language is excellent for large scale and complicated programs.

4. Other programming languages include a wide variety of languages developed for specific applications. The ASSEMBLY language is one used primarily by system programmers to address the hardware. It is rarely used for scientific applications, although it can be, but with difficulty. Currently, high level languages such as C and its extension C++ are satisfying the needs of most programmers, even those requiring interacting with the hardware, and are being used for almost all purposes: scientific, educational, system programming, database, equipment control, games, and many other applications. On the other hand, languages such as CLIPPER are intended for database programming only. Other languages include ALGOL, ADA, COBOL, MODULA, and many others.

The BASIC language has been selected for the models incorporated in this book for the following reasons:

- Simplicity of the language in structure and content
- Ease of learning the language and its interactive use
- Flexibility in programming alteration, updating, and changing
- Universal, popular, interactive, and common use
- Relatively standard language with many versions simple to run in different computers with little or no modification
- Relevance to engineering applications and scientific computations
- Availability of the language program on personal computers
- Immediate detection of errors
- Production of quality graphics and use of sound
- Language requires no hardware knowledge and it shields the user from the operating system of the computer.[11]

Computer programs usually are formed of subprograms (modules) that are easily developed and tested separately. This modular design approach results in short and highly focused programs such as a subroutine, which signifies a set of computer instructions that jointly perform a specific task.

1.3 Fundamentals of the BASIC Language

In BASIC, the instruction can be entered in two ways:

1. Programming mode (indirect), in which the instructions that are entered are those that compose a computer program. Instructions are entered line by line, forming statements.
2. Immediate mode (direct, calculator mode), which signifies the BASIC commands that perform a specific function or task pertaining to the program.

During the stage of program preparation and execution, errors (bugs) may occur. These errors can be grouped as:

- Syntax errors (coding errors, diagnostic) are mistakes made in the usage of the high-level language (for example, misspelling of a key word).
- Run-time errors develop during program execution.
- Logic errors are due to faulty program logic. Logic errors are incorrect order within expressions or incorrect ordering of statements. Usually these types of errors are skipped by the diagnostics.

Debugging and testing should be carried out to produce a reliably operating program. After debugging the program, the next phases are documentation, storage, and maintenance. Documentation signifies addition of English language descriptions to ease implementation of the program by the user. Storage refers to saving the program in a suitable media, while maintenance concerns the upgrading of the program.

BASIC is a simple, reliable language. The instructions resemble elementary algebraic formulae and use plain English words (keywords). Use of the language covers various disciplines such as science, engineering, business, mathematics, finance, database management, commercial, computer games, to name but few.

The language originally was developed by John Kemeny and Thomas Kurtz at Dartmouth College in the early 1960s. Throughout this book QuickBASIC, version 4.5, of Microsoft is adopted for writing programs and models. This version of BASIC has been selected for its ease of use and user-friendly features that facilitate structured programming. The program is menu driven. As such, it is simple to operate and understand and to get help at any stage of programming, as well as making it easy to browse through many topics.

Each program written in BASIC is composed of a sequence of statements that are executed in their order of appearance in the program. The following points summarize the important statements and functions used in QuickBASIC programming and are used in the models in this book.[4,5,7,11–13] They do not cover all the commands in QuickBASIC language; the interested reader is directed to refer to the Microsoft documentation or other publications dealing with this language.

Numeric Constants (Numbers). Numeric data is expressed in BASIC directly as constants and indirectly as variables. Constants refer to values unaltered during program execution. Variables signify symbolic names to which a changing (during execution) value is assigned. Numbers can be positive or negative and can be expressed as integer quantities or decimal (real) quantities. Integer quantities are numbers without a decimal point in the range –32768 to +32767.

In writing numbers the following points need to be considered:

- A number can be preceded by a (+) or a (–) sign.
- A number can contain an exponent, written as xEy, where x is the number and y the exponent. The exponent can be positive or negative but is devoid of a decimal point.
- Letters and symbols such as dollar signs and commas are not allowed in numeric constants.
- Real numbers can be expressed in single or double precision. A single-precision number is a real number stored with 7 significant digits plus a decimal point; a double-precision number is stored with 16 digits plus a decimal point. Both numbers are floating point numbers, i.e., the decimal point floats in its position in the number, depending on the value of the number.

Assignments. These are variable symbolic quantities that can alter during program execution.

Strings. A string is a sequence of characters used to represent non-numeric information. The length of the string depends on the version of BASIC.

Variables. A variable stands for a number or a string. Variables ending in a dollar sign store character information (strings).

Operators. Operators are used to indicate arithmetic operations of addition, subtraction, multiplication, division, and exponentiation as follows:

+ addition
– subtraction
* multiplication
/ division
^ exponentiation

Order of execution gives priority to exponentiations to be followed by multiplication and division, with addition and subtraction to be carried out last. The hierarchy of operation can be altered through the use of pairs of parentheses at proper places in an equation or formula.

The REM Statement ('). The REMark statement adds descriptive comments anywhere into a BASIC program. The computer ignores statements beginning with REM.

The INPUT Statement. The INPUT statement allows entering new data (numerical or string) during the program run. During program execution, the INPUT statement prints a question mark requesting data. Unless the user responds, the program is suspended. Usually a PRINT statement and string constants are used after an INPUT statement to display a message clarifying the required information to be supplied by the user.

The PRINT Statement (Printing Output). The PRINT statement is used to describe numerical or string output information from the computer. Clarification of message can be more descriptive using character strings, semicolon delimiters, blank lines, and TAB functions.

The END Statement. The END statement is used to indicate the end of the program.

The GO TO Statement (Transfer of Control, Unconditional Branching). The GO TO statement is used to transfer the program to some other part of it, changing the sequence of execution.

Intrinsic Functions (Library or Built-in Functions). Table 1.1 outlines some of the functions that usually are included in a BASIC program.

TABLE 1.1
Some of the Built-In Functions in BASIC

Function	Description
ABS(x)	Determines the absolute value of the function x; $y = \|x\|$
SQR(x)	Determines the square root of x, x > 0; $y = \sqrt{x}$
LOG(x)	Determines the natural logarithm of x, x > 0; y = log x
EXP(x)	Determines the exponential of x; $y = e^x$
SIN(x)	Determines the sine of x (x in radian); y = sin(x)
COS(x)	Determines the cosine of x (x in radian); y = cos(x)
TAN(x)	Determines the tangent of x (x in radian); y = tan(x)
ATN(x)	Determines the aractangent of x (x in radian); y = ATN(x)

The LOCATE Statement. The LOCATE statement enables screen design and positions the cursor at any point of the 25 (row)*80 (column) cells on the screen.

The IF-THEN Statement (Conditional Transfer). The IF-THEN statement is used to perform a conditional branching operation. The condition occurring between the IF and the THEN relates one expression (variables, formulas, constants) to another through one of the operators shown in Table 1.2.

TABLE 1.2
Relational Operators Between Expressions

Operator	Relationship
=	Equal
<>	Not equal
<	Less than
>	Greater than
<=	Less than or equal to
>=	Greater than or equal to

READ and DATA Statements. The READ and DATA statements allow incorporation of large amounts of data and information into the program. The DATA statement keeps the actual data and the READ statement assigns the data to a certain variable.

The CASE Construct. The CASE construct allows the selection of one path from a group of alternative paths.

The FOR TO Statement (Building a Loop). The FOR TO statement specifies the number of executions of a loop in the program. The statement includes a running variable whose values change each time the loop is executed.

The NEXT Statement (Closing a Loop). The NEXT statement is used to end a loop.

DO…LOOP. The DO…LOOP construct is a control flow statement that repeats a block of statements while a condition is true or until a condition becomes true.

STRING$. The STRING$ command is a string processing function that returns a string whose characters all have a given ASCII code or whose characters are all the first character of a string expression. By specifying the ASCII code of a graphic character together with the number of such a character, it can be used to draw on the straight lines text screen.

LINE. The LINE statement is a graphics statement that draws a line or box on the graphics screen. The coordinates of the ends of the line have to be known, and, optionally, the command can be used to draw a closed rectangle, if the coordinates of the opposite corners are provided.

PSET and PRESET. The PSET (and PRESET) is a graphics statement that draws a point on the screen. The screen coordinates where the point is to be drawn must be specified, together with the color.

The DIM Statement. The DIM (dimension) statement allows specifying the maximum number of elements (size of array) in an array. An array represents a collection of related variables in a list or table. The individual elements within an array are called subscripted variables.

The DEF Statement (Defining a Function). The DEF statement is used to define a single-line function in the program.

Subroutine. A subroutine is used to structure a sequence of statements. The CALL statement is used to transfer control to the subroutine, and the END SUB statement is used to cause control to transfer back to the main program or to the statement following the point of reference.

The STOP Statement. The STOP statement is used to terminate a job at any point in the program. The STOP statement can appear at any stage in the program except at the very end, unlike the END statement, which only appears at end of program.

The COLOR Statement. The COLOR statement allows the selection of monitor color. The statement specifies the foreground (text) color, the background color, and the border color. Usually the COLOR statement follows a SCREEN statement. The SCREEN statement specifies the mode (text vs. graphics) and color status (enabled or disabled). Table 1.3 outlines the values to be assigned for some colors.

TABLE 1.3
Some of the Screen Colors

Value	Color
0	Black
1	Blue
2	Green
3	Cyan
4	Red
5	Magenta
6	Brown
7	White
8	Gray
9	Light blue
10	Light green
11	Light cyan
12	Light red
13	Light magenta
14	Yellow
15	High intensity white

Apart from using statements for entering data in a program, data may also be introduced using files. A data file is a collection of records maintained on a disk. Data files can be read easily and updated by a BASIC program. Files can be grouped as sequential and direct-access files. Sequential files are those files in which individual items are entered and accessed sequentially, one after another. Direct-access (random-access) files consist of individual data items that are not arranged in any particular pattern or order. The latter files retrieve data more quickly and more efficiently, and individual records can easily be modified. Table 1.4 presents a summary of MS QuickBASIC statements, functions, and commands.

TABLE 1.4
Summary of MS QuickBASIC Statements, Functions, and Commands

Statement/function/command	Purpose	Example
ABS	Keeps absolute value	y = ABS(x)
ATN	Returns aractangent	y = ATN(x)
BEEP	Makes a beeping sound to the user	
CALL	Calls a subroutine	CALL IHR
CHAIN	Passes control to another program	CHAIN "Program 8"
CIRCLE	Draws a circle, an arc, or an ellipse	CIRCLE (100,80),15,1
CLOSE	Closes a file	CLOSE #6
CLS	Clears the screen	CLS
COLOR	Performs color or other screen attributes	COLOR 9,0,1
COS	Determines cosine function	y = COS(x)
DATA	Provides values for variables of the READ statement	READ X,Y,Z,T$ DATA 5,6,–1,7
DATE	Sets the date	DATE$="10-24-95"
DEF FN..	Defines a function	DEF FNB(y)=a*y^3+b*y^2
DIM	Defines arrays	DIM x(120),y$(5,200)
END	Ends program	END
EXP	Determines exponential function	y = EXP(x)
FOR and NEXT	Defines the beginning and end of a FOR-TO loop	For m = 1 TO 39 NEXT m
GOTO	Transfers program to a remote statement	GOTO 3000
IF-THEN	Conditional run	IF y<8 THEN 600
INPUT	Enters data from input device (keyboard)	INPUT x,y,z,t$ INPUT "z=",z
INPUT$	Returns a multi-character string from keyboard	z$ = INPUT$(7)
LET	Assigns values	LET y = a+b–c
LINE	Draws lines	LINE (5,4)–(300,187)
LOCATE	Changes the cursor position	LOCATE 10,45
LOG	Returns the natural logarithm	y = LOG(x)
OPEN	Opens a file	OPEN "L",#3,"IHR"
PRINT	Writes data	PRINT #3,x;y;z;t$
READ	Assigns values in DATA statement to specified variables	READ x,y,z,t$ DATA 5,6,–1,7
REM	Inserts comments and remarks in the program	REM Program listing
RESTORE	Initializes the pointer in DATA statement	RESTORE 45 DATA 5,6,–1,7
SCREEN	Specifies current mode	SCREEN 0

TABLE 1.4 (CONTINUED)
Summary of MS QuickBASIC Statements, Functions, and Commands

Statement/function/ command	Purpose	Example
SIN	Determines the sine function	y = SIN(x)
SQR	Determines square root of a value	y = SQR(x)
STOP	Terminates program	STOP
TAN	Determines the tangent	y = TAN(x)
TIME	Changes the current time	TIME$="12:45"
VAL	Changes a string to a numerical value	y = VAL(x$)

1.4 Programs on the Accompanying CD/ROM

The programs on the CD accompanying this book are all in executable form. Each set of programs carries the name of the chapter in the book to which they pertain. For example, all the programs of Chapter 2 are called CHAPTER2.EXE. These files are in compressed form. To use them, they must be exploded. The following are the steps to be followed to obtain the programs in CHAPTER2.EXE:

1. Insert the CD in the CD drive in your computer.
2. From the CD/ROM prompt (usually D:>), type

   D:\INSTALL

   This command will create a subdirectory on your hard disk named MODELING and will explode all the programs and place them inside the MODELING subdirectory.
3. When the operation is complete, log into the subdirectory CHAPTER2 by typing at the C:> prompt:

   CD CHAPTER2

4. Now you can run any program by simply typing its name. For example, if you want to run program 23.EXE, just type 23 followed by pressing the ENTER key.

In order to use the accompanying CD/ROM, the following system hardware and software specifications are required:

- An IBM-PC computer or compatible machines.
- A CD/ROM drive.
- A hard disk drive with at least 4 MB free space.
- A minimum RAM requirement of 640 kB.
- An adapter such as Color Graphics Adapter (CGA), Enhanced Graphics Adapter (EGA), Video Graphics Adapter (VGA), or compatible video graphics adapter card.
- A video monitor compatible with video graphics adapter.
- A math co-processor chip is recommended to enhance performance.
- An MS-DOS or PC-DOS operating system version 5.00 or above.

1.5 Homework Problems

1. What are the five hardware components of a computer system?
2. What storage medium is most commonly used with microcomputers?
3. Distinguish between RAM and ROM.
4. What is the "brain" of a computer, and what component functions much the same as the human sensory organs?
5. Distinguish between bit and byte.
6. What are the various types of software?
7. Indicate the three basic types of mathematical models.
8. What are the four types of special software available to direct the computer to perform required tasks?
9. Indicate at least five computer languages other than BASIC.
10. Explain the difference between programming language and machine language.
11. What does the term "debugging" mean?
12. Explain the term high-level language (HIL).
13. What are the steps involved in designing a computer program?
14. What is an algorithm?

15. What are the five arithmetic operators and what are their meanings? Give an example of each.
16. What is the order a computer follows when executing an expression?
17. What is a keyword? Give examples.
18. What is the purpose of the keyword REM?
19. What is the purpose of the END statement?
20. What is a constant? Give an example of a statement that causes a constant to be displayed.
21. What is an integer? Give an example.
22. Differentiate between an application program and a utility program.
23. What is a loop?
24. What is an infinite loop?
25. Explain what happens when the computer executes an IF-THEN statement.
26. What is a character string?
27. How is a character variable formed?
28. Explain how the READ and DATA statements work together.
29. What happens when an INPUT statement is encountered in a program?
30. Differentiate between the END statement and the STOP statement.

References

1. McCutcheon, S.C. and French, R.H., *Water Quality Modeling, Vol. 1: Transport and Surface Exchange in Rivers*, CRC Press, Boca Raton, FL, 1989.
2. James, A., *An Introduction to Water Quality Modelling*, John Wiley & Sons, Chichester, U.K., 1984.
3. Welsh, J. and Elder, J., *Introduction to Pascal*, 3rd ed., Prentice-Hall, Englewood Cliffs, NJ, 1988.
4. Davies, G., *Talking BASIC: An Introduction to BASIC Programming for Users of Language*, Cassell Ltd., Maidstone, 1985.
5. Chapra, S.C. and Canale, R.P., *Introduction to Computing for Engineers*, McGraw-Hill, New York, 1986.
6. Confident Computer Corp., *Confident Computer System User's Guide*, Taipei, Taiwan, 1994.

7. Gottfried, B.S., *Schaum's Outline of Theory and Problems of Programming with BASIC*, 3rd ed., Schaum's Outline Series, McGraw-Hill, New York, 1986.
8. Hill, A.E. and Pilkgton, R.D., *A Complete AutoCAD Databook*, Prentice-Hall, Englewood Cliffs, NJ, 1990.
9. Brenner, R.C., *IBM PC Troubleshooting and Repair Guide,* DPB Publications, Delhi, NY, 1985.
10. Calderbank, V.J., *Programming in FORTRAN*, 3rd Ed., Chapman and Hall, London, 1989.
11. Albercht, B., Wiegand, W., and Brown, D., *QuickBASIC Made Easy*, McGraw-Hill, New York, 1989.
12. Lien, D.A., *The BASIC Handbook: Encyclopedia of the BASIC Computer Language*, Compusoft Publishing, San Diego, 1981.
13. Tebbutt, T.H.Y., *BASIC Water and Wastewater Treatment*, Butterworths, London, 1990.

Section II
Water Resources

Chapter 2

Computer Programming For Water and Wastewater Characteristics, Water Resources and Sources, Usage, and Groundwater Flow

Contents

2.1 Introduction

This chapter presents water and wastewater characteristics that can be expressed by mathematical equations or models and then formulated into workable computer programs. Also included in this chapter are computer programs dealing with water resources and sources, usage, and groundwater flow.

The importance of water for all ecosystems is well known, and the availability of water resources has governed the location and evolution of civilizations, their progress and development, and the wellbeing of man's socioeconomic situation (or condition). The importance of water is manifested in the following:[1–14]

1. Existence and evolution of the three kingdoms of plant, animal, and protista
2. Dissolution of organic and inorganic materials
3. Initiation and implementation of agricultural and industrial development projects
4. Initiation, execution, and operation of recreational activities such as boating, sight-seeing, fishing, swimming, golf courses, etc.
5. Transportation of raw materials and navigational activities
6. Conveyance and final disposal of wastes

2.2 Water and Wastewater Characteristics

The characteristics of water or wastewater are important for such purposes as:[14]

- Selecting treatment facilities
- Evaluating operation of existing units
- Evaluating relevance of methods and policies
- Estimating past, present, and future degrees of pollution
- Managing discharge, recreation, and reuse
- Adopting guidelines, regulations, bylaws, and standards

- Employing appropriate surveillance and monitoring systems
- Introducing or updating operational and maintenance and training schemes and programs
- Managing and controlling environmental quality

Classification of water and wastewater characteristics involves physical, chemical, biological and radiological parameters.

2.3 Physical Characteristics

Physical characteristics are those parameters governed by forces of a physical nature. The most important physical parameters include temperature, turbidity, taste, odor, color, conductivity, salinity, solids content, viscosity, and moisture content. Those parameters that are most easily represented by a mathematical equation or formula are presented in this chapter, as well as the computer programs associated with that equation, formula, or model.

2.3.1 Temperature

Water and wastewater may experience variations in temperature due to climatic influences, hot discharges, and industrial discharges. An increase in temperature can pose certain problems; such an increase may:

- Affect the performance of treatment units
- Reduce the concentration of dissolved oxygen
- Accelerate the rate of chemical and biochemical reactions
- Reduce the solubility of gaseous substances
- Increase the rate of corrosion of materials
- Increase the toxicity of dissolved substances
- Increase undesirable growths
- Increase problems of taste and odor

Equations 2.1, 2.2, 2.3, and 2.4 can be used to convert between centigrade (C), Fahrenheit (F), Kelvin (K), and Rankine (R) temperatures

$$°C = \frac{5}{9}(°F - 32) \tag{2.1}$$

$$K = °C + 273.15 \tag{2.2}$$

$$°F = \frac{9}{5}°C + 32 \tag{2.3}$$

$$°R = °F + 459.67 \quad (2.4)$$

Example 2.1

1. Write a short computer program that enables converting temperatures between Fahrenheit, centigrade, kelvin, and Rankine temperatures.
2. Convert 20, 160, and –40°F to centigrade and Kelvin readings. Convert 30, 75, and –10°C to Fahrenheit and Rankine degrees.

Solution

1. For the solution of Example 2.1 (1), see the listing of Program 2.1 below.
2. Solutions for Example 2.1 (2):
   - Given T = 20, 160, and –40°F; T = 30, 75, and –10°C.
   - Determine $°C_{20} = (5/9)*(°F - 32) = (5/9)*(20 - 32) = -6.7$. Similarly, $°C_{160} = 71.1$, $°C_{-40} = -40$.
   - Find $K_{6.7} = °C + 273.15 = -6.7 + 273.15 = 266.45$. Similarly, $K_{71.1} = 344.25$, $K_{-40} = 233.15$.
   - Find $°F_{30} = (9/5)*C + 32 = (9/5)*30 + 32 = 86$. Similarly, $°F_{75} = 167$, $°F_{-10} = 14$.
   - Compute $°R_{86} = 86 + 459.67 = 545.67$. Similarly, $°R_{167} = 626.67$, and $°R_{14} = 473.67$.

Listing of Program 2.1

```
DECLARE SUB box (r1%, c1%, r2%, c2%)          'Draw a frame around the screen
 '***********************************************************************
 'Program 2.1: Convert
 'Converts temperature to different scales
 '***********************************************************************
  DEF FnCen (f) = 5 / 9 * (f - 32)    'Fahrenheit to Centigrade
  DEF FnFah (c) = 9 / 5 * c + 32      'Centigrade to Fahrenheit
  DEF FnKel (c) = c + 273.15          'Centigrade to Kelvin
  DEF FnRan (f) = f + 459.67          'Fahrenheit to Rankine

  DIM ttl$(5)
  ttl$(1) = "Fahrenheit"
  ttl$(2) = "Centigrade"
  ttl$(3) = "Kelvin"
  ttl$(4) = "Rankine"
  ttl$(5) = "Exit"

  DO
    CLS
    CALL box(2, 1, 15, 80) 'Draw a frame around the screen

    LOCATE 3, 15
    PRINT "Program 2.1:Converts temperature to different scales"
    LOCATE 5, 5
    PRINT "Convert temperature from:"
    FOR i = 1 TO 5
      LOCATE i + 5, 5
      PRINT "("; i; ")"; ttl$(i)
```

```
NEXT i
DO
   LOCATE 12, 5
   INPUT "Select an option 1-5 :"; opt
LOOP WHILE opt < 1 OR opt > 5

IF opt < 5 THEN
   CLS
   CALL box(2, 1, 15, 80)  'Draw a frame around the screen
   LOCATE 5, 5
   PRINT "Enter temperature in "; ttl$(opt); " :";
   INPUT t
   ELSE END
END IF

SELECT CASE opt
      CASE 1
         f = t
         c = FnCen(f)
         k = FnKel(c)
         r = FnRan(f)

      CASE 2
         c = t
         f = FnFah(c)
         k = FnKel(c)
         r = FnRan(f)

      CASE 3
         k = t
         c = k - 273.15
         f = FnFah(c)
         r = FnRan(f)

      CASE 4
         r = t
         f = r - 459.67
         c = FnCen(f)
         k = FnKel(c)

   END SELECT
   LOCATE 10, 5
   PRINT "C="; c; "      F="; f; "       K="; k; "       R="; r
   LOCATE 13, 5

   LOCATE 13, 5
   PRINT "press any key to continue....................";
   DO
   LOOP WHILE INKEY$ = ""
LOOP
END

SUB box (r1%, c1%, r2%, c2%) 'This routine is common to all programs in the book
   '*******************************************************************
   'Draws a box with corners at (R1,C1),(R2,C2)
   '***************** DRAW A BOX ***********************************
```

```
  COLOR 15, 1, 4
  CLS
  XL% = c2% - c1% - 1
  'Top line & corners
  LOCATE r1%, c1%
  PRINT CHR$(201);
  PRINT STRING$(XL%, 205);
  PRINT CHR$(187)
  'Side vertical lines
  FOR ROW = r1% + 1 TO r2%
     LOCATE ROW, c1%
     PRINT CHR$(186)
     LOCATE ROW, c2%
     PRINT CHR$(186)
  NEXT ROW
  'Bottom line & corners
  LOCATE ROW, c1%
  PRINT CHR$(200); : PRINT STRING$(XL%, 205); : PRINT CHR$(188)
END SUB
```

2.3.2 Conductivity

Conductivity denotes the ability of an aqueous solution to carry an electric current. This ability is a function of concentration, mobility, valence and relative concentration of ions, and temperature. Generally, solutions of most inorganic acids, bases, and salts are relatively good conductors. The conductivity may also be defined as "the electrical conductance of a conductor or unit length and unit cross-sectional area."[14]

The relationship between conductivity and the concentration of dissolved solids in a solution may be expressed as outlined in Equation 2.5.

$$EC = \frac{TDS}{a} \tag{2.5}$$

where

EC = Electrical conductivity, micromho/cm or microsiemans/cm

TDS = Total dissolved solids, mg/L

a = Constant

Example 2.2

1. Write a computer program that enables the computation of the electrical conductivity of a sample of known concentration of total dissolved solids (TDS), given the total dissolved solids and electrical conductivity (EC) of another sample as TDS_1 and EC_1, respectively.
2. The TDS of a certain sample is recorded as 890 mg/L and its electrical conductivity amounted to 1025 mmhos/cm. Determine the electrical conductivity of another sample that has a TDS of 1450 mg/L.

Solution

1. For the solution of Example 2.2 (1), see listing of Program 2.2 below.
2. Solution for Example 2.2 (2):
   - Find the constant "a" for the sample by: a = TDS/EC, or a = 890/1025 = 0.87.
   - Determine the electrical conductivity for the second sample by: $EC_2 = TDS_2/a$ = 1450/0.87 = 1667 mmohs/cm.

Listing of Program 2.2

```
DECLARE SUB box (r1%, c1%, r2%, c2%)  'Found in Program 2.1
'***************************************************************
'Program 2.2: Conductivity
'Computes electric conductivity of a sample of known TDS concentration
'***************************************************************
CALL box(2, 1, 15, 80)

LOCATE 3, 30
PRINT "Program 2.2"
LOCATE 4, 10
PRINT "Electric conductivity of a sample of known TDS concentration"
row = 6
LOCATE row + 2, 3
INPUT "Enter total Dissolved Solids (TDS) concentration(Sample1) mg/L="; TDS1
LOCATE row + 3, 3
INPUT "Enter electrical conductivity of Sample1   umhos/cm          ="; EC1
LOCATE row + 4, 3
INPUT "Enter total Dissolved Solids (TDS) concentration(Sample2) mg/L="; TDS2
a = TDS1 / EC1
EC2 = TDS2 / a
LOCATE row + 8, 5
PRINT "Electrical conductivity for Sample2="; EC2; "  umhos/cm"
LOCATE 12, 5
DO
LOOP WHILE INKEY$ = ""
END
```

2.3.3 Salinity

Salinity may be described as the total solids in water remaining after all carbonates have been converted to oxides, all bromide and iodide have been replaced by chloride, and all organic matter has been oxidized.[15] The chloride content of natural waters may be augmented from geological formations containing chlorides, through salt water intrusion, or from agricultural, industrial and domestic discharges.[14] Salinity as a function of chlorinity is presented in Equation 2.6.[15]

$$sa = 0.03 + 1.805\ ch \tag{2.6}$$

where

sa = Salinity, g/kg
ch = Chlorinity, g/kg

2.3.4 Solids Content

The term "solids" designates matter that is suspended or dissolved in water and wastewater. Generally, solids are grouped as dissolved, suspended, volatile, fixed, or settleable. Most of the procedures used to quantify solids are gravimetric tests involving measurement of the mass of residues as a function of volume (see Figure 2.1).

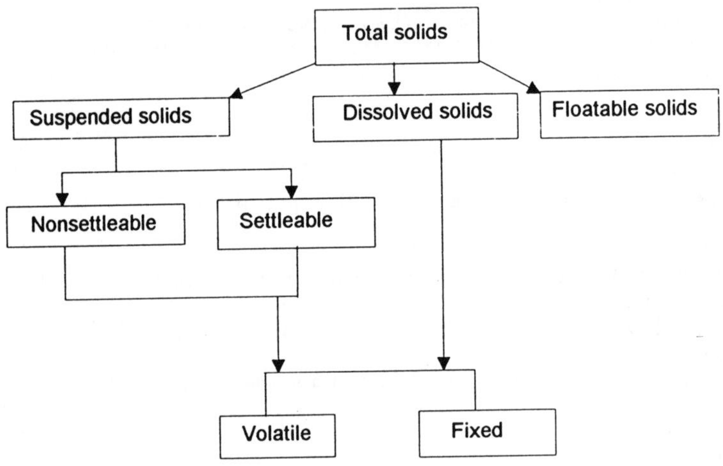

Figure 2.1
Types of solids.

Example 2.3

1. Write a short computer program that enables the computation of the total, volatile, and fixed solids of a certain sample. The following data are given: mass of crucible dish (m_1) in grams, volume of sample (V) in mL, constant mass of dish plus solids dried at 104°C (m_2), and weight of cool dish (after being placed in 550°C furnace) (m_3).
2. Given for a certain sample that $m_1 = 35.6248$ g, $m_2 = 35.6452$ g, $m_3 = 35.6319$ g, and V = 150 mL, find total, volatile, and fixed solids of the sample.

Solution

1. For the solution of Example 2.3 (1), see listing of Program 2.3 below.
2. Solutions for Example 2.3 (2):
   - Given $m_1 = 35.6248$ g, $m_2 = 35.6452$ g, $m_3 = 35.6319$ g, V = 150 mL.

- Evaluate mass of total solids as: m_{solids} = (mass of dish + solids) – mass of dish = $m_2 - m_1$ = 35.6452 – 35.6248 = 0.0204 g = 20.4 mg.
- Find the concentration of total solids: mass of solids (mg)/volume of sample = m_{solids}/V = 20.4*1000/150 = 136 mg/L.
- Determine mass of volatile solids as: $m_{volatile}$ = (mass of dish + volatile solids) – mass of dish = $m_3 - m_1$ = 35.6319 – 35.6248 = 0.00071 g.
- Determine concentration of volatile solids = $m_{volatile}/V$ = 7.1*1000/150 = 47 mg/L.
- Compute concentration of fixed solids as: m_{fixed} = content of total solids – content of volatile solids = $m_{solids} - m_{volatile}$ = 136 – 47 = 89 mg/L.

Listing of Program 2.3

```
DECLARE SUB box (r1%, c1%, r2%, c2%)   'Found in Program 2.1
'*****************************************************************
'Program 2.3: Salinity
'Computes solid contents
'*****************************************************************
CALL box(2, 1, 17, 80)
LOCATE 3, 30
PRINT "Program 2.3: Determines the solids content"
LOCATE 5, 5
INPUT "Enter weight of crucible m1                   (gm) ="; m1
LOCATE 6, 5
INPUT "Enter volume of sample V                      (mL) ="; volume
T1 = 104        'Centigrade
LOCATE 7, 5
INPUT "Enter weight of dish+solids at 104 C            (mg) ="; m2
LOCATE 8, 5
INPUT "Enter weight of dish+ solids after heating to 550 C (mg) ="; m3

msds = m2 - m1                      'solids
csds = msds * 10 ^ 6 / volume          'concentration of solids
mvols = m3 - m1                     'volatiles
cvols = mvols * 10 ^ 6 / volume
cfixs = csds - cvols

LOCATE 10, 5
PRINT "Concentration of total solids    ="; csds; "  mg/L"
LOCATE 11, 5
PRINT "Concentration of volatile solids ="; cvols; "  mg/L"
LOCATE 12, 5
PRINT "Concentration of fixed solids    ="; cfixs; "  mg/L"

LOCATE 22, 3
PRINT "Press a key to continue...."
DO
LOOP WHILE INKEY$ = ""
END
```

2.3.5 Density, Specific Volume, Specific Weight, and Specific Gravity

Density is defined as the mass of a substance per unit volume. Density is temperature dependant. Equation 2.7 may be used to find the density.

$$\rho = \frac{m}{V} \tag{2.7}$$

where:

ρ = Density of the fluid, kg/m^3
m = Mass, kg
V = Volume, m^3

The specific volume is the volume per unit mass, i.e., it is the reciprocal of the density as shown by Equation 2.8.

$$\upsilon = \frac{1}{\rho} \tag{2.8}$$

where:

υ = Specific volume of the fluid, m^3/kg
ρ = Density of the fluid, kg/m^3

The specific weight is the weight of unit volume (see Equation 2.9).

$$\gamma = \frac{mg}{V} = \rho g \tag{2.9}$$

where

γ = Specific weight, N/m^3
m = Mass, kg
g = Gravitational acceleration, m/s^2
V = Volume, m^3
ρ = Density of the fluid, kg/m^3

The specific gravity is the ratio of the density of fluid to the density of water at some specified temperature. Equation 2.10 indicates the specific gravity concept.

$$\text{s.g.} = \frac{\rho}{\rho_{\text{water at }4°\text{C}}} \tag{2.10}$$

Example 2.4

1. Write a computer program that allows the computation of density, specific volume, specific weight, and specific gravity of a particular fluid, given the temperature, mass, and volume of the fluid.
2. The specific weight of water at ordinary temperature and pressure conditions is 9.806 kN/m^3. The specific gravity of mercury is 13.55. Find:

a. Density of mercury
b. Specific weight of mercury
c. Density of water

Solution

1. For the solution to Example 2.4 (1), see listing of Program 2.4 below.
2. Solutions for Example 2.4 (2):
   - Given: $\gamma = 9.806$ kN/m^3, s.g.$_{\cdot Hg} = 13.55$.
   - Determine the density of water as: $\rho = \gamma_{water}/g = 9.806$ kN/m^3/9.81 m/s^2 = 1 Mg/m^3 = 1 g/cm^3.
   - Find specific weight of mercury as: γ_{Hg} = s.g.$_{\cdot Hg}$*γ_{water} = 13.55*9.806 = 133 kN/m^3.
   - Compute density of mercury as: ρ_{Hg} = s.g.$_{\cdot Hg}$*ρ_{water} = 13.55*1 = 13.55 Mg/m^3.

Listing of Program 2.4

```
DECLARE SUB box (r1%, c1%, r2%, c2%)   'Found in Program 2.1
'*************************************************************
'Program 2.4: Density
'Computes Density and specific gravity
'*************************************************************
DIM t$(5)
gamma = 1000          'density of water KN/m3
g = 9.801             'acceleration due to gravity

t$(1) = "1. Compute density"
t$(2) = "2. Compute specific volume"
t$(3) = "3. Compute specific weight"
t$(4) = "4. Compute specific gravity"
t$(5) = "5. Exit program"

start:
CALL box(2, 1, 17, 80)
LOCATE 3, 20
PRINT "Program 2.4:Computes Density and specific gravity"

DO
FOR i = 1 TO 5
LOCATE i + 4, 5
PRINT t$(i)
NEXT i
LOCATE 12, 3
INPUT "Select an option 1..5 :"; opt
LOOP WHILE opt < 1 OR opt > 5

CALL box(2, 1, 17, 80)
SELECT CASE opt
CASE 1
     LOCATE 5, 5
     INPUT "Enter the mass in Kg     ="; m
     LOCATE 6, 5
     INPUT "Enter the volume in m3   ="; V
     Ro = m / V     'the density
     LOCATE 10, 3
     PRINT "The density  Ro="; Ro; "  Kg/m3"
```

```
CASE 2
    LOCATE 5, 5
    INPUT "Enter the mass in Kg      ="; m
    LOCATE 6, 5
    INPUT "Enter the volume in m3    ="; V
    sv = V / m      'specific volume
    LOCATE 10, 3
    PRINT "Specific volume       ="; sv; "  m3/Kg"

CASE 3
    LOCATE 5, 5
    INPUT "Enter the mass in Kg      ="; m
    LOCATE 6, 5
    INPUT "Enter the volume in m3    ="; V
    sw = m * g / V  'specific weight
    LOCATE 10, 3
    PRINT "Specific weight       ="; sw; "  N/m3"

CASE 4
    LOCATE 5, 5
    INPUT "Enter density of material Kg/m3  ="; Ro
    sg = Ro / gamma  'specific gravity
    LOCATE 10, 3
    PRINT "Specific gravity      ="; sg

CASE 5
    END

END SELECT

LOCATE 22, 3
PRINT "press any key to continue.....................";
DO
LOOP WHILE INKEY$ = ""
GOTO start
END
```

2.3.6 Viscosity (Rheological Properties)

Viscosity of a fluid is a measure of its resistance to shear or angular deformation.[16] Viscosity is the property that relates the applied forces to the rates of deformation of the fluid (see Figure 2.2). Viscosity may be found from Newton's law, as presented in Equation 2.11.

$$\tau = \frac{F_\tau}{A} = -\mu \frac{du}{dy} \tag{2.11}$$

where:

τ = Shear stress, $N*m^{-2}$ (see Figure 2.3)
F_τ = Shear force acting on a very wide plate free to move, N
A = Area, m^2
μ = Dynamic (absolute) viscosity, $N*s*m^{-2}$
du/dy = Velocity gradient (rate of angular deformation, rate of shear), s^{-1}

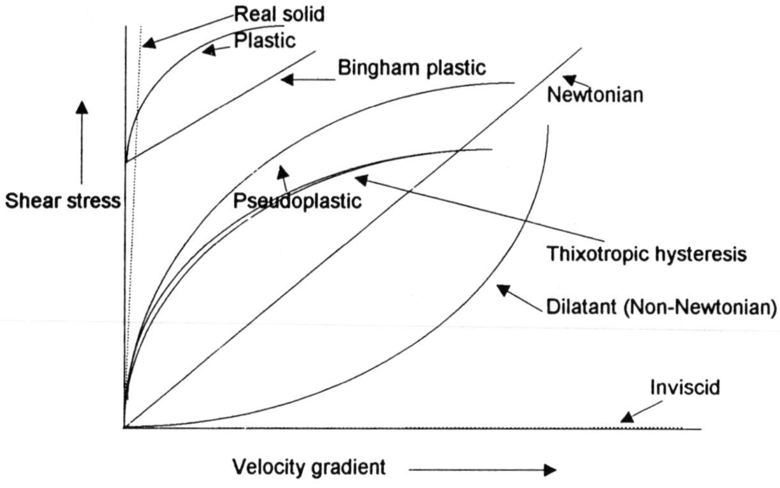

Figure 2.2
Different types of fluids.

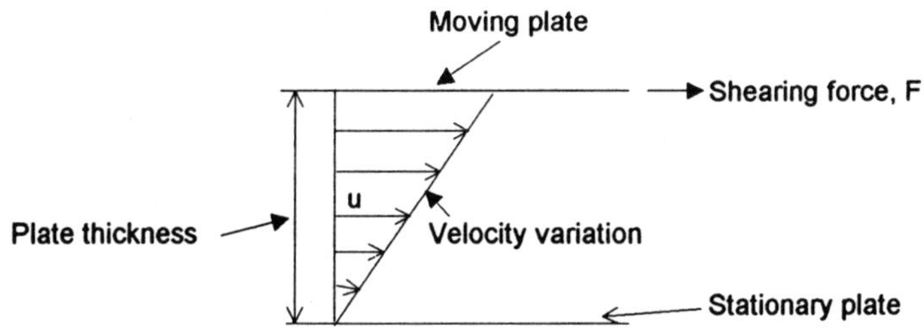

Figure 2.3
Viscosity concept.

Coefficient of dynamic viscosity, μ, may be described as "the shear force per unit area required to drag one layer of fluid with unit velocity past another layer through unit distance in the fluid."[14] Equation 2.12 gives the relationship between viscosity and density.

$$\mu = \rho \nu \quad (2.12)$$

where:

μ = Dynamic (absolute) viscosity, $N*s*m^{-2}$

ν = Kinematic viscosity, m^2*s^{-1} (usually defined as the ratio of dynamic viscosity to mass density)

ρ = Density, $kg*m^{-3}$

For many liquids, the stress-velocity gradient relationship is not a simple ratio. Such liquids are termed non-newtonian, e.g., sewage sludge. Figure 2.2 illustrates the variation of shear stress with velocity gradient for different types of fluids. The factors that influence viscosity include temperature, rate of variation of shearing stress, flow conditions, and characteristics of the fluid.

Example 2.5

1. Write a short computer program to determine the value of the kinematic viscosity of a fluid given its dynamic viscosity, density, and temperature.
2. For a liquid with a dynamic viscosity of $4.5*10^{-3}$ Pa*s and a density of 912 kg/m^3, determine its kinematic viscosity.
3. Use the program developed in (1) to verify data presented in (2).

Solution

1. For the solution to Example 2.5 (1), see listing of Program 2.5 below.
2. Solution for Example 2.5 (2):
   - Given: $\mu = 4.5*10^{-3}$ Pa*s, $\rho = 912$ kg/m^3
   - Determine the kinematic viscosity from $\nu = \mu/\rho = 4.5*10^{-3}$ Pa*s/912 kg/m^3 = 4.9 $*10^{-6}$ m^2/s.

Listing of Program 2.5

```
DECLARE SUB box (r1%, c1%, r2%, c2%)  'Found in Program 2.1
'*****************************************************************
'Program 2.5: Viscosity
'Computes kinematic viscosity from dynamic viscosity
'*****************************************************************
DEF FnKinem = mue / Row        'Row is the density

CALL box(2, 1, 17, 80)
LOCATE 3, 30
PRINT "Program 2.5"
LOCATE 4, 15
PRINT "Computes kinematic viscosity from dynamic viscosity"

LOCATE 8, 5
INPUT "Enter dynamic viscosity u  in  Ns/m2     ="; mue
LOCATE 9, 5
INPUT "Enter density of liquid  Ro in Kg/m3    ="; Row
```

```
LOCATE 13, 3
PRINT "The Kinematic viscosity ="; FnKinem; "  m2/s"

LOCATE 22, 3
PRINT "press any key to continue.....................";
DO
LOOP WHILE INKEY$ = ""
END
```

2.3.7 Surface Tension and Capillary Rise

Surface tension is a property of a liquid that permits the attraction between molecules to form an imaginary film that is able to resist tensile forces at the interface between two immiscible liquids or at the interface between a liquid and a gas.[17] The capillary rise in a small vertical open tube of circular cross-section dipping into a pool of liquid may be obtained[37] from Equation 2.13 (see figure 2.4).

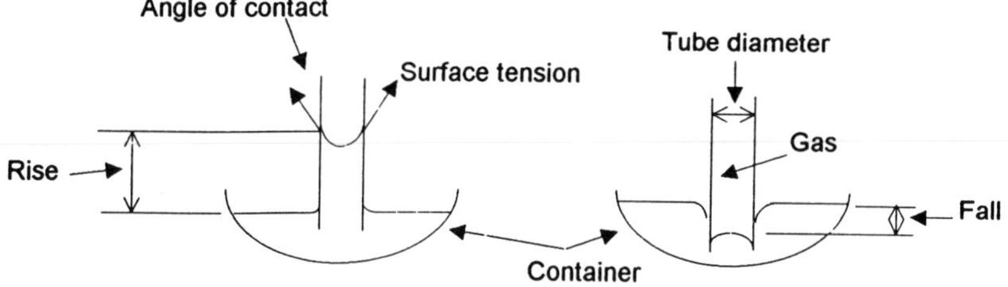

Figure 2.4
Surface tension.

$$h = \frac{4\sigma\cos\phi}{gD(\rho_l - \rho_g)} \tag{2.13}$$

where:

- h = Column height by which liquid rises along capillary tube, m
- σ = Surface tension, N/m
- ϕ = Angle of contact subtended by the heavier fluid and tube (0° for most organic liquids and water against glass, provided the glass is wet with a film of the liquid;[37] 130° for mercury against glass[17])
- ρ_l = Density of liquid, kg/m^3
- ρ_g = Density of gas (or light liquid), kg/m^3
- g = Local gravitational acceleration, m/s^2
- D = Inside diameter of capillary tube, m

Example 2.6

1. Write a computer program to determine the rise of water along capillary tubes for a temperatue range between 0 and 100°C and for tubes with diameters ranging between 10 and 60 mm.
2. Compute the diameter of a clean glass tube required so that the rise of water at 30°C in the tube due to capillary action is not to exceed 2 mm.

Solution

1. For the solution to Example 2.6 (1), see listing of Program 2.6 below.
2. Solution for Example 2.6 (2):
   - Given $h = 2*10^{-3}$ m, T = 30°C.
   - For a temperature of 30°C and from Table A.1 (Appendix A), $\sigma = 0.07118$ N/m and $\gamma = 9.765$ kN/m^3.
   - Compute the diameter as: $d = 4*\sigma*\cos\phi/\gamma*h = [4*0.07118*\cos 0]/[9.765*10^3*2*10^{-3}] = 14.6$ mm.

Listing of Program 2.6

```
DECLARE SUB box (r1%, c1%, r2%, c2%)   'Found in Program 2.1
'*****************************************************************
'Program 2.6: Tension
'Computes capillary heights in surface tension problems
'*****************************************************************
DIM temp(40), sigma(40), density(40), ttl$(5)
'Temperature-Surface tension table for water- in dyne/cm
DATA 74.92,74.78,74.64,74.50,74.36,74.22,74.07,73.93,73.78,73.64
DATA 73.49,73.34,73.19,73.05,72.90,72.75,72.59,72.44,72.28,72.13
DATA 71.97,71.82,71.66,71.50,71.35,71.18,71.02,70.86,70.70,70.53
DATA 70.38,70.21,70.05,69.88
'Temperature-Density table- kg/m3
DATA 999.965,999.941,999.902,999,849,999.781,999.700,999.605,999.498
DATA 999.377,999.244,999.099,998.943,998.774,998.595,998.405,998.203
DATA 997.992,997.770,997.538,997.296,997.044,996.783,996.512,996.232
DATA 995.944,995.646,995.340,995.025,994.702,994.374,994.030,993.680
DATA 993.330,992.296

'constants
pi = 3.142857
g = 9.81            'acceleration due to gravit
phiw = 0            'angle of contact water & glass
phim = 130          'angle of contact mercury & glass
Hgd = 13.55

ttl$(1) = "1. Determine capillary height h"
ttl$(2) = "2. Determine the diameter of the glass tube for a specific h"
ttl$(3) = "3. Determine surface tension"
ttl$(4) = "4. Exit program"
```

```
'Read surface tension table
FOR i = 5 TO 38
temp(i) = i
READ sigma(i)
sigma(i) = sigma(i) / 1000     'convert to N/m
NEXT

'Read density table
FOR i = 5 TO 38
READ density(i)
NEXT

start:
CALL box(2, 1, 20, 80)
LOCATE 3, 7
PRINT "Program 2.6:Computes capillary heights in surface tension problems"

'Print the menu
FOR i = 1 TO 4
LOCATE i + 5, 5
PRINT ttl$(i)
NEXT i
DO
LOCATE 12, 3
INPUT "Select an option (1 to 4): "; opt
LOOP WHILE opt < 1 OR opt > 4

SELECT CASE opt

CASE 1
        'Determine capillary height h
        CALL box(2, 1, 20, 80)
        LOCATE 5, 5
        INPUT "Enter diameter of tube in (mm) (between 10 to 60mm)  ="; diam
        DO
        LOCATE 6, 5
        INPUT "Enter temperature t (C) (5<t<38)                     ="; T
        LOOP WHILE T < 5 OR T > 38
        LOCATE 7, 5
        INPUT "Enter the density of the gas (kg/m3)                 ="; Rog

        'get the value of sigma and density of water corresponding to temperature t
        FOR i = 5 TO 38
        IF T = temp(i) THEN
            s = sigma(i)
            Roww = density(i)
        END IF
        NEXT i
        'compute height h
        h = 4 * s * COS(phi * pi / 180) / (g * diam * (Roww - Rog))
        LOCATE 10, 3
        PRINT "The column height h="; h; " m"
        DO
        LOOP WHILE INKEY$ = ""

CASE 2
        'Determine the diameter of the glass tube for a specific h
        CALL box(2, 1, 20, 80)
```

```
LOCATE 5, 5
INPUT "Enter the density of the liquid or gas, kg/m3  ="; Ro
LOCATE 6, 5
INPUT "Enter surface tension coefficient, N/m         ="; sigma
LOCATE 7, 5
INPUT "Enter angle of contact in degrees              ="; phi
LOCATE 8, 5
INPUT "Enter the exent of the liquid rise in the tube, m ="; h
phi = phi * pi / 180    'in radians
d = 4 * sigma * COS(phi) / (Ro * g * h)
LOCATE 10, 3
PRINT "the required diameter of the tube ="; d; "  m"
DO
        LOOP WHILE INKEY$ = ""

CASE 3
        'Determine sigma
        CALL box(2, 1, 20, 80)
        LOCATE 5, 5
        INPUT "Enter the density of the liquid or gas, kg/m3   ="; Ro
        LOCATE 6, 5
        INPUT "Enter the diameter of the capillary tube,m      ="; d
        LOCATE 7, 5
        INPUT "Enter angle of contact in degrees               ="; phi
        LOCATE 8, 5
        INPUT "Enter the exent of the liquid rise in the tube, m ="; h
        phi = phi * pi / 180    'in radians
        sigma = h * Ro * g * d / (4 * COS(phi))
        LOCATE 10, 3
        PRINT "the surface tension coefficient  ="; sigma; " N/m"
        DO
        LOOP WHILE INKEY$ = ""

CASE 4
        'exit
        END
END SELECT
GOTO start

END
```

2.3.8 Bulk Modulus (Bulk Modulus of Elasticity)

Bulk modulus is a property used to evaluate the degree of compressibility. Equation 2.14 gives the bulk modulus.

$$E_b = -\frac{dP}{dV/V} = \frac{-dP}{\rho/d\rho} \tag{2.14}$$

where:

E_b = Bulk modulus, $N{*}m^{-2}$
dP = Differential change in pressure, Pa

dV = Differential change in volume, m^3

V = Volume, m^3

ρ = Density of fluid, kg/m^3

The negative sign shows that an increase in pressure produces a reduction in volume. Large values of bulk modulus indicate that the fluid is relatively incompressible (i.e., a large pressure change is needed to create a small change in volume).

Example 2.7

1. Write a computer program to compute the bulk modulus given any differential change in pressure needed to create any differential change in volume of a specified volume.
2. A liquid compressed in a cylinder has a volume of 2 L at 1000 kPa and a volume of 1805 cm^3 at 2000 kPa. What is its bulk modulus of elasticity?

Solution

1. For the solution to Example 2.7 (1), see listing of Program 2.7 below.
2. Solution for Example 2.7 (2):
   - Given V_1 = 2000 mL, V_2 = 1805 ml, P_1 = 1000 kPa, P_2 = 2000 kPa.
   - Determine the bulk modulus of the liquid as: $E_b = -\,dP/(dV/V) = -\,(2000 - 1000)kPa/[(1805 - 2000)/2000] = 10.3$ MPa.

Listing of Program 2.7

```
DECLARE SUB box (r1%, c1%, r2%, c2%)  'Found in Program 2.1
'*****************************************************************
'Program 2.7: Bulk
'Computes Bulk modulus given dV and dP
'*****************************************************************

CALL box(2, 1, 17, 80)
LOCATE 3, 15
PRINT "Program 2.7: Computes Bulk modulus given dV and dP"

LOCATE 5, 5
INPUT "Enter initial volume V1 in mL   ="; v1
LOCATE 6, 5
INPUT "Enter final volume V2 in mL     ="; v2
LOCATE 7, 5
INPUT "Enter initial pressure P1 in kPa ="; p1
LOCATE 8, 5
INPUT "Enter final pressure P2 in kPa  ="; p2

Eb = -(p2 - p1) / ((v2 - v1) / v1)
```

```
LOCATE 10, 3
PRINT "The bulk modulus Eb="; Eb; " kPa"
LOCATE 15, 3
PRINT "press any key to continue....................";
DO
LOOP WHILE INKEY$ = ""

END
```

2.4 Chemical Characteristics

Chemical characteristics of significance in the field of water and wastewater include pH, alkalinity, acidity, hardness, dissolved oxygen, dissolved gases, chloride content, nitrogen (ammonia-nitrogen, nitrite, nitrate), nutrients, protein content, oil and grease, carbohydrates, phenols, detergents, toxic metals, BOD, COD, etc. The chemical characteristics that are supported by mathematical equations, formulae, or models are presented here, along with computer programs associated with the equation, formula, or model.

2.4.1 Hydrogen Ion Concentration

The hydrogen ion concentration, pH, is a measure of the acidity or alkalinity of a solution. The pH influences the treatment methods and quality of water supply or wastewater discharge. Equation 2.15 gives a numerical expression for the determination of the pH.

$$pH = -\log[H^+(aq)] = \log\frac{1}{[H^+(aq)]} \tag{2.15}$$

where:

pH = Negative logarithm of the hydrogen ion concentration

$[H^+ (aq)]$ = Concentration of hydrogen (hydroxonium ion, H_3O^+) ions, mole/L

pH ranges from a value of 0 to 14. Solutions with pH of 7 are regarded as neutral, while solutions with pH below 7 constitute acids and those above pH of 7 are alkalies. A strongly acidic solution can, in theory, have a pH less than zero and a strongly alkaline solution can have a pH greater[18] than 14. At a temperature of 298 K, Equation 2.16 is valid.

$$pH + pOH = 14 \tag{2.16}$$

where:

pOH = Negative logarithm of the hydroxyl ion concentration

pH = Negative logarithm of the hydrogen ion concentration

Example 2.8

1. Write a short computer program to determine the pH given one of the following:
   a. Hydrogen ion concentration
   b. Hydroxyl ion concentration or its numerical value
   c. The molar concentration of hydrogen ions
2. Determine which of the following solutions is more acidic or alkaline: a solution with a pH of 7.2, or another containing $1.95*10^{-9}$ g of H^+ per L?
3. Calculate the pH and H^+ concentration, in grams of H^+ per liter, of a solution containing $1*10^{-8.3}$ mol of OH^- per L.
4. Use the program developed in (1) to verify computations conducted on parts (2) and (3).

Solution

1. For the solution to Example 2.8 (1), see listing of Program 2.8 below.
2. Solution to Example 2.8 (2):
   - Given: pH of first solution = 7.2, H^+ of the second solution = $1.95*10^{-9}$ M.
   - Find pH of second solution as: pH = $-\log [H^+] = -\log 1.95*10^{-9} = 8.7$.
   - Since the pH of second solution is higher, then it is more alkaline than the first solution.
3. The dissociation constant for water is $[H^+][OH^-] = 10^{-14}$. Find concentration of hydrogen ions as $[H^+] = 1*10^{-14}/1*10^{-8.3} = 10^{-5.7}$ M. pH = $-\log [H^+] = -\log 1*10^{-5.7} = 5.7$. Hydrogen ion concentration = $[H^+] = 10^{-5.7}*MW = 10^{-5.7}*1.008 = 2.01*10^{-6}$ g H^+/L.

Listing of Program 2.8

```
DECLARE SUB checkph (PH!, T$)
DECLARE SUB box (r1%, c1%, r2%, c2%)   'Found in Program 2.1
'*****************************************************************
'Program 2.8: Ph
'Computes Hydrogyn ion concentration
'*****************************************************************
DEF FnLg10 (x) = -.4342944818# * LOG(x)   'this is log base 10 of x

start:
CALL box(1, 1, 20, 80)
LOCATE 2, 20
PRINT "Program 2.9: Computes Hydrogyn ion concentration"
TTL$(1) = "1. Given Concentration of Hydrogen ions H+ in mg/L or molar"
TTL$(2) = "2. Given Hydroxyl ion concentration in mg/L or molar"
TTL$(3) = "3. Given negative logarithm of hydroxyl ion concentration"
TTL$(4) = "4. Exit program"
DO
FOR i = 1 TO 4
LOCATE i + 2, 3
PRINT TTL$(i)
NEXT i
ROW = 8
LOCATE ROW, 5
```

```
PRINT "Select an option 1-5 then press Enter:";
INPUT opt
LOOP WHILE opt < 1 OR opt > 4

SELECT CASE opt
CASE 1
    LOCATE 10, 5
    INPUT "Enter concentration of Hydrogen ion H+ in mg/L or molar="; H
    PH = FnLg10(H)
    CALL checkph(PH, T$)
    LOCATE 12, 3
    PRINT "PH of the solution    ="; PH; "    "; T$

CASE 2
    LOCATE 10, 5
    INPUT "Enter the Hydroxyl ion concentration in mg/L or molar="; OH
    LOCATE 12, 5
    PH = -FnLg10(10 ^ -14 / OH)
    CALL checkph(PH, T$)
    LOCATE 12, 3
    PRINT "PH of the solution    ="; PH; "    "; T$

CASE 3
    LOCATE 10, 5
    INPUT "Enter the negative logarithm of hydroxyl ion concentration:"; POH
    PH = 14 - POH
    CALL checkph(PH, T$)
    LOCATE 12, 3
    PRINT "PH of the solution    ="; PH; "    "; T$

CASE 4
    END
END SELECT

LOCATE 23, 3
PRINT "press any key to continue....................";
DO
LOOP WHILE INKEY$ = ""
GOTO start
END

SUB checkph (PH, T$)
    IF PH < 7 THEN T$ = "Acidic"
    IF PH = 7 THEN T$ = "Neutral"
    IF PH > 7 THEN T$ = "Alkaline"
END SUB
```

2.4.2 Hardness

Hardness prevents the formation of a soap lather and is generally associated with divalent metallic cations of calcium, Ca^{++}, magnesium, Mg^{++}, strontium, Sr^{++}, ferrous ion, Fe^{++}, and manganous ions, Mn^{++}, as presented in Table 2.1. (Ca^{++} and Mg^{++} are the major ions generally associated with hardness). Hardness may be computed by using Equation 2.17.[15]

TABLE 2.1
Principal Cations and Anions Associated With Hardness[14]

Cations	Anions
Ca^{++}	HCO_3^-
Fe^{++}	NO_3^-
Mg^{++}	$SO_4^=$
Mn^{++}	CL^-
Sr^{++}	$SiO_3^=$

Source: From Rowe, D.R. and Abdel-Magid, I.M., *Handbook of Wastewater Reclamation and Reuse,* Lewis Publishers, Boca Raton, FL, 1995. With permission.

$$Hard = 2.497\,Ca^{++} + 4.118\,Mg^{++} \tag{2.17}$$

where:

Hard = Hardness, milliequivalent $CaCO_3$/L
Ca^{++} = Concentration of calcium ion, mg/L
Mg^{++} = Concentration of magnesium ion, mg/L

When:

- Alkalinity < total hardness, carbonate hardness = alkalinity (mg/L)
- Alkalinity > total hardness, carbonate hardness = total hardness (mg/L)[14]

Advantages and Disadvantages of Hardness in Water

Advantages:

1. Hard water aids in the growth of teeth and bones.
2. Hard water reduces the lead oxide (PbO) toxicity from pipelines made of lead.

Disadvantages:

1. Possible association of soft water with cardiovascular diseases
2. Financial losses due to high consumption of soap
3. Scaling of hot water systems, boilers, domestic appliances, fittings, utensils, bath tubs, sinks, and dishwashers
4. Staining of clothes, household utensils

5. Hard water residues that can remain in the pores of skin
6. Increased laxative effect of hard waters containing magnesium sulfates

Table 2.2 offers a classification of waters according to their degrees of hardness.

TABLE 2.2
Degrees of Hardness[14]

mg/L as $CaCO_3$	Degrees of Hardness
0–75	Soft
75–150	Moderately soft
150–175	Moderately hard
175–300	Hard
300 up	Very hard

Source: From Rowe, D.R. and Abdel-Magid, I.M., *Handbook of Wastewater Reclamation and Reuse,* CRC Press/Lewis Publishers, Boca Raton, FL, 1995. With permission.

Example 2.9

1. Write a computer program that will determine the concentration of anions and cations in milliequivalent per liter and milligram per liter calcium carbonate for the range of ions to be found in hard water. Design the program so that the total, carbonate, and noncarbonate hardness, experimental error, and the degree of water hardness can be calculated as presented in Table 2.2.
2. Laboratory analysis of a water sample yielded the following data:

Cation (mg/)	Anion (mg/L)
$Na^+ = 17.25$	$HCO_3^- = 91.5$
$Ca^{++} = 19$	$SO_4^= = 19.2$
$Mg^{++} = 12.2$	$Cl^- = 28.4$
$Sr^{++} = 3.1$	

   a. Express concentrations of ions in milliequivalent/liter.
   b. Comment about the experimental error, assuming that an error of 10% is acceptable.
   c. Draw a bar graph of the sample.
   d. Determine the total, carbonate, and noncarbonate hardness of the sample.
3. Use program developed in (1) to verify results obtained in (2).

Solution

1. For solution to Example 2.9 (1), see listing of Program 2.9 below.

2. Solutions to Example 2.9 (2):

   a. Given concentration of ions:

      - Determine the concentration of given species in units of milliequivalent/liter by dividing given concentration (mg/L) by the equivalent weight of each substance, i.e., $C_{meq} = C_{mg/L}/EW$.
      - Convert the concentrations (milliequivalent/liter) to units of mg/L as $CaCO_3$. This is achieved by multiplying the concentrations of milliequivalent/liter by the equivalent weight (EW) of $CaCO_3$. Equivalent weight of calcium carbonate = molecular weight of calcium carbonate (MW)/valency (Z), i.e., EW of $CaCO_3$ = MW/Z = (40 + 12 + 16*3)/2 = 50. Tabulate results as in the table shown below:

Constituent	EW (mg/meq)	C (mg/L)	Concentration c = C/EW (meq/L)	c*50 (mg/L$CaCO_3$)
Cations				
Ca^{++}	20	19	0.95	47.5
Mg^{++}	12.2	12.2	1.00	50
Sr^{++}	43.8	3.1	0.07	3.5
Na^{+}	23	17.25	0.75	37.5
Total			2.77	
Anions				
HCO_3^-	61	91.5	1.5	75
$SO_4^=$	48	19.2	0.4	20
Cl^-	35.5	28.4	0.8	40
Total			2.70	

   b. Compute experimental error as: Percent experimental error = (cations – anions)*100/cations = (anions – cations)*100/anions = {(2.77 – 2.7)/2.77}*100 = 3%. Since the computed experimental error is less than 10%, the analysis can be consideted acceptable.

   c. Plot the bar graph as indicated in the following diagram:

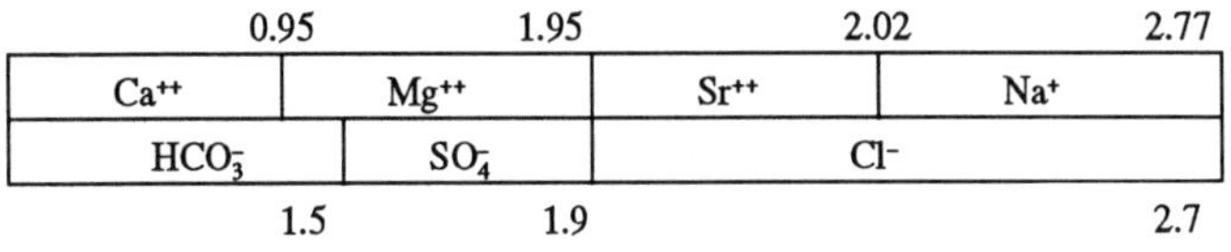

0.95	1.95	2.02	2.77
Ca^{++}	Mg^{++}	Sr^{++}	Na^{+}

HCO_3^-	$SO_4^=$	Cl^-
1.5	1.9	2.7

   d. Determine hardness as: Total hardness = Ca^{++} + Mg^{++} + Sr^{++} = 0.95 + 1.0 + 0.07 = 2.02 meq/L = 2.02*50 = 101 mg/L as $CaCO_3$. Carbonate hardness = HCO_3^- = 1.5 meq/L = 75 mg/L $CaCO_3$. Noncarbonate hardness = total hardness – carbonate hardness = 101 – 75 = 26 mg/L $CaCO_3$.

Listing of Program 2.9

```
DECLARE SUB PART2 ()
DECLARE SUB PART1 ()
DECLARE SUB CONSTANTS ()
DECLARE SUB BOX (R1%, C1%, R2%, C2%)
'**********************************************************
'Program 2.9: Hardness
'Program to compute hardness
'**********************************************************
DIM SHARED EW(12), SYMBOL$(12), CONCENTRATION(12), OPT, COL2(12), COL3(12)
DIM JAMA$(5)

N = 3
J$ = CHR$(13)
CALL CONSTANTS        'constants Ew

CLS : SCREEN 0
COLOR 15

'Define Options:
JAMA$(1) = "(A) Concentration in mg/Litre"
JAMA$(2) = "(B) Concentration in mequivalent/Litre"
JAMA$(3) = "(G) EXIT"

COLOR 13, 0
CALL BOX(1, 1, 20, 80)
LOCATE 2, 30
PRINT "Program 2.10: Computes hardness"
COLOR 14
CALL BOX(3, 18, 16, 70)
CALL CONSTANTS

COLOR 15
'Print Options on the screen:
FOR I = 1 TO N: LOCATE I + 5, 28: PRINT JAMA$(I): NEXT I

'Selections
Y = 1
U$ = CHR$(0) + CHR$(72)
d$ = CHR$(0) + CHR$(80)
R$ = CHR$(0) + CHR$(77)
L$ = CHR$(0) + CHR$(75)

OLDY = N

'Get input from user
90 LOCATE 18, 4
  JAMA$ = UCASE$(INKEY$): IF JAMA$ = "" THEN GOTO 90
'GOTO 90

 OLDY = Y

 IF JAMA$ = CHR$(13) AND Y = 1 OR JAMA$ = "A" THEN
      Y = 1
      OPT = 1
      SCREEN 9
      CALL PART1
      LOCATE 24, 5: INPUT "Press RETURN key to continue.....:"; REP$
      CALL PART2

 END IF
 IF JAMA$ = CHR$(13) AND Y = 2 OR JAMA$ = "B" THEN
```

```
    Y = 2
    OPT = 2
    CALL PART1
    LOCATE 24, 5: INPUT "Press RETURN key to continue.....:"; REP$
    CALL PART2

END IF
    LOCATE 24, 5: INPUT "Press RETURN key to continue.....:"; REP$

IF JAMA$ = CHR$(13) AND Y = 3 OR JAMA$ = "G" THEN
    Y = 3

    END
END IF

END SUB

SUB CONSTANTS
SHARED EW(), SYMBOL$(), CONCENTRATION(), OPT, COL2(), COL3()

EW(1) = 20.0485
EW(2) = 12.1525
EW(3) = 27.9235
EW(4) = 27.4654
EW(5) = 43.81
EW(6) = 22.98977
EW(7) = 30.0246
EW(8) = 61.019
EW(9) = 48.033
EW(10) = 35.453
EW(11) = 62.0049
EW(12) = 38.0419

SYMBOL$(1) = "Ca++"
SYMBOL$(2) = "Mg++"
SYMBOL$(3) = "Fe++"
SYMBOL$(4) = "Mn++"
SYMBOL$(5) = "Sr++"
SYMBOL$(6) = "Na+"
SYMBOL$(7) = "CO3--"
SYMBOL$(8) = "HCO3-"
SYMBOL$(9) = "SO4--"
SYMBOL$(10) = "CL-"
SYMBOL$(11) = "NO3--"
SYMBOL$(12) = "SiO3--"

END SUB

SUB PART1
'*************************************************
SHARED EW(), SYMBOL$(), CONCENTRATION(), OPT, COL2(), COL3()

SCREEN 0
COLOR 15, 1
CLS
BLANK$ = SPACE$(30)
V$ = "####.###        #####.###        ####.###            ####.###"
V1$ = "####.####"

CALL BOX(1, 1, 22, 80)
SCATION = 0
SANION = 0
'LOCATE 3, 10: PRINT BLANK$
ROW = 2
```

```
LOCATE ROW, 3
PRINT "CATIONS:"
FOR I = 1 TO 12
ROW = ROW + 1
IF I = 7 THEN
     LOCATE ROW, 3
     PRINT "ANIONS:"
     ROW = ROW + 1
END IF

LOCATE ROW, 10
     PRINT "Concentration of "; SYMBOL$(I); "       =";
          INPUT CONCENTRATION(I)
     IF OPT = 1 THEN COL2(I) = CONCENTRATION(I) / EW(I)    'C/EW
     IF OPT = 2 THEN COL2(I) = CONCENTRATION(I)
     IF I <= 6 THEN SCATION = SCATION + COL2(I)          'Sum cat.
     IF I > 6 THEN SANION = SANION + COL2(I)             'Sum anions
     COL3(I) = COL2(I) * 50
NEXT I
EERROR = (SCATION - SANION) * 100 / SCATION
FOR I = 3 TO 20
LOCATE I, 3
PRINT SPACE$(60)
NEXT
LOCATE 2, 5
TITLE1$ = "CATION    CONC.(mg/L)    CONC.(mequiv./L)    CONC.(mg/L as CaCO3"
TITLE2$ = "ANION    CONC.(mg/L)    CONC.(mequiv./L)    CONC.(mg/L as CaCo3"
TITLE3$ = "CATION    CONC.(mequiv./L)    CONC.(mg/L as CaCO3"
TITLE4$ = "ANION    CONC.(mequiv./L)    CONC.(mg/L as CaCo3"

IF OPT = 1 THEN PRINT TITLE1$ ELSE PRINT TITLE3$
K = 2
FOR I = 1 TO 12
IF I = 7 THEN
     K = K + 1
     LOCATE K, 5: PRINT SPACE$(40)
     LOCATE K, 5: PRINT "TOTAL": LOCATE K, 30: PRINT SCATION
     K = K + 2
     LOCATE K, 5: PRINT SPACE$(40)
     LOCATE K, 5
     IF OPT = 1 THEN PRINT TITLE2$ ELSE PRINT TITLE4$
END IF
'K = K + 1
LOCATE K + 1, 5: PRINT SYMBOL$(I)
LOCATE K + 1, 15

IF OPT = 1 THEN
     PRINT USING V$; CONCENTRATION(I); COL2(I); COL3(I)
     ELSE
     PRINT USING V$; COL2(I); COL3(I)
END IF
K = K + 1
NEXT I

'K = K + 1
LOCATE K, 5: PRINT "TOTAL"; SPACE$(20); SANION
'K = K + 1
LOCATE K, 5
COLOR 20, 2
PRINT "Total Error          ="; EERROR; "%"

TOTHARD1 = COL2(1) + COL2(2)          'Total hradness as Ca++ + Mg++
TOTHARD2 = COL3(1) + COL3(2)
```

```
COLOR 15, 1
K = K + 1
LOCATE K + 1, 5
PRINT "Tolal Hardness       =";
PRINT USING V1$; TOTHARD1; : PRINT "  meq/L="
LOCATE K + 1, 45
PRINT USING V1$; TOTHARD2;
PRINT "  mg/L as CaCo3"
TOTHARD3 = COL2(7) + COL2(8)           'Carbonate hardness as Co3 + Hco3
TOTHARD4 = COL3(7) + COL3(8)
LOCATE K + 2, 5
PRINT "Carbonate Hardness   =";
PRINT USING V1$; TOTHARD3; : PRINT "  meq/L="
LOCATE K + 2. 45
PRINT USING V1$; TOTHARD4; : PRINT "  mg/L as CaCo3"

LOCATE K + 3, 5
PRINT "NonCarbonate Hardness =";
NONCARB = TOTHARD1 - TOTHARD3
IF NONCARB < 0 THEN NONCARB = 0
PRINT USING V1$; NONCARB; : PRINT "  meq/L="
NONCARB = TOTHARD2 - TOTHARD4
IF NONCARB < 0 THEN NONCARB = 0
LOCATE K + 3, 45
PRINT USING V1$; NONCARB; : PRINT "  mg/L as CaCo3"
'Determine the desgree of hardness
chk = TOTHARD2
IF chk > 0 AND chk <= 75 THEN d$ = "Soft"
IF chk > 75 AND chk <= 150 THEN d$ = "Moderately Soft"
IF chk > 150 AND chk <= 175 THEN d$ = "Moderately Hard"
IF chk > 175 AND chk <= 300 THEN d$ = "Hard"
IF chk > 300 THEN d$ = "Very Hard"
COLOR 20, 2
LOCATE 22, 4
PRINT "The degree of hardness is "; d$
COLOR 15, 1

DO
LOOP WHILE INKEY$ = ""
END SUB

SUB PART2
'*************************************
'PLOTTING ROUTINE
SHARED EW(), SYMBOL$(), CONCENTRATION(), OPT. COL2(), COL3()
DIM P(12)
V$ = "##.##"
SCREEN 9
CALL BOX(1, 1, 22, 80)
SUMCAT = 0
SUMAN = 0
FOR I = 1 TO 12
IF I <= 6 THEN SUMCAT = SUMCAT + COL2(I)
IF I > 6 THEN SUMAN = SUMAN + COL2(I)
P(I) = COL2(I)
NEXT I
'SCALE VALUES
sum1 = 0
FOR I = 1 TO 6
P(I) = INT(P(I) * 70 / SUMCAT)
sum1 = sum1 + P(I)
NEXT
```

```
FOR I = 7 TO 12
P(I) = INT(P(I) * suml / SUMAN)
NEXT

ROW = 10
LOCATE ROW, 3
FOR I = 1 TO 6
COLOR I, 1
IF I = 1 THEN COLOR 15, 1
K = P(I)
PRINT STRING$(K, 219);
NEXT I

X = 3
XSUM = 0
COLOR 15, 1
FOR I = 1 TO 6
IF P(I) > 0 THEN
    LOCATE ROW - 1, X + INT(.5 * P(I))
    PRINT SYMBOL$(I)
    X = X + P(I)
    XSUM = XSUM + COL2(I)
    LOCATE ROW - 2, X
    PRINT USING V$; XSUM
END IF
NEXT I

LOCATE ROW + 1, 3
FOR I = 7 TO 12
COLOR I, 1
K = P(I)
PRINT STRING$(K, 219);
NEXT I

X = 3
XSUM = 0
FOR I = 7 TO 12
IF P(I) > 0 THEN
    LOCATE ROW + 2, X + INT(.5 * P(I))
    PRINT SYMBOL$(I)
    X = X + P(I)
    LOCATE ROW + 3, X
    XSUM = XSUM + COL2(I)
    PRINT USING V$; XSUM
END IF
NEXT I

DO
LOOP WHILE INKEY$ = ""
END SUB
```

2.5 Biological Characteristics

2.5.1 Dissolved Oxygen

Dissolved oxygen (DO) is of paramount importance in aerobic metabolic reactions. The saturation concentration of oxygen may be determined from Equation 2.18.

$$C_s = K_D C_g \tag{2.18}$$

where:

C_s = Saturation concentration, g/m^3
K_D = Distribution coefficient
C_g = Gas concentration in gas phase, g/m^3 (may be calculated from Equation 2.19)

$$Cg = \frac{P_g MW}{RT} \tag{2.19}$$

where:

P_g = Partial pressure of the respective gas in the gas phase, Pa (may be calculated from Equation 2.20)

$$P_g = x_g k_H \tag{2.20}$$

where:

x_g = Mole fraction of gas (may be calculated from Equation 2.21)

$$x_g = \frac{N_g}{N_g + N_w} \tag{2.21}$$

where:

N_g = Moles of gas
N_w = Moles of water
k_H = Henry's constant
MW = Molecular weight of the gas
R = Universal gas constant = 8314.3 J/kg*K 0.082 L.atm/K.mol
T = Absolute temperature, K

Equation 2.22 relates oxygen solubility concentration to pressure.

$$C' = \frac{C_s(P - p_w)}{(760 - p_w)} \tag{2.22}$$

where:

C' = Solubility of oxygen at barometric pressure P and given temperature, mg/L
C_s = Saturation concentration at given temperature, mg/L
P = Barometric pressure, mmHg
p_w = Pressure of saturated water vapor at the temperature of the water, mmHg

Oxygen is slightly soluble in water. The actual content of oxygen that can be found in solution is influenced by different interacting parameters such as solubility of the gas, partial pressure of the gas in the gas phase, temperature, and degree of water purity (e.g., level of salinity, concentration of suspended solids, etc.). Drinking water saturated with oxygen has a pleasant taste, while water devoid of dissolved oxygen has an insipid taste.

Example 2.10

1. Write a computer program to evaluate the saturation concentration of oxygen dissolved in water for any value of pressure and at any temperature and chloride concentration in water.
2. Calculate the concentration of dissolved oxygen in a water sample (with zero salinity) having a dissolved oxygen saturation concentration of 9.2 mg/L at a temperature of 20°C for an atmospheric gauge pressure of 695 mmHg.

3. Use the computer program developed in (1) to verify computations achieved in (2).

Solution

1. For solution to Example 2.10 (1), see listing of Program 2.10 below.
2. Solution to Example 2.10 (2):
   - Given T = 20°C; C′ = 9.2 mg/L; P = 695 mmHg; salinity = 0.
   - From Table A2 in Appendix,[5,22] for T = 20°C find vapor pressure as p_w = 17.535 mmHg.
   - Find the oxygen saturation concentration as: $C_s = C'*(760 - p_w)/(P - p_w) = 9.2*(760 - 17.535)/(695 - 17.535) = 11.08$ mg/L.

Listing of Program 2.10

```
DECLARE SUB BOX (R1%, C1%, R2%, C2%)   'Found in Program 2.1
'************************************************************
'Program 2.10: DO
'Computes Dissolved oxygen concentration
'************************************************************
DIM T(31), Pwi(31)
DATA 5,5,5,6,6,7,7,8,8,9,9,10,11,11,12,13,14,15,16,17,18,19,20
DATA 21,22,24,25,27,28,30,32

'Table of temperature-pressure relationship
FOR i = 1 TO 31
T(i) = i - 1
READ Pwi(i)
NEXT i

CALL BOX(2, 1, 20, 80)
LOCATE 3, 15
PRINT "Program 2.11: Computes Dissolved oxygen concentration"

LOCATE 5, 5
INPUT "Enter Temperature in the range 0 to 30 C        ="; T1
LOCATE 6, 5
INPUT "Enter gauge pressure   in mm Hg                 ="; P
LOCATE 7, 5
INPUT "Enter oxygen solubility concentration  in mg/L  ="; C

'Calcualtions
'Find Pw corresponding to temperature T
FOR i = 1 TO 31
IF T1 <= T(i) THEN
     Pw = Pwi(i)
     GOTO jmp
END IF
NEXT i
jmp:

Ps = 760       'mmHg
Cx = C * (Ps - Pw) / (P - Pw)
```

```
LOCATE 10, 3
PRINT "Oxygen saturation concentration at the given temperature="; Cx; " mg/L"
LOCATE 11, 3
PRINT "Salinity was assumed=0"

DO
LOOP WHILE INKEY$ = ""

END
```

2.5.2 Biochemical Oxygen Demand

The biochemical oxygen demand (BOD) test measures the relative amount of oxygen that is needed to biologically stabilize organic matter present in a sample. The advantages of the test include estimation of the size of treatment units, evaluation of treatment efficiency, and estimation of the relative amount of oxygen required for oxidation of organic pollutants.[14]

The BOD reaction is assumed to follow a first-order reaction and it may be expressed as presented in Equation 2.23.

$$\frac{L_t}{L_0} = e^{-k't} = 10^{-k'_1 t} \tag{2.23}$$

where:

L_t = Amount of first stage BOD remaining in the sample at time t, mg/L
L_o = Initial remaining BOD, total or ultimate first stage BOD present at zero time
k' = Rate constant (to base e), per day
k'_1 = Rate constant (to base 10), per day (= 0.4343*k′)
t = Time, days

Equation 2.24 gives the value of BOD exerted at any time t and temperature T (see Figure 2.5).

$$BOD_t^T = L_0 - L_t = L_0(1 - 10^{-k'_1 t}) \tag{2.24}$$

The usual standard for reporting BOD is based on the 5-d BOD at a temperature of 20°C (BOD_5^{20}). Oxidation of the organic matter is from 60 to 70% complete within the 5-d test period. The 20°C temperature represents an average value for slow-moving streams in temperate climates and is easily duplicated in the laboratory.[21] Table 2.3 gives the wastewater strength in terms of BOD_5 or chemical oxygen demand (COD).

The chemical oxygen demand test is employed to measure the amount of oxygen required to oxidize organic matter by using a strong chemical oxidizing agent in an acidic medium. The standard reagent used in the determination is a boiling mixture of concentrated sulfuric acid and potassium dichromate ($K_2Cr_2O_7$), together with

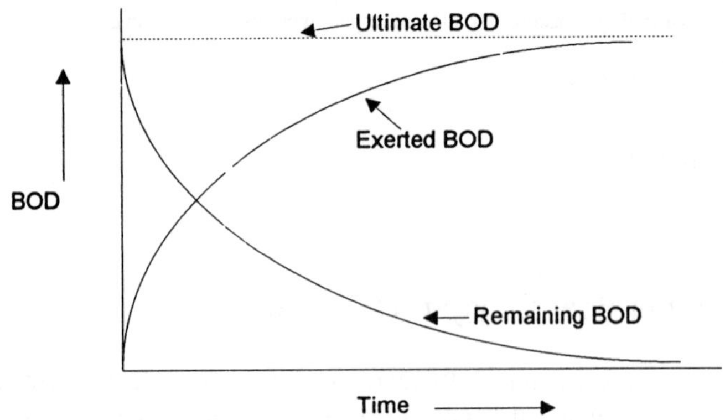

Figure 2.5
Variation of BOD with time.

TABLE 2.3
Sewage Strength in Terms of BOD and COD[14]

Strength	BOD_5 (mg/L)	COD (mg/L)
Weak	<200	<400
Medium	200–350	400–700
Strong	351–500	701–1000
Very strong	>750	>1500

Source: From Rowe, D.R. and Abdel-Magid, I.M., *Handbook of Wastewater Reclamation and Reuse*, CRC Press/Lewis Publishers, Boca Raton, FL, 1995. With permission.

silver sulfate as a catalyst. Generally, the COD of a wastewater is greater than its 5-d BOD. This is for the following reasons:

1. More compounds are likely to be oxidized chemically than can be oxidized biologically.
2. The 5-d BOD does not equal the ultimate BOD.

The advantages of the COD test include:

1. The test can be conducted in a short period of time (2 to 3 h), which is more convenient than the 5-day period required for the determination of the BOD.
2. The test can be used to aid in design, operation, and control of treatment works (if related with BOD).

The Thomas method can be used to determine values for k′ and L_0 from a series of BOD measurements. This method is based on the similarity of two series functions (expansions for $1 - e^{-k'*t}$) and $k'*t*(1 + k'*t/6)^{-3}$)). Thomas developed the approximate linearized formula presented in Equation 2.25.

$$\sqrt[3]{\frac{t}{BOD}} = \sqrt[-3]{k'L_0} + \frac{t\sqrt[3]{(k')^2}}{6\sqrt[3]{L_0}} \qquad (2.25)$$

where:

t = Time, s
BOD = BOD exerted in time t, mg/L
k′ = Reaction rate constant (to base e), d^{-1}
L_0 = Ultimate BOD, mg/L

Equation 2.25 can be put in the form of a straight line, as indicated in Equation 2.26.

$$y = ax + b \qquad (2.26)$$

where:

y = $(t/BOD)^{1/3}$
b = Intercept of the line

$$b = (k'L_0)^{-1/3} \qquad (2.27)$$

Slope (a) of the straight line is given by:

$$a = \frac{\sqrt[3]{k}^{2}}{6\sqrt[3]{L_0}} \qquad (2.28)$$

The reaction rate constant and ultimate BOD may therefore be determined as shown in Equations 2.29 and 2.30.

$$L_0 = \frac{1}{k'b^3} \qquad (2.29)$$

$$k' = \frac{6a}{b} \qquad (2.30)$$

Example 2.11

1. Write a computer program to determine the BOD of a wastewater at any time t, rate constant k, and temperature T. Design the program so that t or k or L_0 can be determined, given the other two parameters. The program should show the strength of the sewage in accordance with Table 2.3
2. Find the amount of the 5-d BOD as compared to the ultimate BOD for a sewage sample which has a rate constant (k_1') value of 0.103/day.

3. A wastewater sample has a 5-d BOD of 324 mg/L and an ultimate BOD of 405 mg/L. Find the rate of biological oxidation for the wastewater.
4. The 3-d BOD of a sample is 255 mg/L at a temperature of 20°C. Determine its 5-d BOD, assuming that the rate constant is 0.22/d (to base 10).

Solution

1. For solution to Example 2.11 (1), see listing for Program 2.11 below.
2. Solution to Example 2.11 (2):
   - Given $k_1 = 0.103/d$.
   - Find the ratio of 5-d BOD to ultimate BOD for the sample as: $BOD_5^{20}/L_0 = (1 - 10^{-k'_1 * t}) = 1 - 10^{-0.103*5} = 0.69$. This shows that the 5-d BOD is approximately 69% of the ultimate BOD.
3. Solution to Example 2.11 (3):
   - Given: $BOD_5^{20} = 324$ mg/L, $L_0 = 405$ mg/L.
   - Determine the reaction rate constant as: $324 = 405(1 - 10^{-5k'_1})$. This yields a value for k'_1 equal to 0.14/d.
4. Solution to Example 2.11 (4):
   - Given: $BOD_3 = 255$ mg/L, $k'_1 = 0.22/d$.
   - Determine ultimate BOD as: $L_0 = BOD/(1 - 10^{-k'_1 * t}) = 255/(1 - 10^{-0.22*3}) = 326.4$ mg/L.
   - Find the 5-d BOD as: $BOD_5 = 326.4*(1 - 10^{-0.22*5}) = 300$ mg/L.

Listing of Program 2.11

```
DECLARE SUB box (r1%, c1%, r2%, c2%)   'Found in Program 2.1
'*****************************************************************
'Program12: BOD-1
'Estimates the BOD for wastewater for a given rate k, temperayture T
' and at any time t
'*****************************************************************
DEF fnlog10 (x) = LOG(x) / LOG(10)
DIM ttl$(5)
ttl$(1) = "1. Determine BOD at any time t"
ttl$(2) = "2. Determine the reaction rate constant k1 to base 10"
ttl$(3) = "3. Determine the reaction rate constant k1 to base e"
ttl$(4) = "4. Determine the ultimate BOD (Lo)"
ttl$(5) = "5. Exit program"
start:
CALL box(1, 1, 20, 80)
LOCATE 2, 10
PRINT "Program 2.12:Estimates the BOD for wastewater"
FOR i = 1 TO 5
LOCATE i + 3, 5: PRINT ttl$(i)
NEXT i
LOCATE 10, 3
DO
INPUT "Select an option (1 to 5): "; opt
LOOP WHILE opt < 1 OR opt > 5
```

```
SELECT CASE opt
CASE 1
        'Determine BOD at any time t
        CALL box(1, 1, 20, 80)
        row = 3
        LOCATE row, 5: INPUT "Enter rate constant k1 (/day) to base 10 ="; k1
        LOCATE row + 1, 5: INPUT "Enter the time t (days), t          ="; t
        LOCATE row + 2, 5: INPUT "Enter the 5-day BOD in mg/L         ="; BOD5
        'First order reaction rate
        BODRATIO = (1 - 10 ^ (-k1 * t))
        LOCATE row + 4, 5
        PRINT "The value of BOD at"; t; " days="; BODRATIO; " of the ultimate BOD"
        B = BODRATIO * BOD5
        LOCATE row + 5, 5
        PRINT "The value of BOD at "; t; " days="; B; "  mg/L"

        'Classify sewage strength
        IF B <= 200 THEN c$ = "Weak"
        IF B > 200 AND B <= 350 THEN c$ = "Medium"
        IF B > 350 AND B <= 500 THEN c$ = "Strong"
        IF B > 750 THEN c$ = "Very Strong"
        LOCATE row + 6, 5
        PRINT "The strength of the sewage is "; c$
        DO
        LOOP WHILE INKEY$ = ""

CASE 2
        'Determine k1 to base 10
        CALL box(2, 1, 20, 80)
        row = 5
        LOCATE row, 3
        INPUT "Enter the value of the ultimate BOD mg/L ="; Lo
        LOCATE row + 1, 5
        INPUT "Enter the value of the 5-day BOD mg/L   ="; BOD5
        k1 = -LOG(1 - BOD5 / Lo) / (t * LOG(10))
        LOCATE row + 3, 5
        PRINT "Rate at which water is being oxidized k1(to base 10)="; k1; " /day"
        DO
```

Example 2.12

1. Write a computer program for the estimation of the ultimate BOD and the rate constant by using the Thomas method for the following data.
2. A BOD test for a wastewater sample, at a temperature of 20°C, gave the following results:

t (days)	0.5	1	1.5	2	2.5	3	3.5	4	4.5	5
BOD (mg/L)	63	112	150	180	202	220	234	241	248	254

Using the Thomas method find:

a. The reaction rate constant
b. Ultimate first stage BOD.
c. The value of k'_1 (to base 10).

Solution

1. For solution to Example 2.12 (1), see listing for Program 2.12 below.
2. Solution to Example 2.12 (2):
   - Use Thomas method to construct the table shown below:

t (d)	BOD (mg/L)	$(t/BOD)^{1/3}$
0.5	63	0.1994695
1	112	0.2074566
1.5	150	0.2154435
2	180	0.2231443
2.5	202	0.231311
3	220	02389092
3.5	234	0.2463868
4	241	0.2550827
4.5	248	0.2627768
5	254	0.2700093

- Plot a graph of $(t/BOD)^{1/3}$ vs. t.
- From the constructed straight line estimate the slope and intercept as: a = slope = $0.015729 = k'^{2/3} / 6L_0^{1/3}$ (1). b = intercept = $0.191745 = (k'^{*}L_0)^{-1/3}$ (2).
- Use Equations 1 and 2 above to find: $k' = 0.492$/day (to base e) and $L_0 = 288.5$ mg/L.
- Determine the reaction rate constant to base 10 as: $k'_1 = 0.4343*k' = 0.4343*0.492185 = 0.214$/d.

Listing of Program 2.12

```
DECLARE SUB box (r1%, c1%, r2%, c2%)   'Fopund in Program 2.1
'****************************************************************
'Program 2.12: BOD-2
'Thomas method for determining BODE
'***************************************************************
DIM T(50), BOD(50), tbod(50), x2(50), xy(50)

CALL box(1, 1, 20, 80)
LOCATE 2, 20
PRINT "Program 2.13: Thomas method for determining BOD"
```

```
ROW = 3
LOCATE ROW, 5

INPUT "Enter number of data points (N)="; N

SXX = 0
SXY = 0
SXN = 0
SYN = 0
v$ = "##.####    ###.####      #.########      #.#######"

FOR I = 1 TO N
LOCATE ROW + I, 5: INPUT "Enter t(Days)="; T(I)
LOCATE ROW + I, 35: INPUT "Enter BOD(mg/L)="; BOD(I)
tbod(I) = (T(I) / BOD(I)) ^ (1 / 3)
SXX = SXX + T(I) ^ 2
SXN = SXN + T(I)
SXY = SXY + T(I) * tbod(I)
SYN = SYN + tbod(I)
NEXT I

'Perform regression on the data to find the slope a and intercept b
'Find statistical properties
XBAR = SXN / N
YBAR = SYN / N
SXX = SXX - SXN ^ 2 / N
SXY = SXY - SXN * SYN / N
a = SXY / SXX
b = YBAR - a * XBAR
kd = 6 * a / b          'Reaction rate constant
Lo = 1 / (b ^ 3 * kd)   'Ultimate first stage BOD
k1 = .4343 * kd         'Reaction rate constant

'Output results
CALL box(2, 1, 20, 80)

ROW = 7
LOCATE ROW, 3
PRINT "RESULTS:"
LOCATE ROW + 1, 5
PRINT "The reaction rate constant K'(to base e)="; kd; " /day"
LOCATE ROW + 2, 5
PRINT "Ultimate first stage BOD            Lo="; Lo; " mg/L"
LOCATE ROW + 3, 5
PRINT "Reaction rate constant  K1(to base 10)  ="; k1; " /day"

LOCATE 23, 3
PRINT "press any key to continue....................";
DO
LOOP WHILE INKEY$ = ""

END
```

2.6 Radioactivity

Radioactivity is a property of unstable atoms. It arises from the spontaneous breaking up of certain heavy atoms into other atoms, which themselves might be radioactive. This phenomenon continues producing a transformation series. The disintegration results in three kinds of radioactive emissions known as alpha particles, beta particles, and gamma rays (Figure 2.6).

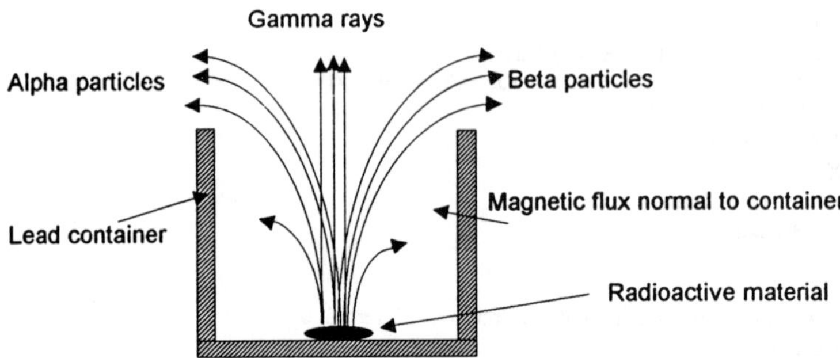

Figure 2.6
Radioactive decay.

Each radioactive substance is characterized by its half life. The half-life period is the time required for half the atoms in any given sample of the substance to decay. The half-life periods are different for each element. The unit of radioactivity is the Curie, which is the number of disintegrations occurring per second in one gram of pure radium.[14]

The rate of decay of any nuclide is directly proportional to the number of atoms present and follows a first order reaction as given by Equation 2.31.

$$\text{Ln}\,\frac{N_t}{N_0} = -\,kt \tag{2.31}$$

where:

N_0 = Number of atoms at time t_0
N_t = Number of atoms found at time t
k = Decay constant
t = Time

Equation 2.32 gives the relationship between decay time and half life.

$$k_t = \frac{0.6932}{t_{1/2}} \tag{2.32}$$

where:

$t_{1/2}$ = Half life of a particular nuclide

Example 2.13

1. Write a computer program for the computation of the number of atoms present after time t for any fraction of the remaining mass. Design the program so as to determine the time needed by the following substances to lose any fraction of their initial mass: polonium, Po^{218} (3.05 min); bromine, Br^{78} (6.4 min); lead, Pb^{214} (26.8 min); sodium, Na^{24} (15 hs); fluorine, F^{131} (8 d); phosphorous, P^{32} (14.3 d); thorium, Th^{234} (24.1 d); cobalt, Co^{60} (5.3 years); hydrogen, H^{3} (12.3 years); cesium, Cs^{137} (30 years); radium, Ra^{226} (1600 years); carbon, C^{14} (5730 years); strontium, Sr^{90} (28.1 years); uranium, U^{234} ($2.48*10^5$ years); potassium, K^{40} ($1.28*10^9$ years); and uranium, U^{238} ($4.51*10^9$ years). (Note: number between parentheses indicates the half life period of the radioactive substance).
2. Radioactive radium has a half life of 1600 years. Find the mass remaining unchanged of 50 units of this radioactive material after 3200, 4800, and 6400 years. Compute the time needed for the substance to lose 10% of its mass.

Solution

1. For solution to Example 2.13 (1), see listing of Program 2.13 below.
2. Solution to Example 2.13 (2):
   - Determine the decay constant as: $k_t = 0.693/t_{1/2} = 0.693/1600 = 4.33*10^{-4}$ year $^{-1}$.
   - Find mass remaining after time t as: $N = N_0*e^{-k*t}$ for t = 1600 $N = 50*e^{-0.000433*1600}$ = 25; for t = 3200, N = 12.5; for t = 4800, N = 6.25; for t = 6400, N = 3.13.
   - Plot graph of mass remaining vs. time and find from graph for a remaining mass of (= 50 – (10*50/100)) 45, the required time = 243 years, or weight lost = 10*50/100 = 5; weight remaining = 50 – 5 = 45.
   - Or use Ln 45/50 = – $4.33*10^{-4}$*t to find t = 243 years.

Listing of Program 2.13

```
DECLARE SUB box (r1%, c1%, r2%, c2%)   'Found in Program 2.1
'*****************************************************************
'Program 2.13: radioactivity
'Computes radioactivity
'*****************************************************************
DIM elem$(16), haf(16), sym$(16)
F1 = 1 / 365.25              'convert days to years
F2 = F1 / 24                 'convert hours to years
F3 = F2 / 60                 'convert minutes to years
DATA "Br78","C14","Co60","Cs137","F131","H3","Pb124","K40","Na24"
DATA "P32","Po218","Ra226","Sr90","Th234","U238","U234"
DATA "Bromine","Carbon","Cobalt","Cesium","Flourine","Hydrogen"
DATA "Lead","Potassium","Sodium",Phosphorous","Polonium"
DATA "Radium","Strontium","Thorium","Uranium","Uranium"
DATA 6.4,5730,5.3,30,8,12.3,26.8,1.28E9,15,14.3,3.05
DATA 1600,28.1,24.1,4.51E9,2.48E5
```

```
'Read data
FOR i = 1 TO 16
READ sym$(i)
NEXT
FOR i = 1 TO 16
READ elem$(i)
NEXT
FOR i = 1 TO 16
READ haf(i)
NEXT
'convert all half times to years
haf(5) = haf(5) * F1
haf(7) = haf(7) * F3
haf(9) = haf(9) * F2
haf(10) = haf(10) * F1
haf(11) = haf(11) * F3
haf(14) = haf(14) * F1
CALL box(1, 1, 20, 80)
LOCATE 2, 25
PRINT "Program 2.8: Radioactivity"
LOCATE 3, 3
FOR i = 1 TO 8
LOCATE 3, 5 + (i - 1) * 9
PRINT i; sym$(i)
LOCATE 4, 5 + (i - 1) * 9
PRINT i + 8; sym$(i + 8)
NEXT
LOCATE 5, 5
PRINT "17 Other material"
LOCATE 6, 3
INPUT "Select a material number from above and press Enter:"; num

IF num = 17 THEN
     CALL box(1, 1, 20, 80)
     LOCATE 2, 25
     PRINT "Program 2.8: Radioactivity/ Screen 2"
     LOCATE 3, 3
     INPUT "Enter the symbol of the material :"; sym$
     LOCATE 4, 3
     INPUT "Enter the half life period        :"; th
END IF
IF num <> 17 THEN th = haf(num)
'Decay constant
kt = .693 / th          'th is the half life period
LOCATE 7, 3
INPUT "Initial mass of material="; no

cycle:
     LOCATE 8, 3
     INPUT "What period of time for decay in years:"; t
     'Mass remaining after any time t for an initial mass no
     n = no * EXP(-kt * t)
     LOCATE 10, 5
     PRINT "Mass remaining after time period t is:"; n
     LOCATE 11, 3
     INPUT "Do you want the mass after another period? Y/N: "; R$
IF UCASE$(R$) = "Y" THEN GOTO cycle
```

```
repeat:
    LOCATE 12, 3
    INPUT "What percentage of mass to loose :"; p

    'to find the time to loose  a percentage p
    n1 = (100 - p) / 100 * no
    t1 = LOG(no / n1) / kt
    LOCATE 14, 5
    PRINT "Time to loose that percentage="; t1; "  years"

    LOCATE 16, 3
    INPUT "Any other percentage ? Y/N :"; ans$
    LOCATE 16, 3
    PRINT SPACE$(30)

IF UCASE$(ans$) = "Y" THEN GOTO repeat
END
```

2.7 Water Resources and Sources, Usage, and Groundwater Flow

2.7.1 Introduction

The sources for a water supply can be grouped into three major categories: rainwater (precipitation), surface water (runoff), and groundwater. Surface water includes both rainwater and groundwater where the water table lies above the ground surface. Surface waters include waters in rivers, streams, lakes, lagoons, ponds, and oceans. Groundwater is contained in the soil or rocks below the water table.[34]

The movement of water on the earth follows a complex cyclic pattern from the sea to the atmosphere and then by precipitation back to the earth, where it collects in rivers, streams and lakes and then runs back to the sea. The cycle may be short-circuited at several stages; there is no uniformity in the time a cycle takes, and the intensity and frequency of the cycle depends on geographical and climatic conditions.[24] (See Figure 2.7.)

2.7.2 Source Selection

The process of selecting a water source must consider the following items:

1. Is the source free from toxic or undesirable chemicals?
2. Does the source contain:
    - Domestic sewage, which may contain disease-causing agents, or undesirable substances that pose a health hazard?
    - Industrial waste, which may contain materials that may render the water unsuitable for the intended use?

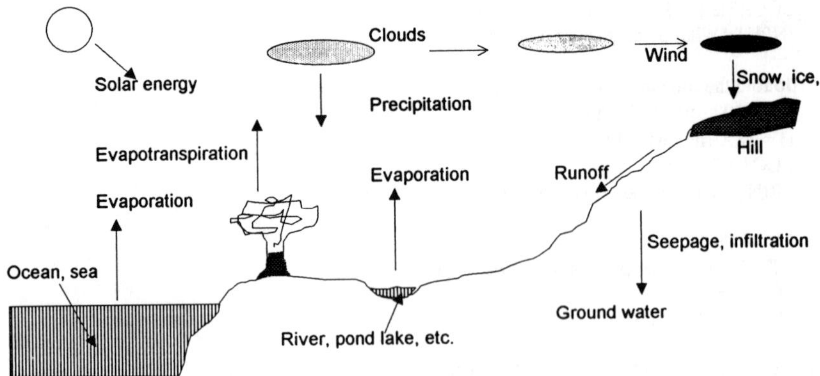

Figure 2.7
Schematic diagram of the hydrological cycle.

- Agricultural waste, which may contain residues of herbicides, insecticides, pesticides, and nutrients?

Figure 2.8 gives an outline that can be followed during the process of source selection.

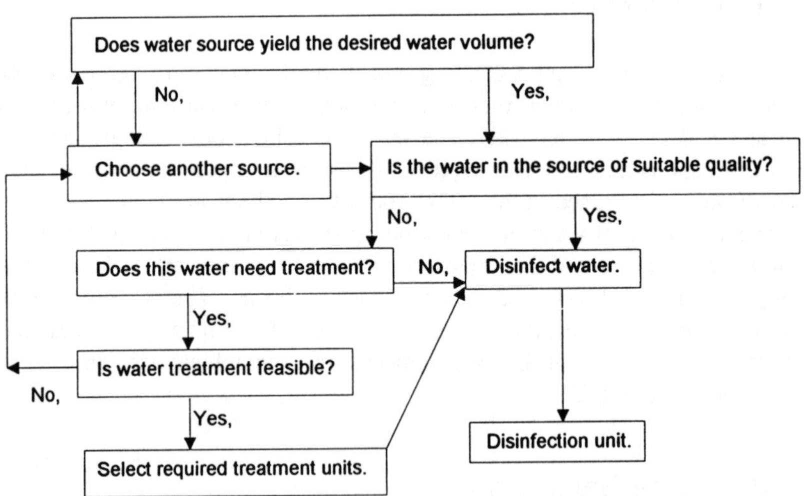

Figure 2.8
Source selection.

Water use and consumption can be grouped into several broad categories: domestic, industrial, commercial and public, agricultural, and recreational. Unaccountable water consumption in general makes up about 10 to 15% of the total water consumption. The total consumption is the sum of all the aforementioned uses and losses. Water use and consumption data are generally expressed in liters per capita per day, or L/c/d.

There is a wide range of uses for water and properly treated and reclaimed wastewater, such as in power plant and industrial cooling water, groundwater recharge, aquaculture (fish farming), silviculture, landscape irrigation, recreational purposes, fire fighting, industrial processes, stock watering, stream flow augmentation, and agricultural irrigation.

2.7.3 Fire Demand

Water is required to extinguish fires and to prevent them from spreading into uncontrollable firestorms. Usually, the quantity of water to be used for firefighting is small; nevertheless, the rate of application is high. For a community of 20,000 or less,[25] Equation 2.33 may be used to estimate the fire demand.

$$Q = 3860 \cdot \sqrt{POP} \cdot (1 - 0.01\sqrt{POP}) \tag{2.33}$$

where:

Q = Fire demand, L/min
POP = Population, in thousands

The U.S. Insurance Service Office (ISO) advocates the use of the formula presented in Equation 2.34 to estimate the fire flow required for a given floor area.

$$Q = 3.7C_f\sqrt{A} \tag{2.34}$$

where:

Q = Required fire flow, L/s
C_f = Coefficient related to the type of construction
A = Total floor area including all stories in the building, but excluding its basement, m^2

The values of the coefficient C_f may be taken as presented in Table 2.4.

TABLE 2.4
Values of the Coefficient C_f[5,11]

Description	C_f
Wood frame construction	1.5
Ordinary construction	1.0
Noncombustible construction	0.8
Fire-resistive construction	0.6

Regardless of the computed value, the fire flow need not exceed 500 L/s for wood-frame or ordinary construction, or 380 L/s for noncombustible or fire-resistive constructions. For normal one-story construction of any type, the fire flow may not exceed 378 L/s. The minimum fire flow may not be less than 32 L/s, and the maximum for all purposes for a single fire is not to exceed 760 L/s. Extra flow may be needed to protect adjacent buildings. For groups of single-family and small two-family residences not exceeding two stories in height, the fire flow outlined in Table 2.5 may be used to evaluate the required fire demand.[11]

TABLE 2.5
Residential Fire Flows[11]

Distance between adjacent units (m)	Required fire flow (L/min.)
>30.5	1890
9.5–30.5	2835–3780
3.4–9.2	3780–5670
≤3	5670–7660[a]

[a] For continuous construction, use 9450 L/min.

The maximum flow required for an individual fire is 45.4 m^3/min for all purposes. In large communities, the possibility of concurrent fires should also be considered. In residential districts, the required fire flow ranges from a minimum of 1.9 m^3/min to a maximum[11] of 9.5 m^3/min. Hydrant spacing is dictated by the required fire flow, since the capacity of a single hydrant is limited. Table 2.6 presents proposed distances for fire hydrants. The required duration for a fire flow is indicated in Table 2.7.

TABLE 2.6
Fire Hydrant Spacing[11]

Hydrant	Distance (m)	Location
Ordinary	60–150	At street intersections where streams can be taken in any direction
High value districts	As close as 30	Additional hydrants may be necessary in the middle of long blocks
Low risk areas	100–150	In towns

Example 2.14

1. Write a computer program that enables the computation of fire demand for communities of varying populations, taking into consideration the varying types of construction and restrictions outlined in Section 2.7.3.
2. Find the fire demand for a town having a population of 0.9 million.
3. Estimate the fire demand for a single-story building made of wood-frame construction, given that the total floor area amounts to 10,000 m^2.

TABLE 2.7
Fire Flow Duration[11]

Required fire flow (L/min)	Duration (h)
<3780	4
3780–4725	5
4725–5670	6
5670–6615	7
6615–7560	8
7560–8505	9
>8505	10

Solution

1. For solution to Example 2.14 (1), see listing of Program 2.14 below.
2. Solution to Example 2.14 (2):
   - Given: POP = 900,000 = 900 in thousands.
   - Determine the fire demand as: $Q = 3860(POP)^{1/2}[(1 - 0.01(POP^{1/2})] = 3860*900^{1/2}*(1 - 0.01*900^{1/2}) = 81\ m^3/min$.
3. Solution to Example 2.14 (3):
   - Given: $A = 10000\ m^2$.
   - Determine the fire flow as: $Q = 3.7*C_f*(A)^{0.5}$.
   - From Table 2.4 for wood-frame construction, the coefficient, $C_f = 1.5$.
   - Compute the fire demand as: $Q = 3.7*1.5*(10,000)^{0.5} = 555$ L/s.
   - Since this is a wood-frame single-story building, however, the maximum fire demand may not exceed 380 L/s.

Listing of Program 2.14

```
DECLARE SUB box (r1%, c1%, r2%, c2%)  'Found in Program 2.1
'**************************************************************
'Program 2.14: Fire
'Computes Fire demand for communities of different population
'**************************************************************

DIM const$(4), ttl$(3), Cw(4), limit(4)
const$(1) = "(1) Wood frame construction"
const$(2) = "(2) Ordinary construction"
const$(3) = "(3) Non-conmbustible construction"
const$(4) = "(4) Fire-resistive construction"
limit(1) = 500
limit(2) = 500
limit(3) = 300
limit(3) = 300
Cw(1) = 1.5
Cw(2) = 1
Cw(3) = .8
Cw(4) = .6
```

```
start:
prompt$ = "Select an option 1 to 4:"
ttl$(1) = "(1) Estimate fire demand for a building"
ttl$(2) = "(2) Estimate fir demand for a whole town"
ttl$(3) = "(3) Exit the program"
CALL box(2, 1, 20, 80)
LOCATE 3, 20
PRINT "Program 2.14:Computes Fire demand for communities"
row = 5
FOR i = 1 TO 3
LOCATE row + i, 5: PRINT ttl$(i)
NEXT i

LOCATE row + 5, 3: PRINT "Select an option 1 to 3:";

INPUT OPT

SELECT CASE OPT
 CASE 1
     CALL box(2, 1, 20, 80)
     row = 5
     FOR i = 1 TO 4
     LOCATE i + row, 5: PRINT const$(i): NEXT i
     LOCATE row + 6, 3: PRINT prompt$;
     INPUT opt1
     LOCATE 13, 5: INPUT "enter Total floor area in sq.m="; A
     Q = 3.7 * Cw(opt1) * A ^ .5
     IF Q < 32 THEN Q = 32
     IF Q > limit(opt1) THEN Q = limit(opt1)
     LOCATE 15, 3: PRINT "Total fire demand for this building="; Q; "  L/sec"

 CASE 2
     CALL box(2, 1, 20, 80)
     LOCATE 5, 5: INPUT "Enter population of town ="; pop
     pop = (pop / 1000) ^ .5
     Q = 3860 * pop * (1 - .01 * pop)
     LOCATE 7, 5: PRINT "FIRE DEMAND FOR THIS TOWN="; Q; "  L/min"

 CASE 3
     END

END SELECT

LOCATE 23, 3
PRINT "press any key to continue.....................";

DO
LOOP WHILE INKEY$ = ""

GOTO start

END
```

2.7.4 Groundwater Flow

Groundwater aquifers are depleted by pumping, natural discharge to surface water systems, and, to a minor extent, evaporation and evapotranspiration. The balance between the depletion rate and the recharge of an aquifer determines the average annual water table level.[26] The principal method for depletion of an aquifer is the pumping of water. The main factors that influence groundwater flow are the following:

- Liquid characteristics, such as density, viscosity, etc.
- Media, through which the liquid flows, such as porosity and permeability
- Existing boundary conditions

Movement of groundwater starts from levels of higher energy to levels of lower energy. Basically, its energy is due to elevation and pressure. The velocity heads are neglected because flow is essentially laminar. The velocity of flow in laminar conditions reaches around 1 cm/sec, and the Reynolds number achieves values ranging between 1 to 10.[27] From the continuity equation, the groundwater flow for a particular aquifer may be determined from Equation 2.35.

$$Q = vA = kjA \tag{2.35}$$

where:

Q = Groundwater flow, m^3/s
v = Average velocity of flow in voids of water bearing material, m/s

$$v = v_a e \tag{2.36}$$

v_a = Actual velocity of flow, m/s
e = Porosity, dimensionless (see Table 2.8)
k = Coefficient of permeability (or conductivity), m/s (see Table 2.9)
j = Hydraulic gradient, dimensionless
A = Cross-sectional area perpendicular to flow direction, m^2

The area may be computed form Equation 2.37.

$$A = Bh \tag{2.37}$$

where:

B = Width of aquifer, m
h = Saturated thickness of aquifer, m

As such, the groundwater flow may be determined as indicated in Equation 2.38.

$$Q = kjBh \tag{2.38}$$

TABLE 2.8
Typical Total Porosities for Selected Materials[35]

Material	Porosity (%)
Unaltered granite and gneiss	0–2
Quartzites	0–1
Shales, slates, mica-schists	0–10
Chalk	5–40
Sandstones	5–40
Volcanic tuff	30–40
Gravels	25–40
Sands	15–48
Silt	35–50
Clays	40–70
Fractured basalt	5–50
Karst limestone	5–50
Limestone, dolomite	0–20

Source: From Nielsen, D.M., Ed., *Practical Handbook of Groundwater Monitoring*, Lewis Publishers, Chelsea, MI, 1991. With permission.

TABLE 2.9
Typical Hydraulic Conductivities for Selected Materials

Geologic Material	Range of K (m/s)
Coarse gravel	10^{-1}–10^{-2}
Sands and gravels	10^{-2}–10^{-5}
Fine sands, silts, loess	10^{-5}–10^{-9}
Clay, shale, glacial till	10^{-5}–10^{-13}
Dolomite limestone	10^{-3}–10^{-5}
Weathered chalk	10^{-3}–10^{-5}
Unweathered chalk	10^{-6}–10^{-9}
Limestone	10^{-3}–10^{-9}
Sandstone	10^{-4}–10^{-10}
Unweathered granite, gneiss, compact basalt	10^{-7}–10^{-13}

Source: From Nielsen, D.M., Ed., *Practical Handbook of Groundwater Monitoring*, Lewis Publishers, Chelsea, MI, 1991.

The transmissibility (TR) or hydraulic conductivity of a certain aquifer may be defined as the capacity of a unit prism of aquifer to yield water. The transmissibility also may be defined as the rate at which water of prevailing kinematic viscosity is transmitted through a unit width of an aquifer under a unit hydraulic gradient. Equation 2.39 may be used for determination of transmissibility of an aquifer.

$$TR = kh \tag{2.39}$$

where:

TR = Transmissibility of aquifer, $m^3/d*m$
k = Permeability coefficient, m/d
h = Saturated thickness of aquifer, m

Using the concept of transmissibility, the groundwater flow may be determined as presented in Equation 2.40.

$$Q = TR \cdot j \cdot B \tag{2.40}$$

where:

Q = Groundwater flow, m^3/s
TR = Transmissibility of aquifer, $m^3/d*m$
j = Hydraulic gradient, dimensionless
B = Width of aquifer, m

Example 2.15

1. Write a computer program to determine the actual velocity of flow in an aquifer, quantity of groundwater flow, transmissibility, given porosity, permeability, hydraulic gradient, saturated thickness, and width of aquifer.
2. An quifer has a porosity of 20% and an average grain size of 1.5 mm. Experimental studies reveal that a tracer needs 540 min to move between two observation wells, 20 m apart. Find the permeability of the aquifer to yield a difference in water surface elevation of 80 cm and a water temperature of 15°C.

Solution

1. For solution to Example 2.15 (1), see listing of Program 2.15 below.
2. Solution to Example 2.15 (2):
   - Given: e = 0.2, d = 1.5×10^{-3} m, L = 20 m, t = 540 min = 9 hr, h = 0.8 m
   - For a temperature of T = 15°C, find from Table A1 in the Appendix the coefficient of viscosity as: $\mu = 1.1447*10^{-3}$ $N*s/m^2$ and density $\rho = 999.099 kg/m^3$.
   - Find the velocity of the traveling tracer by using velocity = distance/time; thus, v_a = l/t = 20/9 = 2.22 m/h.
   - Determine the seepage velocity, v = k*j = k*h/l = k*0.8/20 = 0.04*k.
   - Find the actual velocity as: v = v_a*e = 2.22*(20/100) = 0.444 m/h = $1.235*10^{-4}$ m/s.
   - Compute the coefficient of permeability as: k = 0.444/0.04 = 11.1 m/h = $3.08*10^{-3}$ m/s.
   - Determine the value of the Reynolds number a: Re = $\rho*v*d/\mu$ = 999.1*1.235*10^{-4}*1.5*10^{-3}/1.1447*10^{-3} = 0.162. (Since this value is less than 1, then computations are justifiable).

Listing of Program 2.15

```
DECLARE SUB box (r1%, c1%, r2%, c2%)   'Found in Program 2.1
'**********************************************************************
'Program 2.15: Aquifers
'Computes Actual velocity of flow in aquifers
'**********************************************************************
DIM const$(4), ttl$(4), Cw(4)
ttl$(1) = "1. Determine actual velocity of flow"
ttl$(2) = "2. Determine flow rate"
ttl$(3) = "3. Determine transmissibility"
ttl$(4) = "4. Exit program"

start:
'print the menu
CALL box(1, 1, 20, 80)
LOCATE 3, 15
PRINT "Program 2.15:Computes Actual velocity of flow in aquifers"
FOR i = 1 TO 4
LOCATE i + 5, 5
PRINT ttl$(i)
NEXT i
DO
LOCATE 12, 3
INPUT "Select an option (1 to 4): "; opt
LOOP WHILE opt < 1 OR opt > 4


SELECT CASE opt
CASE 1
CALL box(1, 1, 20, 80)
row = 5
LOCATE row + 1, 5
INPUT "Enter the porosity of the material (as a ratio)  ="; por
LOCATE row + 2, 5
INPUT "Enter the permeability of the material in m3/d.m2="; per
LOCATE row + 3, 5
INPUT "Enter the hydraulic gradient                     ="; j
LOCATE row + 4, 5
INPUT "Enter the starturated thickness of the aquifer m ="; h
LOCATE row + 5, 5
INPUT "Enter the width of the aquifer  in m             ="; B
per = per / (24 * 3600) 'permeability in m/s
TR = per * h            'Transmissibility
Q = TR * j * B          'Groundwater flow
velocity = Q / (B * h)  'velocity of flow
LOCATE row + 7, 3
PRINT "Actual velocity of flow="; velocity; "  m/s"
DO
LOOP WHILE INKEY$ = ""

CASE 2

'Determine flow rate

CALL box(1, 1, 20, 80)

LOCATE 3, 5
INPUT "Enter transmissibilty  (m2/d)   ="; TR
```

```
LOCATE 4, 5
INPUT "Enter hydraulic gradient        ="; j
LOCATE 5, 5
INPUT "Enter width of aquifer (m)      ="; B
Q = TR * j * B
LOCATE 10, 5
PRINT "The flow rate       ="; Q; "  m3/d"
DO
LOOP WHILE INKEY$ = ""

CASE 3

'Determine transmissibility
CALL box(1, 1, 20, 80)

LOCATE 3, 5
INPUT "Enter the permeability  (m/d)         ="; k
LOCATE 4, 5
INPUT "Enter saturated thickness of aquifer (m) ="; h
TR = k * h     'transmissibility
LOCATE 7, 5
PRINT "The transmissibility  ="; TR; "  m3/d*m"

DO
LOOP WHILE INKEY$ = ""

CASE 4
    END

END SELECT

GOTO start
END
```

Field experiments were proposed to find the permeability coefficient of an aquifer by drilling observation wells in the cone of depression of a well and noting the corresponding drawdown in observation wells.[11] By applying the continuity equation and Darcy's law, the steady-state discharge from a well with radial flow in a steady confined aquifer may be predicted by Equation 2.41 (see Figure 2.9).

$$k = \frac{QLn\frac{r_2}{r_1}}{2\pi h(x_1 - x_2)} \tag{2.41}$$

where:

k	= Coefficient of permeability, m/s
Q	= Steady-state discharge from the well (pumping rate from supply well), m^3/s
x_1, x_2	= Drawdowns at observation wells, m
r_1, r_2	= Distance of observation wells from the pumped well (radial direction), m
h	= Depth of aquifer (average thickness of bed at r_1 and r_2 for water table conditions), m

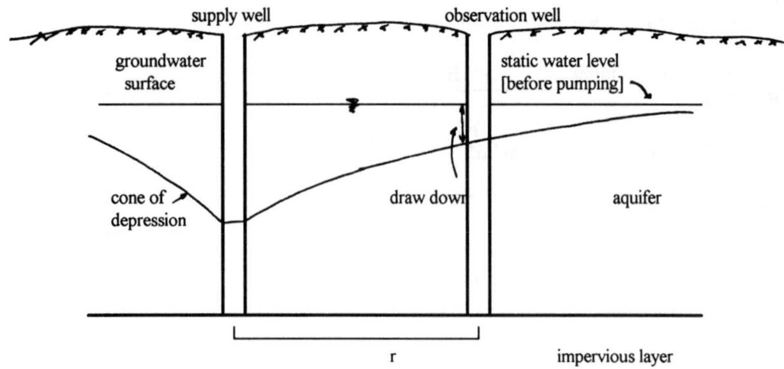

Figure 2.9
Supply and a observation well.

Equation 2.41 is the Thiem formula, or equilibrium equation. The equation may be used to estimate the permeability of the aquifer for measurements around a pumping well.

The specific capacity of a well is defined as the output of the well as divided by the drawdown. This definition is presented mathematically in Equation 2.42.

$$S_c = \frac{Q}{x} \tag{2.42}$$

where

S_c = Specific capacity of well, $m^3/d*m$
Q = Discharge (output) from well, m^3/s
x = Drawdown, m

Example 2.16

1. Write a computer program to find the drawdown, coefficient of permeability, transmissibility, or specific capacity for a steady confined aquifer with known characteristics and pattern of flow between two observation wells.
2. Pumping in an artisan well is conducted at the rate of 0.04 m^3/s. At observation wells that are 100 m and 200 m away, the drawdowns noted are 0.5 and 0.3 m, respectively. The average thickness of the aquifer at the observation wells is 8 m. Compute the coefficient of permeability of the aquifer.
3. A 30-cm well penetrates 25 m below the static water table. After a long period of pumping at a rate of 2500 L/min., the drawdown in wells 20 and 60 m from the pumped well were 1.9 and 0.6 m, respectively. Determine the transmissibility of the aquifer.
4. After prolonged pumping, a well produces 800 m^3/h. For a drawdown from the static level of 75 cm, estimate the specific capacity.

Solution

1. For solution to Example 2.16 (1), see listing of Program 2.16 below.
2. Solution to Example 2.16 (2):
   - $Q = 0.04\ m^3/s$ (see Figure 2.10)
   - Given: $Q = 0.04\ m^3/s$, $r_1 = 100$ m, $r_2 = 200$ m, $x_1 = 0.5$ m, $x_2 = 0.3$ m, h = 8 m.
   - Compute the permeability coefficient as: $k = [0.04]Ln[200/100]/(2\pi*8[0.5 - 0.3]) = 2.8*10^{-3}$ m/s.
3. Solution to example 2.16 (3):
   - Given h = 25 m, Q = 2500 L/min, $r_1 = 20$, $r_2 = 60$ m, $x_1 = 1.9$ m, $x_2 = 0.6$ m.
   - Use Thiem's equation to determine the permeability coefficient as: $k = Q*Ln[r_2/r_1]/\{[h_1]^2 - [h_0]^2\}$.
   - Determine $h_1 = h - x_2 = 25 - 0.6 = 24.4$ m, and $h_0 = h - x_1 = 25 - 1.9 = 23.1$ m.
   - Find permeability as: $k = \{(2500/1000*60)*Ln\ 60/20\}/([24.4]^2 - [23.1]^2) = 7.41*10^{-4}$ m/s = 64 m/d.
   - Compute the value of the transmissibility of the aquifer as: TR = k*h = 64*25 = $1.6*10^6$ Lpd/m.
4. Solution to Example 2.16 (4):
   - Given: $Q = 800\ m^3/h$, x = 0.75 m.
   - Determine the specific capacity of the well as: $S_c = Q/x = 800/75 = 10.7\ m^3/h/cm$.

Listing of Program 2.16

```
DECLARE SUB box (r1%, c1%, r2%, c2%)   'Found in Program 2.1
'******************************************************************
'Program 2.16: Aquifers-2
'Computes Properties of aquifers
'******************************************************************
DIM ttl$(4)
pi = 3.142857
ttl$(1) = "(1) To determine permeability"
ttl$(2) = "(2) To determine Transmissibility"
ttl$(3) = "(3) To determine the specific capacity"
ttl$(4) = "(4) Exit the program"
prompt$ = "Select an option 1-4 :"


start:
    CALL box(1, 1, 20, 80)
    LOCATE 2, 20
    PRINT "Program 2.16:Computes Properties of aquifers"

ROW = 5

FOR i = 1 TO 4
LOCATE ROW + i, 5: PRINT ttl$(i): NEXT i
LOCATE ROW + 6, 3: INPUT "Select an option 1-4 :"; OPT

SELECT CASE OPT
```

```
CASE 1
    CALL box(2, 1, 20, 80)
    ROW = 5
    LOCATE ROW + 1, 5
    INPUT "Enter distance to the first test well r1 (m)    ="; r1
    LOCATE ROW + 2, 5
    INPUT "Enter distance to the 2nd test well r2  (m)     ="; r2
    LOCATE ROW + 3, 5
    INPUT "Enter the draw down at the first test well x1 (m)="; x1
    LOCATE ROW + 4, 5
    INPUT "Enter the draw down at the 2nd test well x2  (m)="; x2
    LOCATE ROW + 5, 5
    INPUT "Enter the average depth of the aquifer h (m)    ="; h
    LOCATE ROW + 6, 5
    INPUT "Enter the pumping flow rate from the well m3/sec ="; Q
    X = r2 / r1
    IF r1 > r2 THEN X = r1 / r2
    K = ABS(Q * LOG(X) / (2 * pi * h * (x1 - x2)))
    LOCATE ROW + 8, 3
    PRINT "Coefficient of permeability K="; K; "  m/sec"

CASE 2
    CALL box(2, 1, 20, 80)
    ROW = 5
    LOCATE ROW + 1, 5
    INPUT "Enter distance to the first test well r1 (m)    ="; r1
    LOCATE ROW + 2, 5
    INPUT "Enter distance to the 2nd test well r2  (m)     ="; r2
    LOCATE ROW + 3, 5
    INPUT "Enter the draw down at the first test well x1 (m)="; x1
    LOCATE ROW + 4, 5
    INPUT "Enter the draw down at the 2nd test well x2  (m)="; x2
    LOCATE ROW + 5, 5
    INPUT "Enter the average depth of the aquifer  h (m)   ="; h
    LOCATE ROW + 6, 5
    INPUT "Enter the pumping flow rate from the well m3/sec ="; Q
    X = r2 / r1
    'IF r1 > r2 THEN X = r1 / r2
    h1 = h - x2
    ho = h - x1
    K = ABS(Q * LOG(X) / (h1 ^ 2 - ho ^ 2))
    LOCATE ROW + 8, 3
    K1 = K * 3600 * 24      'in m/d
    PRINT "Coefficient of permeability K="; K; "  m/sec ="; K1; "  m/d"
    TR = K1 * h
    LOCATE ROW + 9, 3
    PRINT "Transmissibility of the aquifer      ="; TR; " m3/d*m"
CASE 3
    CALL box(2, 1, 20, 80)
    ROW = 5
    LOCATE ROW + 1, 5
    INPUT "Enter flow rate from the well Q (m3/h)          ="; Q
    LOCATE ROW + 2, 5
    INPUT "Enter draw down from static level X (m)         ="; X
    Sc = Q / X
    LOCATE ROW + 5, 3
    PRINT "The specic capacity Sc="; Sc; "  m3/hr/m"
```

```
CASE 4
    END
END SELECT

LOCATE 23, 3
PRINT "press any key to continue....................";
DO
LOOP WHILE INKEY$ = ""
GOTO start

END
```

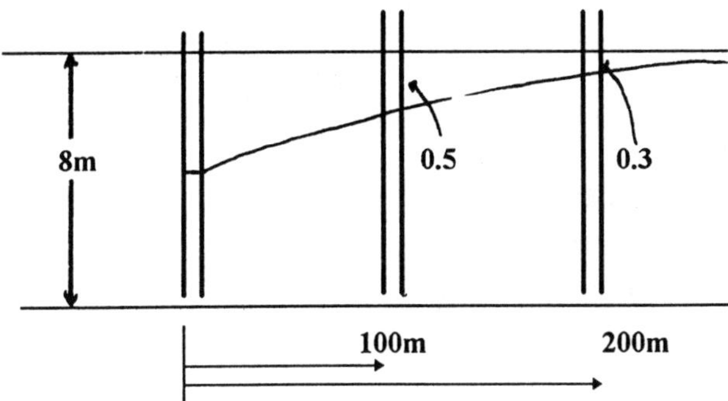

Figure 2.10
Solution to Example 2.16 (2).

For a steady, unconfined flow the discharge of a well may be estimated as presented in Equation 2.43 (see Figure 2.11).

$$Q_0 = \frac{\pi k([h_1]^2 - [h_0]^2)}{Ln\frac{R}{r_0}} \tag{2.43}$$

where:

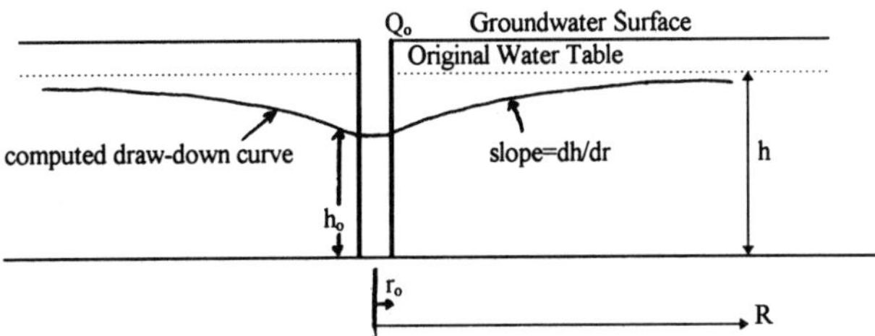

Figure 2.11
Steady unconfined flow.

Q_0 = Steady-state discharge from the well, m^3/s

k = Coefficient of permeability, m/s

h_1 = Depth to original water table when r = R, m

h_0 = Depth to original water table when $r = r_0$, m

R = Radius of influence, m

r_0 = Radius of well, m

Example 2.17

1. Write a computer program to find the radius of zero drawdown, coefficient of permeability, and drawdown in a pumped well for a steady, unconfined aquifer with known characteristics and pattern of flow between two observation wells.
2. A well of diameter 0.3 m contains water to a depth of 50 m before pumping commences. After completion of pumping, the drawdown in a well 20 m away is found to be 5 m, while the drawdown in another well 40 m farther away reached 3 m. For a pumping rate of 2500 L/min, determine:
   a. Radius of zero drawdown
   b. Coefficient of permeability
   c. Drawdown in the pumped well

Solution

1. For solution to Example 2.17 (1), see listing of Program 2.17 below.
2. Solution to Example 2.17 (2):
   a. Given: h = 50 m, r_1 = 20 m, x_1 = 5 m, r_2 = 40 m, x_2 = 3 m, Q = 2500 L/min.
      - Find $h_1 = h - x_1 = 50 - 5 = 45$ m, and $h_2 = h - x_2 = 50 - 3 = 47$ m.
      - Use Equation 2.43 for both observation wells:

$$\left(\pi k \frac{h^2 - h_1^2}{\mathrm{Ln}\frac{R}{r}}\right)_{\text{Firstwell}} = \left(\pi k \frac{h^2 - h_1^2}{\mathrm{Ln}\frac{R}{r}}\right)_{\text{Second well}}$$

      - By substituting given values into the previous equation, then:

$$\frac{50^2 - 45^2}{\mathrm{Ln}\frac{R}{20}} = \frac{50^2 - 47^2}{\mathrm{Ln}\frac{R}{40}}$$

      This yields R = 119.7 m.

   b. Find the permeability coefficient by using the data of one of the wells. For h = 50 m, h_0 = 45 m, r = 20 m, R = 119.7 m, Q = 2500*10^{-3}*60*24 = 3600 m^3/d,

$$k = \frac{3600 \text{Ln} \frac{119.7}{20}}{\pi(50^2 - 45^2)} = 4.32 \text{ m/d}$$

- Depth of the water in the pumped well may be found as:

$$h_1^2 = h^2 - Q \frac{\text{Ln} \frac{R}{r}}{\pi k} = 50^2 - 3600 \frac{\text{Ln} \frac{119.7}{0.15}}{4.32\pi}$$

This yields, h = 27 m.

c. Determine the drawdown at the well as: $x = h - h_1 = 50 - 27 = 23$ m.

Listing of Program 2.17

```
DECLARE SUB box (r1%, c1%, r2%, c2%)   'Found in Program 2.1
'*************************************************************
'Program 2.17: Aquifers-3
'Computes draw down from a steady unconfined aquifer
'with known characteristics
'*************************************************************
pi = 22! / 7!
CALL box(1, 1, 20, 80)
LOCATE 2, 10
PRINT "Program 2.17:Computes draw down from unconfined aquifers"
ROW = 5
LOCATE ROW + 1, 5
INPUT "Enter distance to the first test well r1 (m)              ="; r1
LOCATE ROW + 2, 5
INPUT "Enter distance to the 2nd test well r2   (m)              ="; r2
LOCATE ROW + 3, 5
INPUT "Enter the draw down at the first test well x1 (m)                 ="; x1
LOCATE ROW + 4, 5
INPUT "Enter the draw down at the 2nd test well x2   (m)                 ="; x2
LOCATE ROW + 5, 5
INPUT "Enter the average depth of the aquifer  h (m)             ="; h
LOCATE ROW + 6, 5
INPUT "Enter the pumping flow rate from the well (m3/sec)      ="; Q
LOCATE ROW + 7, 5
INPUT "Enter the diameter of the well (m)                        ="; diam

h1 = h - x1
h2 = h - x2
c = (h ^ 2 - h1 ^ 2) / (h ^ 2 - h2 ^ 2)
R = EXP(1 / (c - 1) * (c * LOG(r2) - LOG(r1)))
LOCATE ROW + 10, 3
PRINT "RESULTS:"
LOCATE ROW + 11, 3
PRINT "Radius of zero draw down    R="; R; "  m"

K = Q * LOG(R / r1) / (pi * (h ^ 2 - h1 ^ 2))
```

```
LOCATE ROW + 12, 3
PRINT "Coefficient of permeability K="; K; " m/sec"
radius = diam / 2
ho = (h ^ 2 - Q * LOG(R / radius) / (pi * K)) ^ .5
x = h - ho
LOCATE ROW + 13, 3
PRINT "The draw down at the well X="; x; " m"
DO
LOOP WHILE INKEY$=""
END
```

Groundwater recharge is useful for a number of reasons: overdraft reduction, surface runoff conservation, and increasing yield of groundwater sources. Groundwater recharge can be placed in two main categories: incidental recharge (e.g., surplus irrigation water) or intentional (deliberate) (e.g., municipal waste, surface water). Artificial or intentional recharge may be defined as augmenting the natural replenishment of groundwater storage by some manmade process, such as surface spreading of water or water well injection.[14,28] Groundwater recharge by reclaimed water has been practiced to augment the groundwater volume and increase its development. The factors that influence this practice include location of natural recharge areas, geological formations, soil structure, hydrological conditions, quantity of water withdrawn, and the degree of wastewater treatment.[14]

2.8 Homework Problems

2.8.1 Discussion Problems

1. Why is water vital for human survival?
2. What benefits are to be gained by acquiring knowledge about the characteristics of water and wastewater?
3. Outline the main physical water-quality parameters relevant in water-quality management.
4. What are the main effects of an increase in temperature of water or wastewater?
5. Define conductivity and indicate units of measurements.
6. Indicate different methods that contribute to increased water salinity.
7. Describe a method for measuring the solids content of a water sample.
8. How can you relate mathematically density, specific volume, specific weight, and specific gravity?
9. What are some of the problems associated with viscous fluids?
10. Give one example for each of the following: dilatant, thixotropic, ideal liquid, ideal solid, non-newtonian, newtonian, plastic, and pseudoplastic fluid.

11. Illustrate differences between kinematic and absolute viscosity coefficient.
12. What is surface tension?
13. What are the practical benefits of surface tension?
14. Define bulk modulus.
15. Describe the differences between alpha, beta, and gamma radiation.
16. Outline the main chemical characteristics relating to water quality.
17. Outline the factors that affect the pH of waters and wastewaters.
18. How can you differentiate between an alkaline and an acidic solution?
19. Define hardness of water. Note the main classes of hardness and discuss cons and pros.
20. Indicate how to differentiate between a hard- and a soft-water sample.
21. Discuss the factors that influence the solubility of a gas in water.
22. What is BOD?
23. What are the advantages of the BOD test?
24. Describe the hydraulic cycle.
25. Outline the preferable water source characteristics.
26. What are the main categories into which water usage may be grouped?
27. What are the factors that influence fire demand.
28. Define the following terms: transmissibility coefficient, confined aquifer, unconfined aquifer, drawdown, and radius of influence.

2.8.2 Specific Mathematical Problems

1. Write a short computer program to illustrate the effect of angle of contact on surface tension.
2. Write a short computer program to enable determination of the temperature when the readings on a Celsius and a Fahrenheit thermometer coincide.
3. Write a simple computer program to find the weight of a liquid, given its volume and specific gravity.
4. Write a short computer program that allows for the conversion of viscosity (dynamic and kinematic) between the SI system of units and the British system.
5. Write a short computer program that allows the determination of the molecular weight and equivalent weights of the materials found in the periodic table.

Test the program by determining the molecular weights of the following compounds: barium acetate ($Ba(C_2H_3O_2)_2$), nitrous oxide (N_2O), ammonium aluminum sulfate $(NH_4)_2Al_2(SO_4)_4$, strontium sulfate ($SrSO_4$), and copper sulfate (blue vitriol) ($CuSO_4*5H_2O$).

6. Write a computer program to determine the normality (N) and molarity (M) of a solution given that: N = (wt/EW)/V and M = (wt/MW)/V, where wt is weight, EW is equivalent weight, MW is molecular weight, and V is volume. Test your program by finding the molarity of a solution that contains 2.5 g of sulfuric acid in 4 L of solution.

7. Write a computer program to determine the percentage of elements found in a compound composed, say, of nine substances: carbon (C), hydrogen (H), nitrogen (N), and oxygen (O). Test your program by determining the percent carbon in fructose, $CH_2OH(CHOH)_3COCH_2OH$, and the percent zinc in the compound zinc dimethyldithiocarbamate, $Zn[S_2CN(CH_3)_2]_2$.

8. Write a computer program that computes the pH of a solution. Design the program so as to indicate whether the solution is acidic or alkaline. The program is to differentiate between the acidity of different solutions given $[H^+]$ or $[OH^-]$ ion concentration.

9. Write a computer program that enables estimating the concentration of a cation or an anion from data of cations involving calcium, magnesium, sodium, strontium, and iron and anions containing chloride, sulfate, nitrate, carbonate, and bicarbonate. Give the total hardness and percent experimental error.

10. Assuming an experimental error of 3%, use the program developed in Problem 9 to find the missing chloride concentration for the following data of a sample of water (all values are in mg/L):

Ca^{++}	35	Cl^-	18
Mg^{++}	40	HCO_3^-	122
Sr^{+++}	9	$SO_4^=$	34
Fe^{+++}	23	NO_3^-	22
		SiO_3	14

11. Write a computer program that enables the computation of the BOD of organic compounds (containing hydrogen, carbon, nitrogen, and oxygen) with a known concentration (in mg/L). Assume reaction products to be carbon dioxide, water, or nitrogen oxide according to reactants entering chemical reaction.

12. Check the validity of the program of problem 11 by determining the total BOD of the following compounds:

    a. 0.1 molar solution of ether, $(CH_3CH_2)_2O$.
    b. 5 g of glycerol, $CH_2OH{\cdot}CHOH{\cdot}CH_2OH$.
    c. 20 mg/L uric acid, $C_5H_4O_3N_4$.

References

1. Abdel-Magid, I.M. and El Hassan, B.M., *Water Supply in the Sudan*, Khartoum University Press, Khartoum, 1986 (Arabic).
2. Abdel-Magid, I.M., *Water Treatment and Sanitary Engineering*, Khartoum University Press, Khartoum, 1986 (Arabic).
3. Barnes, D., Bliss, P. J., Gould, B.W., and Vallentine, H.R., *Water and Wastewater Engineering Systems*, Pitman International, Bath, 1981.
4. Camp, T.R., *Water and Its Impurities*, Reinhold, New York, 1973.
5. Hammer, M.J., *Water and Wastewater Technology*, 2nd ed., Prentice Hall, Englewood Cliffs, NJ, 1986.
6. Husain, S.K., *Textbook of Water Supply and Sanitary Engineering*, 2nd ed., Oxford and IBH Publications, New Delhi, 1981.
7. Lorch, W., Ed., *Handbook of Water Purification*, 2nd ed., McGraw-Hill, New York, 1981.
8. Merritt, F.S., *Standard Handbook for Civil Engineers*, McGraw-Hill, New York, 1976.
9. Peavy, H.S., Rowe, D.R., and Tchobanoglous, G., *Environmental Engineering*, McGraw-Hill, New York, 1985.
10. Salvato, J.A., *Environmental Engineering and Sanitation*, 4th ed., John Wiley & Sons, New York, 1992.
11. McGhee, T.J. and Steel, E.W., *Water Supply and Sewerage*, 6th ed., McGraw-Hill, New York, 1991.
12. Tebbutt, T.H.Y., *Principles of Water Quality Control*, Pergamon Press, Oxford, 1992.
13. Vesilind, P.A. and Peirce, J.J., *Environmental Pollution and Control*, 2nd ed., Butterworth-Heinemann, London, 1990.
14. Rowe, D.R. and Abdel-Magid, I.M., *Handbook of Wastewater Reclamation and Reuse*, CRC Press/Lewis Publishers, Boca Raton, FL, 1995.
15. APHA, *Standard Methods for the Examination of Water and Wastewater*, 18th ed., American Public Health Association, Washington, D.C., 1992.
16. Daugherty, R.L., Franzini, J.B., and Finnemore, E.J., *Fluid Mechanics with Engineering Applications*, McGraw-Hill, New York, 1985.
17. Munson, B.R., Young, D.F., and Okiishi, T.H., *Fundamentals of Fluid Mechanics*, 2nd ed. John Wiley & Sons, New York, 1994.
18. Liptrot, G.F., Thompson, J.J., and Walker, G.R., *Modern Physical Chemistry*, Bell and Hyman Ltd., London, 1985.
19. Berger, B.B., Ed., *Control of Organic Substances in Water and Wastewater*, Noyes Publishing, Park Ridge, NJ, 1987.
20. Abdel-Magid, I.M., *Selected Problems in Wastewater Engineering*, Khartoum University Press, Khartoum, 1986.
21. Metcalf and Eddy, Inc. *Wastewater Engineering: Treatment Disposal Reuse*, 3rd ed., McGraw-Hill, New York, 1991.
22. Whipple, G.C. and Whipple, M.C., Solubility of oxygen in sea water, *J. Am. Chem. Soc.*, 33, 362, 1911.
23. Mara, D., *Sewage Treatment in Hot Climates*, John Wiley & Sons, New York, 1980.
24. Wilson, E.M., *Engineering Hydrology*, 4th ed., MacMillan, New York, 1990.
25. IPHE, *The Public Health Engineering Data Book 1983/84*, Bartlett, R., Ed., Sterling, London, 1984.

26. Viessman, W., Lewis, G.L., and Knapp, J.W., *Introduction to Hydrology*, Harper & Row, New York, 1989.
27. Punmia, B.C., *Environmental Engineering, Vol. 1: Water Supply Engineering*, Standard Book House, New Delhi, 1979.
28. Todd, D.K., *Groundwater Hydrology*, John Wiley & Sons, New York, 1980.
29. Ven Te Chow, Ed., *Handbook of Applied Hydrology: A Compendium of Water Resources Technology*, McGraw-Hill, New York, 1964.
30. Hammer, M.J. and Mac Kichan, K.A., *Hydrology and Quality of Water Resources*, John Wiley & Sons, New York, 1981.
31. Shaw, E.M., *Hydrology in Practice*, D Van Nostrand-Reinhold, New York, 1983.
32. Hofkes, E.H., Huisman, L., Sundaresan, B.B., Netto, J.M.D., and Lanoix, J.N., *Small Community Water Supplies*, John Wiley & Sons, New York, 1986.
33. Van del Leeden, F., *The Water Encyclopedia*, 2nd ed., Lewis, Publishers, Chelsea, MI, 1990.
34. Scott, J.S. and Smith, P.G., *Dictionary of Waste and Water Treatment*, Butterworths, London, 1980.
35. Nielsen, D.M., Ed., *Practical Handbook of Groundwater Monitoring*, Lewis, Chelsea, MI, 1991.
36. Manahan, S.E., *Fundamentals of Environmental Chemistry*, Lewis Publishers, Chelsea, MI, 1993.
37. Perry, R.H., Green, D.W., and Maloney, J.O., Eds., *Perry's Chemical Engineers Handbook*, 6th ed., McGraw Hill, New York, 1984.

Chapter 3

Water Treatment

Contents

This chapter outlines the concepts used in the design of treatment facilities for both unit operations and processes. Unit operations are physical in nature, while unit processes involve chemical and biological principles. Each treatment process, whether physical, chemical, or biological or a combination thereof, is briefly described with emphasis being placed on the design of the units, applying well established formulae, equations, and models that are then used to develop computer programs that can aid in the various design procedures.

3.1 Water Treatment Systems

Generally, water treatment processes are broadly classified as:

1. *Physical treatment, unit operations:* Generally, these are the simplest forms of water treatment. Factors that govern treatment here are physical in nature and include particle size, specific gravity, viscosity, etc. Examples of such treatment units include screening, sedimentation, flotation, aeration, and filtration.
2. *Chemical treatment, unit processes:* These are capable of changing the nature of the pollutants and existing contaminants to by-products which are no longer objectionable. Examples include coagulation, color and odor removal, disinfection, etc.

3. *Biological treatment, unit processes:* These are capable of removing organic matter and soluble and colloidal particles in an engineered system under controlled environmental conditions. An example of such a process is biological filtration.

3.2 Aims of Water Treatment

Municipal water works are required to produce a water supply that is hygienically safe (potable), aesthetically attractive (palatable), and economically satisfactory for its intended uses. Water treatment unit operations and unit processes are both used to provide a water supply that is free of:

- Suspended or floating substances (e.g., leaves, debris, soil particles)
- Colloidal substances (e.g., clay, silt, microorganisms)
- Dissolved solids (e.g., inorganic salts)
- Dissolved gaseous substances (e.g., carbon dioxide, hydrogen sulfide)
- Immiscible liquids such as oils and greases
- Substances that cause corrosion or encrustation
- Odor/taste/color-producing substances
- Pathogenic organisms (bacteria, viruses, amoebas, worms, fly nymphs, cyclops, etc.)
- Compounds (organic or inorganic) that have an adverse effect, acute or long term, on human health.

Some of the first information needed in order to design an economically feasible water treatment facility includes the source of the water supply, the quality and quantity of water available, and the population to be served, as well as the commercial and industrial development in that community so that present demands and further trends can be evaluated. Design capacity must also include social and economic conditions within the community. Estimates must be made as to the design period, design population, design flow, peak flows, design area, and design hydrology.

3.3 Screening

Screening devices are installed to facilitate removal of coarse suspended and floating matter. Presence of coarse solids may damage or interfere with the operation of the pumps and other mechanical equipment. Solids can also obstruct valves and other devices in the plant. Generally, these materials consist of leaves, rags, sticks, vegetable matter, broken stones, tree branches, boards, and other large objects that are relatively inoffensive. The coarse solids are usually removed by simple screening devices.

The minimum velocity, through a screen, that is needed to prevent deposition of sand and suspended matter is related to size and density of the suspended impurities and to flow velocity in the channel from which water or wastewater emerges. The approach velocity of the water in the channel upstream should not be less than 0.3 to 0.5 m/s. This velocity is recommended to prevent settling out of suspended matter. The passing velocity of the water through the openings between the bars should not exceed 0.7 to 1 m/s to prevent fluffy and soft matter from being forced through the openings in the screen.

With the blockage of screen openings, resistance to flow rapidly increases. Flow resistance increases can be evaluated according to Equation 3.1.[1]

$$H_s = (a_0 / a_a)^2 * H_0 \tag{3.1}$$

where:

H_s = Resistance for a clogged screen, m of water

H_0 = Initial resistance for a clean screen, m

a_a = Percentage open area for a clogged screen

a_0 = Percentage open area for a clean screen

The development of this head loss will produce a considerable load on the bar screen, which necessitates an adequate structural design to meet the developing hydraulic load. Frequent cleaning of the screen openings helps to limit the effects of this hydraulic load. Cleaning of openings keeps the maximum value of the resistance below a practical value of 0.5 m. In general, two screens should be provided so that one screen can be shut down while the other is being cleaned or repaired.

3.4 Sedimentation

3.4.1 Introduction

Sedimentation may be defined as the gravitational settling of suspended particles that are heavier than the surrounding fluid. The water to be clarified by sedimentation is held in a tank for a considerable period of time. For tanks with large cross-sectional areas, flow velocities are small. This condition induces a state of virtual quiescence.[1] As such, heavier particles (with a mass density greater than that of the surrounding fluid) start settling under gravitational forces (a process referred to as sedimentation; see Figure 3.1). Lighter particles (with a mass density lower than that of the surrounding fluid) tend to move vertically upwards (a process denoted as flotation; see Figure 3.1). Accordingly, suspended particles are retained either in the scum layer (top layer) at the water surface or in the sludge layer at the tank floor (bottom layer). Removal of settled and floatable particles clarifies the raw water. With the same capacity and tank volume, long, narrow, and shallow basins have the greatest potential for solids removal.

The aims of the sedimentation process include:

- Removal of grit (grit chamber)
- Reduction of content of particulate solids, both settleable and floatable (primary settling)
- Reduction of BOD_5 (primary clarifier)
- Removal of chemical floc (chemical coagulation)
- Removal of biological floc (activated sludge system)
- Separation of solids from mixed liquor (activated sludge)
- Production of a concentrated return sludge flow to sustain biological treatment (secondary clarifiers)
- Increase solids concentration (sludge thickeners and dewatering units)
- Provision of recirculated water flow (high rate trickling filtration)

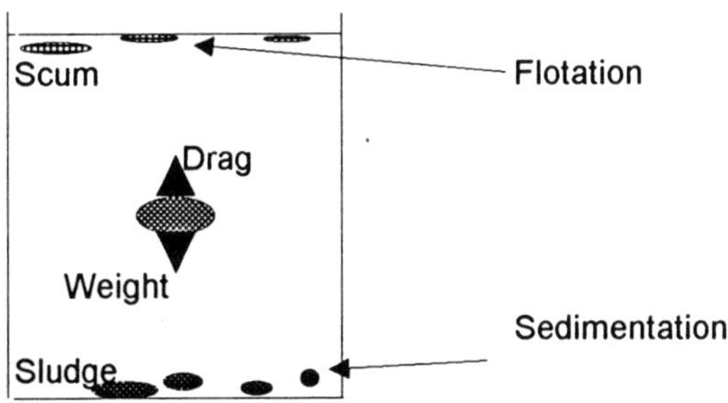

FIGURE 3.1
Sedimentation and flotation.

Sedimentation can be improved by adding coagulant aids, such as lime, soda ash, NaOH, H_2SO_4, clay, bentonite, powdered stone, or polyelectrolytes. Flotation can be facilitated by bubbling air or chlorine gas into the liquid at the bottom of the sedimentation tank.

Sedimentation tanks may be constructed as rectangular, circular, or square in plan. The water in the tank may be at rest or continuously flowing, either vertically or horizontally. Figure 3.2 illustrates two types of sedimentation tanks. Square or circular basins may be preferred, for one or all of the following reasons:

- They make better use of the land area allocated for the project (multiple rectangular tanks require less area than multiple circular tanks).
- They offer savings in cost.
- They use smaller amounts of raw construction materials.
- They allow use of more durable materials (e.g., prestressed concrete).

Best results can be obtained with vertical flow basins, which have large depths and inlet structures that spread the incoming water equally over the entire tank plan area.[2–5]

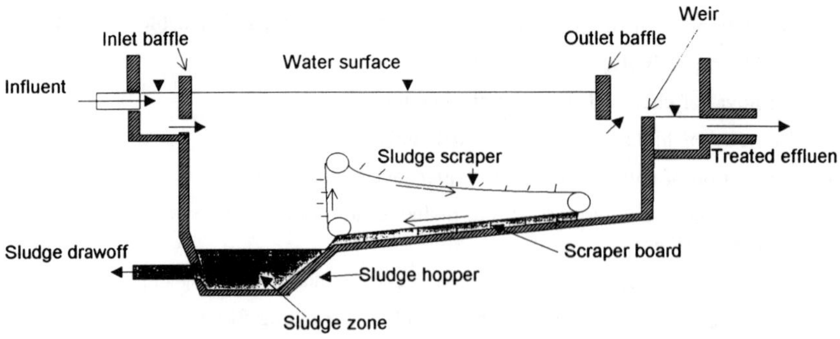

A

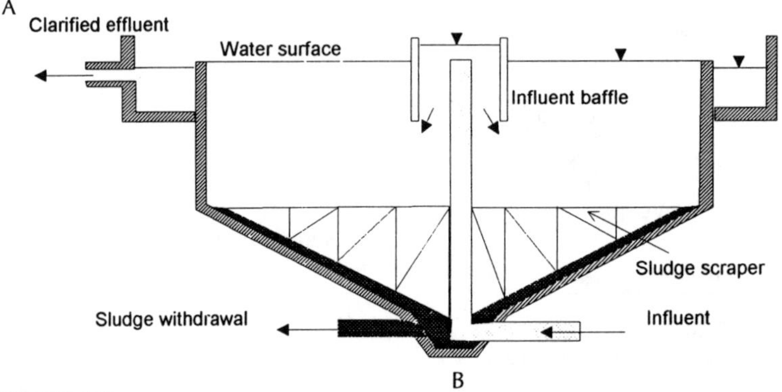

B

FIGURE 3.2
Sedimentation tanks. (A) Schematic diagram of a rectangular horizontal sedimentation tank. (B) A schematic diagram of a circular sedimentation tank.

3.4.2 Types of Settling Phenomenon

The settling phenomenon is governed by the particle shapes, sizes, and characteristics. Particle settling is often classified as follows:

Class I: Discrete Particle Settling. Under discrete particle settling, nonflocculating conditions develop and particles maintain their identity throughout the settling process with a constant velocity, i.e., particles experience no change in their size, shape, or density. General features of discrete settling can be summarized as follows:[6]

- Particles settle as individual entities.
- No significant interaction occurs between adjacent particles.
- Settling is unhindered by the basin walls.

This class of particle settling can be represented by the settling of sand grains in water treatment or the removal of grit in grit chambers in the field of wastewater treatment.

Class II: Flocculent Settling. Under flocculent settling conditions, particles flocculate during sedimentation, i.e., they experience a change in their size, weight, and shape. The main features of this class of particle settling include:

- Particles coalesce together or flocculate.
- Particles are hindered by their neighbors.
- Due to the close proximity of particles, the change in their size is mainly due to particle contact.
- Particles increase in mass with depth and settling time.
- Particles settle rapidly.

Examples of this class of settling occur in alum or iron coagulation.[7]

Class III: Hindered (Zone) Settling. Hindered settling occurs when high density particles interact and individual particles are so close together that the displacement of water by the settling of one affects the relative settling velocities of its neighbors.[8] With the cross-sectional area of a settling tank being fixed, the water displaced by the settling particle must flow in the opposite direction to that of the moving particles. The essential features of this kind of settling include:

- Settling characteristics of adjacent particles is influenced by existence of interparticle forces.
- Moving particles have similar settling velocities.
- Settlement occurs for a mass of particles that are able to settle as one unit.
- Particles remain in fixed places relative to each other.

Examples of this kind of settling may be encountered in lime softening tanks, sludge thickeners, and in activated sludge sedimentation.[7] Compared to the stationary walls of the sedimentation tank, the settling rate under hindered settling conditions is reduced from a value of v to a value of v′ as given by Equation 3.2:[1]

$$v' / v = 1 - (f * C_v)^{2/3} \tag{3.2}$$

where:

v' = Hindered settling velocity, m/s
v = Initial settling velocity of the particle, free settling velocity, m/s
c_v = Volumetric concentration of particles (volume of particles divided by the total volume of the suspension) = $V_{particle}/V_{suspension}$
f = A coefficient that varies with the type of particles

An estimate of the extent to which settling is hindered can also be calculated from Equation 3.3, which is valid for Reynolds numbers less than 0.2, which is generally the situation with hindered settling.[8]

$$v' / v = (1 - C_v)^{4.65} \tag{3.3}$$

where:

v' = Hindered settling velocity, m/s
v = Free settling velocity of the particle, m/s
c_v = Volume of particles divided by the total volume of the suspension

Class IV: Compression Settling. In Compression particle settling, the concentration of particles forms a distinct structure. Compression, under the weight of the particles, governs the settling process. This kind of settling occurs at the bottom of deep secondary settling tanks and takes place under sludge settling conditions.[9] Usually, more than one type of settling occurs in a sedimentation basin.

3.4.3 Class I Settling

When a particle is placed in a liquid of a density lesser than its own, its settling velocity will accelerate until a limiting (terminal) velocity is reached. Beyond this velocity, the settling forces and the submerged weight of the particle, are in balance with the frictional drag forces. This yields turbulent flow conditions. The relationship for this settling velocity is presented in Equation 3.4.

$$v = [3.3 * g * d(s.g - 1)]^{1/2} \tag{3.4}$$

where:

- v = Settling (terminal) velocity of particle, m/s
- g = Gravitational acceleration, m/s^2
- d = Diameter of spherical particle, m
- s.g. = Specific gravity of the particle

The settling velocity of a spherical particle under laminar flow conditions can be expressed by the Stoke's law which is presented in Equation 3.5.

$$v = [*d^2(s.g. - 1)]/(18*v) \tag{3.5}$$

where:

- v = Settling (terminal) velocity of particle, m/s
- g = Gravitational acceleration, m/s^2
- d = Diameter of spherical particle, m
- s.g. = Specific gravity of the particle = ρ_s/ρ
- ρ = Density of fluid, kg/m^3
- ρ_s = Particle density, kg/m^3
- v = Kinematic viscosity, m^2/s

$$v = \mu / \rho \upsilon \tag{3.6}$$

where:

- μ = Dynamic viscosity (absolute), $N*s/m^2$

Example 3.1

1. Write a computer program to determine the settling velocity for spherical discrete particles settling under different flow conditions (laminar or turbulent). Provide for the

program to check the Reynolds number for the computed overflow rate (at different temperatures).

2. Particles are allowed to settle in a settling column test. The particles have a specific gravity of 1.3 and an average diameter of 0.1 mm. Using Stoke's law, determine the settling velocity of the particles in water (the temperature of the water is 20°C).
3. Use the program developed in (1) to verify the computations in (2).

Solution

1. For solution to Example 3.1 (1), see listing of Program 3.1 below.
2. Solution to example 3.1 (2):
   - Given: s.g. = 1.3, d = $0.1*10^{-3}$ m.
   - Find the viscosity coefficient from Table A1 in the Appendix that corresponds to a temperature of T = 20°C as: $\mu = 1.0087*10^{-3}$ $N*s/m^2$, and $\rho = 998.2$ kg/m^3.
   - Use Stoke's law to find the settling velocity of the particles, $v = \rho*g*d^2\ (s.g. - 1) / (18*\mu)$.
   - Substitute the given values in Stoke's equation to find the settling velocity as: $v = 998.2*9.81(0.1*10^{-3})^2(1.3 - 1) / 18*1.0087*10^{-3} = 1.62*10^{-3}$ m/s = 1.62 mm/s.

Listing of Program 3.1

```
DECLARE SUB box (r1%, c1%, r2%, c2%)      'Found in Program 2.1
'*********************************************************************
'Program 3.1: Settling
'Computes settling velocity for spherical discrete particles
'*********************************************************************
DIM temp(25), density(25), viscosity(25)
'density - Appendix C- physical properties of fluids
DATA 999.9,1000,999.7,999.1,998.2,997.1,995.7,994.1,992.2,990.2,988.1
DATA 985.7,983.2,980.6,977.8,974.9,971.8,968.6,965.3,961.9,958.4
'viscosity- same reference
DATA 1.792,1.519,1.308,1.14,1.005,0.894,0.801,0.723,0.656,0.599,0.549
DATA 0.506,0.469,0.436,0.406,0.38,0.357,0.336,0.317,0.299,0.284

'read tables
FOR i = 0 TO 20
temp(i) = i * 5
READ density(i)
NEXT i
FOR i = 0 TO 20
READ viscosity(i)
viscosity(i) = viscosity(i) / 1000
NEXT i

G = 9.81          'Gravity constant
ttl$(1) = "1. Liquid is water"
ttl$(2) = "2. Any other fluid"
ttl$(3) = "3. End program"

START:
CALL box(2, 1, 20, 80)
LOCATE 3, 10
PRINT "Program 3.1: Computes settling velocity for discrete particles"
FOR i = 1 TO 3
```

```
LOCATE i + 3, 5
PRINT ttl$(i)
NEXT i

DO
LOCATE 10, 3
INPUT "Select an option (1 to 3): "; opt
LOOP WHILE opt < 1 OR opt > 3

CALL box(2, 1, 20, 80)

SELECT CASE opt
CASE 1
        LOCATE 5, 5
        INPUT "Enter the temperature(Centigrade)="; T
        LOCATE 6, 5
        INPUT "specific gravity of the particles="; sg
        LOCATE 7, 5
        INPUT "diameter of particles d (m)      ="; d
        'determine the density and the viscosity
        temp(21) = 100
        kr = 100
        FOR i = 0 TO 20
        IF T = temp(i) THEN kr = i
        IF T > temp(i) AND T < temp(i + 1) THEN kr = i
        NEXT i
        Ro = density(kr)
        mue = viscosity(kr)
CASE 2
        LOCATE 6, 5
        INPUT "density of liquid  (Kg/m3)       ="; Ro
        LOCATE 7, 5
        INPUT "specific gravity of the particles="; sg
        LOCATE 8, 5
        INPUT "diameter of particles d (m)      ="; d
        LOCATE 9, 5
        INPUT "Enter viscosity of fluid (Pa.s)   ="; mue
CASE 3
        END
END SELECT

'compute settling velocity (Stoke's law)
velocity = Ro * G * d ^ 2 * (sg - 1) / (18 * mue)

LOCATE 10, 3
PRINT "The settling velocity ="; velocity; "   m/s"

DO
LOOP WHILE INKEY$ = ""
GOTO START
END
```

3.4.4 Settling Characteristics

Laboratory experiments are often conducted to evaluate the settling phenomenon of suspended particles, as well as to provide information needed for the design of sedimentation tanks. The tests are conducted in a settling column apparatus (Figure 3.3), which consists of a cylindrical container having a uniform surface area and tapping points (ports) at preset levels. The cylinder is of a height equal to that of the proposed tank. The temperature of the container can be kept constant by using a water bath. The test procedure may be summarized as follows:

1. The initial suspended solids concentration, C_0, is determined.
2. The settling column is filled with a well mixed suspension of the sample to be tested.
3. Samples are taken at various depths (h_1, h_2,..., etc.) and analyzed for solid concentrations, c_1, c_2, etc. Thus, all particles with a settling velocity less than the design settling velocity (v_{so}) will not be removed during the settling time. Particles with a settling velocity greater than v_{so} will have settled past the sampling point. As such, the ratio of removal for particles with $v < v_{so}$ (i.e., portion of particles x_1 that have settling velocity less than v_1) is given by $x_1 = c_1/c_0$. The entire range of settling velocities present in the system can be obtained by making use, for example, of a sieve analysis plus a hydrometer (wet sieve analysis), or a Coulter counter, or any other appropriate device or technique according to standard procedures.
4. The above procedure is repeated at various time intervals (t_1, t_2, etc.).
5. Data is then plotted to give the cumulative settling characteristics curve for the suspension as presented in Figure 3.4. The overall removal efficiency of the sedimentation tank (degree of removability of particles) can be found through the relationship indicated in Equation 3.7.

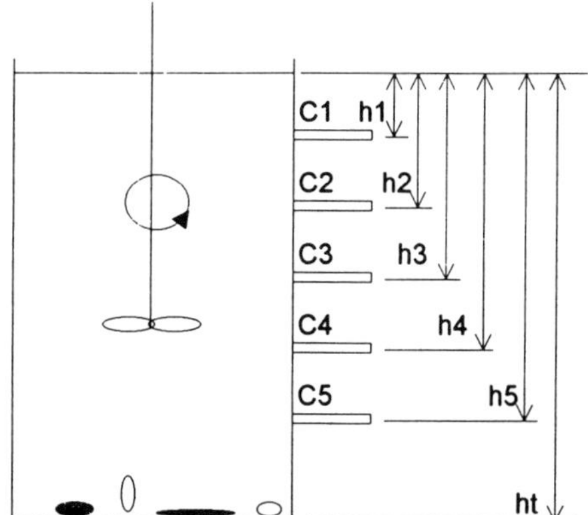

FIGURE 3.3
Settling column test.

$$X_T = 1 - X_0 + \{\int_0^{x_0} v * dx\} / v_{so} \tag{3.7}$$

The settling velocity for a discrete particle that enters the sludge zone at the end of the sedimentation tank can be determined from Equation 3.8.

$$v_{so} = h/t = h/(V/Q) = h * Q/V = Q/(V/h) = Q/A \tag{3.8}$$

where:

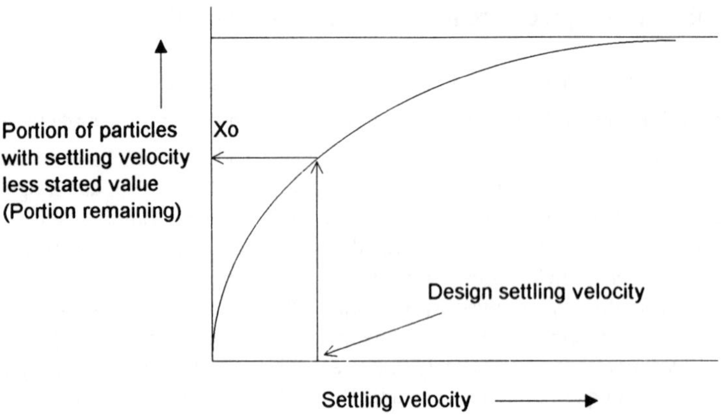

FIGURE 3.4
Cumulative frequency distribution plot.

v_{so} = Settling velocity of the discrete particle, m/s
h = Depth of tank, m
t = Detention time, s (see Equation 3.9)
V = Volume of tank, m^3
Q = Flow rate entering the tank, m^3/s
A = Area perpendicular to direction of flow, m^2

$$t = V / Q \tag{3.9}$$

where:

V = Volume of tank, m^3
Q = Flow rate entering the tank, m^3/s

Equation 3.8 illustrates that for a discrete particle, the solids removal is independent of the depth of the sedimentation tank. To account for the less than optimum conditions encountered in practice, the design settling velocity (v_{so}) obtained from the column test is often multiplied by a factor of 0.65 to 0.85,[9] and the computed detention times are multiplied by a factor of 1.25 to 1.5. Efficiently designed and operated primary clarifiers remove 50 to 65% of the total suspended solids and 25 to 40% of the BOD_5.[10]

Example 3.2

1. Write a computer program to compute the removal efficiency and concentration of suspended solids (SS) in effluent given the relevant column test results.

2. A wastewater treatment plant incorporates four sedimentation tanks for treating a wastewater flow of 1200 m^3/h. Each of the circular sedimentation tanks has a diameter of 21 m. The settling column test conducted for a sample of the wastewater indicated the data tabulated below. Determine the suspended solids removal efficiency of the sedimentation units.

Sampling depth (m)	Concentration of suspended solids removed (mg/L)				
	Sampling time (h)				
	0	1	2	3	4
0.2	190	101	166	183	186
0.5	190	57	85	114	124
1	190	53	58	70	86
1.5	190	50	54	60	66
2	190	44	54	56	57
2.5	190	40	50	51	56

Solution

1. For solution to Example 3.2 (1), see listing of Program 3.2 below.
2. Solution to Example 3.2 (2):
   - Given: Q = 1200 m^3/h, D = 21 m, number of sedimentation tanks = 4, SS concentrations at different depths for different times.
   - Determine the settling velocity as: settling velocity = depth/time = h/t.
   - Find percentage suspended solids remaining, C_r, in the sample for the different time intervals and sampling ports indicated as: C_r = 100 – SS concentration removed = 100 – (C/C_0).
   - Determine percentage SS with settling velocity less than that stated as: $X = C_r/C_0$, as shown in the tabular form below.
   - Plot the percentage SS with settling velocity less than that stated vs. settling velocity to obtain the cumulative frequency distribution curve for the settling particles.
   - Find the flow rate introduced to each tank = total flow/number of tanks = (1200/60*60) / 4 = 0.083 m^3/s.
   - Compute the surface area of each tank, $A = \pi\,(21)^2/4 = 346.4\ m^2$.
   - Determine the surface loading rate for the settling particles, $v_{so} = Q/A = 0.083/346.4 = 2.4*10^{-4}$ m/s = 0.24 mm/s.
   - From the plotted graph and for a design surface loading of 0.24 mm/s, the value of X_0 = 72%.
   - Determine the removal efficiency of the tank by using the following equation:

Depth (mm)	Time (s)	Velocity (mm/s)	Suspended Solids remaining (mg/L)	Suspended Solids with v less than stated (%)
200	3600	0.056	89	46.8
200	7200	0.028	24	12.6
200	10,800	0.019	7	3.7
200	14,400	0.014	4	2.1
500	3600	0.139	133	70
500	7200	0.069	105	55.3
500	10,800	0.046	76	40
500	14,400	0.035	66	34.7
1000	3600	0.278	137	72.1
1000	7200	0.139	132	69.5
1000	10,800	0.093	120	63.2
1000	14,400	0.069	104	54.7
1500	3600	0.417	140	73.7
1500	7200	0.208	136	71.6
1500	10,800	0.139	130	68.4
1500	14,400	0.104	124	65.3
2000	3600	0.556	146	76.8
2000	7200	0.278	136	71.6
2000	10,800	0.185	134	70.5
2000	14,400	0.139	133	70
2500	3600	0.694	150	78.9
2500	7200	0.347	140	73.7
2500	10,800	0.231	139	73.2
2500	14,400	0.174	134	70.5

$$X_T = 1 - X_0 + \{\int_0^{X_0} v * dx\} / v_{so}.$$

- The integral part of the previous equation $\{\int v*dx\}$ may be determined by using a planimeter or by making use of Simpson's rule, by manual means, or by any other appropriate method. In this case, the integral part = $0.76*10^{-3}$.
- Find the removal efficiency as: $X_T = 100 - 72 + (0.76*10^{-3} / 2.4*10^{-4}) = 31\%$.

3. Use the computer program presented in (1) to check the results obtained in (2).

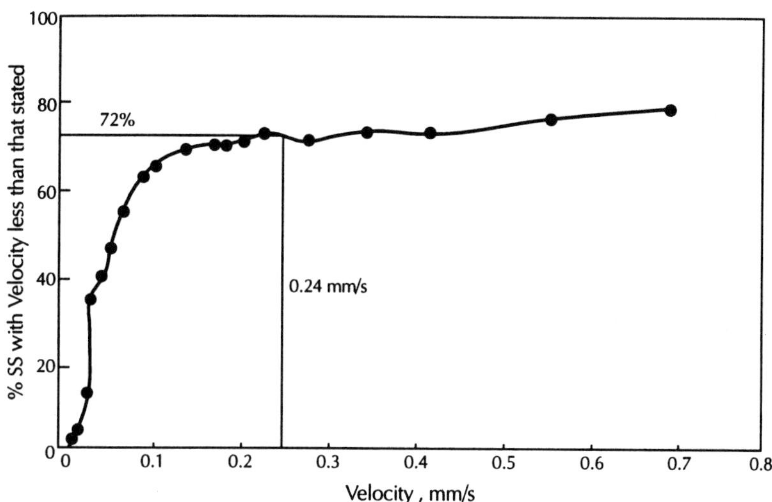

Example 3.2
Cumulative frequency diagram.

Listing of Program 3.2

```
DECLARE SUB integrate (N!, Vs!)
DECLARE SUB waitme ()
DECLARE SUB MIN (X!(), N!, IND!)
DECLARE SUB BOX (r1%, c1%, r2%, c2%)    'Found in Program 2.1
'**************************************************************************
'Program 3.2: Removal
'computes percentage of suspended solids in a settling tank effluent
'**************************************************************************
DIM SHARED depth(20), SS(20, 20), time(20), VELOCITY(50), ssr(50), SSP(50)
DIM ttl$(3)
pi = 3.142857
'Data of example 3.2
DATA 0,1,2,3,4
DATA 0.2,190,101,166,183,186
DATA 0.5,190,57,85,114,124
DATA 1,190,53,58,70,86
DATA 1.5,190,50,54,60,66
DATA 2,190,44,54,56,57
DATA 2.5,190,40,50,51,56

numdepth = 6
numsampletime = 5
ttl$(1) = "1. Input new data"
ttl$(2) = "2. Use the built-in data"
ttl$(3) = "3. Exit program"

start:
CALL BOX(1, 1, 20, 80)
LOCATE 2, 20
PRINT "Program 3.2: computes percentage of suspended solids"
LOCATE 3, 34: PRINT "in a settling tank effluent"
FOR i = 1 TO 3
LOCATE i + 4, 5
PRINT ttl$(i)
```

```
NEXT i
DO
LOCATE 10, 3
INPUT "Select an option (1 to 3): "; opt
LOOP WHILE opt < 1 OR opt > 3

IF opt = 3 THEN END
CALL BOX(1, 1, 20, 80)
LOCATE 3, 5

'read in sampling times
IF opt = 1 THEN
        INPUT "Enter number of sampling times:"; numsampletime
        LOCATE 4, 5: PRINT "Enter sampling times in hours:"
END IF
FOR i = 1 TO numsampletime
IF opt = 1 THEN
LOCATE 5, 5 + (i - 1) * 6
INPUT time(i)
END IF
IF opt = 2 THEN READ time(i)
time(i) = 3600 * time(i)        'time converted to seconds
NEXT i

'Obtain depth and  SS concentration
IF opt = 1 THEN
        LOCATE 6, 5
        INPUT "Enter number of sampling depths:"; numdepth
        CALL BOX(1, 1, 22, 80)
        LOCATE 2, 3
        PRINT "Enter depth & concentration of SS removed (mg/L) for each depth & tim
        LOCATE 3, 3
        PRINT "Depth (m)"
        FOR i = 1 TO numsampletime
        LOCATE 3, 5 + i * 10
        PRINT "time="; STR$(time(i) / 3600);
        NEXT i
        row1 = 3
END IF
FOR i = 1 TO numdepth
        IF opt = 1 THEN
                row1 = row1 + 1
                IF row1 >= 22 THEN
                        CALL BOX(1, 1, 23, 80)
                        LOCATE 3, 3
                        PRINT "Depth (m)"
                        FOR j = 1 TO numsampletime
                        LOCATE 3, 5 + j * 10
                        PRINT "time="; STR$(time(j) / 3600);
                        NEXT j
                        row1 = 3
                END IF
        LOCATE row1, 3: INPUT depth(i)
        END IF
        IF opt = 2 THEN READ depth(i)
        depth(i) = depth(i) * 1000      'convert depth to mm
        FOR j = 1 TO numsampletime
                IF opt = 1 THEN
                        LOCATE row1, 5 + j * 10
                        INPUT SS(i, j)
                END IF
                IF opt = 2 THEN READ SS(i, j)
```

```
        NEXT j
NEXT i

'Computations
CALL BOX(2, 1, 20, 80)
row = 3
LOCATE row, 3
V$ = "####.#   ######.#     ###.####        ####.#          ###.#"
PRINT "depth(mm) time(s)   VELOCITY(mm/s)    SSR(mg/L)       SS%"
k = 0
C$ = " "
FOR i = 1 TO numdepth
FOR j = 2 TO numsampletime
        k = k + 1
        VELOCITY(k) = depth(i) / time(j)
        ssr(k) = SS(i, 1) - SS(i, j)
        SSP(k) = ssr(k) / SS(i, 1) * 100
        row = row + 1
        IF row >= 18 THEN
                DO
                LOCATE 22, 3
                PRINT "Press any key to continue....."
                LOOP WHILE INKEY$ = ""
                CALL BOX(1, 1, 20, 80)
                row = 3
        END IF
        LOCATE row, 3
        PRINT USING V$; depth(i); time(j); VELOCITY(k); ssr(k); SSP(k)
NEXT j
NEXT i
COUNT = k
waitme
CALL BOX(1, 1, 20, 80)

LOCATE 5, 5
INPUT "Flow Rate Q (m3/min)              ="; Q
LOCATE 6, 5
INPUT "Number of tanks                   ="; numTanks
LOCATE 7, 5
INPUT "Diameter of the tank (m)          ="; D

Q = Q / 60                      'flow in m3/sec
Q1 = Q / numTanks               'flow per tank
a = pi * D ^ 2 / 4              'surface area of a tank
Vs = Q1 * 1000 / a    'surface loading rate for settling particles mm/s
'Search for an estimate for SS corresponding to Vs
CALL integrate(COUNT, Vs)
GOTO start
END

SUB integrate (N, Vs)
'*****************************************************************
SHARED depth(), SS(), time(), VELOCITY(), ssr(), SSP()
DIM X(N), Y(N), X1(N), X2(N), Y2(N)

F$ = "###   ##.#####  ##.#####  ###.####"

FOR i = 1 TO N
X(i) = VELOCITY(i)
Y(i) = SSP(i)
NEXT i
```

```
'sort out equations based on x values
FOR i = 1 TO N: X1(i) = X(i): NEXT i
FOR i = 1 TO N
        CALL MIN(X1(), N, IND)
        X2(i) = X(IND)
        Y2(i) = Y(IND)
        X1(IND) = 2E+10
NEXT i
FOR i = 1 TO N
        X(i) = X2(i)
        Y(i) = Y2(i)
NEXT i
'remove repeated equations
kount = 1
X2(1) = X(1)
FOR i = 2 TO N
        IF X(i) <> X(i - 1) THEN
                kount = kount + 1
                X2(kount) = X(i)
                Y2(kount) = Y(i)
        END IF
NEXT i
'Search for SS corresponding to velocity Vs
notav = 0
FOR i = 1 TO kount
IF Vs = X2(i) THEN
        SS = Y2(i)
        GOTO jmp
END IF
IF Vs < X2(i) THEN
        IF i = 1 THEN SS = Y2(1)
        IF i = kount THEN SS = Y2(kount)
        IF i > 1 AND i < kount THEN SS = .5 * (Y2(i) + Y2(i + 1))
        GOTO jmp
END IF
NEXT i
notav = 1
jmp: 'continue
CALL BOX(5, 1, 15, 80)
last = i
LOCATE 6, 3
IF notav = 1 THEN
        PRINT "Velocity out of range, No solution found!!!!"
        waitme
        END
END IF

PRINT "Corresponding SS% TO Vs="; Vs, "  is SS%="; SS
'Find the integration of vdx by trapezoidal rule
sum = .5 * Y2(1) * X2(1)
FOR i = 1 TO last - 1
sum = sum + .5 * (X2(i) + X2(i + 1)) * (Y2(i + 1) - Y2(i)) 'Integration of vdx
NEXT i
LOCATE 8, 3
PRINT "The integral part vdx="; sum
RAV = sum / Vs
Xr = 100 - SS + RAV                         'Removal efficiency
LOCATE 10, 3
PRINT "Rmoval efficiency of the tank="; Xr; " %"
waitme

END SUB
```

```
SUB MIN (X(), N, IND)
XMIN = 1E+10
FOR i = 1 TO N
        IF XMIN > X(i) THEN
                XMIN = X(i)
                IND = i
        END IF
NEXT i
END SUB

SUB waitme
        LOCATE 23, 1
        PRINT "Press any key to continue......"
        DO
        LOOP WHILE INKEY$ = ""
END SUB
```

3.4.5 The Ideal Sedimentation Basin

Settling is assumed to occur in exactly the same way as in a quiescent settling basin in an ideal horizontal flow basin (Figure 3.5). Thus, for optimum performance, the tank ought to have the following characteristics:[1,9,11]

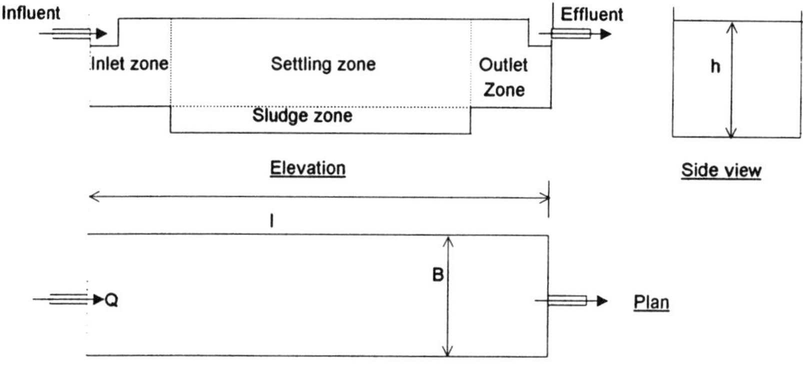

FIGURE 3.5
Schematic diagram of an ideal settling basin.

- Flow within the settling zone is uniform.
- All particles entering the sludge zone are removed, and they remain removed.
- Particles are uniformly distributed in the flow as they enter the settling zone.
- All particles entering the effluent zone leave the tank.
- Concentration of particles at inlet, for each size, is the same at all points in the vertical plane perpendicular to the direction of flow.

3.4.6 Elements that Reduce the Efficiency in the Performance of Sedimentation Basins

The sedimentation efficiency is affected by turbulence, bottom scour, nonuniform velocity distribution, and short-circuiting.

3.4.6.1 Turbulence

With turbulent flow, transverse velocity components exist. This condition scatters the discrete particles and reduces the efficiency of the tank. The horizontal flow occurs under laminar flow conditions when Reynolds numbers (Re) are less than 580 to 2000, depending upon:

- Construction of the inlet zone
- Kind and condition of the sludge removal equipment
- Presence of columns, cross beams, rough walls
- Other obstacles which might interfere with the flow

Reynolds numbers signify the ratio between inertial forces and viscous forces. It can be shown that the Reynolds number may be expressed as shown in Equation 3.10 (see Figure 3.5).

$$Re = v_H * r_H / \nu = Q/(\nu * [B + 2 * h]) = v_s * B * 1/(\nu * [B + 2 * h]) \qquad (3.10)$$

where:

Re = Reynolds number
r_H = Hydraulic radius (= A/w_p = B*h/(B + 2*h)), m
A = Area perpendicular to horizontal flow, m^2
w_p = Wetted perimeter, m
ν = Kinematic viscosity (= μ/ρ), m^2/s
μ = Dynamic (absolute) viscosity, $N*s/m^2$
ρ = Density, kg/m^3
v_H = Velocity of horizontal water movement (= Q/(B*h)), m/s
v_s = Settling rate, displacement velocity, loading rate, overflow rate (= Q/B*1), m/s
Q = Flow rate, m^3/s
B = Tank width, m
h = Tank depth, m
1 = Length of tank, m

In order to decrease Re numbers, one of the following conditions must be considered:

- Reduce flow rate entering tank (less flow)
- Increase width of tank (wider tank)
- Increase tank depth (deeper tank)
- Decrease length of tank (shorter tank)

Therefore, to prevent reduction in basin efficiency by turbulence and bottom scour, the tank must be short, wide, and deep. Equation 3.11 can be used to evaluate Re numbers for various tank conditions.

$$Re = v_{so} * r_H / \nu < 2000 \tag{3.11}$$

v_{so} = Design settling velocity.

3.4.6.2 Bottom Scour

Scouring of settled material starts at a specific velocity due to an increase in the displacement velocity. The scouring velocity develops when the hydraulic shear, between the flowing water and the sludge deposits, equals the mechanical friction of the deposits at the bottom of the tank.

The ratio between scour and settling velocities may be determined for laminar settling conditions as presented in Equation 3.12.

$$v_{sco} / v_s = 36 * \nu * [10 * \rho / (3 * g(\rho_s - \rho) d^3)]^{1/2} \tag{3.12}$$

where:

v_{sco} = Scour velocity, m/s
v_s = Settling velocity, m/s
g = Gravitational acceleration, m/s^2
ρ_s = Density of the particle, kg/m^3
ρ = Density of fluid, kg/m^3

For turbulent flow, Equation 3.13 holds.

$$v_{sco} / v_s = 2 \tag{3.13}$$

The reduction in settling basin efficiency by scour does not create problems as far as the design settling velocity is considered, as long as it is less than the scour velocity (i.e., $v_{so} < v_{sco}$). The ratio between the horizontal water velocity and the settling velocity in a rectangular horizontal flow basin can be calculated from Equation 3.14.

$$v_H / v_{so} = (Q / B * h) / Q / B * l) = l / h \tag{3.14}$$

Equation 3.15 gives the ratio between the length of the tank and its depth.

$$l / h < v_{sco} / v_{so} \tag{3.15}$$

Resuspension of settled matter can be hindered by using baffles. Figure 3.6 illustrates a sketch of round-the-end baffles.

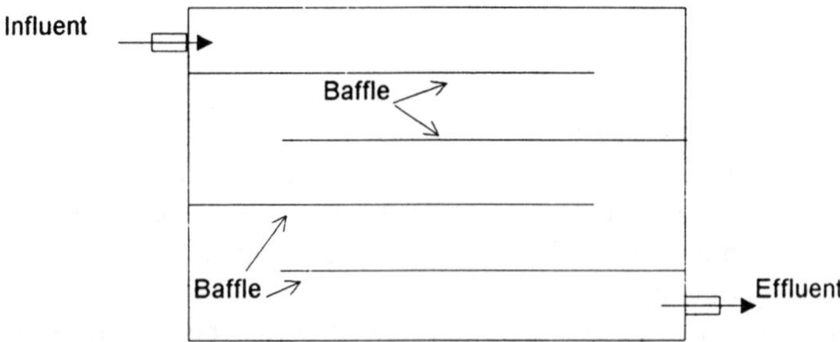

FIGURE 3.6
Round-the-end baffles in a sedimentation tank.

3.4.6.3 Nonuniform Velocity Distribution and Short-Circuiting

The displacement velocity is not constant due to the frictional drag along the walls and floor of the basin which slows down the flow of the water. As such, the velocity is slowed near the walls and floor and has a value larger than average in the center of the basin. Nonetheless, the variation in velocity over the depth of the tank has little effect on the sedimentation efficiency.

Due to the variation in the velocities of the horizontal water movement, some of the particles at the inflow reach the basin outlet in a shorter time than the theoretical detention period, while some other particles take a much longer time period. This phenomenon is referred to as short-circuiting.[1] A direct reduction in the sedimentation basin efficiency takes place due to stagnant areas or eddying currents. This may be induced by an unequal flow distribution of the incoming water or by wind action that produces surface currents. Methods for reducing short-circuiting include:

- Proficient settling tank design
- Equal inflow and outflow over the full width and depth of the basin
- Avoidance of tank inlets with high flow velocities
- Good mixing of tank contents to prevent density currents, etc.

Tank stability can be approached by increasing the ratio between inertia forces and gravity forces, i.e., Froude number. The Froude number, for a rectangular horizontal flow basin, can be expressed mathematically as presented in Equation 3.16.

$$\begin{aligned} Fr &= v_H^2 /(g * r_H) \\ &= Q^2 * [B + 2 * h]/(g * B^3 * h^3) \\ &= (\{v_s^2 * 1^2 * (1 + [2 * h / B]))/ g * h^3 \end{aligned} \tag{3.16}$$

where:

Fr = Froude number, dimensionless
v_H = Average displacement velocity, m/s
g = Gravitational acceleration, m/s^2
r_H = Hydraulic radius, m
Q = Flow rate, m^3/s
B = Breadth of the basin, m
h = Tank depth, m
v_s = Settling velocity, m/s
l = Length of tank, m

High Froude numbers can be achieved if one or all of the following conditions are considered:

- Increasing the incoming flow to the tank
- Reducing the width of the sedimentation tank (narrower tank)
- Reducing the depth of tank (shallower basin)
- Increasing the length of the basin (longer tank)

High Froude numbers are described, but should not be so high as to endanger the basin efficiency by turbulence or bottom scour. Froude numbers should satisfy the parameters in Equation 3.17. Froude numbers are empirical in nature and are based on experimental finding used in designing a settling tank.[1]

$$Fr = v_{so}^2 / g * r_H > 10^{-5} \tag{3.17}$$

Baffles guide and regulate the water flow; any structural material (e.g., wood) may be used for construction of baffles, as the water pressures on both sides of the baffles are the same.

Example 3.3

1. Write a computer program that can be used to check the stability and effects of turbulence (Froude and Reynold's numbers) in a settling basin for horizontal water movement, given the flow rate (m^3/s), tank dimensions, and temperature.
2. A suspension enters a horizontal sedimentation tank at a flow rate of 6 m^3/min. The tank is 4 m wide, 10 m long, and 1.5 m deep. Compute, under ideal settling conditions, the values for the Froude and the Reynolds numbers for the settling basin, given that the kinematic viscosity of the water is $1.004*10^{-6}$ m^2/s.

Solution

1. For solution to Example 3.3 (1), see listing of Program 3.3 below.

2. Solution to Example 3.3 (2):
   - Given: Q = 6/60 = 0.1 m^3/s, B = 4 m, 1 = 10 m, h = 1.5 m, kinematic viscosity = $1.004*10^{-6}$ m^2/s.
   - Determine the hydraulic radius as: $r_H = A/w_p = B*h/[B + 2*h] = 4*1.5/(4 + 2*1.5)$ = 0.857 m.
   - Find the displacement or horizontal tank velocity as: $v_H = Q/B*h = 0.1/4*1.5 =$ 0.017 m/s.
   - Determine Froude's number for horizontal water movement as: $Fr = v_H^2/g*r_H$. $Fr = (0.017)^2/(9.81*0.857) = 3.3*10^{-5}$. This value is greater than 10^{-5}, thus it is all right.
   - Find Reynolds number for the horizontal water movement as: $Re = v_H*r_H/v = 0.017*0.857/1.004*10^{-6} = 14{,}229$, which is greater than 2000, which is not all right. To remedy this situation and avoid erratic tank behavior, baffles should be introduced.

Listing of Program 3.3

```
DECLARE SUB box (r1%, c1%, r2%, c2%)     'Found in Program 2.1
'*****************************************************************
'Program 3.3: Turbulence
' Checks the stability and effects of turbulence
'*****************************************************************
CALL box(2, 1, 20, 80)
LOCATE 3, 15
PRINT "Program 3.3: Checks the stability and effects of turbulence"

ROW = 5
LOCATE ROW + 1, 5
INPUT "Enter the flow rate   Q (m3/s)    ="; Q
LOCATE ROW + 2, 5
PRINT "Enter the dimensions of the tank :"
LOCATE ROW + 3, 5
INPUT "Width of tank B (m)="; B
LOCATE ROW + 4, 5
INPUT "Depth of tank h (m)="; h
LOCATE ROW + 5, 5
INPUT "Length of tank L(m)="; L
LOCATE ROW + 6, 5
INPUT "Kinematic viscosity of water(m2/s)="; u

'Compute hydraulic radius
g = 9.81                     'acceleration due to gravity
Rh = B * h / (B + 2 * h)
VH = Q / (B * h)
Fr = VH ^ 2 / (Rh * g)   'Froude number
Re = VH * Rh / u

LOCATE ROW + 9, 3
PRINT "Froude number               ="; Fr;
        IF Fr >= .00001 THEN
        PRINT "     O.K."
        ELSE PRINT "     NOT O.K., too low Froude number!"
END IF

LOCATE ROW + 10, 3
PRINT "Reynolds number          ="; Re;
```

```
IF Re <= 2000 THEN PRINT "O.K. , No baffles needed"
IF Re > 2000 THEN
        LOCATE ROW + 11, 5
        PRINT "NOT O.K., to avoid erratic tank behavior, use baffles"
END IF
LOCATE 23, 3
PRINT "press any key to continue....................";
DO
LOOP WHILE INKEY$ = ""
END
```

3.4.7 Design of the Settling Zone

The following steps may be used in the design of the settling zone in a sedimentation basin:

- The required surface area can be computed as:

$$A = Q / v_{so} \tag{3.18}$$

where:

A = Surface area, m^2
Q = Amount of water to be treated, m^3/s
v_{so} = Design surface loading to be applied, m/s

- Depth of tank may be computed from the following relationship:
  - For circular tanks:

$$h = 0.17 * A^{1/3} \tag{3.19}$$

where:

h = Depth of tank, m

  - For rectangular tanks:

$$h = 1^{0.8} / 12 \tag{3.20}$$

- The ratio of length of the tank to its width can be calculated from the empirical Equation 3.21.

$$\text{Length/width} = l/B = 6 \text{ to } 10 \tag{3.21}$$

- The flow rate per unit length (weir loading) must not be too high. This is to prevent disturbing the settling of suspended matter near the end of the tank. Settling near the end of the tank occurs due to updraft velocities created at the weir discharge. The allowable weir loading is given by Equation 3.22.

$$Q / B < 5 * h * v \tag{3.22}$$

where:

Q = Flow rate, m^3s
B = Width of tank, m
h = Depth of tank, m
v_{so} = Overflow rate, m/s

With v_{so} = Q/B*1, this requirement is fulfilled in case the ratio of tank length to its depth falls below a value of 5, as shown in Equation 3.23.

$$l / h < 5 \tag{3.23}$$

where:

l = Length of tank, m
h = Depth of the tank, m

3.4.8 Settling of Flocculent Particles

The settling of a suspension containing particles with different velocities of subsidence results in the smaller (lighter) particles being overtaken by the larger sized particles (heavier mass density). This results in a number of collisions and eventually the formation of aggregates of particles.

The removal ratio for flocculent settling increases with a decrease in overflow rate and with an increase in basin depth. Factors affecting flocculent settling efficiency include overflow rate, detention time, and tank depth. The effects of bottom scour and short-circuiting are similar to those for discrete particles. Nonetheless, the effect of turbulence may be neglected due to dispersion. Dispersion prevents part of the suspended matter from reaching the tank bottom, but it augments aggregation of finely divided suspended matter into larger floc that have higher settling velocities. The net effect of turbulence on flocculent settling is minor. Thus, for the design of the sedimentation basin, for a high Froude number and stable flow conditions, the tank ought to be long, narrow, and shallow with the displacement velocity high enough to prevent bottom scour.

Any slight increase in temperature (e.g., from solar heating) results in an increase in the settling velocity of the particles (viscosity effects the settling velocity). However, an increase in temperature can also produce convection currents, which tend to scour and resuspend settled material.

3.4.9 Analysis of Flocculent Settling

The settling column apparatus shown in Figure 3.2 can be used for the determination of the settling characteristics of a suspension of flocculent particles. The procedure for analysis of flocculent settling in the settling column is summarized as follows:

1. See the procedure presented in Section 3.4.4 (Steps 1, 2, 3, and 4).
2. Percentage of SS that have settled past the sampling point is computed as equal to the value of 100 – (C/C_o).
3. The percentage of removal is to be plotted against the time and depth.
4. Curves of equal percentages of removal (iso-concentration lines or iso-removal lines) are to be developed.
5. Overall removal is determined as:

$$\begin{aligned} R_T &= \{\Delta h_1(R_1 + R_2)/2h_t\} + \Delta h_2(R_2 + R_3)/2h_t + \ldots + \{\Delta h_n(R_n + R_{n+1})/2h_t\} \\ &= (1/h_t) * \sum_{i=1}^{i=n} h_i(R_i + R_{i+1}) \end{aligned} \tag{3.24}$$

where

h_i = Sampling port depth i, m
h_t = Height of proposed tank, m
R_T = Total removal, %

3.4.10 General Design Considerations

3.4.10.1 Rectangular Tanks

Rectangular tanks are characterized by their rectangular shapes with horizontal flow. For practical purposes, the following tank dimensions are considered appropriate. A maximum tank length of 100 m, with an average length of 30 m, and a maximum depth of 5 m, with an average depth of 3 m.[12] Such tanks generally are used for primary sedimentation.[13] The sludge hopper acts as a thickener due to accumulation of sludge in it. The bottom slope may be taken as 1% on the average to facilitate sludge sliding into the hopper.[12] In the design of rectangular tanks, the following points should be considered (see Figure 3.7):

- Volume of tank is computed as:

$$V = Q * t \tag{3.25}$$

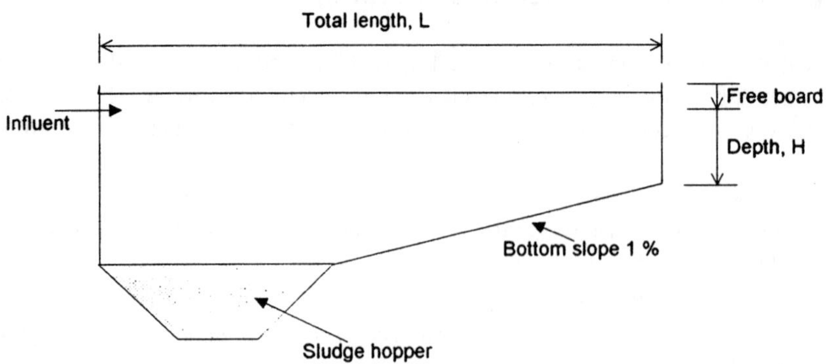

FIGURE 3.7
Rectangular sedimentation tank.

where:

V = Volume of tank, m^3
Q = Design flow, m^3/d
t = Detention time of tank, d (usually taken as $1^1/_2$ to 2 h for maximum flow)

- Minimum number of tanks is 2.
- The length-width ratio may be taken as:[12]

$$4 \leq l / b \leq 1 \tag{3.26}$$

where:

l = Length of tank, m
b = Width of tank, m

- The length-depth ratio may be taken as:[12]

$$25 \geq l / H \geq 1 \tag{3.27}$$

where:

H = Useful clarification depth, m (not to exceed 90 m)[13]

- Daily sludge removal is determined as:

$$V_h Q * C_0 * Eff * (100 / (100 - m.c.)] / \gamma \tag{3.28}$$

where:

V_h = Sludge volume, m^3
Q = Maximum daily flow rate, m^3/d

C_0 = Initial concentration of suspended solids influent to tank, kg/m^3

Eff = Percent suspended solids removal

m.c. = Moisture content of settled sludge, %

ρ = Density of sludge, kg/m^3

3.4.10.2 Circular Settling Tanks (Dorr Settling Tanks)

In tanks with circular shapes, the flow is horizontal. Circular tanks can be used for primary as well as secondary sedimentation purposes. The maximum diameter for circular tanks is considered to be 50 m, with an average diameter of 30 m, and a maximum water depth of 4 m, with an average water depth of 2.5 m.[12] In the design of circular tanks, the following points merit consideration (refer to Figure 3.8):

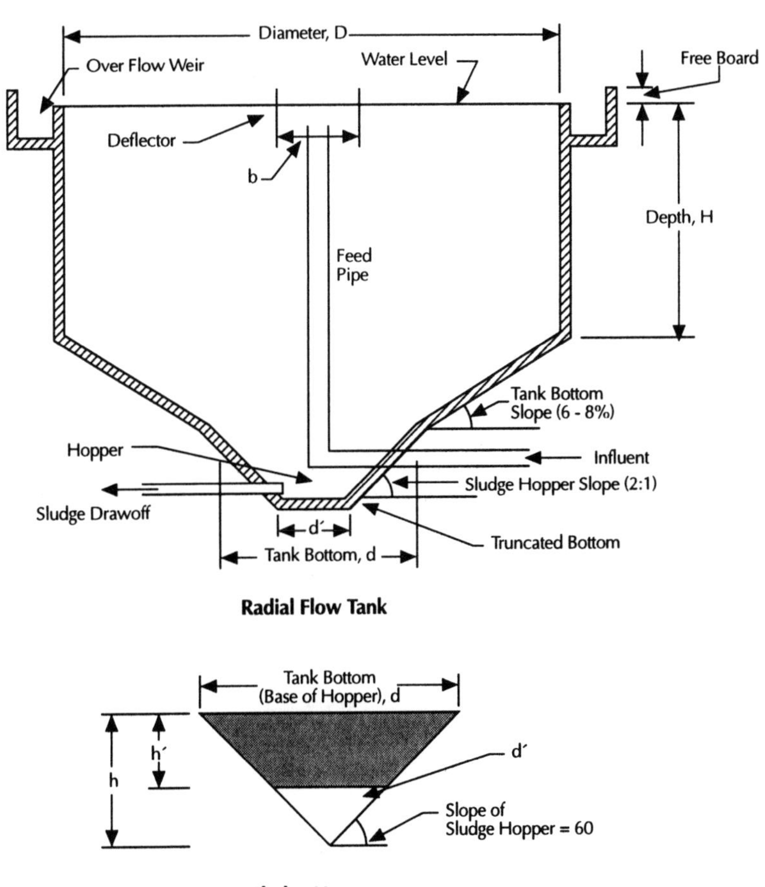

FIGURE 3.8
Radial flow sedimentation tank.

- Ratio of diameter to clarification depth may be found as follows:[12]
  - $6 \le D/H \le 10$ for circular tanks with diameters of 16 to 30 m.
  - $16 \le D/H \le 20$ for circular tanks with diameters of 30 to 50 m.
  - $20 \le D/H \le 25$ for circular tanks serving wastewaters with large amounts of organic matter (high settling velocities).
- Ratio of diameter of circular deflector to diameter of tank may be taken as:[12]

$$b = (10 \text{ to } 20\%) * \tag{3.29}$$

where:

b = Diameter of circular deflector (may be taken as $b \approx 0.15*D$), m.[13]
D = Diameter of circular tank, m

Assuming a single weir in tank: surface loading*area = weir loading*length.

$$\pi(D-b)^2 * v_s / 4 = \pi * D * q_{weir} \tag{3.30}$$

where:

D = Diameter of tank, m
b = Diameter of deflector, m
v_s = Surface loading rate, m/d
q_{weir} = Weir loading (overflow rate), m^3/d/m length

Taking:

$$b = 0.15 * D \tag{3.31}$$

then,

$$D = 5.54 * q_{weir}/v_s$$

- Tank depth may be determined as:

$$H = Q * t / N * A \tag{3.32}$$

where:

A = Effective area of settlement, m^2

$$H = \pi * (D_a - b)^2 / 4 \tag{3.33}$$

where:

N = Number of settling tanks, dimensionless

D_a = Actual diameter of tank, m

Thus,

$$H = 4 * t * Q / \pi * N * (D_a - b)^2 \tag{3.34}$$

- Slope of settling tank bottom, β, may be taken as β = 6 to 8%[12] (or 5 to 10°[13]).
- Slope of sludge hopper, α, may be taken as α = 2:1[12] (≥ 60°[13]).
- Detention time may be taken as 2 h at peak flow.
- Volume of sludge hopper is to hold 12 to 24 h of sludge production, and it may be determined as:

$$V_h = Q * C_0 * Eff / a * C_c \tag{3.35}$$

where:

V_h = Volume of sludge hopper, m^3

Q = Flow rate, m^3/d

C_0 = Concentration of solids entering sedimentation tank, kg/m^3

Eff = Efficiency of solids removal in tank, dimensionless

a = Fraction of flow to be stored in hopper (as sludge)

C_c = Solids content of final sludge, kg/m^3

= (1 – m.c.)*1000, where m.c. = moisture content of sludge (dimensionless).

Considering hopper as a truncated cone, then:

$$V_h = (\pi d^2 * h / 12) - (\pi d'^2 * h' / 1 \tag{3.36}$$

where:

d = Diameter of hopper base (tank floor), m

h = Hopper imaginary depth (depth of total cone), m

d′ = Diameter of truncated cone (hopper floor), m

h′ = Depth of truncated cone, m

Taking a slope of sludge hopper as α = 60, then:

$$(d/2)/h = (d'/2)/h' = \tan 30 \tag{3.37}$$

As such:

$$V_h = 0.227(d^3 - d'^3) \tag{3.38}$$

- The feed pipe is to be located 20 to 30 cm below water level.

Example 3.4

1. Write a computer program for the hydraulic and structural design of a circular sedimentation tank given the relevant design data of rate of flow, Q (m^3/s); loading rate, v_s ($m^3/h/m^2$); influent solids concentration, C_0 (kg/m^3); tank efficiency, Eff (%); sludge moisture content, m.c. (%); sludge production that can be stored by the sludge hopper; weir loading (overflow rate), q_{weir} ($m^3/d/m$ weir length); retention time, t (s) at peak flow; and bottom radius (truncated radius), d′ (m).
2. Design a circular sedimentation tank using the following information:
   - Rate of flow, Q = 15 m^3/min.
   - Loading rate (settling velocity), v_s = 1.4 $m^3/h/m^2$
   - Influent solids concentration, C_0 = 250 mg/L
   - Tank efficiency, Eff = 55%
   - Sludge moisture content, m.c. = 96%
   - Size of sludge hopper sufficient to hold 12 h sludge production
   - Weir loading (overflow rate), q_{weir} = 200 $m^3/d/m$
   - Retention time, t = 2 h at peak flow
   - Bottom radius (truncated radius), d′ = 0.5 m

Solution

1. For solution to Example 3.4 (1), see listing of the following Program 3.4.
2. Solution to Example 3.4 (2):
   - Determine diameter of tank using: $D = 5.54*q_{weir}/v_s = 5.54*200/(1.4*24) = 32.98$ m.
   - Determine required number of sedimentation tanks: Flow from each tank = $\pi*D^2*0.85^2*v_s/4 = \pi*(32.98)^2*85*(1.4*24)/4 = 20738.1$ m^3/d. Flow rate to be treated = 15*60*24 = 21600 m^3/d. Number of needed tanks = 21600/20738.1 ≈ 2.
   - Find actual diameter of tank: $2*\pi*0.85^2*D_a^2*1.4*24/4 = 15*60*24$. Thus, D_a = 23.8 m.
   - Compute useful depth of clarification zone: $H = Q*t/N*D_a^2 = (2/24)*21600/2*23.8^2*(1 - 0.15)^2 = 4.4$ m. Check: D/H = 23.8/4.4 = 5.4 (acceptable falls between 6 and 10).
   - Determine $C_c = (1 - m.c.)*1000 = (1 - 0.96)*1000 = 40$ kg/m^3.
   - Find volume of sludge hopper: $V_h = (21600\ m^3/d*0.25\ kg/m^3*0.55)/[(24/12\ h)*40\ kg/m^3] = 37.125\ m^3$. Since, $V_h = 0.227\ (d^3 - d'^3) = 0.227*[d^3 - (0.5*2)^3] = 37.125$, then d = 5.5 m.

Listing of Program 3.4

```
DECLARE SUB WAITME ()
DECLARE SUB box (r1%, c1%, r2%, c2%)
'*****************************************************************
'Program 3.4: Tanks
'designs a circular sedimentation tank
'*****************************************************************
```

```
CALL box(1, 1, 20, 80)
LOCATE 2, 15
PRINT "Program 3.4:Designs a circular sedimentation tank"

pi = 3.142857
row = 3
LOCATE row + 1, 5
INPUT "Enter flow rate (m3/min)                        ="; Q
LOCATE row + 2, 5
INPUT "Enter loading rate (m3/hr/m2)                   ="; Vs
LOCATE row + 3, 5
INPUT "Enter influent solids concentration (mg/L)  ="; Co
LOCATE row + 4, 5
INPUT "Enter tank efficiency (%)                       ="; Eff
LOCATE row + 5, 5
INPUT "Enter sludge moisture content (%)               ="; mc
LOCATE row + 6, 5
INPUT "Enter period to maintain sludge in hopper (hr) ="; tr
LOCATE row + 7, 5
INPUT "Enter weir loading rate (m3/d/m)                ="; Qweir
LOCATE row + 8, 5
INPUT "Enter retention time at peak flow    (hr)   ="; t
LOCATE row + 9, 5
INPUT "Enter bottom radius (m)                         ="; rdash

Vs = Vs * 24                                     'convert to m/day
Q = Q * 60 * 24                                  'convert to m3/day
Co = Co / 1000                                   'convert to kg/m3
t = t / 24                                       'convert to days
D = 5.54 * Qweir / Vs                            'diameter of tank
Q1 = pi / 4 * D ^ 2 * (1 - .15) ^ 2 * Vs         'flow from each tank
N = CINT(Q / Q1)                                 'Number of tanks

cycle:
D = (Q / (N * pi * .85 ^ 2 * Vs / 4)) ^ .5       'actual diameter
A = pi / 4 * (.85 * D) ^ 2                       'Area of tank
H = Q * t / (N * A)                              'clarification depth
'check D/H
rat = D / H
IF rat > 6 AND rat < 10 THEN
        OK$ = "T"
        t$ = "D/H ratio=" + STR$(rat) + " is O.K."
        ELSE
        OK$ = "F"
        N = N + 1
        GOTO cycle
END IF
'solids content in the final sludge
Cc = (1 - mc / 100) * 1000
'volume of sludge hopper Vh
Eff = Eff / 100                          'efficiency ratio
Vh = Q * Co * Eff / (24 / tr * Cc)
dh = (Vh / .227 + (2 * rdash) ^ 3) ^ (1 / 3)

LOCATE row + 11, 3
PRINT "Number of tanks needed     ="; N
LOCATE row + 12, 3
PRINT "Diameter of tank           ="; D; "  m"
LOCATE row + 13, 3
PRINT "Depth of clarification zone="; H; "  m"; " : "; t$
LOCATE row + 14, 3
PRINT "Diameter of sludge hopper  ="; dh; "  m"
```

```
LOCATE 22, 3
PRINT "press any key to continue....................";
DO
LOOP WHILE INKEY$ = ""
END

SUB WAITME
        LOCATE 22, 1
        PRINT "Press any key to continue......"
        DO
        LOOP WHILE INKEY$ = ""

END SUB
```

3.5 Flocculation and Coagulation

3.5.1 Introduction

In water and wastewater treatment, flocculation and coagulation processes are mainly used for the removal of colloidal substances which cause color and turbidity. In relatively dilute solutions, some particles do not behave as discrete particles but rather they coalesce and combine during settling. As coalescence or flocculation occurs, the mass and settling velocity of the particle increase. The rate of flocculation depends on the opportunity for contact between particles. The factors that influence this process include surface loading, depth of settling tank, velocity gradients, and concentration and size of particles.

To enhance the process of flocculation and coagulation, a small dose of a coagulant (e.g., aluminum or ferric salts) is added to the raw water, after which flocs of coagulated colloidal matter are formed. This enables separation of particles from the fluid by sedimentation and/or filtration. The needed coagulant dose varies with the nature of the suspension. Flocculation occurs at the end of the coagulation phase. It consists of the building up and increase in the flocs with or without the use of flocculation aids or other chemical substances. The major factors that affect and influence the coagulation process include chemical factors (such as the optimum pH; inorganic ions; character of turbidity; nature, type, and concentration of the colloids involved; and type of coagulant) and physical parameters (such as temperature, mixing speed and duration of mixing, and size of colloidal particles).

3.5.2 Electrokinetics of Coagulation

Electrokinetics denotes motion of ions, molecules, or particles along a gradient of an electrical potential.[11,15] Usually, colloidal or suspended particles in water bear a negative electrical charge. This electrical charge induces a positive charge in the vicinity of the neighboring liquid layer (electrical double layer). The electrical double layer provides a difference in potential between the particle and the bulk of the solution. This difference in potential exists across the electrical double layer and is referred to as the electrokinetic potential, zeta potential (ZP), or electrophoretic mobility.

The ZP may be defined as the potential at the surface of a particle that separates the immobile part of the double layer from the diffuse part. Zeta potential is usually employed for rational control of coagulation. The ZP also signifies a measure of the electrokinetic charge that surrounds suspended particulate matter. For most colloids, the electrokinetic potential is in the range of 30 to 60 mV.

Removal of colloids requires reduction of the zeta potential. This is due to the fact that repulsive forces of the electronegative colloid prevent the formation of a floc. The reduction of zeta potential is very effective in the settlement of lyophobic colloids. Equation 3.39 may be used to determine the zeta potential.[11,15]

$$ZP = 4\pi B q^{+} / D_{I} \tag{3.39}$$

where:

ZP = Zeta potential
B = Thickness of the boundary layer
q^{+} = Charge on the colloid
D_I = Dielectric constant

Electrophoretic mobility (EM) refers to the rate of movement of a particle to an electrode under a given electrical potential. For natural water, EM ranges between 2 to $4{*}10^{-4}$ m/s/V/cm. Generally, the higher the ionic strength, the lower the zeta potential. The usual technique for measurement of electrophoretic mobility involves a microscopic viewing of the particle motion under the influence of an electric field. Electrophoretic mobility can be calculated by using Equation 3.40.[15]

$$EM = y * A / t * i * RES_s \tag{3.40}$$

where:

EM = Electrophoretic mobility, μm/s/V/cm
y = Distance covered in t seconds, μm
A = Cross-sectional area, cm^2
i = Current, A
RES_s = Specific resistance of the suspension, Ω*cm

In practice, ZP is assumed to be related to EM as presented in Equation 3.41

$$ZP = 4 * \pi * \mu * EM / D_I \tag{3.41}$$

where:

ZP = Zeta potential

μ = Absolute viscosity, N*s/m^2

EM = Electrophoretic mobility, µm/s/V/cm

D_I = Dielectric constant

Example 3.5

Write a short computer program that can be used to determine the zeta potential and electrophoretic mobility, given viscosity, dielectric constant, cross-sectional area, current density, specific resistance of the suspension, thickness of the boundary layer, and charge on the colloid. Use the relevant equation in each case.

Solution

For solution to Example 3.5, see listing of the following Program 3.5.

Listing of Program 3.5

```
DECLARE SUB box (r1%, c1%, r2%, c2%)    'Found in Program 2.1
'**************************************************************
'Program 3.5: COAGULATION
'Computes the electrokinetics of coagulation
'**************************************************************
pi = 3.14152654
CALL box(2, 1, 20, 80)

LOCATE 3, 15
PRINT "Program 3.5: Computes the electrokinetics of coagulation"
LOCATE 5, 5
INPUT "viscosity coefficient (Ns/m2)     ="; u
LOCATE 6, 5
INPUT "dielectric constant               ="; Di
LOCATE 7, 5
INPUT "Cross sectional area              ="; A
LOCATE 8, 5
INPUT "Current density                   ="; i
LOCATE 9, 5
INPUT "The specific resistance           ="; Rs
LOCATE 10, 5
INPUT "Thickness of boundary layer       ="; B
LOCATE 11, 5
INPUT "Charge on the colloid             ="; q

zp = 4 * pi * B * q / Di          'Zeta potential
'EM = y * A / (t * i * Rs)
EM = zp * Di / (4 * pi * u)       'Electrophoretic mobility

LOCATE 12, 3
PRINT "Zeta potential                ="; zp
LOCATE 13, 3
PRINT "Electrophoretic mobility      ="; EM
```

```
LOCATE 24, 5
PRINT "press any key to continue....................";
DO
LOOP WHILE INKEY$ = ""
END
```

3.5.3 Design Parameters

Parameters important for design of a coagulation system are

1. Flow rate
2. Flocculation time (residence time) in the range of 15 to 30 min.
3. Power input per unit volume (usually from 0.5 to 1.5 kWh/m^3)
4. Mean velocity gradient: For quick circulation in the system the velocity gradient lies between values of 20 and 75 s^{-1}. Equation 3.42 shows the relationship between velocity gradient and power.

$$G = (POW_v / \mu)^{1/2} \tag{3.42}$$

where:

POW_v = Power input per unit volume, W/m^3
G = Velocity gradient, s^{-1}

When impellers are used, velocity gradient may be found from Equation 3.43.

$$G = (2\pi * g * \omega * TOR / 60 * V * \mu)^{1/2} \tag{3.43}$$

where:

G = Velocity gradient, s^{-1}
g = Gravitational acceleration, m/s^2
ω = Impeller rotations, number of rotations/min
TOR = Net torque input, dyne/cm
V = Volume, m^3
μ = Dynamic viscosity, N*s/m^2

5. The power, in kW, requirement to be of values:

$$0.5 < POW < 2.5 \tag{3.44}$$

6. Depth =

$$(Volume)^{1/3}/2 = (flow*detention\ time)^{1/3}/2 = (Q*t)^{1/3}/2 \tag{3.45}$$

7. Surface area =

$$A = Q/v = Q/\{h/t\} = Q*t/h = Q*t/\{(Q*t^{1/3}/2\} = 2*(Q*t)^{2/3} \tag{3.46}$$

3.5.4 Design of Flocculator

The velocity gradients are brought about by rotating paddles in a flocculator; therefore, particles in a rapidly moving stream can overtake and collide with others in a slowly moving stream. The power needed for driving the paddle[16] through the fluid is given by Equation 3.47.

$$POW = \rho * C_D * A * v^3 / 2 \tag{3.47}$$

where:

POW	=	Power needed for driving the paddle through the fluid, J/s
ρ	=	Fluid density, kg/m^3
C_D	=	Drag coefficient
A	=	Paddle area, m^2
v	=	Velocity of paddle relative to fluid, m/s

The velocity gradient initiated as a result of the input power is given by Equation 3.48.

$$G = (POW / \mu * V)^{1/2} \tag{3.48}$$

where:

G	=	Velocity gradient, s^{-1}
μ	=	Viscosity, N*s/m^2
V	=	Tank volume, m^3

The generally accepted design standard requires that G be between 30 and 60 s^{-1}.[17] Time is also an important variable in flocculation. The term G*t often is used in the design (t = hydraulic retention time in a flocculator). Typical G*t values range from 1*10^{14} to 1*10^5.[17]

Example 3.6

1. Write a computer program to determine the power needed for a flocculator system and the area of the paddle required for effective flocculation, given temperature, coefficient of drag for rectangular paddles, paddle tip velocity, velocity gradient, flocculator volume, and the relative velocity of the paddle.
2. A treatment plant incorporates coagulation and sedimentation processes. The existing flocculator has the following properties:

Item	Value
Fluid temperature (T)	20°C
Coefficient of drag for rectangular paddles (C_D)	1.8
Paddle tip velocity (v_p)	0.7 m/s
Velocity gradient (G)	70 s
Flocculator volume (V)	4000 m^3

Using the aforementioned information compute:

a. Power needed for the system
b. Area of the paddle required for effective flocculation (Hint: Use a relative velocity of the paddle equal to 75% of that of the paddle tip velocity.)

Solution

1. For solution to Example 3.6 (1), see Program 3.6 listed below.
2. Solution to Example 3.6 (2):
   - Given T = 20°C, G = 70 s, V = 4000 m^3, v_p = 0.7 m/s, C_D = 1.8
   - Find the viscosity and density from Table A1 in Appendix for a temperature of 20°C as: viscosity $\mu = 1.002*10^{-3}$ N*s/m^2 and density ρ = 998.2 kg/m^3.
   - Determine the theoretical power requirements as: POW = $\mu*G^2*V$.
   - Substitute values given in previous equation: POW = $1.002*10^{-3}*(70)^2*4000$ = 19.6 kW.
   - Determine velocity of the paddle, v = 0.75*0.7 = 0.525 m/s.
   - Find the required paddle area, A, as: A = $2*POW/C_D*\rho*v^3$ = $(2*19.6*10^3)/(1.8*998.2*[0.525]^3)$ = 151 m^2.

Listing of Program 3.6

```
DECLARE SUB box (r1%, c1%, r2%, c2%)    'Found in Program 2.1
'*****************************************************************
'Program 3.6: COAGULATION 2
'Computes the power needed for a flocculator system
'*****************************************************************
'Kinematic viscosity of water u (Ns/m2) versus temp. equation
DATA 0.179E-02,-.654E-04,0.394E-05,-.480E-06,0.431E-07,-.235E-08
DATA 0.801E-10,-.175E-11,0.244E-13,-.210E-15,0.101E-17,-.210E-20

'Density of water Ro (Kg/m3) versus temp (C) equation:
DATA 0.100E+04,-.186E-01,0.301E-01,-.695E-02,0.638E-03,-.333E-04
DATA 0.107E-05,-.220E-07,0.287E-09,-.233E-11,0.106E-13,-.208E-16

pi = 3.141592654
CALL box(2, 1, 17, 80)

LOCATE 3, 15
PRINT "Program 3.6:Computes the power needed for a flocculator system"
LOCATE 5, 5
INPUT "Fluid temperature  T ( Centigrade)="; T
LOCATE 6, 5
```

```
INPUT "Drag coefficient Cd                  ="; Cd
LOCATE 7, 5
INPUT "Paddle tip velocity  Vp (m/s)        ="; vp
LOCATE 8, 5
INPUT "Velocity gradient  G ( /s)           ="; G
LOCATE 9, 5
INPUT "flocculator volume V (cubic m)       ="; v
Rv = .75          'default relative velocity of the paddle
LOCATE 10, 5
INPUT "Relative velocity of the paddle(RATIO)="; Rv

'get the viscosity u (at any temperature T), and density Ro
u = 0
FOR i = 1 TO 12
READ vis
u = u + vis * T ^ (i - 1)
NEXT i

Ro = 0
FOR i = 1 TO 12
READ vis
Ro = Ro + vis * T ^ (i - 1)
NEXT i

LOCATE 11, 3
POW = u * G ^ 2 * v                         'power requirement
vp1 = Rv * vp                               'velocity of the paddle
A = 2 * POW / (Cd * Ro * vp1 ^ 3)           'Required paddle area

LOCATE 12, 3
PRINT "Power needed for the system  POW     ="; POW; "  KW"
LOCATE 13, 3
PRINT "Area of the paddle required          ="; A

LOCATE 24, 5
PRINT "press any key to continue....................";
DO
LOOP WHILE INKEY$ = ""
END
```

3.6 Aeration and Gas Transfer

The process of gas transfer (aeration) signifies the intimate contact between the air and the water for a limited period for the purpose of promoting gas transfer. The advantages of this artificially induced gas process include:[14]

- Addition of oxygen to groundwater
- Addition of oxygen to sewage to assist in biological reactions
- Removal of carbon dioxide, CO_2
- Removal of hydrogen sulfide, H_2S
- Removal of methane, CH_4
- Reduction of taste and odor problems in water
- Removal of volatile oils and similar odor- and taste-producing substances
- Removal of ammonia, NH_3

There are many types of aerators: gravity aerators (e.g., weir aeration and cascades, inclined plane, and vertical stacks of perforated pans or tower cascades), spray aerators, air diffusers (bubble aeration), and mechanical aerators. The factors that influence the solubility of a gas in water include nature of the gas, concentration of the respective gas in the gas phase (i.e., as related to partial pressure of gas in gas phase), temperature of the water (C_s decreases as T increases), and level of impurities contained in the water (C_s decreases with an increase in impurities).

The higher the gas concentration in the gaseous phase, the greater the saturation concentration in the liquid phase. The saturation concentration can be determined by using Henry's law as presented in Equation 3.49.

$$C_s = K_D * MW * P / R * T = k_H * P \tag{3.49}$$

where:

C_s = Saturation concentration, g/m^3
K_D = Distribution coefficient
MW = Molecular weight of the gas
P = Partial pressure of the respective gas in the gas phase, Pa
R = Universal gas constant (8314.3 J/Kg*mol)
T = Absolute temperature, K
k_H = Henry's constant, g/m^3*Pa, g/J (= K_D*MW/R*T)

Solubilities also can be estimated by the Bunsen absorption coefficient, as show in Equation 3.50.

$$C_s = K_b * MW * P / R * T_0 \tag{3.50}$$

where:

K_b = Bunsen absorption coefficient, g/J (= $K_D T_0 T$)
T_0 = Standard temperature, °C

During the process of gas transfer from the gas phase into water, the concentration in the gas phase is decreased. This decrease is due to the absorption of the gas into the liquid phase (process of diffusion, i.e., the tendency of any substance to spread uniformly throughout the space available to it). Some of the theories describing the mechanism of gas transfer include film, penetration, surface renewal, and film-surface renewal.

In some gas transfer operations (e.g., cascades, weir aeration) the efficiency coefficient for the system may be calculated as presented in Equation 3.51.[14]

$$C_e - C_0 = K(C_s - C_0) \tag{3.51}$$

where:

C_e = Effluent concentration, g/m^3
C_0 = Influent concentration, g/m^3
K = Efficiency coefficient of the system
C_s = Saturation concentration, g/m^3

For a cascade aerator, Equation 3.52 can be used to calculate the gas saturation concentration in the liquid phase.

$$(C_s - C_N)/(C_s - C_0) = (1 - k_N)^N \tag{3.52}$$

where:

C_s = Saturation concentration, g/m^3
C_N = Effluent concentration from step N of the cascade aerator, g/m^3
C_0 = Influent concentration, g/m^3
k_N = Efficiency coefficient of the system
N = Number of steps of the cascade aerator

Example 3.7

1. Write a computer program that can be used to determine the efficiency coefficient for a one-step cascade aerator. Also include in the program the procedure to calculate the concentration of gases sorbed into an N-step cascade aerator.
2. Water at a temperature of 20°C has an oxygen content of 15% saturation with zero chloride concentration. A cascade aerator is to be used. The cascade aerator is composed of three steps. Each step can raise the oxygen content of an anaerobic water to 40% saturation. Estimate the oxygen concentration of the effluent from a three-step cascade aerator.
3. Use the computer program developed in (1) to verify the solution of the problem presented in section (2).

Solution

1. For solution to Example 3.7 (1), see Program 3.7 listed below.
2. Solution to Example 3.7 (2):
   - Given influent concentration $C_0 = 0.15*C_s$, T = 20°C, C_0 for each step = 0, C_e for each step = $0.4*C_s$, N = 3 steps.
   - Find the saturation concentration from Table A2 in Appendix corresponding to temperature of 20°C as: $C_s = 9.2$ mg/L.
   - Determine the initial concentration of water as: $C_0 = 0.15*9.2 = 1.38$ mg/L.
   - Find the efficiency coefficient for each step as: $K = (C_e - C_0)/(C_s - C_0) = (0.4*C_s - 0)/(C_s - 0) = 0.4$.
   - Find the effluent oxygen concentration from each step of the 3-step aerator as:

- $C_{e1} = C_0 + K^*(C_s - C_0) = 1.38 + 0.4^*(9.2 - 1.38) = 4.508$ mg/L.
- $C_{e2} = C_{e1} + K^*(C_s - C_{e1}) = 4.508 + 0.4^*(9.2 - 4.508) = 6.385$ mg/L.
- $C_{e3} = C_{e2} + K^*(C_s - C_{e2}) = 6.385 + 0.4^*(9.2 - 6.385) = 7.51$ mg/L.

- Otherwise $(C_s - C_N)/(C_s - C_0) = (1 - k_n)^N$ and $(9.2 - C_{e3})/(9.2 - 1.38) = (1 - 0.4)^3$, which yields, $C_{e3} = 7.51$ mg/L.

Listing of Program 3.7

```
DECLARE SUB box (r1%, c1%, r2%, c2%)    'Found in Program 2.1
'******************************************************************
'Program 3.7: Cascade
'Computes the efficiency coefficient of a cascade aerator
'******************************************************************
DIM temp(31), conc(31), Ce(20)
'data for saturation values od dissolved oxygen at zero chloride
'for different temperatures- Table 2.5
DATA 14.6,14.2,13.8,13.5,13.1,12.8,12.5,12.2,11.9,11.6,11.3,11.1,10.8
DATA 10.6,10.4,10.2,10,9.7,9.5,9.4,9.2,9,8.8,8.7,8.5,8.4,8.2,8.1,7.9
DATA 7.8,7.6
'temperature
FOR i = 1 TO 31
READ conc(i)
temp(i) = i - 1
NEXT i

gammaw = 9.806   'KN/m3
G = 9.81         'Gravity constant

CALL box(2, 1, 20, 80)
LOCATE 3, 5
PRINT "Program 3.7: Computes the efficiency coefficient of a cascade aerator"
LOCATE 5, 5
INPUT "Enter number of steps in the aerator                   ="; N
LOCATE 6, 5
DO
INPUT "Enter the temperature   (C)                            ="; T
LOOP WHILE T < 0 OR T > 30
LOCATE 7, 5
INPUT "Enter influent concentration as percent saturation  ="; PCo
LOCATE 8, 5
INPUT "Enter Chloride concentration in water mg/L          ="; Cc
LOCATE 9, 5
INPUT "Enter percentage of saturation caused by each step  ="; PS

'get saturation concentration at temperature T
FOR i = 1 TO 31
IF T <= temp(i) THEN
        Cs = conc(i)
        GOTO jmp
END IF
NEXT i
jmp:
IF PCo > 1 THEN PCo = PCo / 100 'convert percentage to ratio
IF PS > 1 THEN PS = PS / 100    'convert percentage to ratio
Co = 0
Ce = PS * Cs
K = (Ce - Co) / (Cs - Co)       'for each step
'compute effluent oxygen concentration for each step
```

```
Ce(0) = PCo * Cs                    'initial concentration of water
FOR i = 1 TO N
Ce(i) = Ce(i - 1) + K * (Cs - Ce(i - 1))
LOCATE i + 10, 3
PRINT "Effluent oxygen concentration for step"; i; "="; Ce(i); " mg/L"
NEXT i

LOCATE 22, 3
PRINT "press any key to continue....................";
DO
LOOP WHILE INKEY$ = ""
END
```

3.7 Filtration

3.7.1 Introduction

Filtration is a process used for separating suspended or colloidal impurities from a liquid by passing the liquid through a porous media.[7] Filtration objectives include:[11]

- Partial removal of suspended and colloidal solids
- Alteration of chemical properties of constituents
- Reduction in number of disease-causing agents
- Reduction of color, tastes, odor
- Removal of iron and manganese

Effective filtration depends upon several mechanisms, which include mechanical sieving or straining, precipitation, adsorption, chemical actions, and biological processes.[1,4–6,9,11,18–21]

The porous media preferred for filtration must be[11]

- Inexpensive
- Readily available in sufficient quantities
- An inert material
- Easily cleaned for reuse
- Able to withstand existing pressures

Examples of suitable materials used as filter media include ordinary sand, anthracite coal, broken stones, glass beads, plastic substances, concrete, diatomaceous earth, etc.

In water treatment, sand filters are often used. Sand filters can be classified as single media, dual media, or multimedia filters. Also, they can be classified according to their allowable loading rate, such as slow or rapid sand filters. Another classification is based on water movement through the filter, such as gravity or pressure filters.

Table 3.1 provides a general comparison between slow and rapid sand filters.

TABLE 3.1
Comparison of Rapid and Slow Sand Filters[11]

Parameter	Rapid sand filter	Slow sand filter
Reason for filtration	Removal of SS, reduction of pathogens	Reduction in finalizing treatment
Location in treatment work	After coagulation or sedimentation	Without or after coagulation, or after RSF
Efficiency	Depends on raw water quality and design	Depends on raw water quality and design
Raw water turbidity needed	High	Moderate (<15 NTU)
Design period	10–15 years	10–15 years
Life span	Relatively long	Relatively long
Filtration rate ($m^3/m^2/h$)	5–15	0.1–0.2
Total area (A_t)	Flow rate/filtration velocity	Flow rate/filtration velocity
Area of each filter	(A_t/N –1), (A_t/N – 2)	(A_t/N –1), (A_t/N – 2)
Dimensions[20]		$1 = (2A_t/[N + 1])^{1/2}$ B = (N + 1)L/2N
Effective grain size (mm)	0.4–3	0.15–0.35
Uniformity coefficient	>1.2–1.5	<3–5 (2.5)
Bed thickness (m)	0.6–3	0.8–1.2
Supernatant water level (m)	1–1.5	1–1.5
Minimum depth before resanding (m)	Depends on treatment	0.5
Number of filters[1] (minimum of two filters)	$12(Q)^{1/2}$ (Q in m^3/s)	$15(Q)^{1/2}$ (Q in m^3/s)
Operation period	24 h/day (intermittent not recommended)	24 h/day (intermittent not recommended)
Interval between successive cleanings	12–72 h	20–60 days or more
Filter bed resistance	1.5–4 m	
Method of cleaning	Backwashing (water and/or air)	Scraping top 0.5–2 cm layer
Sludge removal	Manual, mechanical, hydraulic	Manual, mechanical, hydraulic
Filter material	Concrete, brick, plastics, etc.	Concrete, brick, plastics, etc.
Maintenance	Continuous	Continuous
Hazards	Algal growth, change in water quality, clogging	Algal growth, change in water quality, clogging
Control measures	Head loss, flow rate, turbidity	Head loss, flow rate, turbidity
Important quality parameters to be analyzed	Turbidity, bacteriological quality	Turbidity, bacteriological quality

Note: SS = suspended solids. RSF = Rapid sand filter. NTU = Neplometric Turbidity Unit. Q = Flow.
Source: From: Rowe, D.R. and Abdel-Magid, I.M., *Handbook of Wastewater Reclamation and Reuse*, CRC Press/Lewis Publishers, Boca Raton, FL, 1995. (With permission.)

Example 3.8

1. Write a computer program to find the required number of filters to be used, the filtration area, and the required area for each filter, given the filtration rate and the amount of water to be filtered when using:
   - Slow sand filters
   - Rapid sand filters
2. A sedimentation tank effluent is applied to rapid sand filters at a rate of 20 m^3/min. The rate of filtration is to be 200 m^3/m^2/day. Calculate the following:
   - Required number of filters
   - Total filtration area
   - Unit area of each filter
3. Use the program developed in (1) to verify the results obtained in (2).

Solution

1. For solution to Example 3.8 (1), see Program 3.8 listed below.
2. Solution to Example 3.8 (2):
   - Given Q = 20 m^3/min., v_f = 200 m/day.
   - Find required number of filters needed by using the empirical equation: $N = 12*(Q)^{1/2} = 12*(20/60)^{1/2} = 7$ filters (take 9 filters, 2 to serve as standby).
   - Determine total filtration surface area as: $A = Q/v_f = 20*60*24/200 = 144$ m^2.
   - Compute unit filter area for each filter as: $A_n = A/(N - 2) = 144/(9 - 2) = 20.6$ m^2.

Listing of Program 3.8

```
DECLARE FUNCTION ROUNDUP! (x!)
DECLARE SUB box (r1%, c1%, r2%, c2%)     'Found in Program 2.1
'******************************************************************
'Program 3.8: Filters
'Computes number of filters, filtration area, and filter areas
'******************************************************************

CALL box(2, 1, 17, 80)

LOCATE 3, 15
PRINT "Program 3.8:Computes number of filters and filter areas "
LOCATE 6, 5
INPUT "Enter rate of flow in m3/min                    ="; Q
LOCATE 7, 5
```

```
INPUT "Enter the rate of filtration in m3/m2/day       ="; Vf
LOCATE 8, 5
INPUT "Enter the type of the filter, Slow or Rapid(S/R) ="; T$
T$ = UCASE$(T$)
Q = Q / 60       'm3/s

'Number of filters needed
IF T$ = "R" THEN N = ROUNDUP(12 * Q ^ .5) ELSE N = ROUNDUP(15 * Q ^ .5)
IF N > 3 THEN Nadd = 2 ELSE Nadd = 1    'stand by filters
N1 = N + Nadd                           'adding standby filters
Q = Q * 3600 * 24'per day
'Total filtration surface area
A = Q / Vf
'Unit filter area for each filter
An = A / N
LOCATE 10, 3
PRINT "The number of filters needed ="; N;
PRINT "  filters + "; Nadd; " standby="; N1
LOCATE 11, 3
PRINT "Total filtration surface area              ="; A; "  m2"
LOCATE 12, 3
PRINT "unit filter area for each filter           ="; An; "  m2"
LOCATE 24, 5
PRINT "press any key to continue....................";
DO
LOOP WHILE INKEY$ = ""
END

FUNCTION ROUNDUP (x)
'This function rounds up a floating point number
y = INT(x)
IF y < x THEN y = y + 1
ROUNDUP = y
END FUNCTION
```

3.7.2 Filtration Theory

Filter clogging results in an increase in head loss (see Figure 3.9). The Carman-Kozeny equation can be used to estimate the head loss for clean unisized sand; see Equation 3.53.

$$h_L = f * l * (1 - e) * v_f^2 / \phi * d * e^3 * g \tag{3.53}$$

where:

h_L = Head loss through the filter, m

f = Friction factor, dimensionless, where

$$f = \{150(1 - e) / Re\} + 1.75 \tag{3.54}$$

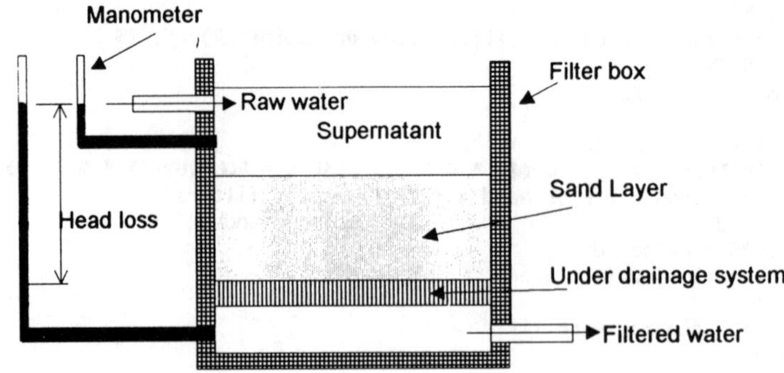

FIGURE 3.9
Schematic diagram of a sand filter.

l = Filter depth, m
e = Porosity of bed, dimensionless
v_f = Filtration velocity, m/s
ϕ = Shape factor, dimensionless
d = Diameter of particle, m
g = Gravitational acceleration, m/s^2
Re = Reynolds number, dimensionless

Another equation that can be used to predict the head loss through a sand filter was developed by Rose and is presented in Equation 3.55.

$$h_L = (1.067 * C_D * v_f^2 * l)/(g * d * \phi * e^4) \qquad (3.55)$$

where:

h_L = Head loss in the filter, m
C_D = Newton's drag coefficient, where

$$C_D = (24/Re) + \{3/(Re)^{1/2}\} + 0.34 \qquad (3.56)$$

v_f = Filtration velocity, m/s
l = Filter depth, m
g = Gravitational acceleration, m/s^2
d = Particle diameter, m
ϕ = Particle shape factor, dimensionless
e = Bed porosity, dimensionless

Example 3.9

1. Write a computer program to compute the head loss by using the Rose or Carman-Kozeny equation. The given information includes the filtration velocity, water viscosity, filter depth, average particle size, specific gravity for sand, the particle shape factor, and the bed porosity. The program needs to be designed so that the head loss can be determined for various temperatures.
2. A tri-filter media (third layer on top) is used for filtering water at 20°C. The properties of the filter are as shown below:

	Sand layers		Anthracite layer
Item	First	Second	Third
Depth (m)	0.6	0.6	0.4
Average particle size (mm)	0.5	1.0	1.4
Specific gravity	2.65	2.0	1.5
Particle shape factor	0.92	0.83	0.9
Porosity of bed (%)	50	45	45
Filtration rate ($m^3/m^2/d$)	200	200	200

   Determine the head loss using both the Carman-Kozeny and Rose equations.

   Find the percent error between the two methods.
3. Use the computer program developed in (1) to verify the results determined in (2).

Solution

1. For solution to Example 3.9 (1), see listing of Program 3.9.
2. Solution to Example 3.9 (2):
   - Given properties of media, $v_f = 200/(24*60*60) = 2.32*10^{-3}$ m/s, T = 20°C, which give from Table A1 in the Appendix, $\mu = 1.0087*10^{-3}$ and $\rho = 998.203$ kg/m³.
   - Rose equation
     - For the first sand layer:
       a. Determine Reynolds number: $Re = \rho v_f*d/\nu = 998.203*2.32*10^{-3}*0.5*10^{-3}/1.0087*10^{-3} = 1.148$.
       b. Compute Newton's drag coefficient as: $C_D = (24/Re) + \{3/(Re)^{1/2}\} + 0.34 = (24/1.148) + \{3/(1.148)^{1/2}\} + 0.34 = 24.05$.
       c. Determine the head loss for the first layer as: $h_L = 1.067*C_D*v_f^2*1/g*d*\phi*e^4$. $h_L = [1.067*24.05*(2.32*10^{-3})^2*0.6\}/\{9.81*0.5 \times 10^{-3}*0.92*(0.5)^4] = 0.293$ m.
     - For the second sand layer:
       a. Determine Reynolds number: $Re = 998.203*2.32*10^{-3}*1*10^{-3} = 2.3$.
       b. Find Newton's drag coefficient as: $C_D = (24/2.3) + \{3/(2.3)^{1/2}\} + 0.34 = 12.75$.
       c. Estimate the head loss for the second layer as: $h_L = [1.067*12.75*(2.32*10^{-3})^2*0.6\}/\{9.81*1*10^{-3}*0.83*(0.45)^4] = 0.132$ m.
     - For the third anthracite layer:
       a. Determine Reynolds number: $Re = 998.203*2.32*10^{-3}*1.4*10^{-3}/1.0087*10^{-3} = 3.21$.

b. Find Newton's drag coefficient as: $C_D = (24/3.21) + \{3/(3.21)^{1/2}\} + 0.34 = 9.49$.

c. Estimate the head loss for the third layer as: $h_L = [1.067*9.49*(2.32*10^{-3})^2*0.4\} /\{9.81*1.4*10^{-3}*0.9*(0.45)^4] = 0.043$ m.

- Determine the total head loss = head loss through the first sand layer + head loss through second sand layer + head loss through anthracite layer = 0.293 + 0.132 +0.043 = 0.468 m.

- Carman-Kozeny equation
  - For the first sand layer:

    a. Determine Reynolds number: $Re = \rho v_f*d/\mu = 998.203*2.32*10^{-3}*0.5*10^{-3}/1.0087*10^{-3} = 1.148$.

    b. Find the coefficient $f = \{150(1 - e)/Re\} + 1.75$. $f = [150(1 - 0.5)/1.148] + 1.75 = 67.08$.

    c. Determine the head loss for first layer as: $h_L = f*1*(1 - e)*v_f^2/\phi*d*e^3*g$. $h_L = 67.08*0.6*(1 - 0.5)*(2.32*10^{-3})^2/[0.92*0.5*10^{-3}*(0.5)^3 9.81] = 0.192$ m.
  - For the second sand layer:

    a. Determine Reynolds number: $Re = 998.203*2.32*10^{-3}*1*10^{-3}/1.0087*10^{-3} = 2.3$.

    b. Find the coefficient $f = \{150(1 - e)/Re\} + 1.75$. $f = [150(1 - 0.45)/2.3] + 1.75 = 37.62$.

    c. Estimate the head loss for the second layer as: $h_L = 37.62*0.6*(1 - 0.45)*(2.32*10^{-3})^2/[0.83*1*10^{-3}*(0.45)^3*9.81] = 0.09$ m.
  - For the third anthracite layer:

    a. Determine Reynolds number: $Re = 998.203*2.32*10^{-3}*1.4*10^{-3}/1.0087*10^{-3} = 3.21$.

    b. Find the coefficient $f = \{150(1 - e)/Re\} + 1.75$. $f = [150(1 - 0.45)/3.21] + 1.75 = 27.45$.

    c. Estimate the head loss for the third layer as: $h_L = 27.45*0.4*(1 - 0.45)*(2.32*10^{-3})^2/[0.9*1.4*10^{-3}*(0.45)^3*9.81] = 0.029$ m.
  - Determine the total head loss = head loss through the first sand layer + head loss through second sand layer + head loss through anthracite layer = 0.192 + 0.09 + 0.029 = 0.311 m.
  - Determine percent error = (0.468 – 0.311)*100/0.468 = 34%.

Listing of Program 3.9

```
DECLARE SUB box (r1%, c1%, r2%, c2%)    'Found in Program 2.1
'*****************************************************************
'Program 3.9: HEAD
'Computes the Head loss in filters
'*****************************************************************
DIM temp(25), density(25), viscosity(25)
'density - Appendix C- physical properties of fluids
DATA 999.9,1000,999.7,999.1,998.2,997.1,995.7,994.1,992.2,990.2,988.1
DATA 985.7,983.2,980.6,977.8,974.9,971.8,968.6,965.3,961.9,958.4
'viscosity- same reference
DATA 1.792,1.519,1.308,1.14,1.005,0.894,0.801,0.723,0.656,0.599,0.549
DATA 0.506,0.469,0.436,0.406,0.38,0.357,0.336,0.317,0.299,0.284
```

```
'Read tables
FOR i = 0 TO 20
temp(i) = i * 5
READ density(i)
NEXT i
FOR i = 0 TO 20
READ viscosity(i)
viscosity(i) = viscosity(i) * .001
NEXT i

CALL box(2, 1, 17, 80)

LOCATE 3, 15
PRINT "Program 3.9: Computes the Head loss in filters"
LOCATE 5, 5
INPUT "number of layers in the filter    ="; N

totalhead = 0
FOR layer = 1 TO N
CALL box(2, 1, 17, 80)
LOCATE 4, 2
PRINT "LAYER NO."; layer
LOCATE 6, 5
INPUT "Filtration velocity (m/s)       ="; v
LOCATE 7, 5
INPUT "Filter depth (m)                  ="; L
LOCATE 8, 5
INPUT "Particle diameter (mm)            ="; d
LOCATE 9, 5
INPUT "Specific gravity                  ="; S
LOCATE 10, 5
INPUT "Particle shape factor (phi)       ="; phi
LOCATE 11, 5
INPUT "Bed porosity e (per cent)         ="; e
LOCATE 12, 5
INPUT "Enter Temperature  in Centigrade  ="; T

'determine the density and the viscosity
temp(21) = 100
kr = 100
FOR i = 0 TO 20
IF T = temp(i) THEN kr = i
IF T > temp(i) AND T < temp(i + 1) THEN kr = i
NEXT i
Ro = density(kr)
vs = viscosity(kr)

'Newton's Drag coefficient Cd
IF e > 1 THEN e = e / 100              'convert to ratio
d = d / 1000                           'convert to m
g = 9.81                               'acceleration due to gravity
Re = Ro * v * d / vs                   'Reynolds number
Cd = 24 / Re + 3 / Re ^ .5 + .34      'drag coefficient
'compute head loss H1
H1 = 1.067 * Cd * v ^ 2 * L / (g * d * phi * e ^ 4)

LOCATE 13, 3
PRINT "Head loss H1       ="; H1; "  m"
totalhead = totalhead + H1
IF layer = N THEN
        LOCATE 15, 3
        PRINT "Total head loss in the filter="; totalhead; "  m"
```

```
END IF

LOCATE 24, 5
PRINT "press any key to continue....................";
DO
LOOP WHILE INKEY$ = ""
NEXT layer
END
```

Example 3.10

1. Write a short computer program that can be used to estimate the filtration rate when given the head loss, viscosity, specific gravity, particle shape factor, and other elements included in the Rose and Carman-Kozeny equations. Plan the computer program so that the suitability of the filtration rate can be evaluated.
2. A single, unisized sand bed filter of depth 1.2 m and a porosity of 0.6 with uniform sand grains of 0.8 mm is used to treat water at 22°C. Find the range of the filtration rate (in m/day) for a head loss of 0.12 m. (Use both the Rose and Carman-Kozeny equations).
3. Use computer programs developed in (1) to verify computations in (2).

Solution

1. For solution to Example 3.10 (1), see Program 3.10 listed below.
2. Solution to Example 3.10 (2):
   - Given: $1 = 1.2$ m, $e = 0.6$, $d = 0.8*10^{-3}$ m, $h_f = 0.12$ m, $T = 22°C$. From Table A1 in Appendix, $\mu = 0.9608*10^{-3}$, $\rho = 997.77$.
     - Rose equation:
       a. Find Reynolds number as: $Re = 997.77*(v/60*60*24)*0.8*10^{-3}/0.9608*10^{-3} = 9.616*10^{-3}v$.
       b. Find Newton's drag coefficient as: $C_D = (24/9.616*10^{-3}/v) + \{3/(9.616*10^{-3}v)^{1/2}\} + 0.34 = 2495.8/v + 30.59/v^{1/2} + 0.34$.
       c. Use the head loss equation: $h_L = 0.12 = 1.067*(2495.8/v + 30.59/v^{1/2} + 0.34)*(v/60*60*24)^2*1.2\}/\{9.81*0.8*10^{-3}*1*(0.6)^4\}$.
       d. Estimate the filtration velocity by trial and error, which yields v = 234 m/day.
     - Carman-Kozeny equation:
       a. Reynolds number as determined before is $Re = 9.616*10^{-3}v$.
       b. Find the coefficient $f = \{150(1-e)/Re\} + 1.75$. $f = [150(1-0.6)/9.616*10^{-3}v] + 1.75 = 6239.6/v + 1.75$.
       c. Estimate the filtration rate from the head loss equation as: $h_L = 0.12 = (6239.6/v + 1.75)*1.2*(1 - 0.6)*(v/60*60*24)^2/[1*0.8*10^{-3}*(0.6)^3*9.81]$, which yields v = 450 m/day.
     - Determine percent difference between the two methods = (450 – 234)*100/450 = 48%.

3. In the development of Program 3.10, the roots of the equation were determined using the numerical method of Newton-Raphson. For a function such as $f(x) = a_0 + a_1x + a_2x^2 + ... + a_nx^n$, if an estimate of the root is chosen to be x_r, an improved estimate can be obtained by taking the derivative at $x = x_r$. Accordingly, the improved estimate for the root can be obtained as $x_{r+1} = x_r + f(x_r)/[df(x_r)/dx]$.

Listing of Program 3.10

```
DECLARE SUB RESULTS (V!, R!)
DECLARE SUB check (f!, fprevious!, X!, m1!, status!)
DECLARE SUB BOX (r1%, c1%, r2%, c2%)    'Found in Program 2.1
'*****************************************************************
'Program 3.10: FILTRATION
'Computes the filtration rate for a given head loss
'REQUIRES ZEROS OF A FUNCTION- USES NEWTON RAPHSON METHOD
'*****************************************************************
DIM temp(40), density(40), viscosity(40)
'Temperature-Density table- kg/m3- see Table 11-1
DATA 999.965,999.941,999.902,999.849,999.781,999.700,999.605
DATA 999.498,999.377,999.244,999.099,998.943,998.774,998.595
DATA 998.405,998.203,997.992,997.770,997.538,997.296,997.044
DATA 996.783,996.512,996.232, 995.944,995.646,995.340,995.025
DATA 994.702,994.371,994.030,993.680,993.330,992.96
'viscosity- same reference
DATA 1.5188,1.4726,1.4288,1.3872,1.3476,1.3097,1.2735,1.2390
DATA 1.2061,1.1748,1.1447,1.1156,1.0875,1.0603,1.0340,1.0087
DATA 0.9843,0.9608,0.9380,0.9161,0.8949,0.8746,0.8551,0.8363
DATA 0.8181,0.8004,0.7834,0.7670,0.7511,0.7357,0.7208,0.7064
DATA 0.6925,0.6791

'Read tables
FOR i = 5 TO 38
temp(i) = i
READ density(i)
NEXT i
FOR i = 5 TO 38
READ viscosity(i)
viscosity(i) = viscosity(i) / 1000
NEXT i

ttl$ = "Program 3.10: Computes the filtration rates in filters"
CALL BOX(2, 1, 17, 80)
LOCATE 3, 15: PRINT ttl$

LOCATE 5, 5
INPUT "Head loss H1 (m)                      ="; H1
LOCATE 6, 5
INPUT "Filter depth  (m)                     ="; L
LOCATE 7, 5
INPUT "Particle diameter (mm)                ="; d
row = 7
row = row + 1: LOCATE row, 5
INPUT "Particle shape factor (phi)           ="; phi
row = row + 1: LOCATE row, 5
INPUT "Bed porosity e (per cent)             ="; e
row = row + 1
DO
LOCATE row, 5
INPUT "Enter Temperature (between 5-38C) ="; T
LOOP WHILE T < 5 OR T > 38
```

```
'determine the density and the viscosity
FOR i = 5 TO 38
IF T = temp(i) THEN kr = i
NEXT i
Ro = density(kr)
vs = viscosity(kr)

IF e > 1 THEN e = e / 100          'convert to ratio
d = d / 1000                       'convert to meter units
g = 9.81                           'acceleration due to gravity
'Compute velocity
k = 60! * 60! * 24!                'conversion factor to /day

'Set up Carman-Kozeny equation
c11 = 1.75
c12 = 1.296E+07 * (1 - e) * vs / (Ro * d)
c13 = -H1 * g * d * phi * e ^ 3 * k ^ 2 / ((1 - e) * L)
Vc = (-c12 + (c12 ^ 2 - 4 * c13 * c11) ^ .5) / (2 * c11)

'Set up Rose equation
B = Ro * d / (k * vs)
c1 = .34
c2 = 3 / B ^ .5
c3 = 24 / B
c4 = H1 * g * d * phi * e ^ 4 * k ^ 2 / (1.067 * L)
m2 = (c4 / c1) ^ .25
m1 = m2 / 10

        'Zeros of function - Search for solution interval
        status = 0
        FOR X = m1 TO m2 STEP m1
                f = c1 * X ^ 4 + c2 * X ^ 3 + c3 * X ^ 2 - c4
                IF X > m1 THEN CALL check(f, fprevious, X, m1, status)
                IF status > 0 THEN GOTO jump
                fprevious = f
                Xprevious = X
                Re = Ro * X ^ 2 * d / vs
        NEXT X
        LOCATE 21, 5: PRINT "NO SOLUTION IN THIS INTERVAL............"

jump:
        CLS
        'Newton Raphson approximation of roots
        Xn = X
        Epsilon = .000001

cycle:
                X = Xn
                f = c1 * X ^ 4 + c2 * X ^ 3 + c3 * X ^ 2 - c4
                fdash = 4 * c1 * X ^ 3 + 3 * c2 * X ^ 2 + 2 * c3 * X
                V = X ^ 2
                deltaX = f / fdash
                'PRINT X, V, f
                Xn = Xn - deltaX  'improved estimate
IF ABS(deltaX) > Epsilon THEN GOTO cycle

CALL BOX(3, 1, 18, 80)
LOCATE 5, 3: PRINT "Solution using Rose equation:"
CALL RESULTS(V, 6)
LOCATE 10, 3: PRINT "Solution using Carman-Kozeny equation:"
CALL RESULTS(Vc, 12)
difference = ABS(Vc - V) / Vc * 100
```

```
LOCATE 16, 10
PRINT "Difference between the two methods="; INT(difference); " %"
LOCATE 22, 5
PRINT "press any key to continue...................."
DO
LOOP WHILE INKEY$ = ""

END


SUB check (f, fprevious, X, m1, status)
        IF X > m1 THEN
             IF SGN(f) <> SGN(fprevious) THEN status = 1
        END IF
END SUB

SUB RESULTS (V, R)
'Output the result of the analysis
LOCATE R, 5
PRINT "The filtration velocity is ="; V; "  m/day"
'Select a suitable type of sand filter:
V = V / 24       'in m/hr
tt$ = "Not suitable for slow nor for rapid sand filter"
IF V >= .1 AND V <= .2 THEN tt$ = "SLOW SAND FILTER"
IF V >= 5 AND V <= 15 THEN tt$ = "RAPID SAND FILTER"
LOCATE R + 1, 5: PRINT "Recommendation :"; tt$

END SUB
```

3.7.3 Clogging of the Filter Bed

The quality of the effluent water from a filter can be evaluated by using Equation 3.57.

$$C_e = C_0 * e^{-\lambda_0 * 1 * L} \quad (3.57)$$

where:

C_e = Concentration of contaminant in the effluent, mg/L
C_0 = Concentration of contaminant at the filter inlet surface, mg/L
λ_0 = Filtration coefficient, m^{-1}
1 = Filter depth, m

The rate of deposition of particulate solids in the filter bed can be estimated from equation 3.58.[1,11]

$$\psi = v * \lambda_0 * C_0 * e^{-l_0\, y} * t \quad (3.58)$$

where:

ψ = The rate of deposition
v = Filtration rate, m/s
λ_0 = Filtration coefficient, m^{-1}
C_0 = Concentration of solid particles at the filter inlet surface, mg/L
y = Depth of deposition of solids from the filter inlet surface, m
t = Time taken for deposition of solids within the filter bed, s

Example 3.11

1. Write a computer program that will estimate the concentration of a contaminant in the effluent from a filter bed as well as the rate of deposition of the contaminant at various depths throughout the filter. Given concentration of contaminant at the filter inlet surface, temperature, filtration coefficient, filter depth, filtration rate, and time taken for deposition of solids within the filter bed. Plan the computer program so that the percent reduction of the contaminant by the filter is also determined.
2. Water at a temperature of 20°C with a contaminant concentration of 200 mg/L is to be treated by a filter bed of 1 m depth with a coefficient of filtration of 5 m^{-1}. Find the concentration of the contaminant in the effluent if the filtration rate is 10 $m^3/m^2/h$.
3. Use the computer program developed in (1) to verify the computations in (2).

Solution

1. For solution to Example 3.11 (1), see listing of Program 3.11 below.

2. Solution to Example 3.11 (2):
   - Given $C_0 = 200$ mg/L, $1 = 1$ m, $\lambda = 5\cdot$/m, $v_f = 10$ m/h $= 2.78*10^{-3}$ m/s.
   - Find the effluent contaminant concentration using Equation 3.57: $C_e = C_0*e^{-\lambda 0L} = 200*e^{-5*1} = 1.3$ mg/L.

Listing of Program 3.11

```
DECLARE SUB box (r1%, c1%, r2%, c2%)    'Found in Program 2.1
'****************************************************************
'Program 3.11: Effluents
'Computes the concentration of impurities of the effluents
'****************************************************************
CALL box(2, 1, 17, 80)

LOCATE 3, 5
PRINT "Program 3.11:Computes the concentration of impurities of effluents"
LOCATE 5, 5
INPUT "Enter concentration of impurities in the effluent (mg/L)  ="; Co
LOCATE 6, 5
INPUT "Enter filtration coefficient  /m        ="; lamda
LOCATE 7, 5
INPUT "Enter filter depth  m                    ="; L
LOCATE 8, 5
INPUT "Enter filtration rate  m/s               ="; v

Cx = Co * EXP(-lamda * L)
LOCATE 12, 3
PRINT "Effluent concentration of impurities     ="; Cx; "  mg/L"
percent = 100 * (Co - Cx) / Co
LOCATE 13, 3
PRINT "Percentage reduction of impurities       ="; percent; " %"
y = L / 2
Cx = Co * EXP(-lamda * y)
LOCATE 14, 3
PRINT "Concentration of impurities at half depth="; Cx; " mg/L"
delta = v * lamda * Co * EXP(-lamda * L)
LOCATE 15, 3
PRINT "Rate of deposition of impurities         ="; delta
LOCATE 24, 5
PRINT "press any key to continue....................";
DO
LOOP WHILE INKEY$ = ""
END
```

3.7.4 Backwashing a Rapid Sand Filter

Rapid filters are cleaned by backwashing. Backwashing refers to reversal of the direction of flow in order to expand the bed at the end of a filter run. Expansion of the bed produces a situation in which the accumulated contaminants are scoured off the sand particle surface. Filter bed expansion may be estimated from the empirical Equation 3.59.

$$l_e / l = (1 - e)/(1 - e_e) = (1 - e)/[1 - (v / v_s)^{0.22}] \tag{3.59}$$

where:

l_e = Expanded bed depth, m

l = Bed depth, m

e = Porosity of bed, dimensionless

e_e = Expanded bed porosity, dimensionless

v = Face velocity of the backwash water (rise rate), m/s

v_s = Sand particles settling velocity, m/s

Example 3.12

1. Write a short computer program to find the filter bed expansion during backwashing, given the rate of water backwashing, sand particle shape factor, the sand bed porosity, the settling velocity of the sand particles, and the depth of the filter.
2. A uniform sand filter bed of depth 0.6 m with a porosity of 0.4 is used for filtering water at a filtration rate of 15 $m^3/m^2/h$ with a measured settling velocity of sand particles of 0.05 m/s. When backwashing is required, it is carried out at a rate of 20 m^3/m^2h. Determine the depth of bed expansion.
3. Use computer program developed in (1) to verify computations in (2).

Solution

1. For solution to Example 3.12 (1), see listing of program 3.12 below.
2. Solution to Example 3.12 (2):
   - Given $l = 0.6$ m, $e = 0.4$, $v = 20$ m/h, $v_s = 0.05$ m/s = 180 m/h.
   - Determine the depth of bed expansion as: $l_e/l = (1 - e)/[1 - (v/v_s)^{0.22}]$. $l_e/0.6 = (1 - 0.4)/[1 - (20/180)^{0.22}]$. This yields a value of $l_e = 0.94$ m.

Listing of Program 3.12

```
DECLARE SUB box (r1%, c1%, r2%, c2%)    'Found in Program 2.1
'*****************************************************************
'Program 3.12: Backwashing
'Computes the depth of bed expansion
'*****************************************************************
CALL box(2, 1, 17, 80)

LOCATE 3, 5
PRINT "Program 3.11:Computes the depth of bed expansion"
LOCATE 5, 5
INPUT "Enter porosity of bed                          ="; e
LOCATE 6, 5
INPUT "Enter depth of bed (m)                         ="; L
LOCATE 7, 5
INPUT "Enter backwashing rate              m/min      ="; V
LOCATE 8, 5
INPUT "Enter particle settling velocity m/min         ="; Vs
```

```
Lx = L * (1 - e) / (1 - (V / Vs) ^ .22)

LOCATE 12, 3
PRINT "The expanded bed depth      ="; Lx; "  m"

LOCATE 24, 5
PRINT "press any key to continue...................";
DO
LOOP WHILE INKEY$ = ""
END
```

3.8 Methods of Desalination

3.8.1 Introduction

Desalination or desalting is a term applied to removal of dissolved salts from waters with the objective of producing water suitable for human consumption.[11,22] Table 3.2 gives an estimate of the dissolved solids concentration of various types of water.

TABLE 3.2
Dissolved Solids Content for Different Waters[11]

Type of water	Total dissolved solids, TDS (mg/L)
Brackish water	1500 to 12,000
Sea water (Middle East)	50,000
Sea water (North Sea)	35,000

Source: From Rowe, D.R. and Abdel-Magid, I.M., *Handbook of Wastewater Reclamation and Reuse*, CRC Press/Lewis Publishers, Boca Raton, FL, 1995. With permission.

Desalination processes include:

1. *Distillation:* This process concerns the boiling of water and the consequent condensation of vapor produced (examples are to be found in solar stills, multistage flash (MSF), multiple effect boiling (MEB), and vapor compression (VC) distillation.
2. *Freezing:* Freezing of a solution of saline water enables the separation of dissolved salts and the production of pure water.
3. *Membrane processes:* Reverse osmosis (RO) applies pressure to force water molecules to flow through a semipermeable membrane while retaining dissolved substances in the solution. Electrodialysis uses ion-selective membranes to separate dissolved salts under the action of an electric current.

Figure 3.10 summarizes the various types of desalination techniques.[11]

3.8.2 Distillation

In distillation, saline water is boiled or evaporated to produce two separate streams: fresh water and concentrate or a brine stream. The first stream has a low concentration

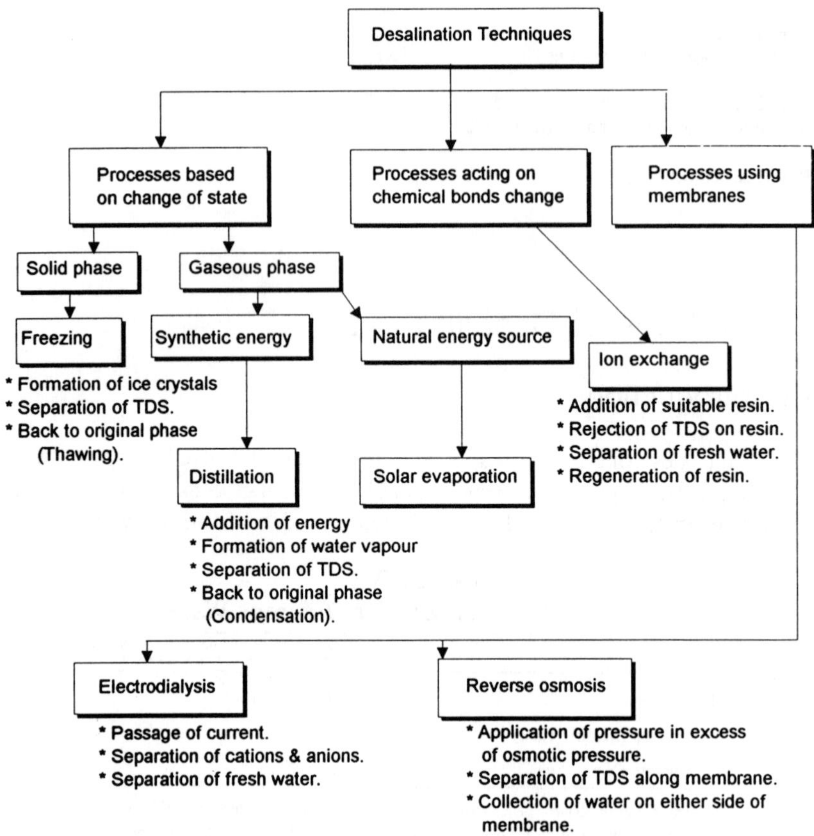

FIGURE 3.10
Summary of various desalination methods.

of dissolved salts, while the second stream contains the remaining dissolved salts. Vapors or steam are then condensed to supply pure water. Benefits of the distillation process include:[11,23]

- Removal of harmful organisms from feed water (e.g., bacteria, viruses)
- Elimination of nonvolatile matter from feed water (e.g., dissolved gases like carbon dioxide, and ammonia)

In the distillation of water, at least two heat exchangers are needed for evaporating the feed water and for aiding vapor in the condensation.

In a multiple-effect distillation unit, the feed water in the first phase is boiled under high pressure while the last phase operates at atmospheric pressure. Figure 3.11 outlines the pressure operation in a multiple-effect distillation process. Heating tubes of the second stage aids the condensation of water vapor emerging from the first phase. The energy released (by latent heat of condensation and by temperature decrease) is employed by the liquid water in the second stage. Cooling water is used in the last stage to finalize the condensation.[11]

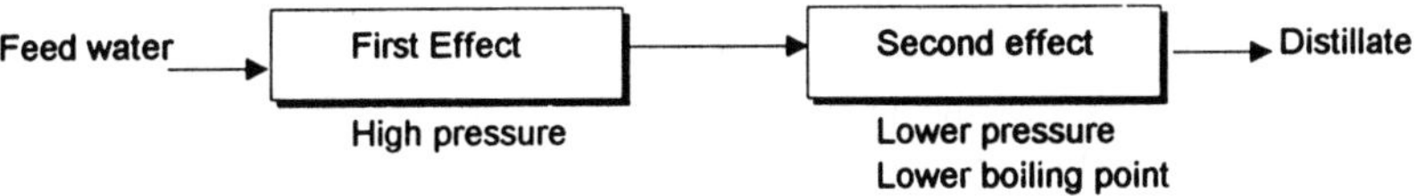

FIGURE 3.11
Pressure regulation within a multiple effect distiller.

Heat exchanged in any phase can be determined by using the following Equation 3.60.

$$HE_i = u_i * A_i * \Delta T_i \tag{3.60}$$

where:

HE_i = Heat exchanged to effect number i, J
u_i = Heat transfer coefficient for heat exchanger number i
A_i = Area of heat exchanger number i, m^2
ΔT_i = Temperature difference between water in effect and steam entering heat exchanger, °C
ΔT_1 = $T_0 - T_1$
T_0 = Temperature of heating steam supplied to first phase, °C
T_1 = Boiling temperature in first phase, °C
ΔT_2 = $T_1 - T_2$
T_2 = Boiling temperature in second phase, etc., °C.

For similar areas of heat exchangers and similar transfer rates of heat in each phase, the relationship between the heat generated in each phase and the temperature difference may be related as illustrated in Equations 3.61, 3.62, and 3.63.

$$HE_i = HE_2 .... = He_i \tag{3.61}$$

$$A_1 = A_2 = A_3 ... = A_i \tag{3.62}$$

$$u_1 * \Delta T_1 = u_2 * \Delta T_2 = ... = u_i * \Delta T_i \tag{3.63}$$

Example 3.13

1. Write a computer program that determines the temperature difference in each phase of a multiphase still composed of N number of stills with a drop in both temperature (from T_1 to T_n) and heat transfer coefficients (from u_1 to u_n); assume the stills have equal areas in the heat exchanger.

2. A triple-phase still of equal areas in the heat exchanger is employed for water distillation. Dry steam enters the first still at a temperature of 120°C and leaves the triple-phase still at a temperature of 40°C with a drop in the pressure in the last phase. Heat transfer coefficients for the three stills are in the ratios of 4:3:2, respectively. Find the temperature difference in each phase in the triple distiller.

Solution

1. For the solution to Example 3.13 (1), see Program 3.13 listed below.
2. Solution to Example 3.13 (2):
   - Given u_1:u_2:u_3 = 4:3:2, T_1 = 120, T_3 = 40.
   - Find the temperature differences for the different phases of the still as: $u_1*\Delta T_1 = u_2*\Delta T_2 = u_3*\Delta T_3$.
   - Determine the total temperature drop in the three phases as equal to $\Delta T_1 + \Delta T_2 + \Delta T_3 = \Delta T = 120 - 40 = 80$.
   - Let $u_1*\Delta T_1 = u_2*\Delta T_2 = u_3*\Delta T_3 = a$, then $\Delta T_1/\Delta T = (a/u_1 + a/u_2 + a/u_3)$.
   - Multiply by u_1/a, then $\Delta T_1/\Delta T = (1)/(1 + u_1/u_2 + u_1/u_3)$.
   - Find temperature differences as: $\Delta T_1/80 = (1)/(1 + (4/3) + (4/2))$. This yields ΔT_1 = 18.5°C.
   - Similarly $\Delta T_2/\Delta T = (1)/(u_2/u_1 + 1 + u_2/u_3)$, which gives ΔT_2 = 24.6°C and $\Delta T_3/\Delta T = (1/)/(u_3/u_1 + u_3/u_2 + 1)$, which gives ΔT_3 = 36.9°C.
3. Use the computer program developed in (1) to verify the manual solution presented in (2).

Listing of Program 3.13

```
DECLARE SUB box (r1%, c1%, r2%, c2%)     'Found in Program 2.1
'***************************************************************
'Program 3.13: Desalination
'Computes the temperature difference in multi-phase still
'***************************************************************
CALL box(2, 1, 20, 80)

LOCATE 3, 5
PRINT "Program 3.13:Computes the temperature difference in multi-phase still"
LOCATE 5, 5
INPUT "Enter number of stills                              ="; N

DIM rat(N), deltaT(N)

LOCATE 6, 5
PRINT "Enter"; N; " heat transfer coefficient ratios    :"
FOR i = 1 TO N
LOCATE i + 6, 10
INPUT "ratio ="; rat(i)
NEXT i

LOCATE 7 + i, 5
INPUT "Enter initial temperature    T1           ="; T1

LOCATE 8 + i, 5
INPUT "Enter final temperature      T3           ="; T3
dT = T1 - T3
```

```
LOCATE 10 + i, 3
PRINT "Temperature difference in the"; N; " stills are as follows:"
LOCATE 11 + i, 3
FOR i = 1 TO N
X = 0
FOR j = 1 TO N
X = X + rat(i) / rat(j)
NEXT j
deltaT(i) = dT / X
PRINT "dT="; deltaT(i); "  ";
NEXT i
PRINT

LOCATE 24, 5
PRINT "press any key to continue....................";
DO
LOOP WHILE INKEY$ = ""
END
```

3.8.3 Osmosis

Osmosis concerns the tendency of a solvent to diffuse through a membrane to a solute. The flow is directed from the more diluted to the more concentrated solution. To prevent diffusion towards the more concentrated solution, osmotic pressure is applied. Figure 3.12 illustrates the concepts of normal and reverse osmosis processes.

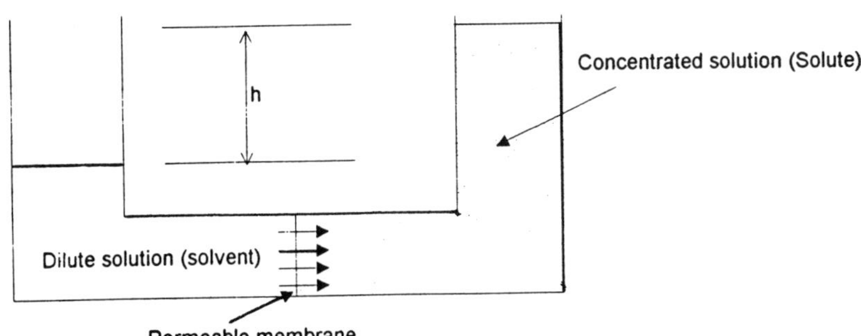

FIGURE 3.12
Osmosis process.

Osmotic pressure, a measure of the force that brings together the solvent molecules, depends upon the number of particles of solute in solution.[11] The passage of the solvent through a membrane produces a driving force, which may be estimated by the difference in vapor pressure of the solvent on the membrane sides. The transfer of solvent through the membrane from the dilute to the concentrated solution continues until hydrostatic pressure exceeds the driving force of the vapor pressure differential.[11,25]

Equation 3.64 may be used to estimate the osmotic pressure at equilibrium for an incompressible solvent.

$$P_{osm} = (R * T / V) * Ln\,(P_0 / P) \tag{3.64}$$

where:

P_{osm} = Osmotic pressure, Pa or atm
R = Universal gas constant (= 0.08206 L*atm/mol*K = 8.314 J/K*mol = 8314 J/kg*K), J/kg*K
T = Temperature, K
P_0 = Vapor pressure of solvent in dilute solution, Pa or atm
P = Vapor pressure of solvent in concentrated solution, Pa or atm
V = Volume per mole of solvent (= 0.018 L for water), m^3 or L

Generally, the existence of a nonvolatile solute in a liquid decreases the vapor pressure of the solution. Raoult's law assumes that for dilute solutions the reduction in vapor pressure of a solvent is directly proportional to the concentration of particles in solution.[11,25] Equation 3.65 signifies Raoult's law for dilute solution relating osmotic pressure to molar concentration of particles in the concentrated solution.

$$P_{osm} = C * R * T \tag{3.65}$$

where:

P_{osm} = Osmotic pr ssure, atm
C = Molar concentration of particles
R = Universal gas constant for all gases, L*atm/mol*K
T = Temperature, K

In the reverse osmosis process, the dissolved solids are separated by a semipermeable membrane. In this case, pressure in excess of the natural osmotic pressure is applied to the feed water (see Figure 3.13).

The advantages of the reverse osmosis process include:[2,11,23,24,26]

- Reduction of content of total dissolved solids (up to 99 reduction)
- Removal of biological and colloidal materials (up to 98% reductions)
- Removal of harmful microorganisms (up to 100% removal)
- Removal of dissolved organic substances (up to 97% removal)
- Production of treated water suitable for human consumption.

Example 3.14

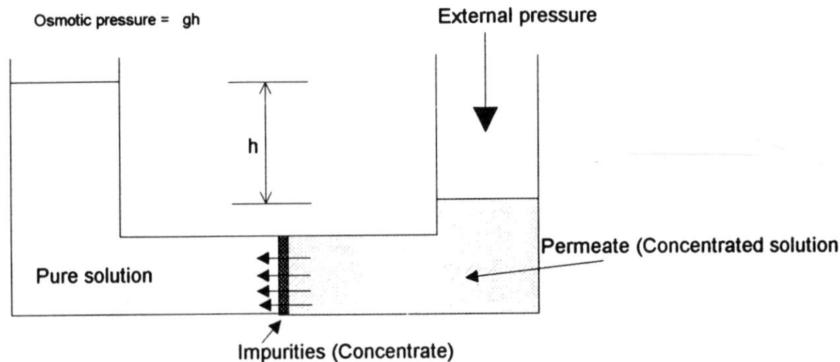

FIGURE 3.13
Reverse osmosis process.

1. Write a short computer program to compute the osmotic pressure per unit volume for a salt solution, given the temperature and the vapor pressure of solvent in dilute and in concentrated solution.
2. A salt solution has a vapor pressure (concentrated solution) of 2.26 kPA and a temperature of 20°C. Assume pure water (dilute solution in this case) has a vapor pressure of 2.34 kPa at this temperature. Estimate the osmotic pressure per unit volume.
3. Use the computer program developed in (1) to verify the manual solution presented in (2).

Solution

1. For solution to Example 3.14 (1), see listing of Program 3.14 below.
2. Solution to Example 3.14 (2):
   - Given T = 20°C, P_0 = vapor pressure of solvent in dilute solution = 2.34 kPa (may be found from Table A1 (Appendix) corresponding to the given temperature in the case of the dilute solution being water). P = vapor pressure of the solvent in the concentrated solution = 2.26 kPa.
   - Determine temperature as: T = 273.16 + 20 = 293.16 K.
   - Find the osmotic pressure as: (P_{osm} = (R*T/V)*Ln (P_0/P). P_{osm} = (8.314 J/K/mol*293.16 K)*Ln (2.34/2.26) = 84.8 Pa/unit volume.

Listing of Program 3.14

```
DECLARE SUB box (r1%, c1%, r2%, c2%)  'Found in Program 2.1
'**************************************************************
'Program 3.14: Osmosis
'Computes the osmotic pressure per unit volume for salt solutions
'**************************************************************
CALL box(2, 1, 20, 80)

LOCATE 3, 10
PRINT "Program 3.14: Computes the osmotic pressure for salt solutions"
LOCATE 5, 5
INPUT "Enter temperature T  (C)                                   ="; T
LOCATE 6, 5
INPUT "Enter vapor pressure of solvent in dilute solution (kPa)   ="; Po
```

```
LOCATE 7, 5
INPUT "Enter vapor pressure of solvent in concentrated solution(kPa)="; P

T = T + 273.16                     'Kelvin
R = 8.314                          'J/K/mol
V = 1                              'unit volume
Posm = (R * T / V) * LOG(Po / P)

LOCATE 12, 3
PRINT "Osmotic pressure            ="; Posm; "  kPa/unit volume"

LOCATE 24, 5
PRINT "press any key to continue...................";
DO
LOOP WHILE INKEY$ = ""
END
```

Example 3.15

1. Write a computer program to find the difference in osmotic pressure across a semipermeable membrane, given the concentrations (in molar, or mg/L) of cations and anions and the prevailing temperature.
2. Analysis of a water sample at a temperature of 23°C indicated the following concentration of ions (in mg ion/L). Cations: Mg^{++} = 0.5, Ca^{++} = 0.8, Na^{+} = 0.3, K^{+} = 0.2. Anions: SO_4^{--} = 1.2, Cl^{-} = 0.1, HCO_3^{-} = 0.25, NO_3^{-} = 0.25. The water sample is to be treated by reverse osmosis. Determine the osmotic pressure across the semipermeable membrane.

Solution

1. For solution to Example 3.15 (1), see Program 3.15 listed below.
2. Solution to Example 3.15 (2):
   - Determine the molar concentration of ions in the water sample as: molar ion concentration = concentration of ion (mg/L) / MW, where MW = molecular weight. Thus (and similarly), Mg^{++} = 0.5/24.3 = 0.0206; Ca^{++} = 0.8/40 = 0.02; Na^{+} = 0.3/23 = 0.013; K^{+} = 0.2/39 = 0.0051; SO_4^{--} = 1.2/96 = 0.0125; Cl^{-} = 0.1/35.5 = 0.0028, HCO_3^{-} = 0.25/61 = 0.0041, NO_3^{-} = 0.25/62 = 0.004.
   - Find the molar ion concentration of the sample of water as: C = 0.0206 + 0.02 + 0.013 + 0.0051 + 0.0125 + 0.0028 + 0.0041 + 0.004 = 0.0821 M.
   - Determine the osmotic pressure as: P_{osm} = C*R*T = 0.0821*0.082*(273.16 + 23) = 1.99 atm.
3. Use the computer program developed in (1) to verify the manual solution presented in (2).

Listing of Program 3.15

```
DECLARE SUB box (r1%, c1%, r2%, c2%)     'Found in Program 2.1
'*****************************************************************
'Program 3.15: Osmosis-2
'Computes difference in osmotic pressure across membranes
'*****************************************************************
```

```
DIM SYM$(8), mw(8), conc(8)
'symbols
SYM$(1) = "Mg++"
SYM$(2) = "Ca++"
SYM$(3) = "Na+ "
SYM$(4) = "K+  "
SYM$(5) = "SO4--"
SYM$(6) = "Cl-"
SYM$(7) = "HCO3-"
SYM$(8) = "NO3"

'molecular weights
mw(1) = 24.3
mw(2) = 40
mw(3) = 23
mw(4) = 39
mw(5) = 96
mw(6) = 35.5
mw(7) = 61
mw(8) = 62

CALL box(1, 1, 20, 80)

LOCATE 2, 5
PRINT "Program 3.15: Computes difference in osmotic pressure across membranes"
LOCATE 3, 5
INPUT "Enter temperature T  (C)                    ="; T
LOCATE 4, 5
mol = 0
PRINT "Enter ion concentrations in mg ion/L :"

LOCATE 5, 2: PRINT "CATIONS:"
FOR i = 1 TO 4
LOCATE i + 5, 10
PRINT SYM$(i)
LOCATE i + 5, 15
INPUT "concentration="; conc(i)
mol = mol + conc(i) / mw(i)                 'molar ion concentration
NEXT i

LOCATE 10, 2: PRINT "ANIONS:"
FOR i = 5 TO 8
LOCATE i + 6, 10
PRINT SYM$(i)
LOCATE i + 6, 15
INPUT "concentration="; conc(i)
mol = mol + conc(i) / mw(i)                 'molar ion concentration
NEXT i

R = .082
Posm = mol * R * (T + 273.16)               'osmotic pressure
LOCATE 17, 3
PRINT "OSMOTIC PRESSURE      ="; Posm; "  atm"
LOCATE 24, 5
PRINT "press any key to continue....................";
DO
LOOP WHILE INKEY$ = ""
END
```

3.8.4 Electrodialysis

In electrodialysis, ions are transferred through an ion-selective membrane from one solution to another under the action of a direct electrical potential.[11] Usually, the electrodialysis device is an array of alternating anion- and cation-selective membranes across which an electric potential is applied (see Figure 3.14).[11,24]

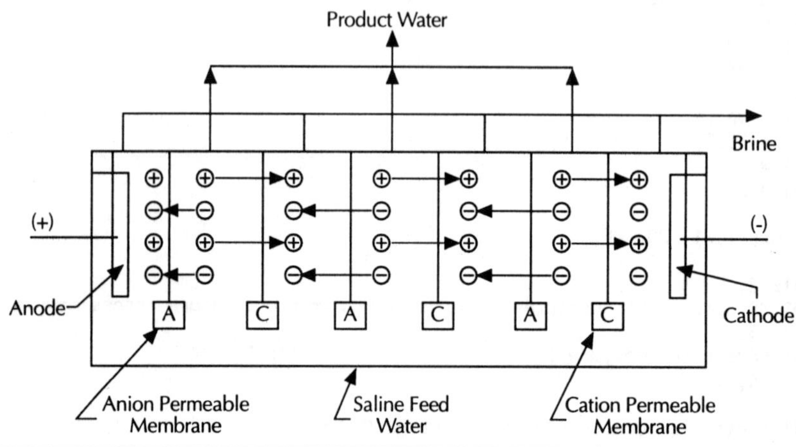

FIGURE 3.14
Schematic representation of an electrodialysis process.

The efficiency of the system to transport counter ions can be determined by using Equation 3.66.[24]

$$i * (Eff' - Eff) / 100 * Far = Diff * (C_0 - C) / B \tag{3.66}$$

where:

- i = Current density, A/cm^2.
- Eff′ = Current efficiency for transport of counter ions through membrane, %
- Eff = Current efficiency for transport of same ions in solution in contact with membrane, %
- Far = Faraday's constant (= 26.8 A*h)
- Diff = Diffusion constant for electrolyte at temperature of electrolysis
- C_0 = Concentration of electrolyte in the depletion compartment
- C = Concentration of electrolyte at the membrane-solution interface
- B = Thickness of diffusion layer at interface

The maximum value of the current density can be determined by using Equation 3.67.[24]

$$i_{max} = (Far * Diff * C_0) * 100 / [B * (Eff' - Eff)] \tag{3.67}$$

where:

i_{max} = Maximum value of the current density, A/cm^2

Problems associated with the electrodialysis process include:[11]

- High costs involved in treating sea water
- Requires expensive pretreatment
- Problems encountered with high-sulfate waters
- Limitation of process of removing only common mineral ions
- Requires skilled maintenance and supervision

Example 3.16

1. Write a computer program to determine the maximum current density, given total dissolved solids (TDS), current efficiency for transport of counter ions through membrane, current efficiency for transport of same ions in solution in contact with the membrane, thickness of diffusion layer at interface, diffusion constant for electrolyte at temperature of electrolysis, and Faraday's constant.
2. Brackish water with a dissolved solids concentration of 1800 mg/L is introduced to an electrodialysis unit. The unit operates under the following conditions:
   - Current efficiency for transport of counter ions through membrane = 80%.
   - Current efficiency for transport of same ions in solution in contact with membrane = 58%.
   - Diffusion constant for electrolyte at temperature of electrolysis = $1.4*10^{-5}$ cm^2/s.
   - Thickness of diffusion layer at interface = 0.5 mm.

   Compute the maximum current density (take the Faraday's constant as equal to 26.8 A*h).
3. Use the computer program developed in (1) to verify the manual solution presented in (2).

Solution

1. For solution to Example 3.16 (1), see listing of Program 3.16 below.
2. Solution to Example 3.16 (2):
   - Given C_0 = 1800 mg/L, Eff = 58%, Eff′ = 80%, Diff = $1.4*10^{-5}$ cm^2/s, B = 0.5 mm = $5*10^{-2}$ cm, Far = 26.8 A*h.
   - Determine the maximum current density as: $i_{max} = 100*Far*Diff*C_0/(B*(Eff' - Eff))$ = $(100*26.8*3600*1.4*10^{-5}*1800*10^{-6})/(5*10^{-2}*(80 - 58))$ = 0.22 A*g/cm^2.

Listing for Program 3.16

```
DECLARE SUB box (r1%, c1%, r2%, c2%)    'Found in Program 2.1
'****************************************************************
'Program 3.16: Electrodialysis
'Computes maximum current density
'****************************************************************
CALL box(1, 1, 20, 80)

LOCATE 2, 20
PRINT "Program 3.16:Computes maximum current density"
LOCATE 5, 5
INPUT "Enter dissolved solids concentration mg/L                =" ; Co
LOCATE 6, 5
INPUT "Enter efficiency for transport of ions through membrane(%)="; eff1
LOCATE 7, 5
INPUT "Enter efficiency for transport of ions in solution(%)    ="; eff2
LOCATE 8, 5
INPUT "Enter diffusion constant for electrolyte   cm2/s         ="; Diff
LOCATE 9, 5
INPUT "Enter thickness of diffusion layer at interface  cm      ="; B
Far = 26.8 * 3600     'Amp*hr
i = Far * Diff * Co * .000001 * 100 / (B * (eff1 - eff2))
i = ABS(i)

LOCATE 12, 3
PRINT "Maximum current density  ="; i; "  Amp.gm/cm2"
LOCATE 24, 5
PRINT "press any key to continue....................";
DO
LOOP WHILE INKEY$ = ""
END
```

3.9 Disinfection Process

3.9.1 Introduction

Disinfection involves the killing of pathogenic organisms with the objective of preventing the spread of water- and wastewater-borne diseases. Disinfection methods can be classified as either physical or chemical. The former methods include heat treatment, exposure to ultraviolet rays, and metal ions (e.g., silver, copper). The latter methods include usage of oxidants such as chlorine gas, ozone, iodine, and chlorinated compounds (e.g., chlorine dioxide) and potassium permanganate.

The preferred characteristics for an effective chemical disinfectant include:[25,29–34]

- Effective in killing pathogenic microorganisms
- Ease of detection and measurement of its concentration in water
- Solubility in water in doses required for disinfection
- Absence of toxicity to humans and animals
- Ease of handling, transporting, and controlling it

- Capable of producing a residual
- Absence of a taste, odor, or color when used
- Available at a reasonable cost

The disinfection process is a function of type and concentration of microorganisms, type and concentration of disinfectant, presence of oxidant-consuming compounds, temperature, dose of chemical, contact time, and pH.

3.9.2 Chlorination

Chlorination is the process of adding chlorine to water with the purpose of killing pathogens. When chlorine is added to water, part of it reacts to produce hypochlorous acid (HOCl), which is the more effective disinfectant (available chlorine); another part forms hypochlorite ion (OCl^-); a third part reacts with ammonia to form chloramines; a fourth part oxidizes inorganic matter (e.g., hydrogen sulfide, iron, manganese); and a fifth part reacts with organic compounds to form trihalomethanes (THMs) and other chlorinated organics. The benefits of using chlorine include effectiveness, reliability, and establishment of a residual in the system.[11]

The rate of kill of organisms is a function of many interacting parameters, such as concentration of disinfectant, ability and time needed to penetrate the bacterial cell wall, distribution of disinfectant and microorganisms, and the contact time between pathogens and disinfectant according to Chick's law.

Equation 3.68 is a mathematical expression for Chick's law which is not valid for all disinfectants nor for all microorganisms.

$$N / N_0 = e^{-k*t} \tag{3.68}$$

where:

N = Number of viable microorganisms of one type at time t
N_0 = Number of viable microorganisms of one type at time t = 0
t = Time, d
k = Constant, d^{-1}

Another relationship between disinfectant concentration and contact time is given in Equation 3.69.

$$C^a * t = k \tag{3.69}$$

where:

C = Concentration of chlorine, mg/L
t = Contact time, or time required for a given percent of microbial kill, min
a, k = Experimental constants that are valid for a particular system

The Van't Hoff-Arrhenius equation can be used to relate the effects of temperature on the disinfection process; see Equation 3.70.[9,27]

$$\mathrm{Ln}\,(t_1 / t_2) = [E'(T_2 - T_1)] / R \quad (3.70)$$

where:

t_1, t_2 = Time required for the given kills, s
E' = Activation energy (see Table 3.3), cal
T_1, T_2 = Temperature corresponding to t_1 and t_2, K
R = Gas constant

TABLE 3.3
Activation Energies (E′) for Aqueous Chlorine[11]

pH	E′,cal
7.0	8200
8.5	6400
9.8	12,000
10.7	15,000

Source: From Rowe, D.R. and Abdel-Magid, I.M., *Handbook of Wastewater Reclamation and Reuse*, CRC Press/Lewis Publishers, Boca Raton, FL, 1995. With permission.

Problems associated with chlorination include:

- Safety considerations during handling, transportation, and storage of chemicals.
- Formation of chlorinated organic compounds (THMs) that can be injurious to human health.

Example 3.17

1. Using Chick's law, write a computer program to find the contact time required for a disinfectant to achieve a percent kill for a microbial system that has a rate constant of k (to base e) or k′ (to base 10).
2. Determine the contact time needed for a disinfectant to achieve a 99.99% kill for a pathogenic microorganism that has a rate constant (to base 10) of 0.1 s^{-1}.

3. Use the program developed in (1) to verify computations of part (2).

Solution

1. For solution to Example 3.17 (1), see listing of Program 3.17 below.
2. Solution to Example 3.17 (2):
   - Given rate of kill = 99.99%, $k' = 0.1\ s^{-1}$.
   - Find the contact time from Chick's law as: $t = -(1/k')*\log(N/N_0)$. $t = -(1/0.1)*\log(100 - 99.99/100) = 40$ s.

Listing of Program 3.17

```
DECLARE SUB box (r1%, c1%, r2%, c2%)     'Found in Program 2.1
'***********************************************************************
'Program 3.17: Chlorination
'Computes the time needed by a disinfectant to achieve a certain kill
'***********************************************************************
CALL box(1, 1, 20, 80)
LOCATE 3, 10
PRINT "Program 3.17: Computes time needed by a disinfectant"
LOCATE 5, 5
INPUT "Enter the rate constant      k     /s  ="; k
LOCATE 6, 5
INPUT "Is it to base 10?                 Y/N  ="; R$
LOCATE 7, 5
INPUT "Enter the percentage kill to achieve ="; N1
IF N1 < 1 THEN N1 = N1 * 100
N2 = (100 - N1) / 100
lgn = LOG(N2) / LOG(10)                  'logarithm base 10 of N2
IF UCASE$(R$) = "N" THEN lgn = LOG(N2)   'logarithm base e of N2
t = -(1 / k) * lgn

LOCATE 10, 3
PRINT "Time needed to achieve this level of kill ="; t; "  seconds"

LOCATE 22, 3
PRINT "press any key to continue....................";
DO
LOOP WHILE INKEY$ = ""
END
```

3.10 Corrosion

3.10.1 Introduction

Water with corrosive characteristics can cause problems in distribution networks and residential plumbing systems. Generally, problems encountered may be classified as follows:[35,37–41]

- *Health problems*: These problems originate from dissolution of certain substances into water from transmission and distribution pipelines or plumbing system.
- *Economic problems*: These problems result from reduced service life of materials due to deterioration within transmission, distribution, and plumbing systems.
- *Aesthetic problems*: These problems are due to dissolution of certain metallic substances into the water.

3.10.2 Corrosion Potential Indicators[35,37–40]

The main corrosion potential indicators include Langelier (LI), Ryznar (RI), and the Aggressiveness (AI) indexes. These indicators are briefly outlined below.

3.10.2.1 Langelier Index

The Langelier index illustrates the tendency for deposition (noncorrosiveness) or dissolution of calcium carbonate scale in pipe. The index is defined as the difference between pH of the water and its saturated pH (when water is in equilibrium with calcium carbonate).[22] Equations 3.71, 3.72, and 3.73 present mathematical expressions for determining Langelier's index.

$$LI = pH_a - pH_s \tag{3.71}$$

where:

LI = Langelier index
pH_a = Actual (measured) pH of the water
pH_s = Saturation pH

$$pH_s = pK_2' - pK_s' + pCa^{++} + pAlk + S \tag{3.72}$$

where:

$pK_2' - pK_s'$ = Dissolution constant estimates based on temperature and total dissolved solids or ionic strength
pCa^{++} = Negative logarithm of the calcium ion concentration (Ca^{++} in moles/L)
$pAlk$ = Negative logarithm of the total alkalinity, (alkalinity in equivalents of $CaCO_3$/L)
S = Salinity correction term

$$S = (2.5\xi^{1/2})/\{1 + 5.3\xi^{1/2} + 5.5\xi\} \tag{3.73}$$

where:

ξ = Ionic strength

Positive LI values indicate a supersaturated condition that will deposit calcium carbonate on the pipe. Negative LI values indicate an undersaturated condition that will dissolve calcium carbonate and will not deposit a protective film of the salt on pipe to stop corrosion.[22]

3.10.2.2 Ryznar Index

The Ryznar index provides an indication of the rate of scale formation or the tendency of the water to be aggressive. Equations 3.74 and 3.75 can be used to determine the Ryznar index.

$$RI = 2\ pH_s - pH_a \tag{3.74}$$

where:

RI = Ryznar index
pH_s = Saturation pH
pH_a = Actual (measured) pH

The relationship between the Ryznar and the Langelier indexes is presented in Equation 3.75.

$$RI = pH_a - 2\ LI \tag{3.75}$$

When:

- RI < 6, calcium carbonate scale deposition increases proportionately.
- RI > 6, corrosion increases.
- RI > 10, conditions are extremely aggressive.

3.10.2.3 Aggressiveness Index

The Aggressiveness index defines water quality that can be transported through asbestos-cement pipes without adverse effects. The relationship between aggressiveness index, pH, total alkalinity, and calcium hardness are presented in Equation 3.76.

$$AI = pH + Log_{10}[(Alk) * (HCa)] \tag{3.76}$$

where:

AI = Aggressiveness index

pH = Negative logarithm of the hydrogen ion concentration

Alk = Total alkalinity, mg/L $CaCo_3$

HCa = Calcium hardness, mg/L $CaCO_3$

Table 3.4 gives a classification of water according to the value of the aggressiveness index.

Example 3.18

1. Write a computer program to determine whether a water sample is corrosive or scale-forming, using the different indexes (Langelier, Ryznar, Aggressiveness). Data given include parameters involved in the computations of each of the relative indexes.
2. Analysis of a water sample revealed the following: calcium hardness = 125 mg/L as $CaCO_3$, alkalinity = 195 mg/L as $CaCO_3$, hydrogen ion concentration of $2.6*10^{-7}$ mol/L. Using the Aggressiveness index, determine whether this water is aggressive to asbestos-cement pipes.
3. Use the computer program involved in (1) to verify computations conducted in (2).

TABLE 3.4
Aggressiveness Index

Value	Description
< 10	Highly aggressive
10 to 11.9	Moderately aggressive
> than 12	Nonaggressive

Solution

1. For solution to Example 3.18 (1), see Program 3.18 listed below.
2. Solution to Example 3.18 (2):
   - Given HCa = 125, Alk = 195, $[H^+] = 2.6*10^{-7}$ mol/L.
   - Determine the pH as: $pH = -\log [H^+] = -\log 2.6*10^{-7} = 6.57$.
   - Find the AI as: $AI = pH + \log_{10} (Alk)*(HCa) = 6.57 + \log (195*125) = 10.96$. Thus, AI = 10.96 < 11.9. This value shows that the water is moderately aggressive to asbestos-cement pipes (Table 3.4).

Listing of Program 3.18

```
DECLARE SUB box (r1%, c1%, r2%, c2%)     'Found in Program 2.1
'*************************************************************************
'Program 3.18: Corrosion
'determines whether a sample is corrosive or precipitative
'*************************************************************************
DEF Fnlog10 (x) = LOG(x) / LOG(10)        'logarithm base 10 of x

CALL box(1, 1, 20, 80)
LOCATE 3, 15
```

```
PRINT "Program 3.18: Checks if a sample is corrosive or precipitative"
LOCATE 5, 5
INPUT "Enter the hydrogen ion concentration(mol/L)="; H
LOCATE 6, 5
INPUT "Enter the total alkalinity (mg/L CaCo3)    ="; ALK
LOCATE 7, 5
INPUT "Enter the calcium hardness (mg/L CaCo3)    ="; HCa

ph = -Fnlog10(H)
AI = ph + Fnlog10(ALK * HCa)
SELECT CASE AI
        CASE IS < 10
                type$ = "Highly aggressive"
        CASE IS > 12
                type$ = "Non-aggressive"
        CASE ELSE
                type$ = "Moderately aggressive"
END SELECT

LOCATE 10, 3
PRINT "Aggressiveness Index ="; AI; SPACE$(10); type$

LOCATE 12, 3
INPUT "Do you want to compute Langelier & Ryznar indices? (Y/N):"; R$

IF UCASE$(R$) = "Y" THEN
        'Compute Langelier & Ryznar Indices
        LOCATE 14, 5
        INPUT "Enter measured pH of water="; pHa
        LOCATE 15, 5
        INPUT "Enter saturation pH       ="; pHs
        LI = pHa - pHs
        RI = 2 * pHs - pHa
        LOCATE 17, 3
        PRINT "Langelier Index="; LI
        LOCATE 18, 3
        PRINT "Ryznar Index="; RI
END IF

LOCATE 22, 3
PRINT "press any key to continue...................";
DO
LOOP WHILE INKEY$ = ""
END
```

3.10.3 Rate of Corrosion

Corrosion may be defined as an attack on the surface of a metal or other solid.[22] Figures 3.15 and 3.16 give a general classification for the corrosion processes. The rate of metal removal due to corrosion can be calculated by using Equation 3.77.[41]

$$R_{mr} = K*wt/A*t*\rho \tag{3.77}$$

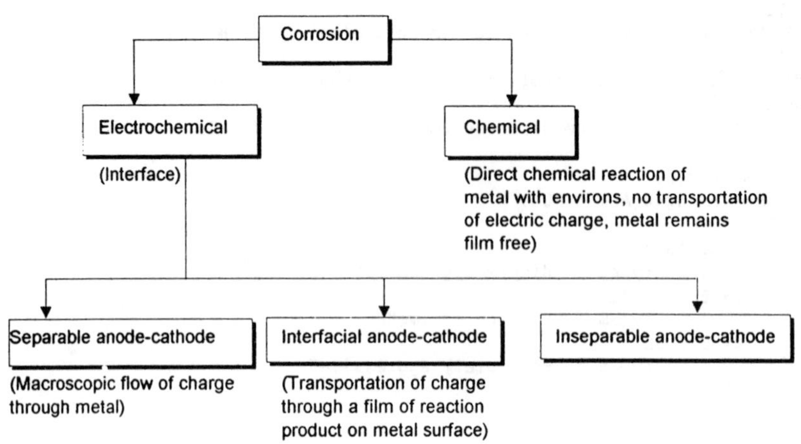

FIGURE 3.15
Types of corrosion.

where:

R_{mr} = Rate of corrosion, m/s
K = Constant
Wt = Weight loss, kg
A = Area of specimen, m^2
t = Time of exposure, s
ρ = Density, kg/m^3

Example 3.19

1. Write a short computer program that determines the rate of corrosion of a metallic surface, given weight loss of material; wt (kg) — or electrochemical equivalent, E_e (mg/Coulomb), current, I (Ampere); area of specimen, A (m^2); time of exposure, t (s); and density of material, ρ (kg/m^3).
2. In an electrochemical cell the rate of corrosion of a 0.5-cm-radius zinc anode is found to be $1.2*10^{-6}$ mm/s for an electric current of 90 mA. Determine the depth to which the anode is immersed in the electrolyte to produce the aforementioned corrosion rate. For zinc, the electrochemical equivalent = 0.3387 mg/C, the density = 7000 kg/m^3, and K = 1.

Solution

1. For solution to Example 3.19 (1), see listing of Program 3.19 below.
2. Solution to Example 3.19 (2):
   - Given r = 0.5 cm, R_{mr} = $1.2*10^{-6}$ mm/s, I = 90 mA, E_e = 0.3387 mg/C, ρ = 7000 kg/m^3.

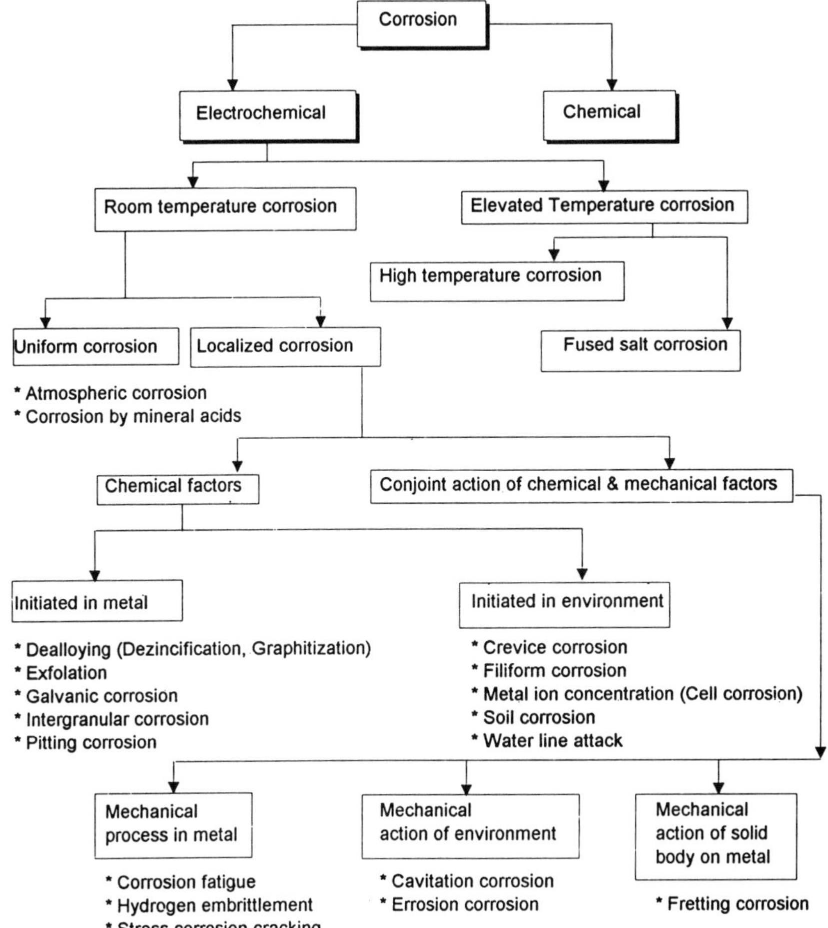

FIGURE 3.16
Classification of corrosion.

- Use corrosion rate Equation 3.77 to find corroded area as: $A = E_e{*}I{*}K/R_{mr}{*}\rho = (0.3387{*}10^{-3}\ g/C{*}90{*}10^{-3}A)/(1.2{*}10^{-7}{*}7 g/cm^3) = 36.29\ cm^2$. Area exposed to corrosion = $\pi d^2/4 + \pi dh$, where d is diameter of anode and h is depth of immersion.
- Find depth of immersion to produce given coronion rate as: $h = (A - \pi{*}r^2)/2\pi r = (36.29 - \pi{*}0.5^2)/2{*}\pi{*}0.5 = 11.3$ cm.

Listing of Program 3.19

```
DECLARE SUB box (r1%, c1%, r2%, c2%)    'Found in Program 2.1
'*************************************************************************
'Program 3.19: Corrosion
'Determines the rate of corrosion of a metallic surface
'PLEASE CHECK EQUATIONS
'*************************************************************************
```

```
DIM ttl$(4)
pi = 3.14159
ttl$(1) = "1. Determine rate of corrosion"
ttl$(2) = "2. Determine the depth of immersion for a given rate"
ttl$(3) = "3. Exit program"

start:
CALL box(1, 1, 20, 80)
LOCATE 3, 10
PRINT "Program 3.19:Determines the rate of corrosion of a metallic surface"
FOR I = 1 TO 3
LOCATE I + 4, 5
PRINT ttl$(I)
NEXT I
DO
LOCATE 10, 3
INPUT "Select an option (1 to 3): "; opt
LOOP WHILE opt < 1 OR opt > 3

CALL box(1, 1, 20, 80)
LOCATE 5, 5
SELECT CASE opt
CASE 1
        INPUT "Enter the weight loss (Kg)                  ="; wt
        LOCATE 6, 5
        INPUT "Enter the area of the specimen (m2)         ="; A
        LOCATE 7, 5
        INPUT "Enter the time of exposure (s)              ="; t
        LOCATE 8, 5
        INPUT "Enter the electrochemical equivalent (mg/L)="; k
        LOCATE 9, 5
        INPUT "Enter the density of the material (Kg/m3) ="; gamma

        Rmr = k * wt / (A * t * gamma)
        LOCATE 12, 3
        PRINT "The rate of corrosion ="; Rmr; "  m/s"
        DO
        LOOP WHILE INKEY$ = ""

CASE 2
        INPUT "Enter the rate of corrosion (m/s)           ="; Rmr
        LOCATE 6, 5
        INPUT "Enter the diameter of the specimen (m)      ="; d
        LOCATE 7, 5
        INPUT "Enter the density of the material (Kg/m3) ="; gamma
        LOCATE 8, 5
        INPUT "Enter the electrochemical equivalent (mg/L)="; k
        LOCATE 9, 5
        INPUT "Enter the electric current  (mA)             ="; I
        LOCATE 10, 5
        INPUT "Enter Ee                                     ="; Ee

        A = Rmr * gamma / (Ee * I * k)               'corroded area
        L = (A - pi * d ^ 2 / 4) / (pi * d)          'depth of immersion
        LOCATE 12, 3
        PRINT "The depth of immersion="; L; "  m"
        DO
        LOOP WHILE INKEY$ = ""

CASE 3
        END
```

```
END SELECT

GOTO start
END
```

3.11 Homework Problems

3.11.1 Discussion Problems

1. What are the various ways in which water treatment can be classified?
2. Discuss the objectives of water treatment.
3. What are the functions of screens at water treatment works?
4. Why should a water suspension have a minimum velocity in order to pass through a screen opening?
5. Define the following terms: sedimentation, flotation, and plain sedimentation.
6. Outline the objectives of the sedimentation process.
7. Why are square and circular settling tanks preferred in water treatment?
8. Discuss the types of settling and outline their main differences.
9. Give examples for each type of settling outlined in Problem 8.
10. Name the factors that affect class I settling.
11. For what purposes are cumulative frequency distribution curves drawn?
12. What are the factors that influence the efficiency of sedimentation basins?
13. Explain the kinetics of the filtration process.
14. Define the terms: aeration, aerator, and saturation concentration.
15. What are the advantages of aeration in water and wastewater treatment?
16. Give some examples of aerators.
17. Outline the factors that affect the solubility of a gas in a liquid.
18. Differentiate between coagulation and flocculation.
19. Indicate how to enhance the process of coagulation and filtration.
20. Define the following terms: electrophoretic mobility, zeta potential, and velocity gradient. State values for each.

21. What are the parameters that govern the design of a flocculator?
22. Define filtration and indicate its objectives in water treatment plants.
23. Differentiate between "slow sand filters" and "rapid sand filters".
24. What are the characteristics of a good filter media?
25. Why is a rapid filter backwashed?
26. What are the main practical desalination methods? Indicate advantages and disadvantages of each method.
27. Define the following terms: osmosis, desalination, distillation, and electrodialysis.
28. What is the purpose of disinfection?
29. What are the desirable properties of an effective disinfectant?
30. Give examples for physical and chemical disinfectants.
31. Present the reactions that can occur if water is chlorinated.
32. What are the problems associated with corrosive waters?
33. Differentiate between a corrosive and a scale-forming water suspension.
34. What are the factors that govern the rate of corrosion of metallic substances.
35. Give a general classification scheme for the types of corrosion expected in a sewerage collection system.

3.11.2 Specific Mathematical Problems

1. Write a short computer program to compute the diameter, D (m), and depth of a circular sedimentation tank, h (m), given influent rate of flow, Q (m^3/s); tank's overflow rate, v_s ($m^3/m^2/s$); and detention time of tank, t (s). Verify your computations for the data given: Q = 38 m^3/h, v_s = 5 $m^3/m^2/d$, t = 2 h.
2. Write a short computer program to compute the diameter, d (m), of spherical particles allowed to settle in an ideal sedimentation tank, given the settling velocity of particles, v_s ($m^3/m^2/s$); specific weight of settling particles, s.g.: and temperature of mixture during the settling process. Assume Stoke's law is valid. Design the program so that a check can be made as to the Reynolds number. Verify your computations for the data given here: v_s = 0.3 m/min, s.g., = 2.6, and T = 20°C.
3. Write a short computer program to evaluate the drag force, F (N), acting on a moving spherical particle with a small Reynolds number, and its terminal settling velocity, given that $F = 6\pi * r * \mu * v$, where μ is the dynamic viscosity ($N*s/m^2$), v is the settling velocity of the sphere (m/s), and r the radius (m).
4. Develop a computer program that can determine the concentration of the suspended solids in the effluent, the Froude and Reynolds numbers for horizontal water movement, the percentage of suspended solids in the effluent having a settling velocity smaller

than a certain given value, and the detention time of the water in the tank. The information given includes: temperature of the settling suspension, T (°C); population number, POP, average per capita domestic water consumption, Q (L/d); suspended solids content of influent, C_0 (mg/L); tank dimensions (i.e., width, B; length l; and depth, h (m)), and settling column test results which showed a straight line relationship for the cumulative frequency distribution of the settling velocities of the particles with characteristics of:

- x_1 percent of the particles having a settling velocity larger than v_1 (mm/s).
- x_2 percent of the particles have a settling velocity smaller than v_2 (mm/s).

Verify your program using the following data: T = 22°C; POP = 71,280; Q = 400 L/d; C_0 = 125 mg/L. The settling column test results:

- 10% of the particles have a settling velocity larger than 0.7 mm/s.
- 10% of the particles have a settling velocity smaller than 0.2 mm/s.
- B = 10 m, l = 60, and h = 1.5 m.

5. Write a computer program to estimate the average grain diameter (mm) needed for a sand filter bed consisting of unisized grains of a porosity, e (%). The bed depth is l (m) operating at a temperature of T (°C) with a filtration velocity of v_f ($m^3/m^2/d$) at a maximum bed head loss of h_f (m). Assume that the head loss has been determined by the Rose's equation. Design the program so that the suitability of the grain diameter is evaluated. Verify your program by using the following data: l = 1 m, e = 50%, v_f = 0.2 $m^3/m^2/h$, h_f = 0.24 m, and $\nu = 1.0105*10^{-6}$ m^2/s.
6. Write a computer program to find the porosity of a filter bed after being backwashed with a filter bed expansion of x (%). Take the filter bed thickness to be l (m), bed porosity of e (%), and a grain size diameter of d (mm). Check the validity of your program for the following data: l = 120 cm, e = 40%, d = 0.95 mm, and x = 25%.
7. Write a computer program to design a slow sand filter intended to treat a flow of Q (m^3/s), using the relationships found in Table 3.1. Use the following data to verify your program: Q = 1000 m^3/d, v_f = 8 m/d.
8. Write a program that provides for the determination of the amount of a disinfectant (in kg/d) needed for disinfecting a water flow of Q (m^3/s) to obtain x mg/L residual, given chlorine contained in the compound as y percent. Use your program to find the amount of chlorine compound needed under the following conditions: Q = 150 m^3/h, x = 0.3 mg/L, y = 65%.
9. The following laboratory data was obtained from a settling column test. The initial solids concentration equals 550 mg/L. Estimate the overall removal efficiency of a settling basin with a depth of 2 m and a detention time of 2 h.

Settling Data

Time (min)	% Removal (initial solids concentration 550 mg/L) Depth (m) 0.5	1	1.5	2
10	20	11	9	8
20	32	22	17	15
30	36	31	26	23
40	46	34	31	30
50	59	38	34	33
60	64	46	38	35
70	70	55	45	34
90	81	66	60	53
110	89	76	89	63
120	91	80	73	69

References

1. Huisman, L., *Sedimentation and Flotation: Sedimentation and Flotation – Mechanical Filtration – Slow Sand Filtration – Rapid Sand Filtration*, Delft University of Technology, Herdruk, The Netherlands, 1977.
2. Barnes, D., Bliss, P. J., Gould, B.W., and Vallentine, H.R., *Water and Wastewater Engineering Systems*, Pitman International, Bath, 1981.
3. Callely, A.G., Forster, C.F.F., and Stafford, D.A., *Treatment of Industrial Effluents*, Hodder and Stoughton, London, 1977.
4. Hammer, M.J., *Water and Wastewater Technology*, John Wiley & Sons, New York, 1986.
5. Huisman, L., Sundaresan, B.B., Netto, J.M.D., Lanoix, J.N., and Hofkes, E.H., *Small Community Water Supplies*, IRC, The Hague and John Wiley & Sons, New York, 1986.
6. Viessman, W. and Hammer, M.J., *Water Supply and Pollution Control*, Harper & Row, New York, 1985.
7. Davis, M.L. and Cornwell, D.A., *Introduction to Environmental Engineering*, Chemical Engineering Series, 2nd ed., McGraw-Hill, New York, 1991.
8. Steel, E.W. and McGhee, T.J., *Water Supply and Sewerage*, 6th ed., McGraw-Hill, New York, 1991.
9. Metcalf and Eddy, Inc., *Wastewater Engineering Treatment Disposal Reuse*, 3rd ed., McGraw-Hill, New York, 1991.
10. Vernick, A.S. and Walker, E.C., *Handbook of Wastewater Treatment Processes*, Marcel Dekker, NY, 1981.
11. Rowe, D.R. and Abdel-Magid, I.M., *Handbook of Wastewater Reclamation and Reuse*, Lewis Publishers, Boca Raton, FL, 1995.
12. Negulescu, M., *Municipal Wastewater Treatment*, Development Water Science 23, Elsevier, Amsterdam, 1985.
13. Wilson, F., *Design Calculations in Wastewater Treatment*, E. & F.N. Spon, London, 1981.

14. Popel, H.J., *Aeration and Gas Transfer*, Delft University of Technology, Herdurk, The Netherlands, 1979.
15. Masschelein, W.J., *Unit Operations*, International Institute for Hydraulic and Environmental Engineering, Delft, The Netherlands, 1977.
16. Punmia, B.C., *Environmental Engineering. Vol. 1: Water Supply*, Standard Book House, Naisarak, Delhi, 1979.
17. Vesilind, P.A. and Peirce, J.J., *Environmental Pollution Control*, Butterworths-Heinemann, London, 1990.
18. Pescod, M.B., Abouzaid, H., and Sundaresan, B.B., *Slow Sand Filtration: A Low Cost Treatment for Water Supplies in Developing Countries*, published for the World Health Organization Regional Office for Europe by the Water Research Center, U.K., in collaboration with the IRC, The Netherlands, Stevenage, Hertfordshire, U.K.
19. Huisman, L., *Slow Sand Filtration*, World Health Organization, Geneva, 1974.
20. *IRC Rep. Intl. Appraisal Meeting, Nagpur, India, Sept. 15–19, 1980: Slow Sand Filtration for Community Water Supply in Developing Countries*, BS 16, The Hague, The Netherlands, 1981.
21. Van Dijk, J.C. and Oomen, J.H.C.M., *Slow Sand Filtration for Community Water Supply in Developing Countries — A Design and Construction Manual*, IRC, Tech. Paper No. 11, The Hague, The Netherlands, 1982.
22. Scott, J.S. and Smith, P.G., *Dictionary of Waste and Water Treatment*, Butterworths, London, 1980.
23. Buros, O.K., *The Desalting ABC's*, International Desalination Association, Topsfield, MA, 1990.
24. Porteous, A., *Desalination Technology: Developments and Practice*, Applied Science, London, 1983.
25. Sawyer, C.N. and McCarty, P.L., *Chemistry for Environmental Engineering*, 3rd ed., McGraw-Hill, New York, 1978.
26. Malik, M.A.S., Tiwari, G.N., Kumar, A., and Sodha, M.S., *Solar Distillation*, Pergamon Press, Oxford 1982.
27. Peavy, H.S., Rowe, D.R., and Tchobanoglous, G., *Environmental Engineering*, McGraw-Hill, New York, 1985.
28. Lacy, R.E., Membrane separation process, *Chem. Eng.*, 4, 56, 1972.
29. El-Hassan, B.M. and Abdel-Magid, I.M., *Environment and Industry: Treatment of Industrial Wastes*, Institute of Environmental Studies, Khartoum University, Khartoum, 1986.
30. Fair, G.M., Geyer, J.C., and Okun, D.A., *Water and Wastewater Engineering*, Vol. I and II, John Wiley & Sons, New York, 1966.
31. Committee Report, Disinfection, *Am. Water Works Assoc.*, 70(4), 219, 1978.
32. Dyer-Smith, P., Brown, Beveri and Co., Water disinfection status and trends, *J. Water Sewage Treat.*, 2(4), 13, 1983.
33. Salvato, J.A., *Environmental Engineering and Sanitation*, 4th ed., John Wiley & Sons, New York, 1992.
34. Degremont, *Water Treatment Handbook*, Vol. I and II, 6th ed., Rueil-Malmaison Cedex, France, 1991.
35. American Water Works Association, *Water Quality and Treatment — A Handbook of Public Water Supplies*, McGraw-Hill, New York, 1971.

36. Fair, G.M., Morris, F.C., Chang, S.L., Weil, I., and Burden, R.A., The behavior of chlorine as a water disinfectant, *J. Am. Water Works Assoc.*, 40, 1051, 1948.
37. American Water Works Association, Determining internal corrosion potential in water supply system, *J. Am Water Works Assoc.*, 76(8), 83, 1984.
38. American Water Works Association, Internal corrosion, *J. Am Water Works Assoc.*, 72(5), 1980 (all articles).
39. Eliassen, R.R.T. and Shrindle, W.B.D., Experimental performance of miracle water conditioner, *J. Am Water Works Assoc.*, 50, 1371, 1958.
40. Hoyt, B.P. et al., Evaluating home plumbing corrosion problems, *J. Am Water Works Assoc.*, 71(12), 720, 1979.
41. Narayan, R.M., *An Introduction to Metallic Corrosion and Its Prevention*, published by Mohan Primlani for Oxford and IBH Pub., New Delhi, 1983.
42. James, A., *An Introduction to Water Quality Modelling*, John Wiley & Sons, New York, 1984.

Chapter 4

Water Storage and Distribution

Contents

4.1 Introduction

The first section of this chapter deals with the various types of water storage reservoirs. Included here is the mass flow curve that can be used to evaluate the volume of water needed in storage to meet the uniform users' demand. Also, accompanying this is a computer program that enables determination of the required volume of water in storage, provided the necessary data is available. The next section deals with the water distribution system and includes patterns and configurations of pipeline networks and fundamental hydraulic principles for both laminar and turbulent flows. Computer programs for both laminar (Poiseuille's equation) and turbulent flow (Darcy's equation) are presented. The hydraulic design and analysis of flow in pipes in the distribution system — such as the equivalent-velocity-head method, the equivalent-pipe-length method, the pipes-in-parallel method, the Hardy cross method, and the analysis of pipe networks by the Finite Element method — are covered. In all of these cases, computer programs are provided, and each computer program incorporates all the elements involved in the design and evaluation of flow of water in pipes in network in water distribution systems. The last section of this chapter deals with the structural design of water storage tanks, circular or rectangular (deep or shallow). Computer programs for the design of both steel and reinforced concrete water tanks are included.

4.1.1 Water Storage

Water storage is needed before and after treatment in order to maintain a reliable, uniform, and constant water supply, to balance out fluctuations in flow, and to furnish large volumes of water during emergencies such as fires. Fire insurance rates generally depend upon water storage capacity available. Water storage in the distribution system is also used to equalize operating pressures as well as to equalize pumping rates. The fluctuations in water demand depend upon many variables, such as:

- Time of day
- Day of the week
- Season of the year
- Weather conditions
- Living standards in the community
- Industrial demands
- Fires
- Main breaks

Benefits of water storage reservoirs include flood control, hydroelectric power, irrigation, water supply, navigation, sailing, preservation of aquatic life, fire protection, emergency needs, recreation, and pollution abatement.[1–4] Reservoirs can be grouped as follows:

1. *Storage (conservation, impounding, direct-supply) reservoirs*: These reservoirs store water in excess of demand, from a natural source in periods of high flow, to be used during periods of dry weather or low flows. Water is stored for periods ranging from a few days to several months or even longer.[2]
2. *Distribution reservoirs*: The distribution reservoirs can be either elevated (see Figure 4.1) or ground reserve water supply under pressure. They store water to provide for varying demands of the community over a period of a day or several days. Water stored in distribution reservoirs is often supplied at a steady rate from a storage reservoir. The distribution reservoirs (elevated storage tanks and towers) provide equalizing storage, provide emergency storage (firefighting, power blackouts, pump station failure), and reduce necessary size and capacity of pipes and treatment facilities.[5] Generally, the volume of water needed to balance or equalize peak hourly flows is about 20% of average daily demand in the service area.[5] When the height of the reservoir exceeds its diameter, it is referred to as stand pipe. Stand pipes are basically tall cylindrical tanks, the upper portion of which constitutes useful storage to produce the necessary head; the lower portion serves to support the structure.[3] A water tower (elevated tank) is a service reservoir or tank (above 15 m in height) raised above ground level.[2,3]
3. *Plant storage reservoirs (clear wells)*: Clear wells are important for the storage of filtered water and to provide operational storage to average out high and low demands. Clear wells with sufficient capacity can prevent the need for varying filtration rates and prevent frequent on/off cycling of water pumps. The clean well also provides storage for filtered water before it is introduced into the distribution system.

The storage volume depends on the water demand as well as the purpose of storage. The amount of storage that needs to be provided can be determined by using Equation 4.1[4]

$$\Delta S = Q_1 - Q_0 \quad (4.1)$$

where:

ΔS = Change in storage volume during a specified time interval
Q_1 = Total inflow volume during a specified time interval
Q_O = Total outflow volume during a specified time interval

The storage volume required above the minimum operating conditions can be determined by:

- Analytical techniques, which are numerical analyses of historical flow records, especially during periods of low flow.
- The graphical method (Rippl mass curve) evaluates the cumulative deficiency between outflow and inflow and selects the maximum value as the required storage.[4]

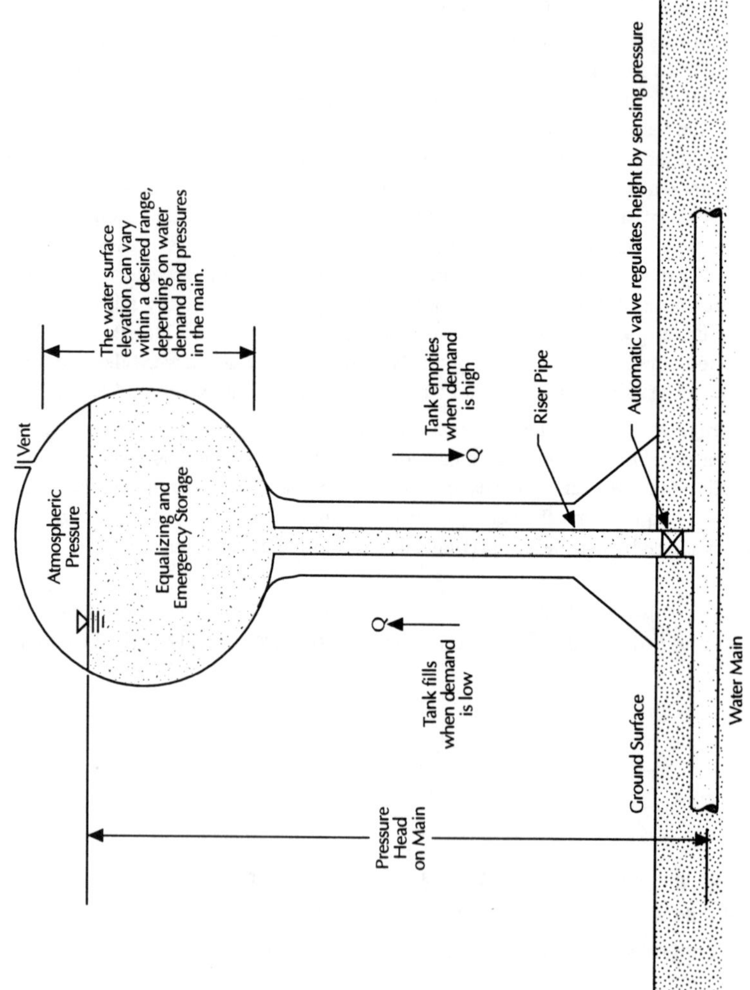

Figure 4.1
An elevated water storage tank.

4.1.2 Mass Curve (Rippl Diagram, S-Curve)

A flow mass curve may be defined as a graph of cumulative values of a hydrologic quantity (e.g., runoff or other flows) plotted as ordinates vs. time on the abscissa. Use of the mass curve technique includes a study of the storage effects on the stream flow conditions. The ordinate of any point in the Rippl diagram indicates total amount of water flowing past a given station during a certain time. Equation 4.2 mathematically represents the flow-mass curve.

$$V = \int_{t_1}^{t_2} Q_t dt \cong \sum_{t_1}^{t_2} Q_t \Delta t \qquad (4.2)$$

where:

- V = Volume of runoff, m^3
- Q = Discharge as a function of time t, m^3/s
- t = Time, s

The capacity of a reservoir needed to maintain a uniform flow in a stream may be estimated by drawing, on the mass curve, a draft line tangent to a point at the beginning of the critical period. The slope of the draft line signifies the uniform regulated discharge. The storage capacity, to meet the uniform demand, is represented by the maximum ordinate between the draft line and the curve. Other ordinates, between the draft line and the mass curve, represent the amount of draft on the reservoir corresponding to other times. Where the draft line intercepts the curve once more, the reservoir is again full, and supply to the reservoir exceeds demand, resulting in overflow (spillage) conditions. The maximum storage required to meet demand also should consider the local rate of evaporation.

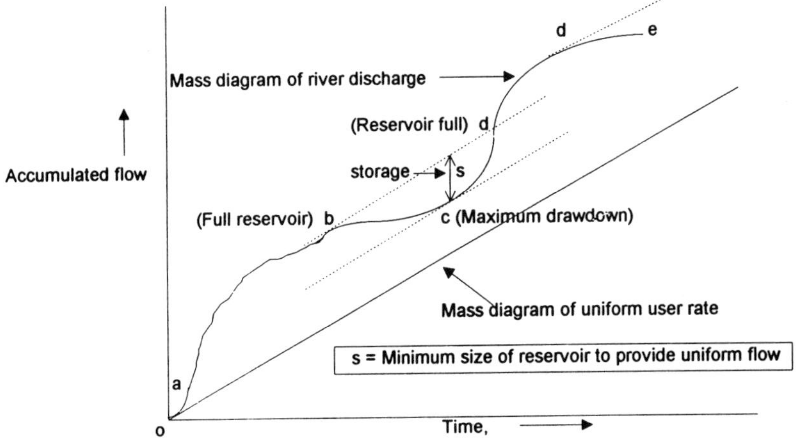

Figure 4.2
Mass diagram.

The mass diagram provides the following information regarding the condition of the reservoir (see Figure 4.2):[7]

1. From a to b, inflow rate exceeds use rate and reservoir is full and overflowing.
2. At b, inflow rate is equal to use rate and reservoir is full but not overflowing.
3. From b to c, use rate or demand exceeds inflow rate and the amount of depletion (drawdown) is increasing.
4. At c, use rate is equal to inflow rate and drawdown is at a maximum.
5. From c to d, inflow rate exceeds use rate and drawdown is decreasing.
6. At d, the reservoir is full.
7. From d to e, conditions are the same as from a to b.
8. The greatest vertical distance(s) between bd and bcd occurs at c. This is the storage required to maintain a use rate during the low flow period from b to c. The largest value, such as s for the entire period of record, is the minimum size of reservoir that would provide this uniform user rate or demand.

Example 4.1

1. Write a computer program that enables determination of the storage requirement to meet a uniform user water drawoff (with no losses from the reservoir). The reservoir is to supply a constant flow rate of Q (in m^3/s) and uses the water flowing from a particular catchment area. The monthly stream flow records are given in total cubic meters.
2. A reservoir is to supply a constant flow rate of 3.1 m^3/s using the amounts of water flowing from a particular catchment area. The monthly stream flow records, in one total cubic meter, are as follow:

Month	J	F	M	A	M	J	J	A	S	O	N	D
m^3 of water (*10^6)	12.8	6	1.2	4.5	9	18.5	46.9	25.1	15	10.5	7.5	7

   Assuming uniform water drawoff and no losses, determine the storage requirement to meet the aforementioned uniform drawoff.

3. Use the program developed in (1) above to verify the computations in section (2).

Solution

1. For solution to Example 4.1 (1), see following listing of Program 4.1.
2. Solution to Example 4.1 (2):
   - Given stream flow records.
   - Determine the cumulative total flow as tabulated below:
   - Draw mass curve for the data given by drawing the cumulative total flow vs. months (see Figure 4.3).

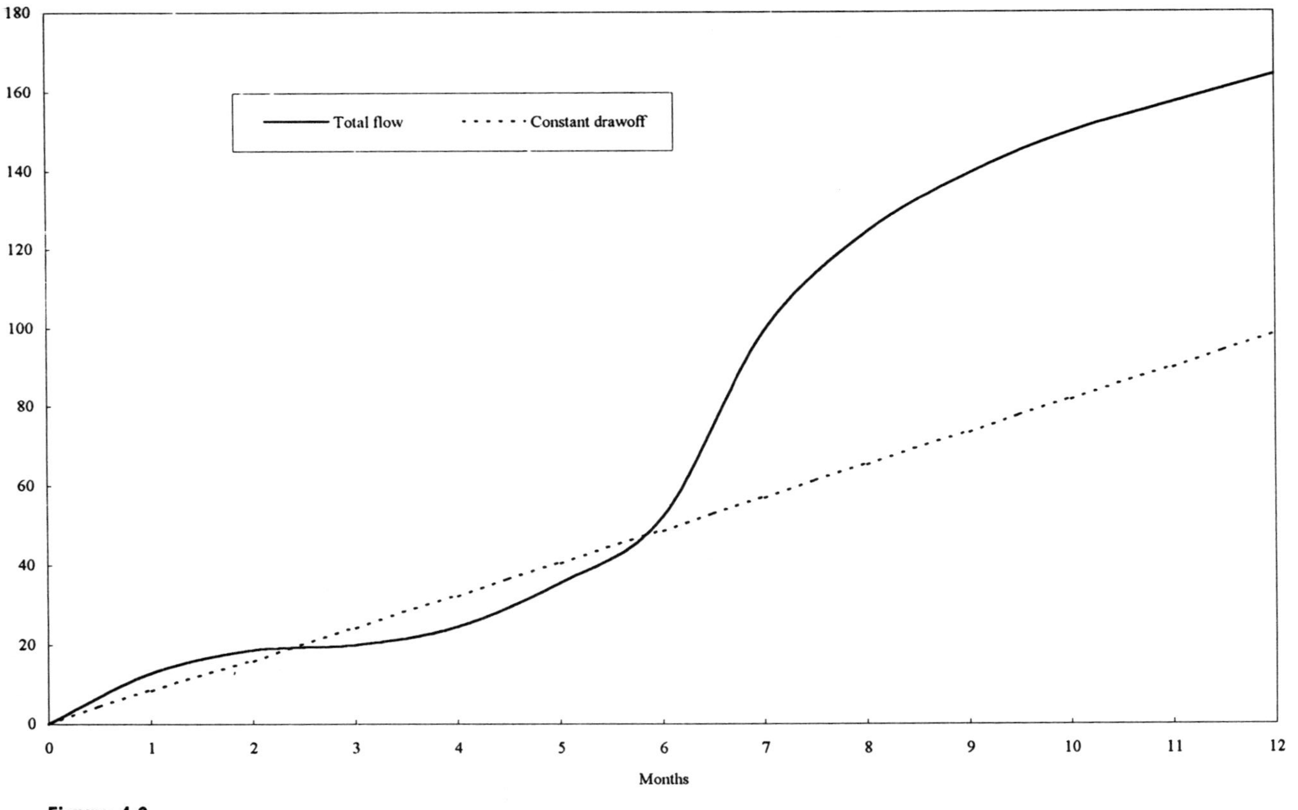

Figure 4.3
Mass diagram for Example 4.1.

Month	Monthly flow	Total flow
1	12.8	12.8
2	6	18.8
3	1.2	20
4	4.5	24.5
5	9	33.5
6	18.5	52
7	46.9	98.9
8	25.1	124
9	15	139
10	10.5	149.5
11	7.5	157
12	7	164

- Determine annual constant draw off (for the month of December) = 3.1 $m^3/s*3600$ s/hr*24 h/d*365 d/yr = $97.8*10^6$ m^3.
- Plot the uniform drawoff as a straight line from the origin to the point of 97.8 Mm^3 on the mass curve. Draw a parallel line from the point where reservoir is full.
- From the mass curve, determine the storage-minimum reservoir required to give demand = 14×10^6 m^3.

Listing of Program 4.1

```
DECLARE SUB BOX (r1%, c1%, r2%, c2%)   'Found in Program 2.1
'*********************************************************************
'Program 4.1: reservoir capacity
'*********************************************************************
DIM ys(12), yd(12), ysd(12), x(12), dif(12)
DATA 12.8,6,1.2,4.5,9,18.5,46.9,25.1,15,10.5,7.5,7
F$ = "####    ##.##^^^^   ##.##^^^^   ##.##^^^^    ##.##^^^^"
CALL BOX(1, 1, 20, 80)

LOCATE 2, 15: PRINT "Program 4.1: reservoir capacity"
LOCATE 3, 10: PRINT "1 Use the data in Example 4.1"
LOCATE 4, 10: PRINT "2 Input new data"
DO
LOCATE 5, 3: INPUT "Select an option (1-2):"; opt
LOOP WHILE opt < 1 OR opt > 2
row = 5

'read built in data
IF opt = 1 THEN q = 3.1
ys(0) = 0: yd(0) = 0: x(0) = 0
IF opt = 2 THEN
CALL BOX(1, 1, 20, 80)
row = 3
LOCATE row, 3: INPUT "Enter the constant flow rate supplied (volume/s):"; q
END IF
Qt = q * 3600! * 24 * 365
```

```
m = Qt / 12!                    'slope of demand curve
FOR i = 1 TO 12
x(i) = i
IF opt = 1 THEN
     READ supply
     supply = supply * 10 ^ 6
END IF
IF opt = 2 THEN
     LOCATE row + i, 3
     PRINT "Enter flow for month "; STR$(i); "=";
     INPUT supply
END IF
ys(i) = ys(i - 1) + supply         'supply
yd(i) = m * i                   'demand
dif(i) = ys(i) - yd(i)
NEXT i

max = 0
mon = 0
FOR i = 1 TO 12
IF dif(i) < 0 THEN GOTO jump
IF dif(i) > max THEN
     max = dif(i)
     mon = i
END IF
NEXT i
jump:

PRINT "slope max at:"; mon; "   ="; max
'parallel line to demand curve
FOR i = 0 TO 12
ysd(i) = yd(i) + max
dif(i) = ys(i) - ysd(i)
'PRINT USING F$; i; ys(i); yd(i); ysd(i); dif(i)
NEXT i

y1 = mon
FOR i = mon + 2 TO 12
IF SGN(dif(i)) <> SGN(dif(i - 1)) THEN GOTO jump2
NEXT i
jump2:
p = i - 1
q = i
maxd = 0
FOR i = mon TO p
IF ABS(dif(i)) > maxd THEN maxd = ABS(dif(i))
NEXT i
SCREEN 12
maxy = ys(12)
IF yd(12) > maxy THEN maxy = yd(12)
miny = -maxy / 15
VIEW (0, 0)-(639, 479), 1, 1
WINDOW (-3, miny)-(12, maxy)
LINE (0, 0)-(12, maxy), 15, B
PSET (0, 0)
FOR x = 0 TO 12
y = ys(x)
LINE -(x, y), 4
```

```
NEXT x
PSET (0, 0)
FOR x = 0 TO 12
y = yd(x)
LINE -(x, y), 15
NEXT x
PSET (0, 0)
FOR x = 0 TO 12
y = ysd(x)
LINE -(x, y), 15
NEXT x


LOCATE 5, 20
PRINT "Capacity="; maxd; "  m3"

DO
LOOP WHILE INKEY$ = ""
```

4.2 Water Distribution

4.2.1 Introduction

Systems used for distribution of water for public water supplies are networks of pipes within networks of streets. The distribution system is designed to serve each individual property, household, commercial establishment, public building, or industrial factory, etc. Flow type and character depend on interrelated factors that include street plan, topography of land, location of supply works, and service storage. The analysis of a water distribution system includes quantifying flows and head losses in the network of pipes and finding the resulting residual pressures.

4.2.2 Patterns of Pipelines in Water Networks

Basically there are two patters of pipelines:[1–5,8,9]

1. *Branching pattern with dead ends*: This system resembles the branching of a tree. The flow of the water continues in the same direction and a single pipe supplies water to a particular area. Advantages of the branching pattern include:
   - Simplicity of water distribution
   - Simplicity in design of pipe network
   - Usage of economical pipe dimensions

   On the other hand, disadvantages include:
   - Accumulation of sediments (due to stagnation of dead ends)
   - Tastes and odors (due to absence of regular flushing)
   - Water shortage (during breakdowns, repair, and maintenance)
   - Occurrence of insufficient water pressure (especially during extension of services)

2. *Grid pattern*: This system serves an area with a high demand for water using a large diameter pipe loop around the area. In this plan, all the pipes are interconnected together with no dead ends. Water can reach any point from more than one direction. The advantages of this system include:
   - Free movement of water in more than one direction
   - Less possibility of stagnation (comparable to the branching system)
   - Continuous water flow (no disconnection during repair or breakdown)
   - Little adverse effect on supply when there are large variations in consumption

   The disadvantages of this system include:
   - Complicated calculations when designing pipes
   - More pipes and fittings required

4.2.3 Viscous Flow in Closed Conduits

A closed conduit may be defined as a pipe or a duct through which fluid flows while completely filling its cross-section. Since the flowing fluid has no free surface, it can be either a liquid or a gas, its pressure may be above or below atmospheric pressure, and this pressure may vary from one cross-section to another along its length.[10] A closed conduit is commonly called a pipe if it has a round cross-section or a duct if it is not round, and they are designed to withstand a considerable pressure difference across their walls without undue distortion of their shape.[11] Fluid flow in a pipe may be laminar or transitional or turbulent (See Table 4.1). The criterion used to determine whether the flow is laminar or turbulent is the Reynolds number presented in Equation 4.3.

TABLE 4.1
Turbulent and Laminar Flow

Criterion	Value of Reynolds number
Laminar	< 2100
Transitional	2100 > Re > 4000
Turbulent	> 4000

$$Re = \rho v D/\mu \tag{4.3}$$

where:

Re = Reynolds number, dimensionless
ρ = Density of fluid, kg/m^3
v = Average velocity in the pipe, m/s
D = Pipe diameter, m
μ = Dynamic viscosity of fluid, $N{*}s/m^2$ or $kg/m{*}s$

4.2.3.1 Fundamental Equations for an Incompressible Flow

For steady, fully developed, incompressible flow (neglecting gravitational effects), the relationship of pressure difference and shear for an element of fluid in a horizontal pipe can be determined by using Equation 4.4 (see Figure 4.4).

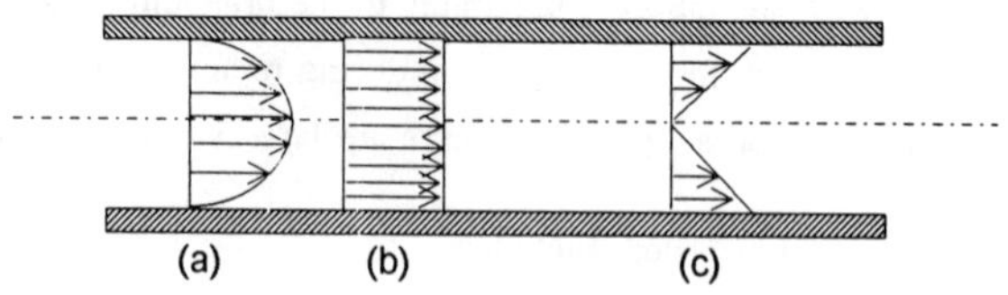

Key
(a) Laminar flow velocity profile
(b) Inviscid Flow velocity profile
(c) Shear stress distribution (laminar or turbulent)

Figure 4.4
Pipe flow velocity profile and shear stress distribution.

$$\Delta P = 4 * 1 * \tau_w / D \tag{4.4}$$

where:

ΔP = Pressure difference, Pa
1 = Length of circular cylindrical element, m
τ_w = Maximum shear at pipe wall (wall shear stress), N/m^2
D = Pipe diameter, m

Equation 4.4 indicates that a small shear stress can produce a large pressure difference if the pipe is relatively long (l/D » 1).

4.2.4 Flow in Pipes

4.2.4.1 Laminar Flow (Hagen-Poiseuille Flow)

For laminar flow of a Newtonian fluid within a horizontal pipe, the velocity and discharge may be obtained from the Poiseuille's law presented in Equations 4.5 and 4.6.

$$v = \Delta P * D^2 / 32\mu * 1 \tag{4.5}$$

where:

v = Average velocity, m/s
ΔP = Pressure difference, Pa
D = Diameter of pipe, m
l = Length, m
μ = Dynamic viscosity, N*s/m²

$$Q = \pi * D^4 * \Delta P /(128\mu * l) \quad (4.6)$$

where:

Q = Flow rate through the pipe, m³/s

To account for a pressure drop in nonhorizontal pipes, the Darcy-Weisbach equation may be used as presented in Equation 4.7

$$h_l = f * (l / D) * (V^2 / 2g) \quad (4.7)$$

where:

f = Friction factor (Darcy friction factor = 64 μ/(v*D*ρ) = 64/Re for laminar flow)
g = Gravitational acceleration, m/s²

Since $\Delta P = \gamma * h_l$ and from Equation 4.7,

$$h_l = 32\ \mu * l * v/(\gamma * D^2) = f*(l/D)*V^2/(2g) \quad (4.8)$$

where:

h_l = Head loss (drop in hydraulic grade line), m*N/N

Example 4.2

1. Write a computer program to determine viscosity of a fluid that flows in a pipe of diameter d (m) and of length l (m) at a rate of flow of Q (m³/s), with the head loss in the pipe given as h_f (m). Let the program indicate whether the flow is laminar or turbulent by checking its Reynolds number.
2. Given values of d = 5 mm, l = 20 cm, Q = 0.95 L/min, h_f = 0.8 m, determine ν, V, and Re.
3. Use the program developed in (1) to check the computations in (2).

Solution

1. For solution to Example 4.2 (1), see following listing of Program 4.2.
2. Solution to Example 4.2 (2):
   - Given d = 5 mm, l = 20 cm, Q = 0.95 L/min, h_f = 0.8 m.

- Find $Q = 0.95*10^{-3}$ m^3/15*60 s = $1.06*10^{-6}$ m^3/s.
- Determine kinematic viscosity as: $\nu = (\pi*D^4*g*h_l)/(128*l*Q) = (\pi*(5*10^{-3})^4*9.81*0.8)/(128*20*10^{-2}*1.06*10^{-6}) = 5.7*10^{-4}$ m^2/s.
- Compute velocity of flow as: $v = Q/A = 1.06*0^{-6}/(\pi*(5*10^{-3})^2/4) = 0.054$ m/s.
- Determine Reynolds number as: $Re = V*d/\nu = 0.054*5*10^{-3}/5.7*10^{-4} = 0.5$.
- Since Reynolds number is less than 2100, then flow is laminar.

Listing of Program 4.2

```
DECLARE SUB box (r1%, c1%, r2%, c2%)
'*******************************************************************
'Program 4.2:Viscous flow in pipes
'*******************************************************************
pi = 3.1415962#
g = 9.81
CALL box(1, 1, 20, 80)
LOCATE 2, 15
PRINT "Program 4.2:Viscous flow in pipes"
row = 5
LOCATE row, 5
INPUT "Enter diameter of pipe (mm)      ="; d
LOCATE row + 1, 5
INPUT "Enter Length of pipe (m)         ="; L
LOCATE row + 2, 5
INPUT "Enter the flow rate (m3/s)       ="; Q
LOCATE row + 3, 5
INPUT "Enter the head loss (m)          ="; hf
d = d / 1000    'convert to metre
nu = pi * d ^ 4 * g * hf / (128 * L * Q)
A = pi * d ^ 2 / 4
V = Q / A
Re = V * d / nu
LOCATE row + 5, 3
PRINT "Kinematic viscosity ="; nu: "  m2/s"
LOCATE row + 6, 3
PRINT "Renolds number ="; Re;
IF Re < 2100 THEN PRINT "  LAMINAR" ELSE PRINT "  TURBULENT"
DO
LOOP WHILE INKEY$ = ""
END
```

4.2.4.2 Turbulent Flow

Turbulent pipe flow actually is more likely to occur than laminar flow in a practical situation.[11,12] The main differences between laminar and turbulent flow occur due to variables such as velocity, pressure, shear stress, and temperature. Shear stress for turbulent flow can be determined by using Equation 4.9.

$$\tau = \xi(du_a / dy) \tag{4.9}$$

where:

ζ = Eddy viscosity, $N.s/m^2$

u_a = Average velocity, m/s

The velocity profile may be obtained by the empirical power-law velocity profile as presented in Equation 4.10

$$u_a / v_c = [1 - (r / R)]^{1/b} \tag{4.10}$$

where:

b = Constant (function of Reynolds number)

v_c = Center-line velocity, m/s

The one seventh power law velocity profile (b = 7) is often used as a reasonable approximation for many practical flows.[11,12]

Using the technique of dimensional analysis of pipe flow, the experimental data and semi-empirical formulas used in analysis of turbulent flow can be grouped in a dimensionless form. For steady, incompressible turbulent flow in a horizontal round pipe, the pressure drop and head loss can be determined by using Equations 4.11 and 4.12.

$$\Delta P_f = f * (1 / D) * (\rho * v^2 / 2) \tag{4.11}$$

where:

f = Friction factor = ϕ(Re, ε/D) = 64/Re for laminar flow (independent of ε/D) = φ(ε/D) for completely (wholly) turbulent flow

ε = Roughness of the pipe wall, m

D = Pipe diameter, m

l = Length of pipe, m

v = Average velocity, m/s

$$h_l = f * (1 / D) * (v^2 / 2g) \tag{4.12}$$

where:

h_l = Head loss, m

It is difficult to estimate the dependence of the friction factor on Reynolds number and relative roughness. Experimental findings have correlated data in terms of relative roughness of commercially available (new and clean) pipe materials and plotting this information in a diagram termed the Moody chart. The Moody chart is valid for steady, fully developed, incompressible pipe flows. The Colebrook formula gives the roughness for the entire nonlaminar range of the Moody chart, as indicated in Equation 4.13.[11,12]

$$1/f^{1/2} = -2\log[(\varepsilon/D)/3.7 + 2.51/(Re^* f^{1/2})] \qquad (4.13)$$

Example 4.3

1. Write a computer program that enables the computation of head loss due to friction and the power needed to maintain flow Q (m^3/s) through a circular horizontal pipe of diameter d (m) and length 1 (m). Use different pipe materials (different roughness coefficients) and different temperatures (different viscosities) in the program.
2. Water flows at the rate of 0.8 L/s through a horizontal, circular 500-m long pipe of diameter 6 cm at a temperature of 20°C. Assuming an absolute roughness of 0.08 mm for the pipe, determine the head loss due to friction using both Poiseuille and Darcy equations. Estimate the power required to maintain the flow in the pipe.
3. Verify your computations in (2) by using the program developed in (1).

Solution

1. For solution to Example 4.3 (1), see following listing of Program 4.3.
2. Solution to Example 4.3 (2):
   - Find the viscosity corresponding to the given temperature from Table A1 in the Appendix as: $\mu = 1.0087*10^{-3}$ N*s/m², $\rho = 998.203$ kg/m3, and $\gamma = 9792$ N/m³.
   - Determine Reynolds number as: $Re = \rho*V*D/\mu = \rho*Q*D/\mu*A = 998.203*0.08*10^{-3}*6*10^{-2}/1.0087*10^{-3}*(\pi*(6*10^{-2})^2/4) = 1680$. Therefore, flow is laminar since Re < 2000.
   - For laminar flow find the loss of head due to friction by using Poiseuille's equation as:

$$\begin{aligned} h_f &= 128*\mu*1*Q/\pi*d^4*\gamma \\ &= 128*1.0087*10^{-3}*500*0.08*10^{-3}/\pi*(6*10^{-2})^4*9792 \\ &= 0.013 \text{ m of water} \end{aligned}$$

   - Determine head loss using Darcy's equation (f = 64/Re) as: $h_f = f*1*v^2/2*D*g = 512*1*Q^2/D^5*g*Re*\pi^2 = 512*500*(0.08*10^{-3})^2/(6*10^{-2})^5*9.81*1680*\pi^2 = 0.013$ m.

Listing of Program 4.3

```
DECLARE FUNCTION log10! (x!)
DECLARE FUNCTION getf! (e!, D!, Re!)
DECLARE SUB box (r1%, c1%, r2%, c2%)   'Found in Program 2.1
'*****************************************************************
'PRINT "Program 4.3:Head loss due to friction in pipes"
'*****************************************************************
pi = 3.1415962#
```

```
g = 9.81
CALL box(1, 1, 20, 80)
LOCATE 2, 15
PRINT "Program 4.3:Head loss due to friction in pipes"
row = 5
LOCATE row, 5
INPUT "Enter diameter of pipe (m)       ="; D
LOCATE row + 1, 5
INPUT "Enter Length of pipe (m)         ="; L
LOCATE row + 2, 5
INPUT "Enter the flow rate (m3/s)       ="; Q
LOCATE row + 3, 5
INPUT "Enter viscosity (Ns/m2)          ="; mu
LOCATE row + 4, 5
INPUT "Enter density (kg/m3)            ="; Ro
gamma = Ro * g
A = pi * D ^ 2 / 4
V = Q / A
Re = Ro * V * D / mu
LOCATE row + 5, 3
PRINT "Renolds number ="; Re;

IF Re < 2000 THEN
     PRINT "LAMINAR"
     hf1 = 128 * mu * L * Q / (pi * D ^ 4 * gamma)  'Poiseuille eqn.
     f = 16 / Re
     hf2 = 4 * f * L * V ^ 2 / (2 * D * g)          'Darcy eqn.
     LOCATE row + 6, 3
     PRINT "hf="; hf1; "  POISEUILLE"; "      hf = "; hf2; "  DARCY"
     ELSE
     PRINT "TURBULENT"
     f = getf(e, D, Re)
     hf = f * L / D * V ^ 2 / (2 * g)
     LOCATE row + 6, 3
     PRINT "hf="; hf
END IF
DO
LOOP WHILE INKEY$ = ""
END

FUNCTION getf (e, D, Re)
FOR i = .000001 TO .001 STEP .000001
RHS = f ^ -.5
LHS = -2 * log10(e / (D * 3.7) + 2.51 / (Re * f ^ .5))
     IF ABS(RHS - LHS) < EPSILON THEN
       getf = i
       EXIT FUNCTION
     END IF
NEXT i
END FUNCTION

FUNCTION log10 (x)
  log10 = LOG(x) / LOG(10)
END FUNCTION
```

4.2.4.3 Minor Losses

Losses (pressure drops) are either major or minor in nature. Major losses are associated with friction in the straight portions of the pipes, while minor losses are due to losses from valves, bends, and tees. Minor head losses may be determined by specifying the loss coefficient as shown in Equation 4.14.

$$h_l = k_l * v^2 / 2g \tag{4.14}$$

where:

k_l = Loss coefficient (depends on geometry of the component and fluid properties) = φ(geometry, Re)

4.2.5 Pipes in Series

4.2.5.1 Equivalent-Velocity-Head Method

The equivalent-velocity-head method concerns a pipe composed of sections of different diameters. In pipes in series, the same fluid flows through all the pipes and the head losses are cumulative (see Figure 4.5). As such, continuity and energy equations yield Equations 4.15 and 4.16.

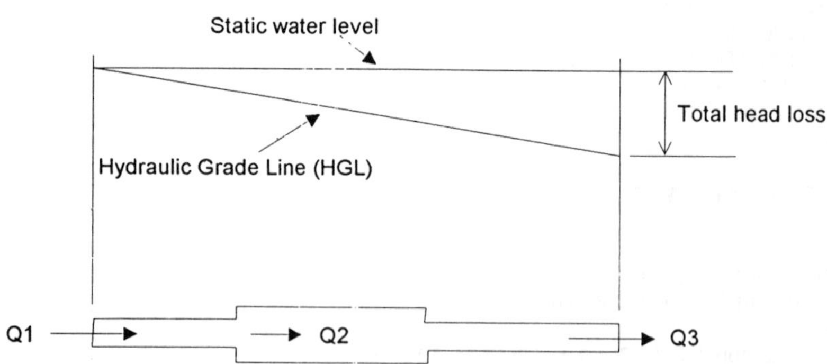

Figure 4.5
Pipes in series.

$$Q = Q_1 = Q_2 = Q_3 = \ldots = Q_N \tag{4.15}$$

$$h_l = h_{l1} + h_{l2} + h_{l3} + \ldots + h_{li} = \sum_{i=1}^{N} h_{li} \tag{4.16}$$

where:

h_{li} = Contribution of head loss from i-th section, m

N = Number of pipes

$$h_1 = (f_1 * l_1 * v_1^2 / 2g * D_1) + (f_2 * l_2 * v_2^2 / 2g * D_2) + \ldots + (f_i * L_i * v_i^2 / 2g * D_i) + k_1 * v^2 / 2g \tag{4.17}$$

where:

D_i = Diameter of i-th pipe, m

Example 4.4

1. Write a computer program that enables determination of flow from a large tank to which two pipes are connected in series, using the equivalent-velocity-head method. The two pipes have the following characteristics: their diameters are D_1 mm and D_2 mm, their lengths are l_1 m and l_2 m, and their friction factors are f_1 and f_2, respectively. The head loss is equal to h_l.
2. Two pipes of diameters 200 mm and 150 mm and of lengths 250 m and 150 m are connected in series from a large tank. For a head loss of 8 m and given friction factors of 0.018 and 0.021, respectively, determine the rate of flow from the tank to the second pipe using the equivalent-velocity-head method.

Solution

1. For solution to Example 4.4 (1) see following listing of Program 4.4.
2. Solution to Example 4.4 (2):
   - Given $d_1 = 200*10^{-3}$ m, l_1 = 250m, f_1 = 0.018 m, d2 = $150*10^{-3}$ m, l_2 = 150 m, h_f = 8 m, f_2 = 0.021 m.
   - Use continuity equation as: $v_2^2 = v_1^2*(D_1/D_2)^4 = v_1^2*(200/150)^4 = 3.16*v_1^2$
   - Use energy equation for the two pipes as: $h_f = (f_1*l_1*v_1^2/2g*D_1) + (f_2*l_2*v_2^2/2g*D_2)$ $8 = (0.018*250*v_1^2/2*9.81*200*10^{-3}) + (0.021*150*v_2^2/2*9.81*150*10^{-3}) = 1.15*v_1^2 + 1.07*v_2^2 = (1.15 + 1.07*3.16)v_1^2 = 4.53*v_1^2$. Thus, v_1 = 1.32 m/s.
   - Determine the flow rate as: $Q = v*A = 1.32*\pi*(200*10^{-3})^2/4) = 0.041$ m³/s.

Listing of Program 4.4

```
DECLARE SUB box (r1%, c1%, r2%, c2%)   'Found in Program 2.1
'*********************************************************************
'Program 4.4:Equivalent Velocity Head Method
'*********************************************************************
pi = 3.1415962#
g = 9.81

CALL box(1, 1, 20, 80)

LOCATE 2, 15
PRINT "Program 4.4:Equivalent Velocity Head Method"
```

```
LOCATE 3, 5
INPUT "Enter how many pipes connected in series ="; N
row = 5
S$ = SPACE$(10)
FOR i = 1 TO N
LOCATE 4, 3
PRINT "PIPE NO:"; STR$(i); ":"
LOCATE row, 48: PRINT S$
LOCATE row, 5
INPUT "Enter diameter of pipe     (m)       ="; d(i)
LOCATE row + 1, 48: PRINT S$
LOCATE row + 1, 5
INPUT "Enter Length of pipe       (m)       ="; l(i)
LOCATE row + 2, 48: PRINT S$
LOCATE row + 2, 5
INPUT "Enter the friction factor  (m)       ="; f(i)
NEXT i

LOCATE row + 5, 5
INPUT "You want to compute discharge for a given head loss? Y/N :"; R$

LOCATE row + 6, 5
IF UCASE$(R$) = "Y" THEN
     INPUT "Enter head loss H (m)                ="; H
     sum = f(1) * l(1) / (2 * g * d(1))
     FOR i = 2 TO N
     R = (d(1) / d(i)) ^ 4
     sum = sum + f(i) * l(i) * R / (2 * g * d(i))
     NEXT i

     V1 = (H / sum) ^ .5           'velocity in pipe 1
     Q = V1 * pi / 4 * d(1) ^ 2     'discharge
     LOCATE row + 9, 3
     PRINT "The flow rate for this head loss  ="; Q; " m3/s"
     ELSE                       'compute head loss in pipes
     INPUT "Enter discharge Q(m3/s)           ="; Q
     H = 0
     FOR i = 1 TO N
     v = 4 * Q / (pi * d(i) ^ 2)
     H = H + f(i) * l(i) * v ^ 2 / (2 * g * d(i))
     NEXT i
     LOCATE 9, 3
     PRINT "Total head loss       ="; H; " m"
END IF
DO
LOOP WHILE INKEY$ = ""
END
```

4.2.5.2 Equivalent-Length Method

In the equivalent-length method, pipes are replaced by equivalent lengths of a selected pipe size. Usually the selected pipe size is one which figures most prominently in the system. An equivalent length may be defined as a length, l_e, of pipe of a certain diameter, D_e, which for the same flow will give the same head loss as the pipe of length l and diameter D under consideration.[13] Thus, pipe friction and continuity equations yield Equation 4.18.

$$l_e = l*(f*v^2/2g*D)/(f_e*v_e^2/2g*D_e) = l*(f/f_e)*(D_e/D)^5 \quad (4.18)$$

where:

l_e = Equivalent length for pipe to be replaced, m

l = Length of pipe to be replaced, m

f = Friction factor of selected pipe, m

v = Velocity of flow through selected pipe, m/s

g = Gravitational acceleration, m/s^2

D = Diameter of pipe to be replaced, m

f_e = Friction factor of selected pipe, m

v_e = Velocity of flow through equivalent pipe, m/s

D_e = Diameter of selected pipe

The equivalent length method is of value when there are minor losses, such as bends, which are expressed in terms of equivalent lengths of pipe.

Example 4.5

1. Write a computer program to solve Example 4.4 using the equivalent length method.
2. Solve Example 4.4 using the equivalent length method.

Solution

1. For solution to Example 4.5 (1), see following listing of Program 4.5.
2. Solution to Example 4.5 (2):
   - Given $d_1 = 200*10^{-3}$ m, $l_1 = 250$ m, $f_1 = 0.018$ m, $d_2 = 150*10^{-3}$ m, $l_2 = 150$ m, $h_f = 8$ m, $f_2 = 0.021$ m.
   - Choose the 200-mm diameter pipe as the standard pipe. The equivalent length for the second pipe may be found as: $l_e = l*(f/f_e)*(D_e/D)^5 = 150*(0.021/0.018)*(200/150)^5 = 737$ m of 200-mm diameter.
   - Determine total effective length = 250 + 737 = 987 m of 200-mm pipe.
   - Use energy equation for the equivalent pipe as: $h_f = (f_1*l_e v_1^2/2g*D_1) = 0.018*987*v_1^2/2*9.81*(200/1000) = 8$, which yields $v_1 = 1.33$ m/s.
   - Determine flow rate as: $Q = v*A = 1.33*(\pi*(200*10^{-3})^2/4) = 0.042\ m^3/s$.

Listing of Program 4.5

```
DECLARE SUB box (r1%, c1%, r2%, c2%)  'Found in Program 2.1
'*****************************************************************
'Program 4.5:Equivalent Length Method
'*****************************************************************
pi = 3.1415962#
g = 9.81

CALL box(1, 1, 20, 80)

LOCATE 2, 15
PRINT "Program 4.5:Equivalent Length Method"

LOCATE 3, 5
INPUT "Enter how many pipes connected in series ="; N
row = 5
S$ = SPACE$(10)
MAXD = 0
FOR i = 1 TO N
LOCATE 4, 3
PRINT "PIPE NO:"; STR$(i); ":"
LOCATE row, 48: PRINT S$
LOCATE row, 5
INPUT "Enter diameter of pipe     (m)          ="; d(i)
IF MAXD < d(i) THEN
      MAXD = d(i)
      EQP = i
END IF
LOCATE row + 1, 48: PRINT S$
LOCATE row + 1, 5
INPUT "Enter Length of pipe       (m)          ="; l(i)
LOCATE row + 2, 48: PRINT S$
LOCATE row + 2, 5
INPUT "Enter the friction factor  (m)          ="; f(i)
NEXT i
'determine the equivalent length
sum = 0
FOR i = 1 TO N
sum = sum + l(i) * f(i) / f(EQP) * (MAXD / d(i)) ^ 5
NEXT i

LOCATE row + 5, 5
INPUT "You want to compute discharge for a given head loss? Y/N :"; R$

LOCATE row + 6, 5
IF UCASE$(R$) = "Y" THEN
      INPUT "Enter head loss H (m)                 ="; H
      V1 = (2 * g * MAXD * H / (sum * f(EQP))) ^ .5   'velocity in pipe
      Q = V1 * pi / 4 * d(1) ^ 2      'discharge
      LOCATE row + 9, 3
      PRINT "The flow rate for this head loss ="; Q; " m3/s"
      ELSE                       'compute head loss in pipes
      INPUT "Enter discharge Q(m3/s)              ="; Q
      v = 4 * Q / (pi * MAXD ^ 2)
      H = f(EQP) * sum * v ^ 2 / (2 * g * MAXD)
```

```
    LOCATE 9, 3
    PRINT "Total head loss      ="; H; " m"
END IF

DO
LOOP WHILE INKEY$ = ""

END
```

4.2.6 Pipes in Parallel (Pipe Network)

A pipe network is formed by a group of pipes which are interconnected to allow flow of a fluid from a particular inlet to a certain outlet through different directions. A loop is defined as a string of connected pipes. Loops may be classified as follows:[14]

1. *Closed loop*: Last pipe in the closed loop connected to the first pipe in it.
2. *Pseudo loop*: Contains a series of pipes that are not closed, but whose starting and ending nodes have fixed hydraulic grade line elevations, e.g., a loop connecting two reservoirs.
3. *Open loop*: Loop that does not end at its starting node, or a pseudo loop that does not have fixed-head end nodes.

Generally, the number of loops is equal to the number of pipes less the number of nodes less 1 as shown in Equation 4.19.

$$\text{Nloops} = \text{Npipes} - \text{Nnodes} - 1 \tag{4.19}$$

In parallel pipes, head losses are the same in any of the lines while discharges are cumulative. Equations 4.20 and 4.21 can be formulated from Figure 4.6 (size of pipes, fluid properties, and roughnesses are assumed to be known).

$$Q = Q_1 + Q_2 + Q_3 + \ldots + Q_i = \sum_{i=1}^{N} Q_i \tag{4.20}$$

where:

Q =Total flow through a node, m/s^3

Q_i = Flow through pipe number i, m/s^3

N = Number of pipes

$$h_1 = h_{11} + h_{12} + h_{13} + \ldots + h_{1i} = \sum_{i=1}^{N} h_{1i} \tag{4.21}$$

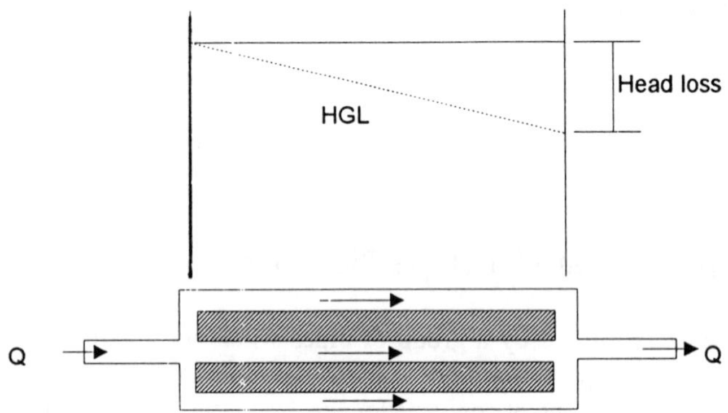

Figure 4.6
Pipes in parallel.

where:

h_l = Total head loss, m
h_{li} = Head loss at pipe number i, m

The cases that can be met are:

1. Unknown discharge for known elevation of hydraulic grade line
2. Unknown direction and quantity of flow and head loss in each line for known discharge

Analysis of a network of pipes by Bernoulli's and continuity equations would result in formulation of a large volume of simultaneous equations which are tedious to solve. Instead of this lengthy procedure, the method of successive approximations may be employed. This method assumes values for flow in each pipe or head losses at each node, and it checks that the chosen values satisfy the following requirements:[10]

1. The head loss between any two junctions is the same for all routes between these junctions.
2. The inflow to each junction equals the outflow from that junction.

Assumptions are corrected by successive approximations until they satisfy requirements within the required degree of accuracy.[10] The procedure used in the corrections is the Hardy Cross (relaxation) method. This method considers the network as a combination of simple loops (circuits). Each loop in the network is balanced to maintain compatible flow conditions in the system.

The procedure adopted for individual loops is summarized as follows:

1. Continuity equation is satisfied at all junctions. This indicates that at every junction the total quantity of water entering is equal to the algebraic sum of water leaving, including any water removed from or added to the system at that junction, i.e.,

$$\sum_{i=1}^{N} Q_i = 0 \tag{4.22}$$

where:

Q_i = Flow at note i, m^3/s

N = Number of nodes

Equation 4.22 is referred to as Kirchoff's node law.

2. Law of conservation of energy is valid. Energy loss is the same for all routes through which water passes. This indicates that around any closed loop of pipes the algebraic sum of the energy losses must be zero. Furthermore, the sum of the pipe head losses (in a series of pipes connecting two constant head sources) must equal the difference in head between the two sources, i.e., in every loop the algebraic sum of head losses through any chosen path is zero.

$$(\Sigma h)\ \text{loop} = 0 \tag{4.23}$$

Equation 4.23 is termed Kirchoff's loop law.

For solving a network problem the following methods can be used:

1. Balancing heads by correcting assumed flows
2. Balancing flows by correcting assumed heads

The Hardy Cross method analyzes the network by using the former concept of assuming initial flows. The imbalance in the energy equations is then determined, and flows in each loop are corrected accordingly. Corrections are carried out until convergence is achieved, when the largest correction is less than some tolerance level. Energy losses may be found using any one of the standard energy loss equations, neglecting similar minor losses and kinetic energy. All energy loss equations can be expressed as indicated in Equation 4.24.

$$h = k * Q^n \tag{4.24}$$

where:

h = Head loss, m

k = Resistance coefficient, numerical constant (a function of pipe geometry, diameter, length, material and age of conduit, and fluid viscosity)

Q = Volume rate of flow in the pipe, m^3/s

n = Constant exponent for all pipes. If the Darcy-Wesisbach equation is used ($h = f*l*v^2/2D*g$), n = 2; if Manning's formula is used ($v = r_H^{2/3} * j^{1/2}/n$), n = 2 (turbulent flow). If the Hazen-Williams formula is used ($v = 0.894*c*(r_H)^{0.63}*j^{0.54}$, n = 1.85; c is the Hazen-Williams roughness factor (depends on the material and age of the conduit). Generally, v = constant*$D^x(h_f/l)^y$, where x and y are constants depending upon the equation used.

The relationship of the discharges and corrections may be determined from Equation 4.25.

$$Q_2 = Q_1 + \Delta Q_1 \qquad (4.25)$$

where:

Q_2 = Second set of discharge
Q_1 = Initial set of assumed values of discharge
ΔQ_1 = First order correction to the discharge

The sum of the energy losses around a loop based on the first assumption is as indicated in Equation 4.26.

$$h_1 = \Sigma k * Q_1^{\,n} \qquad (4.26)$$

The sum of the energy losses when the first correction is applied yields Equation 4.27.

$$h_2 = \Sigma\{k(Q_1 + \Delta Q_1)^n\} \qquad (4.27)$$

Equation 4.28 is obtained by expanding Equation 4.27 into a series and neglecting small values.

$$h_2 = \Sigma\{k(Q_1^{\,n} + n * Q_1^{\,(n-1)} * \Delta Q_1)\} \qquad (4.28)$$

where:

h_2 = Algebraic sum of the losses around a loop (= 0)

Putting Equation 4.28 equal to zero and solving for Q gives Equation 4.29.

$$\Delta Q_1 = -\Sigma h / \{n\Sigma(h/Q)\} \qquad (4.29)$$

The negative sign illustrates that the positive (clockwise) discharges ought to be reduced, and the negative (anticlockwise, Tawaf-wise) discharges ought to be increased. The process is iterative and must be continued until the desired degree of accuracy is achieved.[10]

In summary, the Hardy Cross method involves the following steps:

1. Geometric configuration of the network is established.

2. Appropriate flows are assumed in each pipe (continuity must be satisfied at each junction, clockwise flows are positive, and they produce positive head losses).
3. For each elementary loop in the system:
    - Sign convention is established.
    - Head loss in each pipe is estimated.
    - Algebraic sum of the head loss around the loop is evaluated.
    - Correction factor is determined for each loop.
    - Flow adjustment is conducted for the loop.
4. The aforementioned step is applied to each circuit in the network, with flow correction established to each pipe, and the procedure is repeated to fulfill the required accuracy. It is to be noted that a common element in two different loops receives two different flow rate corrections.
5. For large complex networks, the analysis needs to be conducted by using a computer program.

In modeling a water supply network, the necessary data needed includes system maps or pipe and valve grid maps; geometric date, such as pipe diameters, lengths of pipes, materials, pipe junction elevations, pump characteristic curves, valve types, and sets; operational data, such as total water production, consumption and system losses, control valve set points, and reservoir levels.

Example 4.6

1. Write a computer program to compute the flow through each pipe of the network of the three pipes (A, B, and C) in parallel as shown in Figure 4.7A, given rate of flow to system Q, water temperature T, pipe diameters D_a, D_b, D_c; friction factor f_a, f_b, f_c; length l_a, l_b, l_c; pressure at point 1, with point 1 at a level of z_1; and point 2 at elevation of z_2. Let the program determine the pressure at point 2. Use Darcy-Weisbach equation.
2. Solve for flow through each pipe of the network discussed in (1) given values of D_a = 150 mm, D_b = 250 mm, D_c = 200 mm, f_a = 0.001, f_b = 0.0013, f_c = 0.0015, l_a = 600 m, l_b = 1000 m, l_c = 800 m, P_1 = 140 kPa, z_1 = 15 m, and z_2 = 9 m, Q = 36 m^3/min., and temperature = 20°C.
3. Use the program written in (1) to verify computations in (2).

Solution

1. For solution to Example 4.6 (1), see following listing of Program 4.6.
2. Solution to Example 4.6 (2):
    - Given D_a = 150 mm, D_b = 250 mm, D_c = 200 mm, f_a = 0.001, f_b = 0.0013, f_c = 0.0015, l_a = 600 m, l_b = 1000 m, l_c = 800 m, P_1 = 140 kPa, z_1 = 15 m, and z_2 = 9 m, Q = 36 m^3/min.
    - Use Darcy-Weisbach equation for the flow as: $h = f*l*v^2/2D*g = 8*f*l*Q^2/(\pi^2*D^5*g) = k*Q^2$.
    - Assume flow passing through each pipe (long, narrow pipe offers more resistance to flow), and take clockwise flow (anti Tawaf-wise) to be positive.

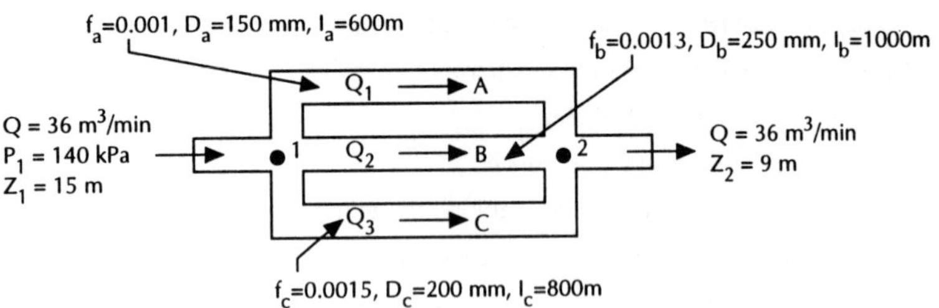

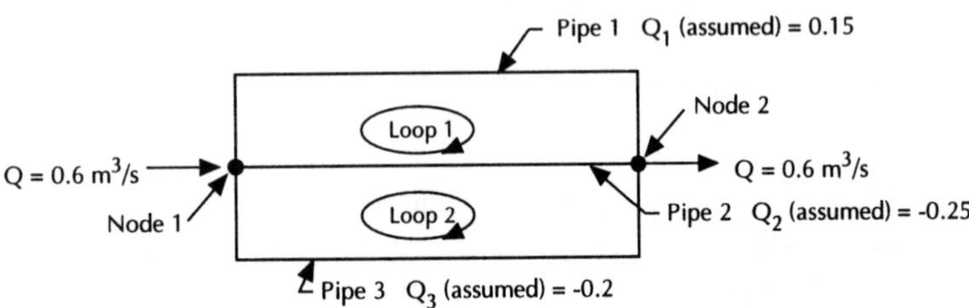

Figure 4.7
Example 4.6.

- Find the values of k (= $8*f*l/(\pi^2*D^5*g)$), h, and h/Q for each pipe in a loop.
- Compute the correction factor for each loop, $\Delta Q = -\Sigma h/\{2\Sigma(h/Q)\}$, and make adjustments of flow as shown in Table 4.2
- Repeat previous steps, taking in consideration the correction factor for each pipe in a loop and especially treating the common element 1B2 as show in Table 4.2.
- Adjustments are carried on until a desirable conversion is achieved.
- Values of flow in pipes are $Q_{1A2} = 0.123$ m³/s, $Q_{1B2} = 0.229$ m³/s, $Q_{1C2} = 0.178$ m³/s.
- Pressure at point 2 can be determined by using Bernoulli's equation between points 1 and 2 as: $(P_1/\gamma) + (v_1^2/2g) + z_1 = P_2/\gamma) + (v_2^2/2g) + z_2 = h_{f1B2}$.
- The head loss through pipe 1B2 can be found as: $h_{f1B2} = k*Q^2 = 109.9*(0.299)^2 =$ 9.8 m.
- For the same pipe: $v_1 = v_2$; therefore, $P_2 = P_1 + \gamma(z_2 - z_1) - \gamma h_{f1B2} = 140*10^3 +$ $998.2*(15 - 9) - 998.2*9.8 = 136.2$ kPa.

TABLE 4.2
First Iteration

Pipe	f	l	d	k	Q	h	h/Q	ΔQ
[1A2]	0.001	600	0.15	652.3289506	0.15	14.67740139	97.84934259	−0.031152405
[1B2]	0.0013	1000	0.25	109.9043816	−0.25	−6.86902385	+27.4760954	
Result						7.808377539	125.325438	
[1B2]	0.0013	1000	0.25	109.9043816	0.25	6.86902385	+27.4760954	0.030845955
[1C2]	0.0015	800	0.2	309.6014355	−0.2	−12.38405742	+61.92028711	
Result						−5.515033572	89.39638251	

Second Iteration

Pipe	f	l	d	k	Q	h	h/Q	ΔQ
[1A2]	0.001	600	0.15	652.3289506	0.118847595	9.21398393	77.52772709	0.006637748
[1B2]	0.0013	1000	0.25	109.9043816	−0.31199836	−10.69841963	+34.28998678	
Result						−1.484435698	111.8177139	
[1B2]	0.0013	1000	0.25	109.9043816	0.3119836	10.69841963	34.28998678	−0.010614809
[1C2]	0.0015	800	0.2	309.6014355	−1.69154045	−8.858654037	+52.37033518	
Result						1.839765591	86.66032196	

Third Iteration

Pipe	f	l	d	k	Q	h	h/Q	ΔQ
[1A2]	0.001	600	0.15	652.3289506	0.125485344	10.27194444	81.85772253	–0.00316841
[1B2]	0.0013	1000	0.25	109.9043816	–0.294745802	–9.547952797	+32.39385509	
Result						0.723991643	114.2515776	
[1B2]	0.0013	1000	0.25	109.9043816	0.294745802	9.547952797	32.39385509	0.002597301
[1C2]	0.0015	800	0.2	309.6014355	–0.179768854	–10.00534037	+55.65669539	
Result						–0.457387574	88.05055048	

Fourth Iteration

Pipe	f	l	d	k	Q	h	h/Q	ΔQ
[1A2]	0.001	600	0.15	652.3289506	0.122316934	9.759775429	79.79087707	0.00073294
[1B2]	0.0013	1000	0.25	109.9043816	–0.300511513	–9.9251536	+33.02753198	
Result						–0.165378172	112.818409	
[1B2]	0.0013	1000	0.25	109.9043816	0.300511513	9.9251536	33.02753198	–0.001176825
[1C2]	0.0015	800	0.2	309.6014355	–0.177171553	–9.718314548	+54.85256726	
Result						0.206839053	87.88009924	

TABLE 4.2 (continued)

Fifth Iteration

Pipe	f	l	d	k	Q	h	h/Q	ΔQ
[1A2]	0.001	600	0.15	652.3289506	0.123049873	9.877089656	80.26899483	–0.000343475
[1B2]	0.0013	1000	0.25	109.9043816	–0.298601748	–9.799404796	+32.81764044	
Result						0.07768486	113.0866353	
[1B2]	0.0013	1000	0.25	109.9043816	0.298601748	9.799404796	32.81764044	0.000275132
[1C2]	0.0015	800	0.2	309.6014355	–0.178348379	–9.847847101	+55.21691407	
Result						–0.048442305	88.03455451	

Sixth Iteration

Pipe	f	l	d	k	Q	h	h/Q	ΔQ
[1A2]	0.001	600	0.15	652.3289506	0.122706398	9.822025831	80.04493615	7.9799E-05
[1B2]	0.0013	1000	0.25	109.9043816	–0.299220355	–9.840049319	+32.8856281	
Result						–0.018023487	112.9305642	
[1B2]	0.0013	1000	0.25	109.9043816	0.299220355	9.840049319	32.8856281	–0.000128172
[1C2]	0.0015	800	0.2	309.6014355	–0.178073246	–9.81748662	+55.1317327	
Result						0.022562698	88.0173608	

Listing of Program 4.6

```
DECLARE SUB waitme ()
DECLARE FUNCTION checkrow! (X!)
DECLARE SUB box (r1%, c1%, r2%, c2%)   'Found in Program 2.1
'**************************************************************************
'Program 4.6: Hardy Cross method
'**************************************************************************
CALL box(1, 1, 20, 80)
row = 2
LOCATE row, 10
PRINT "Program 4.6: Analysis of Pipe Networks by Hardy Cross method"

NITER = 100           'limit number of iterations
EPSILON = .00001
iter = 0

LOCATE row + 1, 3
INPUT "Number of loops="; NLOOPS
LOCATE row + 1, 25
INPUT "Number of pipes="; NPIPES
LOCATE row + 1, 50
INPUT "Number of boundary pipes="; NBOUNDS
LOCATE row + 2, 3
INPUT "Are you using Metric units? Y/N :"; an$
IF UCASE$(an$) = "Y" THEN unit$ = "Metric" ELSE unit$ = "fps"

DIM NSIDE(NLOOPS), ISIDE(NLOOPS, 10), DONE(NPIPES), COMON(NPIPES)
DIM NBC(NBOUNDS), Qo(NPIPES), D(NPIPES), XL(NPIPES), CW(NPIPES), DH(NPIPES)
DIM ERRORQ(NLOOPS)

LOCATE row + 2, 3
PRINT "LOOP DATA:"
row = row + 2
FOR i = 1 TO NLOOPS
     row = checkrow(row)
     LOCATE row, 3
     PRINT "Loop No."; STR$(i); ":"
     row = row + 1
     LOCATE row, 4
     INPUT "Enter number of sides in the loop ="; NSIDE(i)
     FOR j = 1 TO NSIDE(i)
     row = checkrow(row)
     LOCATE row, 4
     PRINT "side "; STR$(j); "=";
     INPUT ISIDE(i, j)
     NEXT j
NEXT i

'Boundary pipes
FOR i = 1 TO NBOUNDS
row = checkrow(row)
LOCATE row, 5
INPUT "Enter boundary pipe:"; NBC(i)
NEXT i

' Pipes data
CALL box(1, 1, 20, 80)
t1$ = "Pipe  Diameter(m)   Length(m)   Roughness(m)  Initial Flow(m3/s)"
```

```
t2$ = "Pipe  Diameter(ft)  Length(ft)    Roughness(ft)  Initial Flow(cfs)"
IF unit$ = "Metric" THEN
      t$ = t1$
      hfactor = .08256
      ELSE
      t$ = t2$
      hfactor = .025153
END IF
row = 2
FOR i = 1 TO NPIPES
COMON(i) = 1
DONE(i) = 0
row = checkrow(row)
IF row = 3 THEN
      LOCATE row, 3
      PRINT t$
      row = row + 1
END IF
LOCATE row, 2: PRINT i
LOCATE row, 10: INPUT D(i)
LOCATE row, 23: INPUT XL(i)
LOCATE row, 38: INPUT CW(i)
LOCATE row, 53: INPUT Qo(i)
NEXT i


'Processing
FOR i = 1 TO NBOUNDS
k = NBC(i)
COMON(k) = 0
NEXT i

start:                'Iteration loop
iter = iter + 1
FOR iloop = 1 TO NLOOPS
sum1 = 0!
sum2 = 0!
FOR ISIDE = 1 TO NSIDE(iloop)
ipipe = ISIDE(iloop, ISIDE)
IF (DONE(ipipe) = 1) AND (COMON(ipipe) = 1) THEN Qo(ipipe) = -Qo(ipipe)
Q = Qo(ipipe)
Q1 = ABS(Q)
r = 1!
IF Q < 0! THEN r = -1!
h = r * hfactor * CW(ipipe) * XL(ipipe) * Q1 ^ 2 / D(ipipe) ^ 5
DH(ipipe) = h
hoq = h / Q
sum1 = sum1 + h
sum2 = sum2 + hoq
NEXT ISIDE
dq = -sum1 / (2 * sum2)
FOR i = 1 TO NSIDE(iloop)
ipipe = ISIDE(iloop, i)
Qo(ipipe) = Qo(ipipe) + dq
DONE(ipipe) = 1
NEXT i
ERRORQ(iloop) = dq
NEXT iloop
```

```
z = 0!
FOR i = 1 TO NLOOPS
IF ABS(ERRORQ(i) > ABS(z)) THEN z = ERRORQ(i)
NEXT i

IF iter > NITER THEN GOTO finish
IF ABS(z) > EPSILON THEN GOTO start

finish:
CALL box(1, 1, 20, 80)
LOCATE 2, 3: PRINT "Final Results:"
LOCATE 3, 3
t$ = "PIPE NO.   Head Loss       Q"
PRINT t$
F$ = "  ###      ##.##^^^^    ##.##^^^^"
row = 4
FOR i = 1 TO NPIPES
row = row + 1
IF row > 19 THEN
  CALL box(1, 1, 20, 80)
  LOCATE 3, 3
  PRINT t$
  row = 4
END IF
LOCATE row, 3
PRINT USING F$; i; DH(i); Qo(i)
NEXT i
LOCATE row + 1, 3
PRINT "TOTAL ITERATIONS USED ="; iter
waitme
END

FUNCTION checkrow (X)
     X = X + 1
     IF X >= 18 THEN
       CALL box(1, 1, 20, 80)
       X = 3
       LOCATE 3, 3
     END IF
     checkrow = X
END FUNCTION

SUB waitme
DO
LOOP WHILE INKEY$ = ""
END SUB
```

Using Program 4.6

Using the data of Example 4.6, the nodes are numbered 1 and 2, the loops are numbered 1 and 2, and the pipes are numbered 1 and 2 as shown in Figure 4.7B. The input data is then arranged as follows:

Number of loops = 2

Number of pipes = 3

Number of boundary nodes = 2

Loop 1: Number of sides in the loop = 2, side 1 = 1, side 2 = 2

Loop 2: Number of sides in the loop = 2, side 1 = 2, side 2 = 3

Boundary pipe 1 = 1

Boundary pipe 2 = 3

Pipes data:

Pipe 1: Diameter = 0.15, length = 600, roughness = 0.001, initial discharge = 0.15

Pipe 2: Diameter = 0.25, length = 1000, roughness = 0.0013, initial discharge = –0.25

Pipe 3: Diameter = 0.2, length = 800, roughness = 0.0015, initial discharge = –0.2

The output of the program gives the same final results as obtained by manual computations.

4.2.6.1 Disadvantages of the Hardy Cross Method[15,18–22,24,25]

1. A lot of time and tedious work is exhausted in assuming initial flows.
2. Limitations in usage for large flows result in convergence problems.
3. Direction of flow is assumed incorrectly occasionally.
4. There are complications in the method for complex systems such as reservoirs, interior pumps, valves, etc.

4.2.6.2 Analysis of Pipe Networks by the Finite Element Method

The finite element method for the analysis of water distribution networks may be used with the aid of personal computers, which make tedious, iterative calculations more amenable to quick solutions.[15–18,24,25] Application of the finite element method to a structural problem demands the subdivision of the structure into a number of discrete elements. Each of these elements must satisfy three conditions:[15,24,25]

1. Equilibrium of forces
2. Compatibility of strains
3. Force-displacement relationship specified by the geometric and elastic properties of the discrete element

An equivalent set of conditions for a pipe network exists; hence, the ability to draw the analogy:[24,25]

1. The algebraic sum of the flows at any joint or node must be zero.
2. The value of the piezometric head at a joint or node is the same for all pipes connected to that joint.
3. The flow-head relationship (such as Darcy-Weisbach or Hazen-Williams) must be satisfied for each element or pipe.

For a direct application of the finite element method involving a matrix solution, a linear relationship is required to define the element or pipe (see Figure 4.8). Hence, there is a relationship of the form indicated in Equation 4.30.

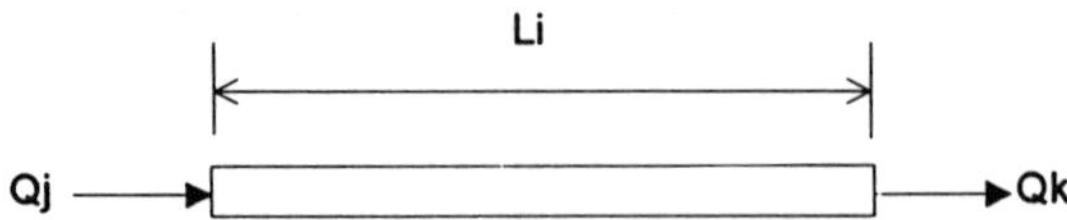

Figure 4.8
Pipe element. (From Abdel-Mageid, H., Hago, A., and Abdel-Magid, I.M., *J. Water Intl.*, 16 (2), 96–101, 1991. With permission.)

$$q = c * h \tag{4.30}$$

where:

q = Flow, m^3/s
h = Head loss, m
c = Hydraulic properties of the pipe (to be assumed)

The solution technique can be subdivided into three steps:[24,25]

1. An initial value of the pipe coefficient, c, is selected for each pipe and is then combined to yield the system matrix coefficient (C). The system matrix is then solved for the value of the piezometric head at each joint.
2. The individual pipe flows, q, are computed by means of Equation 4.30 using the difference between the determined piezometric heads. These flows are then substituted in the Darcy-Weisbach equation to calculate the pipe head losses. If the pipe head losses obtained from the Darcy Weisbach equation correspond to those obtained from the matrix solution, then the unique solution, satisfying both the Darcy-Weisbach equation and the linear Equation 4.30, has been found.
3. If there is a difference between the values of head loss calculated by the two methods, the values of c are changed to cause the problem to converge to a solution.

In analysis of a pipe network, Figure 4.9 is used[24,25] to show the application of the finite element method. Nodes and pipes are numbered for identification purposes. As a sign convention, any external flow at a joint will be positive when fluid is input and negative when actual consumption occurs. Applying the equilibrium flow-criteria at each node then:

$$
\begin{aligned}
Q_1 &= q_1 + q_2 \\
Q_2 &= q_1 + q_3 + q_4 + q_5 \\
Q_3 &= q_2 + q_3 + q_6 + q_8 \\
Q_4 &= q_4 + q_6 + q_7 + q_9 \\
Q_5 &= q_5 + q_7 + q_8 \\
Q_6 &= q_9
\end{aligned}
\tag{4.31}
$$

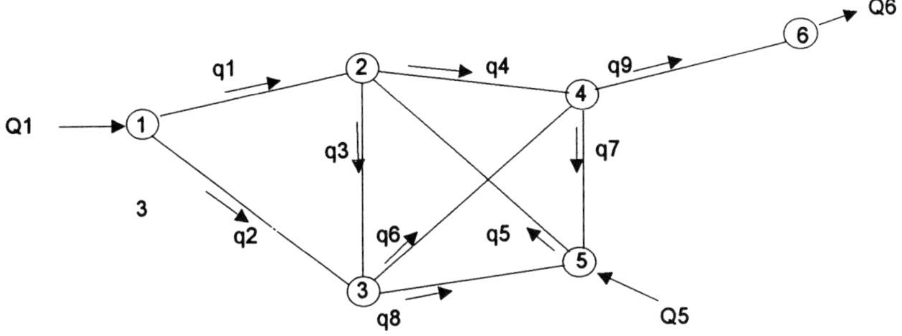

Figure 4.9
Pipe network. (From Abdel-Mageid, H., Hago, A., and Abdel-Magid, I.M., *J. Water Intl.*, 16 (2), 96–101, 1991. With permission.)

But the pipe flows in individual elements are given by:

$$
\begin{aligned}
q_1 &= c_1(H_1 - H_2) \\
q_2 &= c_2(H_1 - H_3) \\
q_3 &= c_3(H_2 - H_3) \\
q_4 &= c_4(H_2 - H_4) \\
q_5 &= c_5(H_5 - H_2) \\
q_6 &= c_6(H_3 - H_4) \\
q_7 &= c_7(H_4 - H_5) \\
q_8 &= c_8(H_3 - H_5) \\
q_9 &= c_9(H_4 - H_6)
\end{aligned}
\tag{4.32}
$$

Applying the sign convention on the flow in the element, it will be noted that, while the flow at one end, j, of the element will be input (i.e., positive), it is an output (i.e., negative) at the other end, k, of the element. It follows that, if the flow in element m, whose ends are nodes j and k, is given by $q_m = c_m (H_j - H_k)$ at end j, the same flow when considering end k i will be $q_m = c_m (H_k - H_j)$. Thus, substituting Equations 4.32 in Equations 4.31, observing the sign convention and writing in matrix form, we get Equation 4.33.

$$\begin{bmatrix} Q \\ Q \\ Q \\ Q \\ Q \\ Q \end{bmatrix} = \begin{bmatrix} (C + C_2) & -C_1 & -C_2 & 0 & 0 & 0 \\ -C_1 & (C_1 + C_3 + C_4 + C_5) & -C_3 & -C_4 & -C_5 & 0 \\ -C_2 & -C_3 & (C_2 + C_3 + C_6 + C_8) & -C_6 & -C_8 & 0 \\ 0 & -C_4 & -C_6 & (C_4 + C_6 + C_7 + C_8) & -C_7 & -C_9 \\ 0 & -C_5 & -C_8 & -C_7 & (C_5 + C_7 + C_8) & 0 \\ 0 & 0 & 0 & -C_9 & 0 & C_9 \end{bmatrix} = \begin{bmatrix} H_1 \\ H_2 \\ H_3 \\ H_4 \\ H_5 \\ H_6 \end{bmatrix} \qquad (4.33)$$

Equation 4.33 may be put in a compact form as shown in Equation 4.34.

$$Q = C * H \qquad (4.34)$$

where:

Q = Network consumption vector
C = Network characteristics matrix
H = Network head vector

The input data in the finite element procedure consists of the following:

1. The number of nodes, elements, known consumption nodes, known head nodes, and fluid properties
2. The element number, its diameter, its length, and the number of its first and second node connectivities
3. The node number and its known head or consumption.

The Darcy-Weisbach relationship may be used in the analysis of head loss, as presented in Equation 4.35.

$$h_f = [f * L / 2g * D * A^2)]q^2 = C_1 * q^2 \qquad (4.35)$$

where:

h_f = Head loss, m
f = Friction factor
L = Length of pipe, m

D = Diameter of pipe, m
A = Cross sectional area of pipe, m^2
g = Acceleration of gravity, m/s^2
q = Flow discharge, m^3/s

The friction factor may be determined from Equation 4.36.[18,23–25]

$$f = 0.094K^{0.225} + 0.53K + 88.0K^{0.44}\,Re^{-a} \tag{4.36}$$

where:

f = Friction factor.
Re = Reynolds number.

$$K = \varepsilon / D, \text{ where } \varepsilon = \text{Roughness coefficient} \tag{4.37}$$

a = Constant to be found from Equation 4.38.

$$a = 1.62K^{0.134} \tag{4.38}$$

The initial value of pipe coefficient, C_0, was chosen to correspond to Re = 200,000 in each pipe, a typical value for practical problems.[15] Accordingly, and since $Re = qD/(Av)$, the initial flow in the pipe becomes:

$$q_0 = 200,000v * A / D \tag{4.39}$$

where:

A = Area of the pipe, m^2
D = Diameter, m
v = Kinematic viscosity of the fluid, m^2/s

The value of head loss, h_0, corresponding to the flow, q_0, can be calculated from Equation 4.35. The pipe coefficient is then found from Equation 4.30 as indicated in Equation 4.40.

$$c_0 = q_0 / h_0 \tag{4.40}$$

This initial value of the pipe coefficient, c_0, for each pipe was then combined, according to the geometry of the network, to obtain the initial network characteristic matrix, $[C_1]_0$. Using a standard finite element procedure, the matrix assembly process to form the global matrix can be summarized. Starting with a zero matrix, the following operations are performed for each element:

1. Add coefficient $-C_1$ to position (k,k) and (j,j).

2. Add coefficient $-C_1$ to position (k,j) and (j,k), according to the connectivity table. Once all elements are considered, the [C] matrix is assembled.

Before the total system of equations can be solved, it is necessary to introduce proper boundary conditions for at least some of the nodes of the network. The two possible types of boundary conditions involve specifying either head or consumption for any given node. The introduction of the boundary conditions for prescribed heads can be implemented by performing the following steps:

1. Add the contribution of the prescribed unknown H_j to the vector of nodal consumption [Q].
2. Zero the jth column and jth row of the matrix [C}.
3. Make the jth coefficient of the vector nodal consumption equal to H_j.

A boundary condition of the second type, where the discharge rather than the head is prescribed, is handled by simply placing the value of the prescribed discharge or consumption in the proper position in the vector of nodal consumption. The solution of the simultaneous algebraic equations can be done using the Gauss elimination technique.[26] The solution of the system of equations could provide the values of the previously unknown nodal heads. With these, discharges could be computed for every element using Equation 4.30.

During the checking procedure, the flow, q_c, for each pipe calculated via Equation 4.30 and the matrix solution may be used to determine the head loss, h_c, from the Darcy-Weisbach equation.[15,24,25] The first step in the development of the program is to obtain the correction for the c value for each pipe by assuming that the point (h_c, q_c) is the unique solution. Thus, the correct linear relationship is defined by the straight line joining this point to the origin and is defined by Equation 4.41.

$$h = (q_c / h_c)q \tag{4.41}$$

The new value of c is then set equal to q_c/h_c. When all pipe coefficients are corrected in a similar way, the flow distribution obviously is altered, and this method can be overcorrection when the matrix is resolved. To dampen this overcorrection effect, an averaging technique is introduced. The corrected value of c is taken to be the mean of the c value defined by Equation 4.41 and the value of c used to obtain the matrix solution.[24,25]

Example 4.7

Write a computer program to predict flows in a network that consists of E number of elements and N number of nodes, given input and withdrawals at nodes, pipe characteristics for each pipe in the system, and water level in tank at a fixed node.

Listing of Program 4.7

```
DECLARE SUB GAUSS (NNODE, Z(), Q(), H())
DECLARE SUB WAITME ()
```

```
DECLARE FUNCTION checkrow! (X!)
DECLARE SUB BOX (X1%, Y1%, X2%, Y2%)      'Found in Program 2.1
'**************************************************************
'Program 4.7: FEM Analysis of Pipe Networks
'**************************************************************
CALL BOX(1, 1, 20, 80)
LOCATE 2, 15
PRINT "Program 4.7: FEM Analysis of Pipe Networks"
LOCATE 10, 3
INPUT "Is data stored in a file? Y/N :"; an$
an$ = UCASE$(an$)
LOCATE 11, 3
IF an$ = "Y" THEN
        INPUT "Enter name of data file :"; fin$
        OPEN fin$ FOR INPUT AS #1
        INPUT #1, NELEM, NNODE, met$
        INPUT #1, NVFIX, NPFLOW, ROUGHNESS
ELSE
        INPUT "Enter name of file to store data:"; fout$
        OPEN fout$ FOR OUTPUT AS #2
        CALL BOX(1, 1, 20, 80)
        LOCATE 2, 15
        PRINT "Program 4.7: FEM Analysis of Pipe Networks"
        LOCATE 3, 5
        INPUT "Enter number of elements              ="; NELEM
        LOCATE 4, 5
        INPUT "Enter number of nodes                 ="; NNODE
        LOCATE 5, 5
        INPUT "Are you using Metric units? Y/N       :"; met$
        LOCATE 6, 5
        INPUT "Enter number of nodes with known head  ="; NVFIX
        LOCATE 7, 5
        INPUT "Enter number of nodes with known flow  ="; NPFLOW
        LOCATE 8, 5
        INPUT "Enter roughness(Enter zero if unknown)  ="; ROUGHNESS
        PRINT #2, NELEM, NNODE, CHR$(34); met$; CHR$(34)
        PRINT #2, NVFIX, NPFLOW, ROUGHNESS
END IF

N = NELEM
IF NNODE > NELEM THEN N = NNODE
DIM DIAM(N), XLEN(N), AK(N), LNODS(N, 2), HEAD(N), H(N), Q(N), C(N), F(N)
DIM D(N), NODEH(N), Z(N, N), DELH(N), QO(N), CO(N), R(N), DH(N)

FOR I = 1 TO NNODE
H(I) = 0!
Q(I) = 0!
NEXT I

IF an$ = "Y" THEN
        FOR I = 1 TO NELEM
        INPUT #1, DIAM(I), XLEN(I), LNODS(I, 1), LNODS(I, 2)
        NEXT I
        FOR I = 1 TO NPFLOW
        INPUT #1, NODEQ, Q(NODEQ)
        NEXT I
        FOR I = 1 TO NVFIX
        INPUT #1, NODEH(I), HEAD(NODEH(I))
```

```
        NEXT I
        CLOSE #1
END IF

'Constant values
IF UCASE$(met$) = "Y" THEN unit$ = "Metric" ELSE unit$ = "fps"
Re = 200000!        'Assumed initial Reynolds number
EPSILON = 10 ^ -3    'iteration control factor
IF ROUGHNESS = 0! THEN
        ROUGHNESS = .004                    'in feets
        IF unit$ = "Metric" THEN ROUGHNESS = .0013    'in metres
END IF
PI = 22 / 7!
viscous = 1.059 * 10 ^ -5
G = 32.2                  'gravitational acceleration
CNVRT = 1!
IF unit$ = "Metric" THEN
        viscous = 9.76E-07
        G = 9.807
        CNVRT = 3600! * 24     'convert to m3/day
END IF

'get input from keyboard
IF an$ = "N" THEN
CALL BOX(1, 1, 20, 80)
row = 2
LOCATE row, 3
PRINT "Pipes data:"
u$ = "(ft)"
IF unit$ = "Metric" THEN u$ = "(m )"
t$ = "Element No.  Diameter" + u$ + "  Length" + u$ + "  End1  End2"
FOR I = 1 TO NELEM
row = checkrow(row)
IF row = 3 THEN
        LOCATE 3, 3
        PRINT t$
        row = row + 1
END IF
LOCATE row, 7: PRINT I
LOCATE row, 20: INPUT DIAM(I)
LOCATE row, 30: INPUT XLEN(I)
LOCATE row, 44: INPUT LNODS(I, 1)
LOCATE row, 50: INPUT LNODS(I, 2)
PRINT #2, DIAM(I), XLEN(I), LNODS(I, 1), LNODS(I, 2)
NEXT I

'Input external nodal flows
CALL BOX(1, 1, 20, 80)
LOCATE 2, 3: PRINT "External nodal flows:"
row = 2
LOCATE row, 3
t$ = "Node Number      External flow"
FOR I = 1 TO NPFLOW
row = checkrow(row)
IF row = 3 THEN
        LOCATE 3, 3
        PRINT t$
        row = row + 1
```

```
END IF
LOCATE row, 8: INPUT NODEQ
LOCATE row, 28: INPUT Q(NODEQ)
PRINT #2, NODEQ, Q(NODEQ)
NEXT I

'Input known head at nodes
CALL BOX(1, 1, 20, 80)
row = 2
LOCATE row, 3
t$ = "Node Number        Known Head"
FOR I = 1 TO NVFIX
row = checkrow(row)
IF row = 3 THEN
          LOCATE 3, 3
          PRINT t$
          row = row + 1
END IF
LOCATE row, 8: INPUT NODEH(I)
LOCATE row, 28: INPUT HEAD(NODEH(I))
PRINT #2, NODEH(I), HEAD(NODEH(I))
NEXT I
CLOSE #2
END IF        'end of input from keyboard
FOR I = 1 TO NNODE
QO(I) = Q(I)
NEXT I

'Calculate pipe coefficient C(I)
FOR I = 1 TO NELEM
AK(I) = ROUGHNESS / DIAM(I)
ff = -1.62 * AK(I) ^ .134
bk = 88! * AK(I) ^ .44 * Re ^ ff
F(I) = .094 * AK(I) ^ .225 + .53 * AK(I) + bk
Q1 = Re * viscous * PI / 4! * DIAM(I)
ct = 8! * F(I) * XLEN(I) / (PI ^ 2 * G * DIAM(I) ^ 5)
H1 = ct * Q1 ^ 2
CO(I) = Q1 / H1
C(I) = CO(I)
NEXT I

'Assembly of global matrix z(i,j)
ITER = 0
start:
ITER = ITER + 1
FOR I = 1 TO NELEM
Q(I) = QO(I) / CNVRT
FOR J = 1 TO NNODE
Z(I, J) = 0!
NEXT J
NEXT I

FOR I = 1 TO NELEM
L1 = LNODS(I, 1)
L2 = LNODS(I, 2)
Z(L1, L1) = Z(L1, L1) + C(I)
Z(L2, L2) = Z(L2, L2) + C(I)
Z(L1, L2) = Z(L1, L2) - C(I)
```

```
Z(L2, L1) = Z(L2, L1) - C(I)
NEXT I

'Impose boundary conditions
FOR INODE = 1 TO NVFIX
J = NODEH(INODE)
Z(J, J) = Z(J, J) * 1E+08
Q(J) = HEAD(J) * Z(J, J)
NEXT INODE

'Solve the equations by Gaussian elimination
CALL GAUSS(NNODE, Z(), Q(), H())
'Compute headlosses and discharges
ICHECKER = 0
FOR I = 1 TO NELEM
L1 = LNODS(I, 1)
L2 = LNODS(I, 2)
DELH(I) = ABS(H(L1) - H(L2))
D(I) = C(I) * DELH(I)
R(I) = 4! * D(I) / (PI * viscous * DIAM(I))
EE = -1.62 * AK(I) ^ .134
WW = 88! * AK(I) ^ .44 * R(I) ^ EE
F(I) = .094 * AK(I) ^ .225 + .53 * AK(I) + WW
DH(I) = XLEN(I) * D(I) ^ 2 / DIAM(I) ^ 5
DH(I) = DH(I) * 8! * F(I) / (G * PI ^ 2)
DIF = ABS(DH(I) - DELH(I))
IF DIF > EPSILON THEN
        ICHECKER = ICHECKER + 1
        C(I) = .5 * (C(I) + (D(I) / DH(I)))
END IF
NEXT I
IF ICHECKER > 0 THEN GOTO start

'Results
CALL BOX(1, 1, 20, 80)
row = 2
LOCATE row, 3
t$ = "Element No.   Discharge    HEAD LOSS"
FOR I = 1 TO NELEM
row = checkrow(row)
IF row = 3 THEN
        LOCATE 3, 3
        PRINT t$
        row = row + 1
END IF
LOCATE row, 8: PRINT I
LOCATE row, 17: PRINT D(I) * CNVRT
LOCATE row, 31: PRINT DELH(I)
NEXT I
WAITME
CALL BOX(1, 1, 20, 80)
row = 2
LOCATE row, 3
t$ = "Node Number     Head"
FOR I = 1 TO NNODE
row = checkrow(row)
IF row = 3 THEN
        LOCATE 3, 3
```

```
        PRINT t$
        row = row + 1
END IF
LOCATE row, 8: PRINT I
LOCATE row, 22: PRINT H(I)
NEXT I
PRINT
PRINT "TOTAL ITERATIONS="; ITER
WAITME
END

FUNCTION checkrow (X)
        X = X + 1
        IF X >= 18 THEN
        WAITME
          CALL BOX(1, 1, 20, 80)
          X = 3
          LOCATE 3, 3
        END IF
        checkrow = X
END FUNCTION

SUB GAUSS (NNODE, Z(), Q(), H())
'***************************************************************
'Solves a system of linear equations by Gaussian elimination
'***************************************************************
'Forward Elimination
FOR I = 1 TO NNODE - 1
ZZ = Z(I, I)
Q(I) = Q(I) / ZZ
FOR J = 1 TO NNODE
Z(I, J) = Z(I, J) / ZZ
NEXT J
FOR K = I + 1 TO NNODE
COF = -Z(K, I)
Q(K) = Q(K) + COF * Q(I)
FOR J = 1 TO NNODE
Z(K, J) = Z(K, J) + COF * Z(I, J)
NEXT J
NEXT K
NEXT I

'BACKSUBSTITUTION

H(NNODE) = Q(NNODE) / Z(NNODE, NNODE)
FOR I = NNODE - 1 TO 1 STEP -1
H(I) = Q(I)
FOR J = I + 1 TO NNODE
H(I) = H(I) - Z(I, J) * H(J)
NEXT J
NEXT I

END SUB

SUB WAITME
```

```
DO
LOOP WHILE INKEY$ = ""

END SUB
```

4.2.6.3 Advantages of the Finite Element Method

The major advantages of the finite element method over the Hardy Cross method include:[24,25]

1. The speed of convergence and the apparent lack of convergence problems
2. No need for an initial guess of flow distribution
3. Flexibility in applying boundary conditions
4. Ease of modifying and extending network without interrupting the whole system
5. Absence of artificial loops
6. The choice of flow-head loss relationships
7. Ease of input data
8. The ability to account for temperature effects
9. The unlimited network size (depending only on computer storage capacity)

The effect of pumps, boosters, pressure-reducing valves, nonreturn valves, etc., can easily be incorporated in the program if their actual head-flow relationships are known.

4.3 Structural Design of Storage Tanks

The structural design procedure involves two steps:

1. *Analysis of forces*: Finding the distribution of the vertical and horizontal bending moments and the axial forces in the tank.
2. *Design of sections*: Proportioning the sections and determining the necessary amounts of steel required.

4.3.1 Analysis of Forces

4.3.1.1 Circular Tanks

Assuming that the thickness of the walls of the tank is small in comparison to the overall dimensions of the tank, the tank can be analyzed as a cylindrical shell. Considering a tank with a uniform wall thickness under lateral fluid pressure, the governing equilibrium equation in terms of the lateral deflection of the tank w is given by the differential Equation 4.42.[27]

$$\frac{d^4w}{dx^4} + 4\beta^4 w = -\gamma(H - x)/D \quad (4.42)$$

where:

w = Lateral deflection of the tank
x = Distance along height of the tank measured from its base
H = Height of tank
γ = Density of liquid, KN/m^3
β = A constant given by Equation 4.43.

$$\beta^4 = 3(1 - v^2)/(a^2t^2) \quad (4.43)$$

v = Poisson's ratio of the material of the tank
a = Radius of the tank
t = Thickness of the walls of the tank
D = Flexural rigidity of the tank, given by Equation 4.44.

$$D = Et^3/(12(1 - v^2)) \quad (4.44)$$

E = Young's modulus of the material of the tank, KN/m^2

For a tank with its walls fixed to the base, a solution to the differential Equation 4.42 is given by Equation 4.45.

$$w = e^{-\beta x}(C_1 \cos\beta x + C_2 \sin\beta x) - \frac{(H - x)a^2}{Et} \quad (4.45)$$

The constants C_1 and C_2 are obtained by applying the boundary conditions and are given by Equation 4.46.

$$C_1 = \gamma a^2 H/(Et) \quad (4.46)$$

$$C_2 = \gamma a^2(H - 1/\beta)/(Et) \quad (4.47)$$

Using the solution given by Equation 4.46, together with the strain displacement relationship and Hook's law, the bending moments in the walls of the tank can be found as shown in Equation 4.48.

$$M_x = -D\, d^2w/dx^2 = \gamma aHt[-\varsigma + (1 - 1/(\beta H))\theta]/(12(1 - v^2))^{0.5} \quad (4.48)$$

where:

M_x = Vertical moment on the wall per unit width, KNm

$$\varsigma = e''^{x} \sin\beta x \quad (4.49)$$

$$\theta = e^{-x} \cos\beta x \quad (4.50)$$

The horizontal moment M_y on the walls per unit width is given by Equation 4.51.

$$M_y = vM_x \quad (4.51)$$

where:

v = Poisson's ratio of the material of the tank

The axial thrust on the walls of the tank per unit width N_y, which acts in the horizontal direction, is given by Equation 4.52.

$$N_y = -Etw/a \quad (4.52)$$

The maximum shear force at the foot of the walls will be given by Equation 4.53:

$$Q_0 = -aHt(2\beta - 1/H)/(12(1 - v^2)) \quad (4.53)$$

Example 4.8

1. Write a computer program to compute the bending moments, axial thrust, and shearing forces in a circular tank of height H, radius R, and wall thickness t. The tank retains a fluid with density γ.
2. A cylindrical tank with radius 9 m and height of 8 m has a wall thickness of 0.35 m and retains a liquid with a density of 9.99 KN/m³. Assuming Poisson's ratio of 0.25, compute the bending moments, axial thrust, and shearing forces at mid-height and at the base of the walls.

Solution

1. For solution to Example 4.8 (1), see following listing of Program 4.8.
2. Using the above data with Program 4.8, the following results are obtained:
   - At a depth of 4m: Mx = 4.17 kNm/m, My = –1.04 kNm/m, N = 390.41 kN/m, Q = 5.39 KN/m.
   - At a depth of 8m: Mx = 62.2 kNm/m, My = 15.5 kNm/m, N = 0, Q = 100.15 KN/m.

Listing of Program 4.8

```
DECLARE SUB box (r1%, c1%, r2%, c2%)  'Found in Program 2.1
'*******************************************************************
'Program 4.8: Analysis of Circular Tanks
'*******************************************************************
```

```
CALL box(1, 1, 20, 80)
LOCATE 2, 15
PRINT "Program 4.8: Analysis of Circular Tanks"
row = 2
pi = 3.4156503#

LOCATE row + 1, 5
INPUT "Enter the height of tank              H (m) ="; H
LOCATE row + 2, 5
INPUT "Enter the thickness of wall           t (m) ="; t
LOCATE row + 3, 5
INPUT "Enter the diameter of the tank         Dt(m) ="; Dt
LOCATE row + 4, 5
LOCATE row + 5, 5
INPUT "Enter the density of the liquid       (KN/m3)="; gama
LOCATE row + 6, 5
INPUT "Enter Poissons ratio of the material of tank  ="; mu
row = 8
cycle:
LOCATE row + 1, 5
INPUT "Enter depth at which forces are required, x(m)="; y
x = H - y
R = Dt / 2
b = (3 * (1 - mu ^ 2)) ^ .25 / (R * t) ^ .5
bh = b * H
c0 = (12 * (1 - mu ^ 2)) ^ .5
N = gama * R * H * (1 - x / H - EXP(-b * x) * COS(b * x) - (1 - 1 / bh) * EXP(-b * x) * SIN(b * x))
M = gama * R * H * t * (-EXP(-b * x) * SIN(b * x) + (1 - 1 / bh) * EXP(-b * x) * COS(b * x)) / c0
c1 = 1 / bh
c2 = -(2 - 1 / bh)
V = gama * R * H * t / c0 * (b * EXP(-b * x)) * (c1 * SIN(b * x) + c2 * COS(b * x))
LOCATE row + 2, 5
PRINT "At a depth ="; y; "  m from the top of the tank:"
LOCATE row + 3, 5
PRINT "Vertical moment         Mx="; M; "  KNm/m"
LOCATE row + 4, 5
PRINT "Horizontal moment        My="; mu * M; "  KNm/m"
LOCATE row + 5, 5
PRINT "Axial ring force        N="; N; "  KN/m"
LOCATE row + 6, 5
PRINT "Shearing force          V="; V; "  KN/m"
LOCATE row + 8, 3
INPUT "Any other section? Y/N:"; R$
IF UCASE$(R$) = "Y" THEN
CALL box(1, 1, 20, 80)
row = 5
GOTO cycle
END IF
END
```

4.3.1.2 Rectangular Tanks

Rectangular tanks can be divided into two types: deep and shallow tanks.

Deep Tanks: Here the height of the walls H is greater than twice the lateral dimensions, i.e., $H > 2L_x$ and $H > 2L_y$, where L_x and L_y are the plan length and width of the tank. Such tanks will resist the applied forces primarily horizontally. The moments in the horizontal strips of the walls can be found using the equation of three moments as shown in Equations 4.54, 4.55, and 4.56.

At the center of the longer side L_x:

$$M_{x1} = (0.5a^2 + a - 1)PL_y^2 / 12 \tag{4.54}$$

At the center of the shorter side L_y:

$$M_{x2} = (0.5 + 5a - a^2)PL_y^2 / 12 \tag{4.55}$$

where:

M_{x1}, M_{x2} = Horizontal moments at the center of the long and the short sides of the tank, respectively

a = Sides ratio = L_x/L_y

P = Fluid pressure at any depth z from the top of the tank = γz

γ = Density of the retained material

z = Depth at which the moments and forces are calculated

The horizontal moment at the junction between the walls of the tank M_s will be opposite in direction to both M_{x1} and M_{x2} and will be given by Equation 4.56.

$$M_s = (1 - a - a^2)PL_y^2 \tag{4.56}$$

The moment on the vertical strips of the walls of the tank M_y per unit width of the walls can be calculated from Equations 4.57 and 4.58:

$$M^-_{ymax} = -\gamma HL_y^2 / 24 \tag{4.57}$$

and

$$M^+_{ymax} = +\gamma HL_y^2 / 12 \tag{4.58}$$

The axial force in the wall acts horizontally and is equal to the reaction on the other wall. For the side with length L_x, the axial force is as given by Equation 4.59:

$$N_{x1} = PL_y / 2 \quad (4.59)$$

and that on the side with length L_y is given by Equation 4.60:

$$N_{x2} = PL_x / 2 \quad (4.60)$$

where:

P = Pressure of the fluid at the depth under consideration

Shallow Tanks: Here the depth $H < 2L_x$ and $H < 2L_y$. In such a case, the tank walls act as two-dimensional two-way plates. The analysis of the walls in this case requires the solution of the equilibrium equation for plates, which is Equation 4.61.[27]

$$\partial^4 w / \partial x^4 + \partial^4 w / \partial x^2 \partial^2 + \partial^4 w / \partial y^4 = -q / D \quad (4.61)$$

where:

w = Lateral deflection of the tank walls
q = Lateral pressure on the walls
D = Flexural rigidity of the walls of the tank.

The usual procedure is to assume each plate to be completely fixed to its neighboring plate, while its top can be free or simply supported. The moments are then found by solving Equation 4.61. Various methods exist for the solution which include the approximate elasticity methods described by Timoshinko and Krieger[27] or the numerical approximate methods of the finite differences[27] or the finite elements.[28] The moments found in this way at the edges of the plates will not be in equilibrium and can be distributed using the stiffness of the mating walls in the normal way.[29] Details of the method of solution for Equation 4.61 using the Finite Difference method are given in Chapter 5. The method is suitable for both shallow and deep tanks (see Program 5.5).

Example 4.9

1. Write a computer program to analyze a rectangular tank, given its plan dimensions L_x and L_y, the wall thickness t, its depth H, and the density of the retained material γ.
2. Use the program developed in (1) above to analyze a rectangular tank of plan dimensions 3 m × 2 m with a depth of 8 m and a wall thickness of 0.2 m, at a depth of 4 m and at a depth of 8 m. The material retained has a density of 9.81 KN/m^3. Take Poisson's ratio = 0.25

Solution

1. For the solution to Example 4.9 see the listing of the following Program 4.9.

2. Using the following data with the program: H = 8 m, t = 0.2 m, Lx = 3 m, Ly = 2 m, γ = 9.81KN/m^3, and Poisson's ratio = 0.25, we get the following results: at a depth of 4 m: Mx = 21.255 KNm/m, My = –3.27 KNm/m, Mc = –35.97 KNm/m, NX = 39.24, Ny = 58.86 KN/m at a depth of 8 m, Mx = 42.51 KNm/m, My = –6.54 KNm/m, Mc = –71.48 Knm/m = 78.48 KN/m, Ny = 117.72 KN/m.

Listing of Program 4.9

```
DECLARE SUB box (r1%, c1%, r2%, c2%)   'Found in Program 2.1
'*****************************************************************
'Program 4.9: Analysis of Rctangular Tanks
'*****************************************************************
CALL box(1, 1, 20, 80)
LOCATE 2, 15
PRINT "Program 4.8: Analysis of Rectangular Tanks"
ROW = 2
LOCATE ROW + 1, 5
INPUT "Enter the height of tank                h (m) ="; h
LOCATE ROW + 2, 5
INPUT "Enter the thickness of wall              t (m) ="; t
LOCATE ROW + 3, 5
INPUT "Enter the length of the tank           Lx  (m)="; Lx
LOCATE ROW + 4, 5
INPUT "Enter the width of the tank            Ly (m)="; Ly
LOCATE ROW + 5, 5
INPUT "Enter the density of the liquid       (KN/m3)="; gama
LOCATE ROW + 6, 5
INPUT "Enter Poissons ratio of the material of tank  ="; mu

cycle:
ROW = 8
LOCATE ROW + 1, 5
INPUT "Enter depth at which forces are required, x(m)="; y
a = Lx / Ly
P = gama * y
Mx1 = (.5 * a ^ 2 + a - 1) * P * Ly ^ 2 / 12
Mx2 = (.5 + a - a ^ 2) * P * Ly ^ 2 / 12
Mc = (1 - a - a ^ 2) * P * Ly ^ 2 / 12
Nx = P * Ly / 2
Ny = P * Lx / 2
LOCATE ROW + 2, 5
PRINT "At a depth ="; y; "  m from the top of the tank:"
LOCATE ROW + 3, 5
PRINT "Horizontal moment at the centre of the long side Mx1="; Mx1; "  KNm/m"
LOCATE ROW + 4, 5
PRINT "Horizontal moment at the centre of the short side Mx2="; Mx2; "  KNm/m"
LOCATE ROW + 5, 5
PRINT "Horizontal moment at the joints of the sides      Mc ="; Mc; "  KNm/m"
LOCATE ROW + 6, 5
PRINT "Horizontal thrust on the long sides              Nx="; Nx; "  KN/m"
LOCATE ROW + 7, 5
PRINT "Horizontal thrust on the short sides             Ny="; Ny; "  KN/m"
LOCATE ROW + 8, 3
INPUT "Any other section? Y/N:"; R$
IF UCASE$(R$) = "Y" THEN GOTO cycle
END
```

4.3.2 Design of the Section

For a section subjected to a bending moment M, the amount of steel needed can be computed from Equation 4.62:

$$Ast = M/(0.87f_y Z) \tag{4.62}$$

where:

Ast = Area of steel needed in mm^2
M = Applied bending moment, in Nmm.
f_y = Yield strength of the steel, in N/mm^2.
Z = The lever arm, mm

For a certain moment M, the lever arm Z can be obtained from Equation 4.63:

$$Z = d[0.5 + \sqrt{(0.25 - K/0.9)}] \tag{4.63}$$

where:

d = Effective depth of the section = total depth minus the cover to steel, in mm
K = $M/(f_{cu}*b*d^2)$, where
M = Bending moment, N*mm
b = Breadth of the section, mm
d = Effective depth, mm
f_{cu} = Concrete strength, N/mm^2

If the section is subject to an axial tensile force N in addition to the bending moment M (as is the case with the horizontal strips in tanks), the total area of steel can be obtained form Equation (4.64):

$$Ast = N/fst + [M - N*(d - 0.5*t)]/(fst*Z) \tag{4.64}$$

where:

N = Axial tensile force on the section, N
fst = The design steel stress, N/mm^2
M = Bending moment on the section, Nmm
d = Effective depth of the section, mm
t = Thickness of the section, mm
Z = Lever arm given by Equation 4.63, mm

To cover the effects of thermal and shrinkage, a check for cracks can be made. However, most codes of practice recommend that the area of the steel in any section must not be less than Astmin, given by Equation 4.65.

$$\text{Astmin} = F * b * t \tag{4.65}$$

where:

- b = Breadth of section, mm
- t = Thickness of walls
- F = 0.0024 for mild steel reinforcement (i.e., grade 250) = 0.0015 for high yield steel reinforcement (i.e., grade 460)

Example 4.10

1. Write a computer program to design the reinforcement for a section of a concrete tank for a certain moment M and axial tension N, given the wall thickness t, concrete strength (grade of concrete) fcu, steel yield strength fy.
2. Use the program developed in (1) to design a concrete tank section subject to a vertical bending moment Mx = 4.17 Knm/m, horizontal bending moment of 1.04 Knm/m, an axial (ring) tension of 390.41 KN/m, and a shearing force of 5.39 KN/m. Take fcu = 30 N/mm^2, fy = 250 N/mm^2 and a wall thickness of 0.2 m, assuming severe exposure conditions.

Solution

1. For solution to Example 4.10 (1), see following listing of Program 4.10.
2. Input for the program will be in the following form: fcu = 30, fy = 250, Mh = 1.04, Mv = 4.17, N = 390, V = 5.39, t = 0.2, severe exposure conditions. The following results will be obtained: cover to steel = 40 mm, horizontal steel = 1739.1 mm^2, vertical steel = 300 mm^2.

Listing of Program 4.10

```
DECLARE SUB box (r1%, c1%, r2%, c2%)    'Found in Program 2.1
'*****************************************************************
'Program 4.10: Structural design of tanks
'*****************************************************************
DIM d(4), A(4), spac(4), NB(4)
CALL box(1, 1, 20, 80)
row = 2
LOCATE row, 15
PRINT "Program 4.10: Structural design of tanks"
LOCATE row + 1, 5
INPUT "Enter grade of concrete      fcu(N/mm2)="; fcu
LOCATE row + 2, 5
INPUT "Enter the steel yield stress  fy(N/mm2)="; fy
LOCATE row + 3, 5
INPUT "Enter the horizontal moment  Mh (KNm/m)="; Mh
```

```
LOCATE row + 4, 5
INPUT "Enter the vertical moment    Mv (KNm/m)="; Mv
LOCATE row + 5, 5
INPUT "Enter the ring tension        N (KN/m)="; N
LOCATE row + 6, 5
INPUT "Enter the maximum shear       V (KN/m)="; V
DO
LOCATE row + 7, 5
INPUT "Enter the thickness of the wall   t(m)="; t
LOOP WHILE t < .2

'Determine the cover to steel reinforcement
GOSUB COVER

PI = 22 / 7
br = 1000                    '1m width
t = t * 1000                 'Convert to mm
d = t - CV - 10               'Effective depth, assuming 20mm bars
Astmin = .0015 * br * t       'Minimum area of steel for grade 250 steel
IF fy > 250 THEN Astmin = .0024 * br * t       'For grade 460 steel
Mh = ABS(Mh) * 10 ^ 6          'Convert to KNmm
Mv = ABS(Mv) * 10 ^ 6          'Convert to KNmm
N = ABS(N) * 10 ^ 3           'Convert to KN
V = ABS(V) * 10 ^ 3           'Convert to KN
fst = .87 * fy               'Design steel stress

RECYCLE:
     'Horizontal steel
     maxMu = .156 * fcu * br * d ^ 2
     DO
       d = d + 100
       t = t + 100
     LOOP WHILE Mh > maxMu
     K = Mh / (fcu * br * d ^ 2)
     z = d * (.5 + (.25 - K / .9) ^ .5)
     IF z > .94 * d THEN z = .94 * d
     Ast = N / fst
     Ast2 = (Mh - N * (d - t / 2)) / (fst * z)
     IF Ast2 > 0 THEN Ast = Ast + Ast2
     IF Ast < Astmin THEN Ast = Astmin

     'Vertical steel
     DO
  d = d + 100
  t = t + 100
LOOP WHILE Mv > maxMu
K = Mv / (fcu * br * d ^ 2)
z = d * (.5 + (.25 - K / .9) ^ .5)
IF z > .94 * d THEN z = .94 * d
Astv = Mv / (fst * z)
IF Astv < Astmin THEN Astv = Astmin

'Check for shear stress
Ra = 100 * Ast / (br * d)
IF d > 400 THEN d = 400
Vc = .79 * Ra ^ (1 / 3) * (400 / d) ^ .25 / 1.25
IF fu > 40 THEN fu = 40
IF fu > 25 THEN Vc = Vc * (fu / 25) ^ (1 / 3)
```

```
    stress = V / (br * d)
    IF stress > Vc THEN
        d = d * V / Vc * 1.1
        t = d + CV + 10
        GOTO RECYCLE
    END IF

    'Check for cracking at serviceability limit state
    N = N / 1.5                 'Service ring tension
    Mmax = Mh
    Asm = Ast
    dr = 1
    IF Mmax < Mv THEN
        Mmax = Mv
        Asm = Astv
        dr = 2
    END IF
    Mmax = Mmax / 1.5           'Service maximum moment
    Ec = 4500 * fcu ^ .5        'Youngs modulus of concrete
    Ec = .5 * Ec                'Service Youngs modulus
    Es = 2 * 10 ^ 5             'Youngs modulus for steel
    alpha = Es / Ec             'modular ratio
    r = alpha * Asm / (br * d)  'steel ratio
    x = d * r * ((1 + 2 / r) ^ .5 - 1)
    z = d - x / 3               'lever arm
    fs = Mmax / (Asm * z)       'Service stress in steel
    fsa = .58 * fy              'Allowable service stress
    IF fs > fsa THEN
        'Cracking is excessive, so increase area of steel
        fact = fs / fsa
        IF dr = 1 THEN Ast = Ast * fact ELSE Astv = Astv * fact
    END IF

LOCATE row + 9, 3
PRINT "The cover to reinforcement   ="; CV; " mm"
LOCATE row + 10, 3
PRINT "Area of horizontal steel    ="; Ast; "  sq.mm /m"
LOCATE row + 11, 3
PRINT "Area of vertical steel      ="; Astv; "  sq.mm /m"
LOCATE row + 12, 3
PRINT "Shear stress                ="; stress; " N/mm2"
LOCATE row + 13, 3
PRINT "Service steel stress        ="; fs; " N/mm2; O.K.: No Cracking"
LOCATE row + 14, 3
PRINT "Final depth of wall required ="; t; " mm"
DO
LOOP WHILE INKEY$=“”
END

COVER:
CALL box(1, 1, 20, 80)
row = 2
LOCATE row, 15
PRINT "Program 4.10: Structural design of tanks"
LOCATE row + 1, 5: PRINT "THE EXPOSURE CONDITION"
LOCATE row + 2, 5: PRINT "(1)  Mild"
LOCATE row + 3, 5: PRINT "(2)  Moderate"
LOCATE row + 4, 5: PRINT "(3)  Severe"
```

```
LOCATE row + 5, 5: PRINT "(4) Very severe"
LOCATE row + 6, 5: PRINT "(5) Extreme"
LOCATE row + 7, 5: INPUT "Select your choise"; EXC$

SELECT CASE EXC$
CASE "1"
    IF fcu <= 30 THEN CV = 25
    IF fcu >= 35 THEN CV = 20

CASE "2"
    IF fcu <= 35 THEN CV = 35
    IF fcu = 40 THEN CV = 30
    IF fcu = 45 THEN CV = 25
    IF fcu = 50 THEN CV = 20

CASE "3"
    IF fcu <= 40 THEN CV = 40
    IF fcu = 45 THEN CV = 30
    IF fcu = 50 THEN CV = 25

CASE "4"
    IF fcu <= 40 THEN CV = 50
    IF fcu = 45 THEN CV = 40
    IF fcu = 50 THEN CV = 30

CASE "5"
    IF fcu <= 45 THEN CV = 60
    IF fcu = 50 THEN CV = 50
END SELECT
RETURN
```

4.4 Homework Problems

4.4.1 Discussion Problems

1. Why is water storage needed?
2. Outline the classification used for water storage reservoirs.
3. What are the differences between an elevated water storage tower and a stand pipe?
4. What parameters govern the size of storage reservoirs?
5. Define the following terms: mass curve and water supply network.
6. Show how to estimate the required reservoir storage capacity in order to maintain uniform user flow.
7. What are the main types of pipeline patterns? Indicate advantages and disadvantages of each system.

8. What is the difference between a pipe and a duct?

9. Differentiate between laminar, transitional, and turbulent flow.

10. Define the following numbers: Reynolds number, Froude number, and Weber number. Show that each number is a dimensionless quantity. Show examples and the use of each of these numbers.

11. Differentiate between steady-state and nonsteady-state flow.

12. What are the differences between compressible and incompressible flows?

13. Define the following: shear wall, eddy viscosity, dimensional analysis, and network loop.

14. What are the advantages of the Moody chart? Indicate the range of the validity of the Moody chart.

15. What is the value of the equivalent-velocity-head method in analysis of pipe flow? Indicate differences between this method and the method of equivalent length.

16. Outline the classification of loops in pipe networks. What are the differences between them?

17. What are the disadvantages of analyzing networks by the Bernoulli and the continuity equations?

18. Define the method of successive approximations and finite element method. Indicate their main differences.

19. Indicate the assumptions made in the method of successive approximations. Show how to correct these assumptions.

20. Briefly describe the Hardy Cross method of network analysis. Indicate its disadvantages and its limitations.

21. What is the required data needed for appropriate modelling of a water supply network?

22. What are the steps involved in the structural design of a water storage tank?

23. How can you structurally differentiate between a cylindrical shell, deep rectangular tank, and a shallow rectangular tank?

24. Indicate the assumptions made in the structural design of a water storage tank.

25. Briefly describe the following methods for determining moments: approximate elasticity method, numerical approximate methods of finite differences, and finite elements method.

26. Why is it important to check a structural design for cracks?

4.4.2 Specific Mathematical Problems

1. Write a computer program to determine the pressure difference in a horizontal circular pipe, given its length, 1 (m); diameter, D (m); and the maximum shear that can develop at pipe wall, τ_w (N/m^2). Check the validity of your program, given D = 300 mm and τ_w = 0.4 kPa per unit length.
2. Write a short program to estimate minor losses in a piping system, h_l (m), given loss coefficient k_l, and velocity of flow, v (m/s). Use the relationship: $h_l = k_l*v^2/2g$. Check the validity of your program for the following data: v = 15 m/s, k_l = 0.5.
3. Write a computer program that enables the determination of the pressure drop in a pipe using Poiseuille's law ($\Delta P = 32\mu lv/D^2$). Let the program also find the pressure drop through the Darcy-Weisbach equation ($\Delta P = fl\rho v^2/2D$) with the friction factor found either for laminar flow (f = 64/Re) or for turbulent flow, f determined from Colebrook's equation: $1/(f)^{1/2} = -2\log[((\varepsilon/D)/3.7) + (2.51/(Re*(f)^{1/2}))]$ and given roughness of pipe wall, ε (m); pipe diameter, D (m); velocity of flow through pipe, v (m/s); and fluid properties (density, ρ (kg/m^3), dynamic viscosity, μ ($N*s/m^2$), kinematic viscosity, ν (m^2/s)). Let the program determine whether the flow is laminar, transition, or turbulent. Check the validity of your program for a gas flowing through a tube at v = 10 m/s, D = 5 mm, ε = 0.002 m, 1 = 0.2 m, ρ = 1.33 kg/m^3, μ = $2*10^{-5}$ $N*s/m^2$, and ν = $1.5*10^{-5}$ m^2/s.

References

1. Davis, M.L. and Cornwell, D.A., *Introduction to Environmental Engineering*, 2nd ed., Chemical Engineering Series, McGraw-Hill, New York, 1991.
2. Scott, J.S. and Smith, P.G., *Dictionary of Waste and Water Treatment*, Butterworths, London, 1980.
3. Henry, J.G. and Heinke, G.W., *Environmental Science and Engineering*, Prentice-Hall, Englewood Cliffs, NJ, 1989.
4. Viessman, W. and Hammer, M.J., *Water Supply and Pollution Control*, Harper & Row, New York, 1985.
5. Nathanson, J.A., *Basic Environmental Technology: Water Supply, Waste Disposal and Pollution Control*, John Wiley & Sons, New York, 1986.
6. Bruce, J.P. and Clark, R.H., *Introduction to Hydrolometeorology*, Pergamon Press, Oxford, 1966.
7. Wisler, C.O. and Brater, E.F., *Hydrology*, 2nd ed., John Wiley & Sons, New York, 1959.
8. Fair, G.M., Geyer, J.C., and Okun, D.A., *Water and Wastewater Engineering*, Vol. I and II, John Wiley & Sons, New York, 1966.
9. Abdel-Magid, I.M. and El-Hassan, B.M., *Water Supply in the Sudan*, Khartoum University Press, Sudan National Council for Research, Khartoum, 1986 (Arabic).
10. Douglas, J.F., Gasiorek, J.M., and Swaffield, J.A., *Fluid Mechanics*, Longman Scientific and Technical, New York, 1994.
11. Munson, B.R., Young, D.F., and Okiishi, T.H., *Fundamentals of Fluid Mechanics*, 2nd ed., John Wiley & Sons, New York, 1994.
12. Streeter, V.L. and Wylie, E.B., *Fluid Mechanics*, McGraw-Hill, New York, 1988.
13. Daugherty, R.L., Franzini, J.B., and Finnemore, E.J., *Fluid Mechanics with Engineering Applications*, McGraw-Hill, New York, 1985.

14. CECOMP, *Micro Hardy Cross Manual*, version 4.21, Granada Hills, CA, 1988.
15. Collins, A.G. and Johnson R.L., Finite element method for water distribution networks, *J. Am. Waterworks Assoc.*, 67(7), 385–389, 1975.
16. Gientke, F.J., Finite Element Solution for Flow in Noncircular Conduits, Proc. Am. Soc. Civil Engineers, *J. Hydraulics Div.*, 100 (HY3), 425–442, 1974.
17. Lam, C.F. and Wolla, M.L., Computer analysis of water distribution systems, Part I. Formulation of equations, *Proc. Am. Soc. Civil Engineers, J. Hydraulics Div.*, 98 (HY2), 335–344, 1972.
18. Wood, D.J. and Charles, C.O., Hydraulic network analysis using linear theory, *Proc. Am. Soc. Civil Engineers, J. Hydraulics Div.*, 98 (HY7), 1157–1170, 1972.
19. Shamir, U. and Howard, C.D., Water distribution systems analysis, *Proc. Am. Soc. Civil Engineers, J. Hydraulics Div.*, 94 (HY1), 219–234, 1968.
20. Epp, R. and Fowler, A.G., Efficient code for steady state flows in networks, *Proc. Am. Soc. Civil Engineers, J. Hydraulics Div.*, 96 (HY1), 43–56, 1970.
21. McCormich, M. and Bellamy, C.J., A computer program for the analysis of networks of pipes and pumps, *J. Inst. Eng.*, 38 (3), 51–58, 1968.
22. Liu, K.T., *The Numerical Analysis of Water Supply Networks by a Digital Computer*, Proc., IAHR 13th Congress, Kyoto, Japan, 31 Aug.–5 Sept. 1969, pp. 35–42.
23. Wood, D.J., Slurry flow in pipe networks, Proc. Am. Soc. Civil Engineers, *J. Hydraulics Div.*, 102 (HY1), 57–70, 1980.
24. Abdel-Magid, H., Analysis and Design of Networks Using Finite Element Method, M.Sc. thesis, Department of Civil Engineering, University of Khartoum, 1987.
25. Abdel-Magid, H., Hago, A., and Abdel-Magid, I.M., Analysis of Pipe Networks by the Finite Element Method, *J. Water Int.*, 16 (2), 96–101, 1991.
26. Henrici, P., *Elements of Numerical Analysis*, John Wiley & Sons, New York, 1967.
27. Timoshenko, S.P. and Woinowsky-Krieger, S., *Theory of Plates and Shells*, 28th ed., McGraw-Hill, New York, 1989.
28. British Standard BS8110, *Structural Use of Concrete, Part I*, British Standards Institution, 1985.
29. Kong, K. and Evans, R.H., *Reinforced and Prestressed Concrete*, Van Nostrand-Reinhold, New York, 1989.

Section III
Wastewater Engineering

Chapter 5

Wastewater Collection System

Contents

5.1 Introduction

The collection and disposal of wastewater from its sources (domestic, commercial, industrial, agricultureal, etc.) is of paramount importance to safeguarding public health and hygiene. Sewage ought to be collected and conveyed as soon as possible from the point of production to a treatment facility or an approved final disposal location with a minimum of cost. This chapter presents information dealing with the terminology used in the wastewater collection field, such as the classification of sewers (sanitary, combined, and storm), and the advantages and disadvantages of each of these systems. The methods used to estimate the design or peak wastewater flows including extraneous water entering the sewers due to infiltration and inflow are included. The Rational method to estimate peak flows from storm water runoff is covered, and a computer program to estimate the flow and the required storm sewer size is presented. Also discussed are the fundamentals of sewer design considering the following items: uniform flow, incompressible flow, Bernoulli's equation, application of the momentum equation, the continuity equation, and determination of the hydraulic radius. The most frequently used equations, formulas, and models used in the design of sewers such as Darcy-Weisbach, Chezy, Manning's, Kutter-Manning's, Hazen-Williams, Prandtl-Colebrook, Stricker, and Scimemi are documented.

This information lays the foundation for the development of computer programs for each of these models. The computer programs can then be used to make rapid and repeated calculations for design and evaluation of the various elements involved in sanitary sewer design. Equations to calculate the self-cleaning velocities in sewers are presented and a computer program is included that can be used for the same purpose. A summary of the information needed and the steps involved in the design, layout, construction, and installation of sanitary sewers are documented. The basic information needed includes topographical, geological, aerial, and developmental maps. The next step in the process is the sizing and required gradients for the sewers as well as the location for manholes. Other important elements in the design of sanitary sewers are included here.

Equations that can be applied in the evaluation of sulfide buildup in sewers flowing both full and partially full are presented, and a computer program is incorporated for rapid and easy determinations for the various variables involved. The last section of this chapter deals with the design, operation, and construction of waste disposal systems for rural areas. Septic tanks are the most often used systems in this situation; however, they are also the avenue of last resort. A computer program for septic tank sizing and layout that incorporates the structural design of septic tanks is displayed. A brief discussion of Imhoff tanks concludes this chapter.

5.2 Sewers and Sewerage Systems

A sewer is a pipe or conduit, generally closed but normally not flowing full, for conveying sewage. The functions of a sewer include:

- Collection of wastewater from its sources
- Transportation of wastewater to treatment works and points of final disposal
- Protection of public health and welfare, especially in areas of concentrated population and/or development

Socioeconomic factors of the system affect the selection of one of the following sewer systems:

1. *Separate sanitary sewer*: This system is used for collection and conveyance of wastes from domestic, commercial, and industrial establisments. In this system, surface waters or runoff are disposed of in a storm sewer system, while domestic sewage, commercial, and industrial wastes are carried in another set of sewers.
2. *Pseuo-separate (partially separate) system*: It is a combination of the separate and combined systems. One system receives the sewage and a part of the storm water (the runoff from the roofs of the buildings which have a sanitary connection to the system); the other system takes care of the remaining storm water.
3. *Combined sewer system*: In this system the same sewer serves to carry domestic, commercial, and industrial wastes, as well as surface and storm water flow.

Here the terms sewage and wastewater are often used interchangeably. In some cases, however, the term sewage indicates that human body wastes are present. The term wastewater is more inclusive and indicates not only the presence of discharges from individual homes but also commercial, institutional, and industrial discharges.

5.2.1 Advantages and Disadvantages of Sewer Systems

The following points outline the main advantages and disadvantages of sewer systems:

5.2.1.1 Separate System

Advantages of a separate sewer system include:

- System is economical since it is smaller in size.
- There is a lower risk of stream pollution, as storm overflows are excluded.
- Quantity of sewage to be treated is smaller.
- Pumping of sewage (when needed) is less costly than in the combined sewer system.

Disadvantages of the separate sewer system include:

- System requires flushing (expensive), since it is difficult to assure a self-cleansing velocity in a sewer (unless laid on a steep gradient or slope).
- Double house-plumbing (two sewers in a street) leads to obstruction of traffic during maintenance and repairs.
- Maintenance costs of the two systems may be greater than those for one.

5.2.1.2 Combined System

Advantages of the combined sewer system include:

- Rain water keeps sewage fresh, making it easier and more economical to treat.
- Sewage is diluted.
- Water provides automatic flushing.
- Easier cleaning since sewers are bigger in size.
- House plumbing is economical.

Disadvantages of the combined system:

- With much higher flows, a larger and more expensive wastewater treatment facility is needed.
- Operational problems occur when the capacity of the treatment facility is exceeded (lowered efficiency).
- Grit, sand, and other debris interfere with operation of the treatment facility.
- With increased flows during storms, the sewer lines can be surcharged. Sewer lines normally are not designed to handle flows under pressure.
- Washout of solids in the secondary treatment system can wipe out the efficiency of the treatment plant and result in pollution of the receiving rivers or streams.
- Oil, grease, rubber tire debris, and toxic trace metals can be present in the runoff from roads and streets.

5.2.1.3 Other Considerations

Misuse of a sanitary sewer system may create problems, such as explosions and creation of fire hazards, clogging (e.g., grease, bed load, other debris), physical damage (e.g., discharge of corrosive or abrasive wastes), overloading (e.g., improper connections), pollution of waterways, and interference with treatment (e.g., peak flows, unbiodegradable wastes). Investigations carried out during sewer design include:[1] Topographical surveys (e.g., maps, photographs), surface and subsurface investigations (e.g., runoff coefficient, permeability, depth of water table, infiltration), existing sewerage system (e.g., shortcomings, difficulties, percent loaded), water supply (e.g., water consumption, percentage returned to sewer), public services (e.g., layout of distribution networks, electricity, telephone, gas), existing structures and their elevations, development plan of the area, industries, population (e.g., maxima, minima, average and distribution of rainfall, river flow, sea level, current prevailing wind, temperature, evaporation), history of the town, political information (e.g., ordinances and laws affecting sewer connections regulations, rates, sewerage authority), financial information, and miscellaneous (e.g., tourism, reuse of effluent for irrigation). Figure 5.1 shows the various phases in a sanitary sewer project.

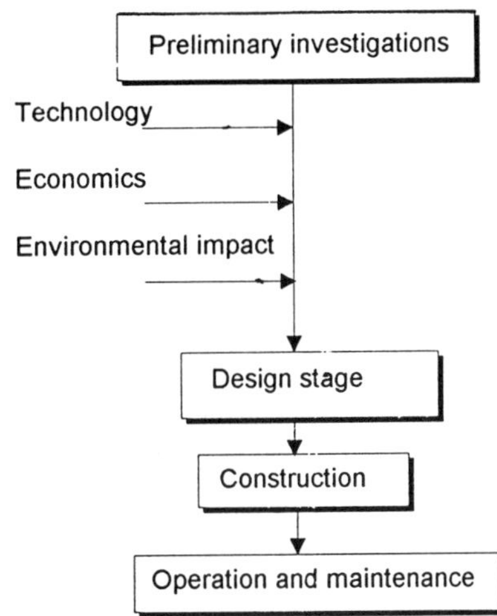

Figure 5.1
Phases of projects of a sanitary sewer.

5.2.2 Flow Rates of Sanitary Wastewater

The rate of flow of wastewater varies through the day. Relatively steady flow from industrial and commercial sources occurs mostly during the day. The design should be based on flows estimated at some future time (design period), usually taken to be between 25 and 50 years.[2]

When estimating the design sewage flows for a sewer, allowance must be made for extraneous water entering the sewer as infiltration and inflow. Infiltration is water other than wastewater that enters the sewerage system through cracked pipes, defective joints, and cracks in the walls of manholes. Inflow is also water other than wastewater entering the sewerage system from sources such as roof basement, yard, and foundation drains; from springs and swampy areas and perforated manhole covers; through cross connections between storm and sanitary sewers, cooling towers, street washwater, and surface runoff.

Velocities and hydraulic forces should not exceed maximum allowable flow rates as dictated by the sewer material. A minimum velocity ought to be maintained in the sewer to prevent settling of grit and suspended solids and accumulations of grease and slime on sewer walls. For most sanitary sewers, flows occur under the action of gravitational forces. Sanitary sewers should be designed for peak flow rates and for areas with a saturated density of people. Estimations of flow rates may be based on maximum water use (assume average sewage flow = average water consumption rates) anticipated through population density, or conducted in accordance with the number of structures or extent of the area served. The peak wastewater flow may be estimated from Equation 5.1.

$$Q_p = (2 \text{ to } 3)Q_a \tag{5.1}$$

where:

Q_p = Peak flow, m^3/s
Q_a = Average flow, m^3/s

5.2.3 Storm Water

The primary source of storm water flow is precipitation. Factors that may affect the production of storm water include intensity and duration of storms; distances traveled by water before reaching sewers; topography and slope of drainage area; or shape and size of drainage area. The Rational method (Lloyd Davies method) may be employed to estimate the rate of runoff. The method is valid for suburban areas up to 15 km^2.[1,3] The mathematical presentation of this method is provided in Equation 5.2.

$$Q_p = 0.278 * c * I * A \tag{5.2}$$

where:

Q = Peak flow rate at point, m^3/s
A = Drainage area upstream of point, km^2
I = Mean rainfall intensity for a duration equal to the time of concentration (the time required for the entire tributary area to contribute runoff from heavy rainfall to the flow at the point, i.e., flow time from the most remote part of the area to the point), mm/h
c = Coefficient of runoff, dimensionless ($0 < c < 1$, see Table 5.1)

In storm sewers, greater velocities are required than in sanitary sewers because of the heavy sand and grit which is washed into them. The minimum allowable velocity is 0.75 m/s, and 0.9 m/s is desirable. Because of abrasive character of the solids, excessively high velocities should be avoided, 2.4 m/s being considered the desirable upper limit.[5]

Example 5.1

1. Write a computer program to estimate the diameter of a sewer for a suburb of any area, A (km^2), given peak flow rate at point, Q (m^3/s), drainage area upstream of point, A (km^2), mean rainfall intensity, I (mm/h), and coefficient of runoff, c (as presented in Table 5.1).

TABLE 5.1
Typical Values of c for the Rational Method[3,4,9,17–19]

Area	c
City and commercial	0.70 – 0.95
Industrial	
Light areas	0.50 – 0.80
Heavy areas	0.60 – 0.90
Residential	
Single family	0.30 – 0.50
Multi-units, detached	0.40 – 0.60
Multi-units, attached	0.60 – 0.75
Suburban	0.25 – 0.40
Parks and undeveloped land	0.10 – 0.30
Pavement	
Asphaltic	0.70 – 0.95
Concrete	0.80 – 0.95
Brick	0.70 – 0.85
Playgrounds	0.20 – 0.35
Lawns	0.05 – 0.35
Watertight roofs	0.70 – 0.95
Woodland areas	0.01 – 0.20

2. A circular storm sewer is required to serve an area of 900 ha; the average ground slope is 1 in 500. Determine the size of the sewer at point of discharge knowing that runoff coefficient is 0.4, time of concentration is 1000 s, and n = 0.013. Take I = 750/(t + 10).
3. Use the computer program developed in (1) to support manual computations conducted in (2).

Solution

1. For solution to Example 5.1 (1), see the following listing of Program 5.1.
2. Solution to Example 5.1 (2):
   - Given $A = 900\ ha = 9\ km^2$, $j = 1/500 = 0.002$, $c = 0.4$, $t = 1000\ s = 1000/60 = 16.7\ min$.
   - Find rainfall intensity as: $I = 750/(16.7 + 10) = 28.1\ mm/h$.
   - Find $Q = 0.278*c*I*A = 0.278*0.4*28.1*9 = 28.1\ m^3/s = 1687.3\ m^3/min$.
   - The diameter of the sewer and flow rate through it may be determined from a nomograph corresponding to the specified slope.[1,3,5–8,10–12] Alternatively, the following procedure can be used: $v = (l/n)(D/4)^{2/3}*j^{1/2}$, and with $Q = (\pi/4)*D^2*v$, it gives: $D = (4^{5/3}*Q*n/\pi*j^{1/2})^{3/8} = (4^{5/3}*28.1*0.013/\pi*0.002^{1/2})^{3/8} = 3.4$ (take 3.5) m, which gives a discharge of: $Q = (\pi/4)*(3.4)^2*(3.4/4)^{2/3}*(0.002)^{1/2}/0.013 = 28\ m^3/s$.

Listing of Program 5.1

```
DECLARE SUB box (r1%, c1%, r2%, c2%)      'Found in Program 2.1
'*****************************************************************
'Program 5.1: Water Collection
'*****************************************************************
pi = 3.1415962#
g = 9.81

CALL box(1, 1, 20, 80)

LOCATE 2, 15
PRINT "Program 5.1: Water Collection"
row = 5
LOCATE row, 5
INPUT "Enter drainage area    A(Km2)          ="; A
LOCATE row + 1, 5
INPUT "Enter the value of n                   ="; n
LOCATE row + 2, 5
INPUT "Enter coefficient of run off c         ="; c
LOCATE row + 3, 5
INPUT "Enter average ground slope j           ="; j
LOCATE row + 4, 5
INPUT "Enter time of concentration    t(sec)  ="; t

t = t / 60             'convert to minutes
I = 750 / (t + 10)     'rainfall intensity
Q = .278 * c * I * A
D = (4 ^ (5 / 3) * Q * n / (pi * j ^ .5)) ^ (3 / 8)
Q = pi / 4 * D ^ 2 * (D / 4) ^ (2 / 3) * j ^ .5 / n
LOCATE row + 6, 3
PRINT "The required diameter of sewer      ="; D; " m"
LOCATE row + 7, 3
PRINT "This delivers a discharge of  Q     ="; Q; " m3/s"

DO
LOOP WHILE INKEY$ = ""
END
```

5.2.4 Hydraulics of Sewers

In the design of sanitary sewers the following assumptions are made:

- One-dimensional flow conditions (velocity of flow is uniform across each section of flow)
- Incompressible flow (fluid motion with negligible changes in density), where pressure flow totally fills a closed conduit, except for the possible occurrence of water hammer in pressure conduits.
- Steady-state flow (flow conditions at any point in fluid do not change with time) with constant rate of flow; actually flow is quasi-steady (varies significantly from hour to hour)
- Valid continuity principle (Q = A*v)

- Valid Bernoulli equation: Total energy (at a point in a flowing fluid) per unit mass = Σ (potential energy + pressure energy + kinetic energy), or

$$H = (Z) + (P/\gamma) + (v^2/2g) \tag{5.3}$$

where:

H = Total energy (total head, energy head), m
Z = Elevation of point on a streamline above horizontal datum, m
P = Pressure at point, Pa
γ = Fluid specific weight, N/m^3
v = Average velocity for all streamlines, m/s

- Momentum principle applied as presented in Equation 5.4:

$$\Sigma F = \gamma[Q(v_2 - v_1))/g \tag{5.4}$$

where:

ΣF = Sum of all forces acting on fluid contained between two cross-sections (pressure, weight, constrains-friction), N
v = Average velocity, m/s
γ = Density

- Hydraulic radius given by:

$$r_H = A/w_p \tag{5.5}$$

where:

r_H = Hydraulic radius (= D/4 for circular closed conduits), m
A = Area, m^2
w_p = Wetted perimeter of flow cross sections, m

5.2.5 Flow Friction Formulas

The most frequently used flow equations are briefly outlined below.[1,3–13]

5.2.5.1 Darcy-Weisbach equation for Pipe Flow

$$h_f = f * v^2 * L/2g * D \tag{5.6}$$

where:

h_f = Friction head, m
f = Friction factor
v = Velocity of flow, m/s
L = Length of conduit, m
g = Gravitational acceleration, m/s^2
D = Diameter of circular conduit, m

5.2.5.2 Chezy Equation

$$v = c_f (r_H * j)^{1/2} \tag{5.7}$$

where:

c_f = Friction coefficient, or coefficient of DeChezy, $m^{1/2}/s$
r_H = Hydraulic radius, m
j = Slope of invert or bed, m/m

5.2.5.3 Manning's Formula

Metric units:

$$v = (r_H)^{2/3} * j^{1/2} / n \tag{5.8a}$$

English units:

$$v = 1.49 * r_H^{2/3} * j^{1/2} / n \tag{5.8b}$$

where:

n = Manning's coefficient.

5.2.5.4 Kutter-Manning's formula

$$v = \left[\frac{23 + (0.00155 / j) + (1/n)}{1 + (23 + (0.0015 / j)(n / \sqrt{rH})}\right] * \sqrt{rH * j} \tag{5.9}$$

where:

v = Mean velocity of flow, m/s
j = Slope of energy grade line, dimensionless
n = Kutter coefficient of roughness = Manning's coefficient for types of pipe commonly used in sewer construction

5.2.5.5 Hazen-Williams formula

Metric units:

$$v = 0.85 * c * r_H^{0.63} * j^{0.54} \tag{5.10a}$$

English units:

$$v = 1.32 * c * r_H^{0.63} * j^{0.54} \tag{5.10b}$$

where:

c = Coefficient that depends on roughness of conduit, $100 < c < 140$

5.2.5.6 Prandtl-Colebrook Formula

$$v = -2 * \log(\{2.51\nu / [D[2g * D * j]^{1/2}] + (\{k/(3.71 * D)\}) * [2g * D * j]^{1/2}) \tag{5.11}$$

where:

- ν = Kinematic viscosity, m^2/s
- D = Diameter of sewer, mm
- K = Factor, usually taken as 0.025 mm
- v = Velocity of flow, m/s
- j = Slope, m/km

5.2.5.7 Strickler Formula (Manning's)

$$v = c * r_H^{2/3} j^{1/2} \tag{5.12}$$

where:

c = Factor (= 100)

5.2.5.8 Scimemi Formula (Hazen)

$$v = c * D^{0.68} * j^{0.56} \tag{5.13}$$

where:

c = Factor (= 61.5)

The flow velocity should not exceed a maximum value to avoid excessive erosion and dissipation of energy at the point of discharge. Following are recommended limits:

- For clear water, velocities up to 12 m/s are recommended.
- For sanitary sewers, velocities up to 3 m/s are recommended (possibly up to 5 m/s).
- For storm sewers, velocities up to 5 m/s are allowed (possibly upt to 10 m/s).

A minimum velocity should be maintained to avoid accumulation of deposits and their biodegradation (to avoid production of hydrogen sulfide, which promotes crown corrosion).

Example 5.2

1. Write a computer program that would allow the computation of velocity (m/s) of flow in a pipe using one of the following formulas, given the relevant data for each: (a) Darcy-Weisbach equation, (b) Chezy equation, (c) Manning's formula, (d) Kutter formula, (e) Hazen-Williams formula, (f) Prandtl-Colebrook formula, (g) Strickler formula, or (h) Scimemi formula.
2. Using Manning's equation, find the velocity of flow and sewer capacity for a concrete sewer of diameter 1.52 m laid at a slope of 0.0008, when sewer is flowing full. (Take Manning's friction coefficient equal to 0.013).

Solution

1. For solution to Example 5.2 (1), see the following listing of Program 5.2.
2. Solution to Example 5.2 (2):
   - Given D = 1.52 m, j = 0.008.
   - Determine the hydraulic radius as: $r_H = (\pi D^2/4)/(\pi D) = D/4 = 1.52/4 = 0.38$ m.
   - Find the average velocity of flow in the pipe as: $v = r_H{}^{2/3} * j^{1/2}/n = 0.38^{2/3} * 0.0008^{1/2}/0.013 = 1.14$ m/s.
   - Determine the capacity of the sewer as: $Q = v*A = 1.14*(\pi*1.52^2)/4 = 2.07$ m³/s.

Listing of Program 5.2

```
DECLARE FUNCTION log10! (x!)
DECLARE SUB box (r1%, c1%, r2%, c2%)      'Found in Program 2.1
'*************************************************************
'Program 5.2: Friction Flow in Pipes
'*************************************************************
pi = 3.1415962#
g = 9.81

m$(1) = "Darcy-Weisbach equation"
m$(2) = "Chezy equation"
```

```
m$(3) = "Manning's equation"
m$(4) = "Kutter-Mannings formula"
m$(5) = "Hazen-Williams formula"
m$(6) = "Prandtl-Colebrook formula"
m$(7) = "Strickler formula"
m$(8) = "Scimemi formula (Hazen)"
m$(9) = "Exit the program"

start:
CALL box(1, 1, 20, 80)
LOCATE 2, 15
PRINT "Program 5.2: Friction Flow in Pipes"
LOCATE 5, 3: PRINT "To compute the velocity of flow using:"

row = 5
FOR I = 1 TO 9
LOCATE row + I, 10
PRINT I; m$(I)
NEXT I
DO
LOCATE row + 11, 3
INPUT "Select an option (1-9): "; opt
LOOP WHILE opt < 1 OR opt > 9
IF opt = 9 THEN END

CALL box(1, 1, 20, 80)
LOCATE 2, 15
PRINT "Program 5.2: Friction Flow in Pipes"
LOCATE 3, 3: PRINT m$(opt)

row = 4
LOCATE row, 5
INPUT "Enter the diameter of the pipe    (m)="; D
RH = D / 4

SELECT CASE opt
CASE 1
      'Darcy-Weisbach equation
      LOCATE row + 1, 5
      INPUT "Enter the head loss hf (m)         ="; hf
      LOCATE row + 2, 5
      INPUT "Enter length of pipe   (m)         ="; L
      LOCATE row + 3, 5
      INPUT "Enter friction factor f            ="; f
      v = (hf * D * 2 * g / (L * f)) ^ .5 /n
CASE 2
      'Chezy equation
      LOCATE row + 1, 5
      INPUT "Enter the slope of the pipe        ="; j
      LOCATE row + 2, 5
      INPUT "Enter Chezy coefficient Cf         ="; Cf
      v = Cf * (RH * j) ^ .5
CASE 3
      'Manning's equation
      LOCATE row + 1, 5
      INPUT "Enter Manning coefficient n        ="; n
      LOCATE row + 2, 5
      INPUT "Enter the slope of the pipe        ="; j
```

```
        LOCATE row + 3, 5
        INPUT "Using Metric units? Y/N          ="; r$
        r$ = UCASE$(r$)
        fact = 1.49
        IF r$ = "Y" THEN fact = 1
        v = fact * RH ^ (2 / 3) * j ^ .5 /n
CASE 4
        'Kutter Manning's formula
        LOCATE row + 1, 5
        INPUT "Enter the slope of the pipe      ="; j
        LOCATE row+2,5
        INPUT "Enter Manning's coefficient n    ="; n
        v = ((1 / n + 23 + .00155 / j) / (1 + n * (23 + .00155 / j) / RH ^ .5)) * RH ^ (2 / 3) * j ^ .5 / n
CASE 5
        'Hazen-Williams formula
        LOCATE row + 1, 5
        INPUT "Enter the slope of the pipe      ="; j
        LOCATE row + 2, 5
        INPUT "Using Metric units? Y/N          ="; r$
        r$ = UCASE$(r$)
        fact = 1.32
        IF r$ = "Y" THEN fact = .85
        LOCATE row + 3, 5
        INPUT "Enter Hazen-William coefficient c ="; c
        v = fact * c * RH ^ .63 * j ^ .54
CASE 6
        'Prandtl-Colebrook formula
        LOCATE row + 1, 5
        INPUT "Enter the slope of the pipe      ="; j
        LOCATE row + 2, 5
        INPUT "Enter viscosity                  ="; vis
        LOCATE row + 3, 5
        INPUT "Enter the factor k               ="; k
        z = (2 * g * D * j) ^ .5
        v = -2 * log10(2.51 * vis / (D * z) + k / (3.71 * D)) * z
CASE 7
        'Strickler formula
        LOCATE row + 1, 5
        INPUT "Enter the slope of the pipe      ="; j
        v = 100 * RH ^ (2 / 3) * j ^ .5
CASE 8
        'Scimemi formula (Hazen)
        LOCATE row + 1, 5
        INPUT "Enter the slope of the pipe      ="; j
        v = 61.5 * D ^ .68 * j ^ .56
CASE 9
        END
END SELECT

LOCATE row + 5, 3
PRINT "The velocity  ="; v
DO
LOCATE 22, 1: PRINT "Press any key to continue............"
LOOP WHILE INKEY$ = ""

GOTO start
END
```

```
FUNCTION log10 (x)
  log10 = LOG(x) / LOG(10)
END FUNCTION
```

5.2.6 Design Computations

Design computations include conduit capacity, self cleansing velocities, maximum or minimum slopes, significant water-surface elevations, and changes in conduit size.

5.2.6.1 Capacity of Flow Estimates

1. For pipes flowing full, the size and slope may be determined by using the aforementioned pipe flow equations.
2. For partially full pipes, the hydraulic element chart (see Diagram A4 in the Appendix) gives: Q/Q_f; v/v_f; A/A_f; r_H/r_{Hf}; n/n_f, and D/D_f; otherwise, the following equations may be used (see Figure 5.2).

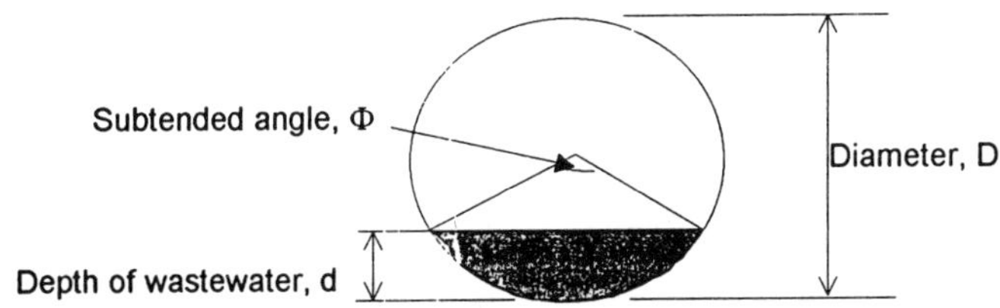

Figure 5.2
Cross-section of a sewer.

In Figure 5.2:

D = Diameter of sewer

d = Depth of flow

ϕ = Angle subtended between the two radii formed at the ends of chord of partial depth

$$\phi = 2\cos^{-1}(1 - 2d/D) \tag{5.14}$$

j = Slope

From Figure 5.2 the following equations can be developed:

1. For a pipe flowing full:

   Area: $A_f = \pi D^2/4$

   Wetted perimeter: $(w_p)_f = \pi D$

Hydraulic radius: $(r_H)_f = D/4$ (5.15)

Velocity of flow: $v_f = (l/n)r_H^{2/3}j^{1/2} = (l/n)(D/4)^{2/3}j^{1/2}$ (5.16)

Flow: $Q_f = A_f v_f = (l/n)(D/4)^{2/3}j^{1/2}*\pi D^2/4 = (l/n)j^{1/2}\pi D^{8/3}/4^{5/3}$ (5.17)

2. For a pipe flowing partially full:

Area: $A_p = ((\pi D^2/4)\phi/360) - (D/2 - d)(Dd - d^2)^{1/2} = D^2((\pi\phi/1440) - (0.5 - (d/D))(d - (d/D)^2)^{1/2}$ (5.18)

Wetted perimeter: $(w_p)_p = \pi D\phi/360$ (5.19)

Hydraulic radius: $(r_H)_p = A_p/(w_p)_p = (D/4) - (360D/\pi\phi)*(0.5 - (d/D))*((d/D) - (d/D)^2)^{1/2})$

Velocity of flow $= v_p = (1/n)(r_H)_p^{2/3}j^{1/2}$ (5.20)

Flow: $Q_p = A_p * v_p$

The hydraulic elements can be specified or formulated by the following equations:

$$(r_H)_p / D = \tfrac{1}{4} - (360/\pi\phi) * (0.5 - (d/D)) * ((d/D) - (d/D)^2)^{1/2}) \quad (5.21)$$

$$A_p / A_f = (\phi/360) - (4\pi) * (0.5 - (d/D)) * ((d/D) - (d/D)^2)^{1/2}) \quad (5.22)$$

$$(r_H)_p /(r_H)_f = 1 - (1440\pi\phi) * (0.5 - (d/D)) * ((d/D) - (d/D)^2)^{1/2}) \quad (5.23)$$

$$\begin{aligned} v_p / v_f &= ((r_H)_p /(r_H)_f)^{2/3} \\ &= (1 - (1440/\pi\phi) * (0.5 - (d/D)) * ((d/D) - (d/D)^2)^{1/2}))^{2/3} \end{aligned} \quad (5.24)$$

$$\begin{aligned} Q_p / Q_f &= v_p A_p / v_f A_f \\ &= (1 - (1440/\pi\phi) * (0.5 - (d/D)) * ((d/D) - (d/D)^2)^{1/2}))^{2/3} * \\ &\quad ((\phi 360) - (4\pi) * (0.5 - (d/D)) * ((d/D) - (d/D)^2)^{1/2})) \end{aligned} \quad (5.25)$$

Table 5.2 can be developed from the above equations.

Example 5.3

1. Write a computer program to determine the velocity, gradient (slope), and rate of flow for a sewer flowing partially full at depth, d (m), given the sewer diameter, D (m), and velocity of sewer when flowing full, v_f(m/s).
2. A sewer of diameter 150 mm is to flow at a depth of 40% on a grade that enables self-cleaning, similar to that at full depth at a velocity of 0.85 m/s. Find the needed grades, associated velocities, and flow rates at full depth and at 40% depth. Take Manning's coefficient of friction to be equal to 0.013.

TABLE 5.2
Design Computations for Partial Pipe Flow

Depth ratio (d/D)	Hydraulic radius/dia $(r_H)_p/D)$	Area ratio (A_p/A_f)	Hydraulic radius ratio $(r_H)_p/(r_H)_f$	For same n: Velocity ratio (V_p/V_f)	Discharge ratio (Q_p/Q_f)
0.1	0.06352	0.052023	0.254081	0.401157	0.020869
0.11	0.069522	0.059825	0.278086	0.426042	0.025488
0.12	0.075458	0.067945	0.301833	0.449964	0.030573
0.13	0.08133	0.076363	0.32532	0.473014	0.036121
0.14	0.087136	0.08506	0.348546	0.495268	0.042128
0.15	0.092878	0.094022	0.37151	0.51679	0.04859
0.16	0.098553	0.103234	0.394212	0.537633	0.055502
0.17	0.104162	0.112682	0.416649	0.557845	0.062859
0.18	0.109705	0.122353	0.438821	0.577464	0,070654
0.19	0.115182	0.132237	0.460727	0.596526	0.078882
0.2	0.120591	0.142321	0.482365	0.61506	0.087536
0.21	0.125934	0.152597	0.503735	0.633094	0.096608
0.22	0.131209	0.163054	0.524835	0.650652	0.106091
0.23	0.136416	0.173683	0.545664	0.667755	0.115977
0.24	0.141555	0.184475	0.566221	0.684422	0.126259
0.25	0.146626	0.195422	0.586503	0.70067	0.136927
0.26	0.151628	0.206517	0.606511	0.716516	0.147973
0.27	0.156561	0.21775	0.626242	0.731973	0.159387
0.28	0.161424	0.229116	0.645696	0.747054	0.171162
0.29	0.166218	0.240606	0.66487	0.761771	0.183287
0.3	0.170941	0.252214	0.683764	0.776135	0.195752
0.31	0.175594	0.263933	0.702375	0.790156	0.208549
0.32	0.180176	0.275757	0.720703	0.803842	0.221665
0.33	0.184686	0.287679	0.738745	0.817203	0.235092
0.34	0.189125	0.299693	0.7565	0.830244	0.248819
0.35	0.193492	0.311793	0.773967	0.842975	0.262834
0.36	0.197786	0.323973	0.791143	0.855401	0.277127
0.37	0.202007	0.336228	0.808026	0.867528	0.291687
0.38	0.206154	0.348551	0.824616	0.879362	0.306502
0.39	0.210227	0.360937	0.84091	0.890908	0.321561
0.4	0.214226	0.37338	0.856905	0.90217	0.336852
0.41	0.21815	0.385875	0.872601	0.913154	0.352363
0.42	0.221999	0.398417	0.887995	0.923862	0.368082
0.43	0.225771	0.411	0.903085	0.934299	0.383997
0.44	0.229467	0.423619	0.917868	0.944467	0.400094
0.45	0.233086	0.436269	0.932343	0.954371	0.416362
0.46	0.236627	0.448944	0.946506	0.964012	0.432787

TABLE 5.2 (CONTINUED)
Design Computations for Partial Pipe Flow

				For same n	
Depth ratio	Hydraulic radius/dia	Area ratio	Hydraulic radius ratio	Velocity ratio	Discharge ratio
0.47	0.240089	0.46164	0.960356	0.973393	0.449357
0.48	0.243473	0.474351	0.973891	0.982517	0.466058
0.49	0.246776	0.487072	0.987106	0.991385	0.482876
0.5	0.25	0.499799	1	1	0.499799
0.51	0.253142	0.512525	1.01257	1.008362	0.516811
0.52	0.256203	0.525247	1.024812	1.016474	0.533899
0.53	0.259181	0.537958	1.036725	1.024336	0.551049
0.54	0.262076	0.550654	1.048304	1.031949	0.568246
0.55	0.264886	0.563329	1.059546	1.039313	0.585475
0.56	0.267612	0.575979	1.070448	1.04643	0.602722
0.57	0.270251	0.588598	1.081005	1.0533	0.61997
0.58	0.272804	0.601181	1.091216	1.059922	0.637205
0.59	0.275269	0.613723	1.101074	1.066296	0.65441
0.6	0.277644	0.626218	1.110577	1.072422	0.67157
0.61	0.27993	0.638661	1.119719	1.0783	0.688668
0.62	0.282124	0.651047	1.128497	1.083927	0.705688
0.63	0.284226	0.66337	1.136905	1.089305	0.722612
0.64	0.286234	0.675624	1.144938	1.09443	0.739423
0.65	0.288148	0.687804	1.15259	1.099301	0.756104
0.66	0.289964	0.699904	1.159857	1.103917	0.772636
0.67	0.291683	0.711918	1.166732	1.108275	0.789001
0.68	0.293302	0.72384	1.173209	1.112372	0.80518
0.69	0.29482	0.735664	1.17928	1.116207	0.821153
0.7	0.296235	0.747383	1.184939	1.119774	0.836901
0.71	0.297544	0.758991	1.190177	1.123072	0.852402
0.72	0.298747	0.770482	1.194986	1.126096	0.867636
0.73	0.299839	0.781847	1.199358	1.12884	0.882581
0.74	0.300821	0.793081	1.203282	1.131301	0.897213
0.75	0.301687	0.804175	1.206748	1.133473	0.911511
0.76	0.302436	0.815123	1.209745	1.135349	0.925448
0.77	0.303065	0.825915	1.21226	1.136922	0.939001
0.78	0.30357	0.836544	1.21428	1.138184	0.952141
0.79	0.303947	0.847001	1.21579	1.139128	0.964842
0.8	0.304193	0.857276	1.216773	1.139742	0.977074
0.81	0.304303	0.867361	1.217211	1.140015	0.988805
0.82	0.304271	0.877245	1.217084	1.139936	1.000003
0.83	0.304092	0.886916	1.216369	1.139489	1.010631
0.84	0,30376	0.896364	1.215041	1.138659	1.020653

TABLE 5.2 (CONTINUED)
Design Computations for Partial Pipe Flow

Depth ratio	Hydraulic radius/dia	Area ratio	Hydraulic radius ratio	For same n: Velocity ratio	For same n: Discharge ratio
0.85	0.303267	0.905575	1.213068	1.137427	1.030026
0.86	0.302605	0.914537	1.210419	1.13577	1.038704
0.87	0.301763	0.923235	1.207054	1.133664	1.046638
0.88	0.300731	0.931653	1.202925	1.131077	1.053771
0.89	0.299495	0.939772	1.197978	1.127975	1.060039
0.9	0.298037	0.947575	1.192147	1.124311	1.065369
0.91	0.296337	0.955037	1.185347	1.120032	1.069672
0.92	0.294369	0.962135	1.177476	1.115068	1.072846
0.93	0.292099	0.968838	1.168397	1.109329	1.07476
0.94	0.289482	0.975111	1.157926	1.102691	1.075247
0.95	0.286452	0.980912	1.145806	1.094983	1.074082
0.96	0.282912	0.986186	1.131647	1.085944	1.070942
0.97	0.278702	0.99086	1.114807	1.075143	1.065316
0.98	0.273515	0.994827	1.094058	1.061762	1.056269
0.99	0.266576	0.997906	1.066304	1.043728	1.041542
1	0.25	0.999598	1	1	0.999598

Solution

1. For solution to Example 5.3 (1), see the following listing of Program 5.3.
2. Solution to Example 5.3 (2):
   - Given D = 0.15 m, d = 0.4*D, v_f = 0.85 m/s, n = 0.013.
   - Find the slope as: $j = [v_f*n/r_H^{2/3}]^2 = [0.85*0.013/(0.15/4)^{2/3}]^2 = 9.7*10^{-3}$.
   - Compute $Q_f = \pi*D^2*v_f/4 = (\pi*0.15^2/4)*0.85 = 0.015$ m³/s.
   - Find central angle = ϕ = 2 cos –1 [(D/2) – d]/(D/2) = (1 – 2d/D) = 2 $\cos^{-1}$ (1 – 2*0.4) = 156.93°.
   - Find ratio of areas at partial and full flow as: $A_p/A_f = (\phi/360) - (\sin\phi/2\pi) = (156.93/360) - (\sin 156.93/2\pi) = 0.37$
   - Determine $A_p = (\pi*0.15^2/4)*0.37 = 6.5*10^{-3}$ m², or $A_p = (D^2/4)(\pi*\phi/360 - (\sin\phi)/2) = (0.15^2/4)(\pi*156.93/360 - ((\sin 156.93)/2) = 6.6*10^{-3}$.
   - Determine hydraulic radius as: $r_H = (D/4)*(1 - 360\sin\phi/2\pi\phi) = (0.15/4)[1 - (360*\sin 156.93)/2\pi*156.93] = 0.032$. Thus, velocity $v_p = (0.032)^{2/3}*(9.7*10^{-3})^{1/2}/0.013 = 0.76$ m/s.

Listing of Program 5.3

```
DECLARE SUB box (r1%, c1%, r2%, c2%)     'Found in Program 2.1
'*******************************************************************
'Program 5.3:Capacity of flow estimates
'*******************************************************************
```

```
pi = 3.1415962#
g = 9.81

CALL box(1, 1, 20, 80)

LOCATE 2, 15
PRINT "Program 5.3:Capacity of flow estimates"
row = 3
LOCATE row, 5
INPUT "Enter the diameter of the sewer (m)      ="; D
LOCATE row + 1, 5
INPUT "Enter the velocity of full flow vf(m/s)  ="; vf
LOCATE row + 2, 5
INPUT "Enter depth of flow in the pipe h (m)    ="; h
LOCATE row + 3, 5
INPUT "Enter the manning coefficient n          ="; n

rh = D / 4                          'hydraulic radius
j = (vf * n / rh ^ (2 / 3)) ^ 2       'the slope
Q = pi * D ^ 2 / 4 * vf               'the flow
r = D / 2
z = r - h
x = (r ^ 2 - z ^ 2) ^ .5
phi = 2 * ATN(z / x)                  'central angle
z = (180 - phi * 180 / pi)            'in degrees
rat = z / 360 - SIN(phi / (2 * pi))    'ratio of areas
Ap = pi / 4 * D ^ 2 * rat             'partial area of flow
rh = D / 4 * (1 - 360 * SIN(phi) / (2 * pi * z))
vp = rh ^ (2 / 3) * j ^ .5 / n

LOCATE row + 5, 3: PRINT "RESULTS:"
LOCATE row + 6, 3
PRINT "The slope                            ="; j
LOCATE row + 7, 3
PRINT "The full flow Qf                      ="; Q; " m3/s"
LOCATE row + 8, 3
PRINT "The central angle phi                  ="; z; " degrees"
LOCATE row + 9, 3
PRINT "Ratio of areas at partial and full flow  ="; rat
LOCATE row + 10, 3
PRINT "Area of partial flow  Ap               ="; Ap; "  m2"
LOCATE row + 11, 3
PRINT "The hydraulic radius                  ="; rh; "  m"
LOCATE row + 12, 3
PRINT "The velocity of flow                  ="; vp; "  m/s"

DO
LOOP WHILE INKEY$ = ""
END
```

5.2.6.2 Self-Cleansing Velocities

Self-cleaning velocity is referred to as that velocity sufficient to prevent deposition of suspended matter. The velocity needed to carry sediment in a pipe flowing full is given by Equation 5.26.

$$
\begin{aligned}
v_{sc} &= (r_H)^{1/6}[b'(s.g.-1)*d]^{1/2}/n \\
&= (8b'*g(s.g.-1)*d/f)^{1/2}
\end{aligned}
\tag{5.26}
$$

where:

v_{sc} = Self-cleansing velocity, m/s
r_H = Hydraulic radius, m
b' = Dimensionless constant (sediment-scouring characteristics, see Table 5.3)
s.g. = Specific gravity, dimensionless
d = Diameter of sediment grain, m
n = Roughness factor, Manning and Kutter factor, $m^{1/6}$
f = Friction (Darcy-Weisbach) factor, dimensionless (usually 0.02 to 0.03)[5]

TABLE 5.3
Values of b′[5]

Value of b′	Significance
0.04	To start motion of clean granular particles
0.8	For adequate self-cleansing of cohesive material

A minimum velocity is often accepted as the design criterion. As such, the recommended velocities of Table 5.4 may be adopted. Generally, it is desirable to have a minimum velocity of 0.91 m/s or more whenever possible. As such, 0.6 m/s < minimum design velocity < 3.5 m/s at peak flow.

TABLE 5.4
Recommended Minimum Velocities

Sewer	$(V_{sc})_{min}$ (m/s)
Sanitary	< 0.61
Storm	< 0.75

5.2.7 Summary of Sewer System Design[1,5,21]

1. The first information needed to design and lay out a sanitary sewer system is a topographical map (contour map), not only for the immediate area under consideration, but also the surrounding area. With this information regarding the various drainage basins, it is then possible to evaluate the number and location for the lift stations. A geological map indicating the soil type and rock formations is also necessary. Aerial photographs are very helpful if available.
2. Details as to the zoning in the design and layout area are required, indicating which areas are industrial, commercial, and residential. Essential information such as lot sizes, population density, and street layouts must be provided.

3. With the information indicated in (1) and (2), the size and gradient of the sewers can be determined. The sewer must be on an adequate slope to prevent deposition of solids (a slope of 2% for house connections, with a minimum slope of 1% being used). Sanitary sewers with diameters of 375 mm (15") are designed to flow half full; large sewers are designed to flow three quarters full.
4. The sanitary sewers must be installed deep enough to receive flows from the tributary areas. Sufficient depth must be provided to prevent freezing in cold climates. In most cases, a minimum depth from the ground surface to the crown of the pipe should not be less than 1.5 m. The sanitary sewer also should be at least 1 m below the first floor elevation of any adjacent dwellings.
5. Manhole locations at:
   - Junctions of sanitary sewers
   - Changes in gradient or gradient or alignment (except in curved alignments)
   - Locations that provide easy access to the sewer for preventive maintenance and emergency service

   A minimum diameter for a manhole is 1.22 m. A straight sewer is to be used between manholes. Large accessible sewers may be curved and manholes placed at 100- to 200-m intervals. The spacing for manholes is in the range of 90 to 150 m and 150 to 300 m for large diameter sewers.
6. Sanitary sewers should not be installed in the same trench as water mains in order to protect public health.[21]
7. No sewer line should pass under a building.
8. Sewer lines should be installed down the center line of the lanes rather than the center line of the streets. In this way, one lane is always available to keep traffic moving. In wide streets, sewers can be laid outside the curb between the curb and the sidewalk.
9. Coordination between utilities responsible for power poles, gas meters, telephone juction boxes, television wires, etc., is necessary to plan for construction and layout of the sewer lines. This planning can help avoid conflicts and problems that may arise among the various organizations involved.
10. Planting of trees and shrubs and construction of fences or retaining walls that interfere with access to the sewer lines should be controlled or prohibited.
11. Integrity of a sewer line (leakage) is tested for by a low-pressure air test. Water testing generally has been replaced by low-pressure air testing.[21]
12. Combined sewers should now be prohibited in order to prevent present and future wastewater treatment problems (see Section 5.2.1 for disadvantages).

5.2.8 Corrosion in Sanitary Sewers

Corrosion in a sanitary sewer is caused by the production of hydrogen sulfide, H_2S. Following are some of the problems encountered with H_2S:[1]

- Generates odors
- Creates health risk to maintenance workers
- Develops corrosion on unprotected sewer pipes made of concrete materials or metals

- Interferes with treatment processes (it affects activated sludge and increases prechloride demand, and public complaint may arise due to generation of odor issuing from influent structure)
- H_2S toxicity (numerous workers are killed each year by entering manholes without the proper precautions)

Generation of sulfides in sanitary sewers depends on wastewater temperature, dilution rate, hydraulic flow conditions, and agitation of deposited waste solids. When insufficient dissolved oxygen is present in the wastewater, anaerobic bacteria produce sulfides. Generation of sulfide is encountered on the interior walls of sanitary sewers. Sulfide in municipal wastewater may exist, in part, as insoluble sulfides of various metals; however, the concentration of metal sulfide is generally low.[1] The major part of sulfides is normally retained in solution as a mixture (referred to as dissolved sulfide) of hydrogen sulfide and ionic HS^-.

5.2.8.1 Sulfide Buildup Estimates

An indicator of the potential of sulfide buildup in relatively small gravity sewers (not over 600 mm or 24" in diameter) can be estimated by using Equation 5.27.[1]

$$f = (BOD_e * w_p)/(j^{1/2} * Q^{0.33} * B) \tag{5.27}$$

where:

f = Defined function (see Table 5.5)
BOD_e = Effective biochemical oxygen demand, mg/L
w_p = Wetted perimeter, ft
j = Hydraulic slope, dimensionless
Q = Discharge volume, ft^3/s
B = Surface width, ft

TABLE 5.5
Sulfide Condition for Different Values of f

f value	Sulfide condition
$f < 5000$	Sulfide rarely generated
$5000 \leq f \leq 10000$	Marginal condition of sulfide generation
$f > 10000$	Sulfide generation common

Source: From Joint Task Force of the American Society of Civil Engineers and the Water Pollution Control Federation, *Gravity Sanitary Sewer Design and Construction*, ACSE Manuals and Reports on Engineering, Practice Number 60, ASCE WPCF, New York, 1982. Reprinted by permission of the publisher, ASCE.

Example 5.4

1. Write a computer program to estimate the likelihood of sulfide buildup in a sewer, given pipe diameter of sanitary sewer, D (mm); flow rate, Q (m^3/s); slope, j (m/m); wetted perimeter, w_p (ft); surface width of flow, B (ft); and effective biochemical oxygen demand, BOD_e (mg/L). Let the program indicate the level of sulfide generation based on Table 5.5.
2. A sanitary sewer of pipe diameter 600 mm (24") carries a flow = 2.2 m^3/min. at a slope = 0.0015. Taking wetted perimeter = 0.82 m and surface width of flow = 1 m, give an indicator of the likelihood of sulfide buildup in the sewer, knowing that the effective biochemical oxygen demand = 295 mg/L.

Solution

1. For solution to Example 5.4 (1), see the following listing of Program 5.4.
2. Solution to Example 5.4 (2):
   - Given d = 600 mm, Q = 2.2 m^3/min = 2.2*35.3147/60 = 1.295 cfs, j = 0.0015, w_p = 0.82 m = 0.82*3.2808 = 2.69 ft, width = B = 1 m = 1*3.2808 = 3.2808 ft, BOD = 295 mg/L.
   - Determine the indicator of sulfide buildup as: $f = BOD_e * w_p/(j^{1/2} * Q^{0.33} * B) = 295*2.69/[(0.0015)^{1/2}*1.295^{0.33}*3.2808] = 5734.6$.
   - By comparing this value with values outlined in Table 5.5, it can be concluded the value 5735 indicates a marginal condition for sulfide generation.

Listing of Program 5.4

```
DECLARE SUB box (r1%, c1%, r2%, c2%)       'Found in Program 2.1
'******************************************************************
'Program 5.4: Sulfide Build Up Indicator
'******************************************************************
pi = 3.1415962#
g = 9.81

CALL box(1, 1, 20, 80)

LOCATE 2, 15
PRINT "Program 5.4: Sulfide Build Up Indicator"
row = 3
LOCATE row, 5
INPUT "Enter the diameter of the sewer (m)     ="; D
LOCATE row + 1, 5
INPUT "Enter the flow rate in the sewer Q(m3/s) ="; Q
LOCATE row + 2, 5
INPUT "Enter the slope of the pipe          S  ="; S
LOCATE row + 3, 5
INPUT "Enter the wetted perimeter  Wp(m)       ="; Wp
LOCATE row + 4, 5
INPUT "Enter the surface width of flow b(m)    ="; B
LOCATE row + 5, 5
INPUT "Enter effective biochemical demand(mg/L) ="; BOD
```

```
D = D * .3048          'convert to ft
Q = Q / .3048 ^ 3      'convert to cfs
Wp = Wp / .3048        'convert to ft
B = B / .3048          'convert to ft

Z = BOD * Wp / (S ^ .5 * Q ^ .33 * B)
r$ = "Sulfide rarely generated"
IF Z > 5000 AND Z < 10000 THEN r$ = "Marginal condition of sulfide generation"
IF Z > 10000 THEN r$ = "Sulfide generation common"

LOCATE row + 9, 3: PRINT "RESULT:"
LOCATE row + 10, 5
PRINT "Sulfide build up indicator Z="; Z
LOCATE row + 11, 5
PRINT r$
DO
LOOP WHILE INKEY$ = ""
END
```

5.2.8.2 Filled Pipe Conditions

Equation 5.28 may be used to estimate the level of sulfide buildup in completely filled sewer pipes (force mains).[1]

$$dS_u / dt = [3.28 * a * BOD_e((1 + 0.12 * D_i) / 0.12 * D_i)] / 0.25 * D_i \quad (5.28)$$

where:

dS_u/dt = Increase in sulfide concentration, mg/L*h
a = Coefficient (= 0.3 mm/h)[1]
BOD_e = Effective biochemical oxygen demand = BOD $(\theta)^{T-20}$ (5.29)
D_i = Internal diameter of sewer pipe, mm
BOD = Climate BOD, mg/L
T = Temperature, °C
θ = Temperature coefficient (may be taken as equal to 1.07)[1]

5.2.8.3 Partially Filled Pipe Conditions

Equation 5.30 may be used to estimate the sulfide buildup in gravity flow in a partially filled sanitary sewer.

$$dS_u / dt = (a' * BOD_e / r_H) - [b(j * v)^{0.375} * S_u / D_m] \quad (5.30)$$

where:

dS_u/dt = Rate of change of total sulfide concentration, mg/L*h

a′ = Effective sulfide flux coefficient (a′ = $0.4*10^{-3}$ m/h for DO < 0.5 mg/L; a′ = 0 when DO is high) m/h[1]

r_H = Hydraulic radius of stream, m

b = Empirical coefficient, sulfide loss coefficient (= 0.64 (conservative) – 0.96 (less conservative value))[1]

S_u = Total sulfide concentration, mg/L

j = Energy gradient of stream

v = Velocity of flow, m/s

D_m = Mean hydraulic depth, ft

Equation 5.30 should not be used for systems when DO > 0.5 mg/L or for sanitary sewers with an effective slope > 0.6%.[1] Sulfide concentrations approach a limiting value (S_{ulim}) when Equation 5.30 approaches zero.[1] This can be represented by Equation 5.3.1, taking the value of 0.96 for the sulfide loss coefficient, b.[1]

$$\begin{aligned} S_{ulim} &= [0.33*10^{-3} BOD_e * D_m)/[j*v)^{0.375} r_H) \\ &= [0.52*10^{-3} BOD_e * w_p)/[j*v)^{0.375} B) \end{aligned} \tag{5.31}$$

Sulfide generation in a long reach of sewer line with uniform slope and flow can be determined from Equation 5.32.[1]

$$S_{u2} = S_{ulim} - [(S_{ulim} - S_{ul}) / \log^{-1}\{(j*v)^{0.375} * \Delta t / 1.15 D_m\}] \tag{5.32}$$

where:

S_{ul} = Sulfide concentration at upstream location, mg/L

S_{u2} = Sulfide concentration at downstream location, mg/L

Δt = Time of flow from upstream location to downstream location, h

5.3 Wastewater Disposal for Rural Inhabitants

Small communities in rural areas often need inexpensive, robust, compact sewage treatment works that require little maintenance and can be installed and operated with unskilled labor. Examples of these units include septic tanks, Imhoff tanks, ventilated improved pit (VIP) latrines, etc. The most widely used system is the septic tank. Following is information that can be used in the design, construction, and operation of septic tanks.

5.3.1 Septic Tanks

5.3.1.1 Introduction

A septic tank is an underground (covered) box.[2,3,7,8,11,13–15] The tank receives wastewater and treats it in a horizontal flow pattern. The tank removes solids and promotes partial deposition. Functions taking place in a septic tank include settling of solids, flotation of grease, anaerobic decomposition of solids, and storage of sludge. The retention of large solids and grease is essential to avoid plugging final disposal localities, such as absorption fields. The liquor within the tank has a considerable detention time to maximize solid deposition. For large installations serving multiple families or institutions, a shorter detention time may be adhered to.

Figures 5.3 and 5.4 illustrate the zones that are found in a septic tank.

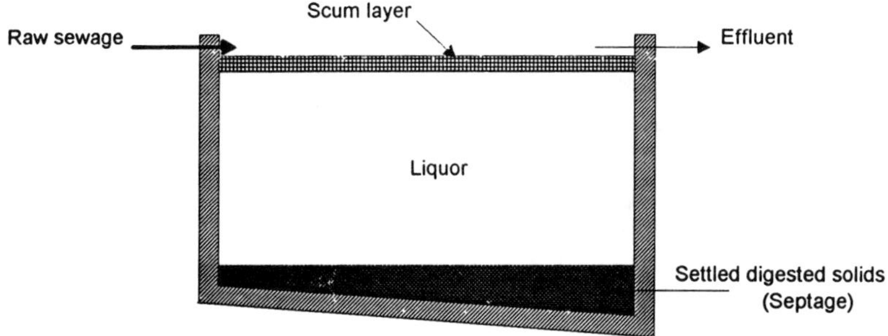

Figure 5.3
Regions in a septic tank.

- The scum layer zone, which forms a crust on surface of tank
- The liquor layer zone, where there is solids deposition and diffusion of solubilized material from the third layer
- The settling and digestion-of-solids zone

The materials that usually are used in the construction of a septic tank include concrete, perforated reinforced concrete, concrete blocks, steel, brick, stone masonry, and fiberglass. The septic tank may be of a rectangular or circular (prefabricated concrete pipes) shape.

5.3.1.2 Designing Septic Tanks

In designing a septic tank the following points must be considered:

- Selection of the location of the tank[20a]
  - To be located down hill
  - To be located at least 15 m from the nearest water supply system (including neighbors)
  - To be located at least 3 m from the nearest building

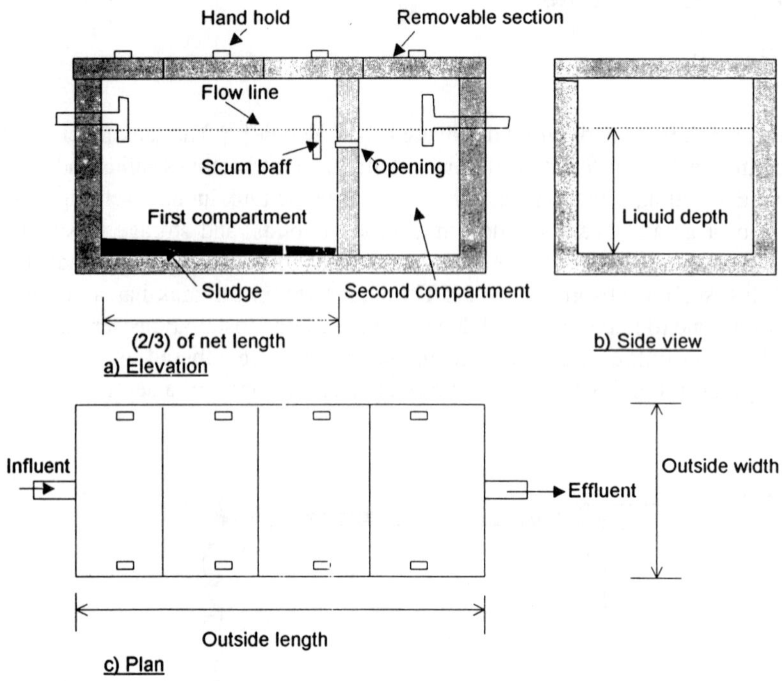

Figure 5.4
A schematic diagram of a septic tank.

- Rain or surface water not to flow over or stand on the tank
- Vehicles not to drive over the tank

- Determination of tank capacity and dimensions
- Determination of necessary labor, materials, and tools
- Maintenance of the septic tank
    - Sludge removed every 2 to 5 years (under normal loading conditions)
    - Sludge discharged to a nearby wastewater treatment plant for treatment or pumped into the soil (avoiding groundwater pollution)
    - Use of strong bactericidal cleansers by householders or use of unnecessary amounts of detergents discouraged
    - With good maintenance a tank can last for 20 years or more
    - Tank inspected at least once a year to determine if it needs cleaning; tank needs to be cleaned if depth of sludge exceeds one third liquid depth or when bottom of scum layer is within 75 mm of bottom of outlet pipe "T" fitting[20a]
    - Removing crust (black-colored digesting sludge) to be avoided when desludging the tank.

Following is general information that can be helpful in designing a septic tank:

1. Tank should receive all sewage from the building to be served (this includes excreta and washwater and excludes rainwater, surface water, or subsurface drainage). Determine amount of sewage entering the tank during each 24-h period = Q.
2. Determine desired retention time t.
   - A minimum value of t of 1 day is taken (a smaller tank implies less initial cost).
   - A maximum value of t of 3 days is taken (a large tank signifies high initial cost and smaller cleaning intervals, and it treats more sewage, thus increases life of subsurface absorption system).
3. Determine capacity of tank V = Q*t. Otherwise, tank capacity may be determined from Equations 5.33a, 5.33b, and 5.33c.
   - Volume of septic tank:

$$V = 180 * P + 2000 \tag{5.33a}$$

where:

V = Minimum volume of septic tank, Liter.

P = Number of people served.

   - Where garbage grinders are used, Equation 5.33b can be used to determine needed capacity.

$$V = 250 * P + 2000 \tag{5.33b}$$

   - When septic tank is used for serving schools and intermittent uses, its volume can be determined from Equation 5.33c.

$$V = 90 * P + 2000 \tag{5.33c}$$

4. Determine tank dimensions as follows:
   - Take length:width ratio = 3:1 to 2:1.

$$L / B = 3:1 \text{ to } 2:1 \tag{5.34}$$

where:

L = Internal length of tank, m.

B = Internal width of tank, m (width of tank not to be less than 1 m)

   - Find inside and outside tank dimensions. Recommended liquid depth in tank = 1.2 m (usually varies from 1.1 to 1.8 m). For a tank serving up to 50 people, use an outlet liquid depth of 1.2 m; when tank serves up to 150 people, use a liquid depth of 1.5 m; and use a liquid depth of 1.8 when tank serves a larger number of people.
   - Inlet and outlet pipes should be fitted with open T sewer pipe fittings.
   - Bottom of inlet pipe 300 mm should be below top of tank.

- Bottom of outlet pipe 75 mm should be below bottom of inlet (375 mm below top of tank).
- Downward T connection of inlet should be 20% of outlet liquid depth, while that of the outlet should be 40% of liquid depth.

5. The tank floor can be sloped towards the inlet at a slope of 10%.
6. When two compartments are considered, one compartment (near inlet) can be twice the capacity of the second compartment. This yields a first compartment length as determined from Equation 5.35.

$$L_c = 2L/3 \quad (5.35)$$

where:

L_c = Internal length of first compartment, m

L = Internal length of tank, m

7. Walls must be watertight (25-mm thick inside coating of cement plaster is applied, usually two coats of 12 mm each are applied).[20a]
8. Floor of tank should be of reinforced concrete (usually 100 to 150 mm thick) and should rest on a bed of gravel or sand 75 mm thick.[20a]
9. Top should be watertight (usually reinforced concrete is used). Generally, all or most of the sections can be removed to clean tank, and one or two sections over outlet can be removed to inspect tank.
10. Inlet and outlet pipes should be fitted with open T sewer pipe fittings.
11. A proper plan view (top, side, and/or end views) of the tank is needed.

5.3.1.3 Septic Tank Effluents

The effluent from a septic tank is offensive and potentially dangerous (BOD_5, 120 to 270 mg/L; SS_{mean}, 44 to 69 mg/L). Usually the effluent is discharged to a subsurface tile field (leaching field, drain field), absorption and evapotranspiration mounds (beds of imported porous soil covered with a top soil that is planted with vegetation to appear as a part of the landscape), seepage pits, or intermittent sand filters. The ability of the ground to absorb the effluent ought to be determined. Percolation tests, used to measure the suitability of ground for the fields, are conducted as follows.[13]

1. A hold about 100 mm (4") or more in diameter is dug as deep as the proposed tile field trench.
2. The hole sides are scratched and all loose soil is removed.
3. 500 mm (2") of fine gravel or coarse sand are placed in bottom of pit.
4. The hole is filled with water to a depth of 300 mm and allowed to stand overnight (at least 4 h).
5. Next day, pit is filled with water to about 150 mm (6") above gravel, and the drop in water level in 30 min. is measured.
6. The percolation rate is calculated as mm/min. (t).

7. Tables are used to determine the required trench area, or it can be computed according to Equation 5.36.

$$Q = 204 * t_{25} \tag{5.36}$$

where:

Q = Flow that can be applied per unit area, L/m²*d

t_{25} = Time required for the water surface to fall 25 mm = 25/subsidence rate, min.

Example 5.5

1. Write a computer program to design a septic tank for any number of people. Design the program so that the necessary structural design of the tank is included.
2. Design a septic tank to serve a family dwelling of 20 people.
3. Use the above data with the program developed in (1) above in order to produce the full design of the tank.

Solution

1. For solution to Example 5.5 (1), see the following listing of Program 5.5.
2. Solution to Example 5.5 (2):
   - Given P = 20.
   - Determine the tank volume using: V = 180*P + 2000 = 180*20 + 2000 = 5600 L.
   - Take outlet liquid depth d = 1.2 m.
   - Take a length-width ratio of 2.5, i.e., L/B = 2.5, and determine the plan area as: Area = volume/depth

     $V/d = 5600*10^{-3}/1.2 = 4.67\ m^2 = L*B = 2.5*B,^2$ which yields a width B = 1.36 m; therefore, L = 2.5*B = 3.4 m.
   - Determine liquid depth at inlet (D) from relationship of slope: Slope = (D – d)/L, thus D = length*floor slope + outlet depth = 0.1*3.4 + 1.2 = 1.54 m.
   - Determine total depth of tank as: Total depth = liquid depth at inlet + distance of bottom of inlet pipe from liquid surface + distance between inlet and tank cover = 1.54 + 0.075 + 0.3 = 1.92 m.
   - Find distance of bottom of T sewer inlet pipe from top of tank = 0.3 + 0.075 + 20% of liquid depth = 0.3 + 0.075 + 0.2*1.2 = 0.62 m.
   - Find distance of bottom of T sewer outlet pipe from top of tank = 0.3 + 0.75 + 40% of liquid depth = 0.3 + 0.075 + 0.4*1.2 = 0.86 m.
   - When considering a two-compartment tank, then the length of the first compartment can be determined as: L_c = 2*L/3 = 2*3.4/3 = 2.27 m
3. As an example, the following data can be taken to run the program: number of people = 20, no garbage grinder wastes, not a public place, grade of concrete = 30, yield stress for steel = 250 N/mm². Severe exposure is assumed, as is a modulus of subgrade soil 0.

The results of the program will be as follows:

- Vertical wall thickness = 200 mm; steel in all directions will be 1040 mm^2 everywhere.
- Base slab thickness = 200 mm; steel in both directions will be 1300 mm^2
- Cover slab thickness = 150 mm; steel in both directions will be 780 mm^2

Listing of Program 5.5

```
DECLARE FUNCTION MAX! (A!(), n!)
DECLARE FUNCTION MAX2! (x!, y!)
DECLARE SUB Designsection (t, cv, Mh, Mv, n, V, fcu, fy, Astx, Astv)
DECLARE SUB Get.Cover (fcu, cv)
DECLARE SUB GetBasePlateMoments (t, Lx, Ly, H, fcu, Mx, My)
DECLARE SUB GetCoverPlateMoments (Lx!, Ly!, udl!, Mx!, My!)
DECLARE SUB GetWallsMoments (Lx!, Ly!, tw, fcu, MAXMX, MAXMY, MAXSX, MAXSY, MAXTN)
DECLARE SUB SOLVE (GSTIF!(), GCON!(), HEAD!(), n!)
DECLARE SUB box (r1%, c1%, r2%, c2%)      'Found in Program 2.1
DECLARE SUB waitme ()
'***************************************************************
'Program 5.5: Design of Septic Tanks
'***************************************************************

pi = 3.1415962#
g = 9.81

CALL box(1, 1, 20, 80)

LOCATE 2, 15
PRINT "Program 5.5: Design of Septic Tanks"
row = 3
LOCATE row, 5
INPUT "Enter the number of people to be served          ="; P
LOCATE row + 1, 5
INPUT "Will garbage grinder wastes be used?    Y/N   ="; r$
LOCATE row + 2, 5
INPUT "Is the tank in a school/public utility?  Y/N   ="; S$
LOCATE row + 3, 5
INPUT "Enter the grade of concrete      N/mm2          ="; fcu
LOCATE row + 4, 5
INPUT "Enter the yield stress of steel  N/mm2          ="; fy
r$ = UCASE$(r$)
S$ = UCASE$(S$)
V = 180 * P + 2000                    'volume of tank in liters
IF r$ = "Y" THEN V = 250 * P + 2000
IF S$ = "Y" THEN V = 90 * P + 2000
V = V / 1000                          'convert to m3

'compute liquid depth at outlet (m)
IF P <= 50 THEN d0 = 1.2
IF P > 50 AND P <= 150 THEN d0 = 1.5
IF P > 150 THEN d0 = 1.8

'compute depth at inlet (m) pipe
sp$ = SPACE$(60)
DO
FOR I = 8 TO 19
```

```
LOCATE I, 3: PRINT sp$
NEXT I
row = 3
10 : LOCATE row + 5, 5: PRINT sp$: LOCATE row + 6, 5
INPUT "Select a sides ratio (L/B) between 2 and 3 ="; rat
IF rat < 2 OR rat > 3 THEN GOTO 10
Ly = (V / (d0 * rat)) ^ .5     'width of tank
Lx = Ly * rat                  'length of tank

'compute liquid depth at inlet di
S = .1                  'slope of base
di = d0 + S * Lx        'liquid depth at inlet pipe
dt = di + .075 + .3     'total depth of tank at inlet
dti = .375 + .2 * d0    'depth from top of tank to bottom of t-sewer at inlet
dt0 = .375 + .4 * d0    'depth from top of tank to bottom of t-sewer at inlet
Lc = 2 * Lx / 3         'length of the second compartment

'output results of overall dimensioning
row = 9
LOCATE row + 1, 3
PRINT "OVERALL DIMENSIONS:"
LOCATE row + 2, 5
PRINT "Length of tank          ="; INT(Lx * 100) / 100; " m"
LOCATE row + 3, 5
PRINT "Width of tank           ="; INT(Ly * 100) / 100; " m"
LOCATE row + 4, 5
PRINT "Depth at inlet          ="; INT(dt * 100) / 100; " m"
LOCATE row + 5, 5
PRINT "Depth at outlet         ="; INT(d0 * 100) / 100; " m"
LOCATE row + 6, 5
PRINT "Depth of T-sewer inlet pipe from top        ="; INT(dti * 100) / 100; " m"
LOCATE row + 7, 5
PRINT "Depth of T-sewer outlet pipe from top       ="; INT(dt0 * 100) / 100; " m"
LOCATE row + 8, 5
PRINT "Length of the second compartment  ="; INT(Lc * 100) / 100; " m"
LOCATE row + 11, 5
INPUT "Are these dimensions acceptable? Y/N :"; r$
LOOP WHILE UCASE$(r$) = "N"

'Structural design of the tank walls
'Assume thicknesses as follows:
tw = 200                    'Thickness of walls in mm
tb = 200                    'Thickness of base slab in mm
tc = 150                    'Thickness of cover slab in mm
H = dt                      'Total depth of tank in mm
CALL Get.Cover(fcu, cv)     'Determine cover to steel

'Design the walls
CALL GetWallsMoments(Lx, H, tw, fcu, M11, M12, M13, M14, TN1)
CALL GetWallsMoments(Ly, H, tw, fcu, M21, M22, M23, M24, TN2)
'midspan section
Mx = MAX2(M11, M21)
My = MAX2(M12, M22)
TN = MAX2(TN1, TN2)

CALL Designsection(tw, cv, Mx, My, TN, TN, fcu, fy, Ast1, Ast2)
'Section at the joint of the walls
Mx = MAX2(M13, M23)
```

```
My = MAX2(M14, M24)
TN = MAX2(TN1, TN2)
CALL Designsection(tw, cv, Mx, My, TN, TN, fcu, fy, Ast3, Ast4)

'Design the bottom base slab
CALL GetBasePlateMoments(tb, Lx, Ly, H, fcu, Mxb, Myb)
TN = -10 * dt   'tension in the base slab made -ve for identification only
CALL Designsection(tb, cv, Mxb, Myb, TN, 0, fcu, fy, Ast5, Ast6)

'Design the cover slab
udl = 1.6 * 1.5 + 1.4 * 24 * tc / 1000  'Ultimate design load in KN/m2
CALL GetCoverPlateMoments(Lx, Ly, udl, Mxc, Myc)
V1 = udl * Lx / 2                'Shear on short support
V2 = udl * Ly / 2                'Shear on long support
V = MAX2(V1, V2)
CALL Designsection(tc, cv, Mxc, Myc, 0, V, fcu, fy, Ast7, Ast8)

'Output results of the structural design
CALL box(1, 1, 20, 80)
row = 2
LOCATE row, 15: PRINT "Structural Design of Septic Tanks"
F$ = "##.##^^^^  "
LOCATE row + 1, 3: PRINT "Walls Forces: ";
PRINT USING F$; M11; M12; M13; M14; TN1
LOCATE row + 2, 3: PRINT "Walls Forces: ";
PRINT USING F$; M21; M22; M23; M24; TN2
LOCATE row + 3, 3: PRINT "Bottom slab : ";
PRINT USING F$; Mxb; Myb
LOCATE row + 4, 3: PRINT "Cover slab  : ";
PRINT USING F$; Mxc; Myc

row = 6
LOCATE row + 1, 3: PRINT "Vertical walls:"
LOCATE row + 2, 5: PRINT "Thickness of walls          ="; INT(tw); " mm"
LOCATE row + 3, 5: PRINT "Vertical steel              ="; INT(Ast1); " mm2/m"
LOCATE row + 4, 5: PRINT "Horizontal steel            ="; INT(Ast2); " mm2/m"
LOCATE row + 5, 5: PRINT "Horizontal steel at support ="; INT(Ast3); " mm2/m"
LOCATE row + 6, 5: PRINT "Vertical steel at support   ="; INT(Ast3); " mm2/m"

LOCATE row + 7, 3: PRINT "Base Slab:"
LOCATE row + 8, 5: PRINT "Thickness of base           ="; INT(tb); " mm"
LOCATE row + 9, 5: PRINT "Long direction steel        ="; INT(Ast5); " mm2/m"
LOCATE row + 10, 5: PRINT "Short direction steel      ="; INT(Ast6); " mm2/m"

LOCATE row + 11, 3: PRINT "Cover Slab:"
LOCATE row + 12, 5: PRINT "Thickness of cover slab    ="; INT(tc); "  mm"
LOCATE row + 13, 5: PRINT "Long direction steel       ="; INT(Ast7); " mm2/m"
LOCATE row + 14, 5: PRINT "Short direction steel      ="; INT(Ast8); " mm2/m"

CALL waitme
END

SUB Designsection (t, cv, Mh, Mv, n, V, fcu, fy, Astx, Astv)
'***********************************************************************
'Structural design of a section
'***********************************************************************
br = 1000              '1m width
D = t - cv - 10        'Effective depth, assuming 20mm bars
```

```
ft = 1.3                    'tensile strength of concrete for crack control
rcrit = ft / fy             'critical steel percentage for crack control
fst = .87 * fy              'Design steel stress
Astmin = rcrit * br * t     'Minimum area of steel for grade 250 steel

RECYCLE:
     'X-direction steel
     MaxMu = .156 * fcu * br * D ^ 2 / 10 ^ 6      'in KNm
     DO WHILE Mh > MaxMu
       D = D + 25
       t = t + 25
       MaxMu = .156 * fcu * br * D ^ 2 / 10 ^ 6

LOOP
k = Mh * 10 ^ 6 / (fcu * br * D ^ 2)
Z = D * (.5 + (.25 - k / .9) ^ .5)
IF Z > .94 * D THEN Z = .94 * D
TN = ABS(n)
Ast = TN * 10 ^ 3 / fst
Ast2 = ABS((Mh * 10 ^ 6 - TN * 10 ^ 3 * (D - t / 2)) / (fst * Z))
Astx = Ast + Ast2
IF Astx < Astmin THEN Astx = Astmin

'Y-direction steel
DO WHILE Mv > MaxMu
  D = D + 25
  t = t + 25
  MaxMu = .156 * fcu * br * D ^ 2 / 10 ^ 6
LOOP
k = Mv * 10 ^ 6 / (fcu * br * D ^ 2)
Z = D * (.5 + (.25 - k / .9) ^ .5)
IF Z > .94 * D THEN Z = .94 * D
'for base slab, there is tension in this direction as well
IF n < 0 THEN TN = ABS(n) ELSE TN = 0!
Ast = TN * 10 ^ 3 / fst
Ast2 = ABS((Mv * 10 ^ 6 - TN * 10 ^ 3 * (D - t / 2)) / (fst * Z))
Astv = Ast + Ast2
IF Astv < Astmin THEN Astv = Astmin

'Check for shear stress
Ast = MAX2(Astx, Astv)
stress = V * 10 ^ 3 / (br * D)
Ra = 100 * Ast / (br * D)
d1 = D
IF d1 > 400 THEN d1 = 400
Vc = .79 * Ra ^ (1 / 3) * (400 / d1) ^ .25 / 1.25
IF fcu > 25 THEN Vc = Vc * (fcu / 25) ^ (1 / 3)
IF stress > Vc THEN
     D = D * stress / Vc * 1.1
     t = D + cv + 10
     GOTO RECYCLE
END IF

'Check for cracking at serviceability limit state
n = n / 1.5                 'Service ring tension
Mmax = Mh
Asm = Ast
dr = 1
```

```
IF Mmax < Mv THEN
     Mmax = Mv
     Asm = Astv
     dr = 2
END IF
     Mmax = Mmax / 1.5              'Service maximum moment
     Ec = 5500 * fcu ^ .5           'Young's modulus of concrete N/mm2
     Ec = .5 * Ec                   'Service Young's modulus of concrete
     Es = 2 * 10 ^ 5                'Young's modulus for steel
     Alpha = Es / Ec                'modular ratio
     r = Alpha * Asm / (br * D)     'steel ratio
     x = D * r * ((1 + 2 / r) ^ .5 - 1)
     Z = D - x / 3                  'lever arm
     fs = Mmax / (Asm * Z)          'Service stress in steel
     fsa = .58 * fy                 'Allowable service stress
     IF fs > fsa THEN
          'Cracking is excessive, so increase area of steel
          fact = fs / fsa
          IF dr = 1 THEN Ast = Ast * fact ELSE Astv = Astv * fact
     END IF
END SUB

SUB Get.Cover (fcu, cv)
'**********************************************************
'Determine the cover for a septic tank
'**********************************************************
CALL box(1, 1, 20, 80)
row = 2
LOCATE row, 15
PRINT "Program 5.5: Structural design of tanks"
LOCATE row + 1, 5: PRINT "THE EXPOSURE CONDITION"
LOCATE row + 2, 5: PRINT "(1)  Mild"
LOCATE row + 3, 5: PRINT "(2)  Moderate"
LOCATE row + 4, 5: PRINT "(3)  Severe"
LOCATE row + 5, 5: PRINT "(4)  Very severe"
LOCATE row + 6, 5: PRINT "(5)  Extreme"
LOCATE row + 7, 5: INPUT "Select your choice"; EXC$

SELECT CASE EXC$
CASE "1"
     IF fcu <= 30 THEN cv = 25
     IF fcu >= 35 THEN cv = 20

CASE "2"
     IF fcu <= 35 THEN cv = 35
     IF fcu = 40 THEN cv = 30
     IF fcu = 45 THEN cv = 25
     IF fcu = 50 THEN cv = 20

CASE "3"
     IF fcu <= 40 THEN cv = 40
     IF fcu = 45 THEN cv = 30
     IF fcu = 50 THEN cv = 25

CASE "4"
     IF fcu <= 40 THEN cv = 50
     IF fcu = 45 THEN cv = 40
     IF fcu = 50 THEN cv = 30
```

```
CASE "5"
    IF fcu <= 45 THEN cv = 60
    IF fcu = 50 THEN cv = 50
END SELECT
END SUB

SUB GetBasePlateMoments (t, Lx, Ly, H, fcu, Mx, My)
'*****************************************************************
'Routine: PLATES ON ELASTIC FOUNDATIONS.
'file: QB\BASIC\ELASTPL.BAS
'written by Dr. A.W. Hago, Muscat, April 1995
'*****************************************************************
CALL box(1, 1, 20, 80)
LOCATE 2, 15
PRINT "STRUCTURAL DESIGN OF TANK BASE SLAB"
LOCATE 10, 5
INPUT "Enter the modulus of subgrade soil (KN/m2) (0 if unknown) ="; kmodulus
IF kmodulus = 0 THEN kmodulus = 39000   'Assumed clay soil
pi = 22! / 7!
E = 5500000! * (fcu) ^ .5      'Young's modulus of concrete, KN/m2
mu = .16                       'Poisson's ratio for concrete
udl = 10 * H                   'Uniform Pressure on the base slab

D = E * t ^ 3 / (12 * (1 - mu ^ 2))

x = .5 * Lx: y = .5 * Ly       'coordinates of center point
w = 0!
wxx = 0!
wyy = 0!
m1 = 15
n1 = 15
FOR m = 1 TO m1 STEP 2
FOR n = 1 TO n1 STEP 2
c = 16 * udl / (m * n * pi ^ 2)
cx = SIN(m * pi * x / Lx)
cy = SIN(n * pi * y / Ly)
cd = pi ^ 4 * D * ((m / Lx) ^ 2 + (n / Ly) ^ 2) ^ 2 + kmodulus
w = w + c * cx * cy / cd
wxx = wxx - c * (m * pi / Lx) ^ 2 * cx * cy / cd
wyy = wyy - c * (n * pi / Ly) ^ 2 * cx * cy / cd
NEXT n
NEXT m
Mx = -D * (wxx + mu * wyy)
My = -D * (wyy + mu * wxx)
END SUB
SUB GetCoverPlateMoments (Lx, Ly, udl, Mx, My)
'****************************************************
'Routine to Compute moments in a simply supported slab
'Using the deflection method (Dr. Marcus)
'****************************************************
Alpha = Ly / Lx        'Sides ratio
B = Alpha ^ 4
qx = udl * B / (1 + B) 'Load carried by x strips
qy = udl / (1 + B)     'Load carried by y strips
Mx = qx * Lx ^ 2 / 8
My = qy * Ly ^ 2 / 8
END SUB
```

```
SUB GetWallsMoments (Lx, Ly, H, fcu, MAXMX, MAXMY, MAXSX, MAXSY, MAXTN)
'*******************************************************************
'Routine to compute the moments on the walls of a tank, Using the
'Finite difference Method. The side walls are assumed as
'FIXED ON THREE SIDES, SIMPLY SUPPORTED ON THE FOURTH TOP SIDE,
'Under linearly varying Hydrostatic Pressure of a liquid.
'*******************************************************************
NDIVX = 4                   'Four divisions of one half for symmetry
NDIVY = 4                   'Four divisions on full height
Nx = NDIVX
NY = NDIVY - 1
DX = .5 * Lx / NDIVX          'Consider half wall for symmetry
DY = Ly / NDIVY
NDOF = Nx * NY                'Total number of degrees of freedom
udl = 10! * Ly               'Maximum liquid pressure in KN/m2
mu = .16                    'Poisson's ratio for concrete
E = 5500000! * fcu ^ .5       'Young's modulus for concrete, KN/m2

RIGIDITY = E * H ^ 3 / (12 * (1 - mu ^ 2))     'Flexural rigidity
pload = udl / RIGIDITY

MM = NDOF + 1

DIM S!(NDOF, MM), w(NDOF), Mx(NDOF), My(NDOF), MXV(NY), MYH(Nx), AXIAL(NDIVY)
DIM Q(NDOF)

L3 = DY ^ 3
L4 = DX ^ 4
Z = (DX / DY) ^ 2
A = (6 + 6 * Z ^ 2 + 8 * Z) / L4
B = -4 * (1 + Z) / L4
c = -4 * Z * (1 + Z) / L4
D = 2 * Z / L4
E = Z ^ 2 / L4
F = 1 / L4

FOR I = 1 TO NDOF
FOR J = 1 TO NDOF + 1
S!(I, J) = 0!
NEXT J
NEXT I

'CONNECTIVITY
k = 0
FOR I = 1 TO Nx
FOR J = 1 TO NY
K13 = 0: K14 = 0: K15 = 0: K16 = 0
k = k + 1
K1 = k - NY
K2 = k + NY
K3 = k - 1
K4 = k + 1
K5 = k + NY + 1
K6 = k - NY + 1
K7 = k - NY - 1
K8 = k + NY - 1
K9 = k - 2 * NY
K10 = k + 2 * NY
```

```
K11 = k - 2
K12 = k + 2
'special cases
IF I = 1 THEN
     K1 = 0
     K6 = 0
     K7 = 0
     K9 = k
END IF
IF I = 2 THEN K9 = 0
IF I = Nx THEN
     K2 = k - NY
     K5 = K2 + 1
     K8 = K2 - 1
     K10 = k - 2 * NY
END IF
IF I = Nx - 1 THEN K10 = k

IF J = 1 THEN
     K7 = 0
     K3 = 0
     K8 = 0
     K11 = k
END IF
IF J = 2 THEN K11 = 0
IF J = NY - 1 THEN
     K4 = 0
     K5 = 0
     K6 = 0
     K12 = k
END IF

S(k, k) = S(k, k) + A
IF K1 > 0 THEN S(k, K1) = S(k, K1) + B
IF K2 > 0 THEN S(k, K2) = S(k, K2) + B
IF K3 > 0 THEN S(k, K3) = S(k, K3) + c
IF K4 > 0 THEN S(k, K4) = S(k, K4) + c
IF K5 > 0 THEN S(k, K5) = S(k, K5) + D
IF K6 > 0 THEN S(k, K6) = S(k, K6) + D
IF K7 > 0 THEN S(k, K7) = S(k, K7) + D
IF K8 > 0 THEN S(k, K8) = S(k, K8) + D
IF K9 > 0 THEN S(k, K9) = S(k, K9) + F
IF K10 > 0 THEN S(k, K10) = S(k, K10) + F
IF K11 > 0 THEN S(k, K11) = S(k, K11) + E
IF K12 > 0 THEN
     IF J = NY - 1 THEN S(k, K12) = S(k, K12) - E
     IF J < NY - 1 THEN S(k, K12) = S(k, K12) + E
END IF

'load vector
Q(k) = pload * (1 - J / NDIVY)          'Hydrostatic pressure
NEXT J
NEXT I

CALL SOLVE(S(), Q(), w(), NDOF)
'Compute moment Mx on interior points
FOR I = 1 TO NDOF
FOR J = 1 TO NDOF
```

```
S(I, J) = 0!
NEXT J
NEXT I

r1 = 2 * RIGIDITY * (1 / DX ^ 2 + mu / DY ^ 2)
r2 = -RIGIDITY / DX ^ 2
r3 = -mu * RIGIDITY / DY ^ 2
k = 0
FOR I = 1 TO Nx
FOR J = 1 TO NY
k = k + 1
K1 = k - NY
K2 = k + NY
K3 = k - 1
K4 = k + 1
'special cases
IF I = 1 THEN K1 = 0
IF I = Nx THEN K2 = k - NY
IF J = 1 THEN K3 = 0
IF J = NY - 1 THEN K4 = 0
S(k, k) = S(k, k) + r1
IF K1 > 0 THEN S(k, K1) = S(k, K1) + r2
IF K2 > 0 THEN S(k, K2) = S(k, K2) + r2
IF K3 > 0 THEN S(k, K3) = S(k, K3) + r3
IF K4 > 0 THEN S(k, K4) = S(k, K4) + r3
NEXT J
NEXT I

FOR I = 1 TO NDOF
mmt = 0!
FOR J = 1 TO NDOF
mmt = mmt + S(I, J) * w(J)
NEXT J
Mx(I) = mmt
NEXT I

'Compute support moment MXV of the Vertical side Ly
FOR I = 1 TO NY
MXV(I) = -2 * RIGIDITY * w(I) / DX ^ 2
NEXT I

'Compute moment My on interior points
FOR I = 1 TO NDOF
FOR J = 1 TO NDOF
S(I, J) = 0!
NEXT J
NEXT I
r1 = 2 * RIGIDITY * (1 / DY ^ 2 + mu / DX ^ 2)
r2 = -mu * RIGIDITY / DX ^ 2
r3 = -RIGIDITY / DY ^ 2
k = 0
FOR I = 1 TO Nx
FOR J = 1 TO NY
k = k + 1
K1 = k - NY
K2 = k + NY
K3 = k - 1
K4 = k + 1
```

```
'special cases
IF I = 1 THEN K1 = 0
IF I = Nx THEN K2 = k - NY
IF J = 1 THEN K3 = 0
IF J = NY - 1 THEN K4 = 0

S(k, k) = S(k, k) + r1
IF K1 > 0 THEN S(k, K1) = S(k, K1) + r2
IF K2 > 0 THEN S(k, K2) = S(k, K2) + r2
IF K3 > 0 THEN S(k, K3) = S(k, K3) + r3
IF K4 > 0 THEN S(k, K4) = S(k, K4) + r3
NEXT J
NEXT I

FOR I = 1 TO NDOF
mmt = 0!
FOR J = 1 TO NDOF
mmt = mmt + S(I, J) * w(J)
NEXT J
My(I) = mmt
NEXT I

'Compute support moment MYH on horizontal side Lx
FOR I = 1 TO Nx
k = (I - 1) * NY + 1
MYH(I) = -2 * RIGIDITY * w(k) / DY ^ 2
NEXT I

'Compute axial tension
FOR I = 1 TO NDIVY
AXIAL(I) = 10 * Ly * (1 - I / NDIVY) * Lx / 2!
NEXT I

'Get maximum values
MAXMX = MAX(Mx(), NDOF)        'Maximum +ve x moment
MAXMY = MAX(My(), NDOF)        'Maximum +ve y moment
MAXSX = MAX(MXV(), NY)         'Maximum -ve x moment
MAXSY = MAX(MYH(), Nx)         'Maximum -ve y moment
MAXTN = MAX(AXIAL(), NDIVY)    'Maximum tension
END SUB

FUNCTION MAX (A(), n)
'*****************************************************
'Function to determine the maximum value in an array
'*****************************************************
x = 0
FOR I = 1 TO n
A(I) = ABS(A(I))
IF x < A(I) THEN x = A(I)
NEXT I
MAX = x
END FUNCTION

FUNCTION MAX2 (x, y)
'**********************************************************
' This function returns with the maximum of two numbers
'**********************************************************
Z = ABS(x)
```

```
IF x > y THEN Z = x ELSE Z = y
MAX2 = Z
END FUNCTION
SUB SOLVE (GSTIF(), GCON(), HEAD(), n)
'*******************************************************
'     SOLVING EQUATIONS BY GAUSSIAN ELIMINATION
'*******************************************************
DIM O(n), S(n), x(n)

'********** ORDERING THE equations
FOR I = 1 TO n
     O(I) = I
     S(I) = ABS(GSTIF(I, 1))
     FOR J = 2 TO n
       IF ABS(GSTIF(I, J)) > S(I) THEN
         S(I) = ABS(GSTIF(I, J))
       ELSE
       END IF
     NEXT J
    NEXT I

'elimination *******
   FOR k = 1 TO n - 1
   PIVOT = k
   BIG = ABS(GSTIF(O(k), k) / S(O(k)))
   FOR II = k + 1 TO n
      DUMMY = ABS(GSTIF(O(II), k) / S(O(II)))
      IF DUMMY > BIG THEN
      BIG = DUMMY
      PIVOT = II
      ELSE
      END IF
   NEXT II
   DUMMY = O(PIVOT)
   O(PIVOT) = O(k)
   O(k) = DUMMY

    FOR I = k + 1 TO n
        FACTOR = GSTIF(O(I), k) / GSTIF(O(k), k)
        FOR J = k + 1 TO n
          GSTIF(O(I), J) = GSTIF(O(I), J) - FACTOR * GSTIF(O(k), J)
        NEXT J
       GCON(O(I)) = GCON(O(I)) - FACTOR * GCON(O(k))
     NEXT I
    NEXT k

'********** BACK SUBSTITUTION ********
  x(n) = GCON(O(n)) / GSTIF(O(n), n)
  HEAD(n) = x(n)
  FOR I = n - 1 TO 1 STEP -1
     SUM = 0
     FOR J = I + 1 TO n
       SUM = SUM + GSTIF(O(I), J) * x(J)
      NEXT J
      x(I) = (GCON(O(I)) - SUM) / GSTIF(O(I), I)
      HEAD(I) = x(I)
   NEXT I
END SUB
```

```
SUB waitme
DO
LOCATE 22, 1: PRINT "Press a key to continue............"
LOOP WHILE INKEY$ = ""
END SUB
```

5.3.2 Imhoff Tanks

An Imhoff tank is a form of a septic tank. It is a two-story tank which performs the same functions of sedimentation and anaerobic digestion of sludge. The two processes are achieved in separate compartments. The incoming wastewater flows through the upper compartment, allowing solids to settle to the bottom of the chamber, which is in the shape of a hopper. At the bottom of the hopper, the solids pass through a baffled outlet into the lower chamber in which anaerobic digestion takes place. Since there is no intimate contact between sewage and the digesting sludge, a better effluent is obtained. Gases from the sludge compartment are directed away from the falling solids so as not to hinder their descent.[16] Although Imhoff tanks function better than septic tanks (especially in warm climates), nonetheless they are more elaborate and more costly to operate.

Advantages of the Imhoff tank:

- It yields good results without skilled attention.
- It has minimum problems with sludge disposal.

Disadvantages of the Imhoff tank:

- Greater depth means greater costs.
- Unsuitable when high acidic conditions prevail.
- No adequate control over its operation.
- Insufficient area of vents causes the gas to lift the fluffy scum, a condition known as "foaming".

For the design of an Imhoff tank the following points must be considered:

- Detention time 1 to 2 h.
- Length-to-width ratio between 2:1 to 4:1.
- Depth should be kept shallow, 10 to 12 m.
- Vent area for gas release and accumulation of scum should be 20 to 30% of total Imhoff tank surface area.

5.4 Homework Problems

5.4.1 Discussion Problems

1. What are the various sources of wastewater?
2. Outline the problems associated with sewer systems.

3. Define the following terms: sewer, sewerage system, and storm sewer.
4. What are the functions of sewers?
5. Indicate the classes of sewer systems. Point out advantages and disadvantages of each system. Which system would you recommend in your community? State your reasons.
6. What is the difference between the terms sewage and wastewater?
7. Briefly discuss the investigations that need to be carried out during sewer design.
8. Why do industrial and commercial rates of flow of wastewatere experience diurnal variations?
9. Indicate factors influencing design of sewers.
10. What are the benefits of maintaining a minimum flow rate in a sewer?
11. Outline factors affecting the volume of a stormwater that must be controlled.
12. Point out the limitations of the Rational method.
13. Why is there a need to have greater velocities in storm sewers compared to sanitary sewers?
14. Briefly explain the assumptions used in the design of sanitary sewers.
15. Define the following terms: one-dimensional flow, uniform flow, incompressible flow, steady flow, quasisteady flow, continuity equation, Bernoulli's equation, momentum principle, hydraulic radius, laminar flow, and self-cleaning velocity.
16. What are the serious effects of generation of sulfides in a sewer operations?
17. Indicate the factors that govern the generation of sulfide in a sanitary sewer.
18. Show how to avoid and control sulfide buildup ina sanitary sewer.
19. Write briefly about methods of wastewater disposal in rural areas.
20. Define the following terms: septic tank, septage, Imhoff tank, Imhoff cone, and V.I.P. latrine.
21. Outline the differences between a septic tank and an Imhoff tank.
22. Discuss the following statement, "The effluent of a septic tank is offensive and potentially dangerous."
23. What are the factors that affect the design of percolation fields?
24. State which system gives a better effluent, a septic tank or an Imhoff tank? Why? State your reasons.

25. Discuss the factors that affect the structural design of a reinforced concrete septic tank.

26. What are the pros and cons of using an Imhoff tank?

5.4.2 Specific Mathematical Problems

1. Using the equations mentioned in this chapter, write a computer program to find the limiting total sulfide, amount, and dissolved sulfide concentration that can develop in a sanitary sewer. Data given: diameter of pipe (ft), length of sanitary sewer (ft), flow rate (cfs), relative depth of flow, slope, Manning's coefficient, wastewater characteristics, BOD (mg/L), climatic temperature (°C), pH, insoluble sulfide (mg/L), total sulfide (mg/L), hydraulic characteristics — surface width of flow (ft), wetted perimeter (ft), exposed perimeter (ft), velocity of flow (fps), and mean hydraulic depth (ft).

2. Write a computer program to design a two-compartment septic tank. The volume of the tank can be computed from the following equation: V = 180*POP + 2000, where V is volume of tank in liters, POP is the number of people to be served by the unit.

3a. Write a computer program to design a suitable percolation field for a septic tank serving a certain number of people.

3b. Determine the size of a septic tank and the percolation field for a mobile home park which has 187 residents. Percolation tests indicate an average percolation rate of 6 mm/min.

4. Expand the computer program of Example 5.5 to enable plotting the plan, side view, and elevation of the septic tank.

References

1. Joint Task Force of the American Society of Civil Engineers and the Water Pollution Control Federation, *Gravity Sanitary Sewer Design and Construction*, ACSE Manuals and Reports on Engineering, Practice Number 60, ASCE, WPCF, New York, 1982.
2. Metcalf and Eddy, Inc. *Wastewater Engineering Treatment Disposal Reuse*, 3rd ed., McGraw-Hill, New York, 1991.
3. Abdel-Magid, I.M. *Selected Problems in Wastewater Engineering*, Khartoum University Press (KUP), National Council for Research (NRC), Khartoum 1986.
4. Davis, M.L. and Cornwell, D.A., *Introduction to Environmental Engineering*, 2nd ed., Chemical Engineering Series, McGraw-Hill, New York, 1991.
5. Steel, E.W. and McGhee, T.J., *Water Supply and Sewerage*, 5th ed., McGraw-Hill, New York, 1979.
6. Barnes, D., Bliss, P.J., Gould, B.W., and Vallentine, H.R., *Water and Wastewater Engineering Systems*, Pitman Publishing, Marshfield, MA, 1981.

7. Fair, G.M., Geyer, J.C., and Okun, D.A., *Water and Wastewater Engineering*, Vol. I and II, John Wiley & Sons, New York, 1966.
8. Frederick, S., Ed., *Standard Handbook for Civil Engineers*, McGraw-Hill, New York, 1976.
9. Hammer, M.J., *Water and Wastewater Technology*, John Wiley & Sons, New York, 1977.
10. Husain, S.K., *Textbook of Water Supply and Sanitary Engineering*, Oxford and IBH Publishers, New Delhi, 1981.
11. Merritt, F.S., *Standard Handbook for Civil Engineers*, McGraw-Hill, New York, 1976.
12. Peavy, H.S., Rowe, D.R., and Tchobanoglous, G., *Environmental Engineering*, McGraw-Hill, New York, 1988.
13. Vesilind, P.A. and Peirce, J.J., *Environmental Pollution Control*, Butterworths-Heinemann, London, 1990.
14. Mara, D., *Sewage Treatment in Hot Climates*, John Wiley & Sons, New York, 1980.
15. Salvato, J.A., *Environmental Engineering and Sanitation*, 4th ed., John Wiley & Sons, New York, 1992.
16. Scott, J.S. and Smith, P.G., *Dictionary of Waste and Water Treatment*, Butterworths, London, 1980.
17. Viessman, W. and Lewis, G.L., *Introduction to Hydrology*, Harper & Row, New York, 1989.
18. Viessman, W. and Hammer, M.J., *Water Supply and Pollution Control*, Harper & Row, New York, 1985.
19. Nathanson, J.A., *Basic Environmental Technology: Water Supply, Waste Disposal and Pollution Control*, John Wiley & Sons, New York, 1986.
20. U.S. Agency for International Development, *Water for the World*, Technical notes produced under contract to the U.S. Agency for International Development by National Demonstration Water Project, Institute for Rural Water and National Environmental Health Association, 1982. (a) *Designing Septic Tanks*, Technical Notes No. SAN2.D.3. (b) *Operating and Maintaining Septic Tanks*, Technical Note No. SAN.2.0.3.
21. Health Education Services, *Recommended Standards for Wastewater Facilities: Policies for the Design, Review, and Approval of Plans and Specifications for Wastewater Collection and Treatment Facilities*, report of the Wastewater Committee of the Great Lakes-Upper Mississippi River Board of State Public Health and Environmental Managers, Albany, 1990 Edition.

Chapter 6

Wastewater Treatment Technology and Disposal

Contents

6.1 Introduction

Wastewater may be described as, "A combination of the liquid or water-carried wastes removed from residences (domestic), institutions, commercial, and industrial establishments together with such groundwater, surface water, and storm water as may be present."[1,5] Wastewater is a complex heterogeneous solution which can pollute or contaminate our environment namely the water, air, food, and soil. Such conditions demand appropriate collection, proper treatment, and final disposal of wastewater in order to safeguard the environment and protect human health and welfare.[5] The first section in this chapter deals with the environmental problems associated with the discharge of inadequately treated wastewater. The sources of wastewater, as well as an evaluation and selection of the wastewater design flow rates (minimum, maximum, average and peak), are presented. Computer programs are developed to calculate these various flow rates. A discussion of the relationship between BOD_5 and the population equivalent (PE) is accompanied by a computer program that correlates these two factors.

The next section in this chapter gives the reasons for treating wastewater and the classification and selection of treatment processes (physical, chemical, and biological). The basic principles involved in the design of grit chambers and a computer program that can be used for this purpose are provided. The most commonly used biological processes such as activated sludge (and its many modifications), trickling filters (high and low rates), and waste stabilization ponds are presented. In all cases, computer programs to aid in the evaluation and design of each system are formulated. The operation and design of systems that can be used to render sludge suitable for disposal or use are considered. A computer program for the design of anaerobic digesters is presented here. Also discussed is the specific resistance test used to evaluate the dewatering characteristics of a sludge, which is accompanied by a computer program for determination of both specific resistance and the compressibility coefficient for a sludge. The design and scaling-up of centrifugal sludge dewatering units are considered, and computer programs that can be used for this purpose are developed. The last section of this chapter is devoted to the ultimate disposal of treated wastewater, including "dilution" (which, of course, in no solution to pollution) and discharge to surface water such as rivers, lakes, reservoirs, and estuaries. Computer programs that can be used to evaluate the effects of wastewater discharges in each of these situations is included.

The problems associated with inadequately treated wastewater discharges include:[1–6]

- Transmission of diseases (disease-causing agents such as bacteria, protozoa, viruses, and helminths) and other public health long-term physiological effects (such as newly created organic substances)
- Accumulation of highly persistent detergents, pesticides, and other toxic substances
- Generation of odors (e.g., hydrogen sulfide, ammonia, mercaptans, and other trace gases containing sulfurs, hydrogen, and nitrogen)
- Pollution of bathing sites with oil, grease, and other undesirable debris
- Establishment of eutrophic conditions (enrichment of water by plant nutrients, etc.)
- Production of objectionable and dangerous levels of solids in wastewater along the banks of rivers and streams, which leads to degradation of water quality both for groundwater and surface water
- Destruction of fish and wildlife
- Contamination of water supplies
- Aesthetic objections

6.2 Sources and Evaluation of Wastewater Flow Rates

Sources of generation of wastewater are as follows:

- *Domestic wastewater*: discharges from residences, commercial, institutional, and related units
- *Industrial wastewater*: discharges from factories and industrial establishments

- *Storm water*: water that results from rainfall runoff
- *Infiltration/inflow*: extraneous water that enters the sewer from the ground via different routes and storm water that is discharged from sources such as roof leaders, foundation drains, and storm sewers[5]

The domestic wastewater rate of flow is normally estimated by finding the dry weather flow, DWF, which is defined as the total average discharge of sanitary sewage and is the normal flow in a sewer during the dry weather.[5] Dry weather flow also may be described as the average daily flow in the sewer after several days during which rainfall has not exceeded 2.5 mm in the previous 24 h.[7] The DWF for a certain locality can be determined from Equation 6.1.

$$DWF = (POP*Q) + I_r + T_w - E_v \tag{6.1}$$

where:

DWF = Dry weather flow, L/d
POP = Number of people served by the sewer, dimensionless
Q = Average water consumption, L/c*d
I_r = Average infiltration into the sewer, L/d (also estimated as L/d/km length of sewer for different ages of the sewer). Usually, I_r varies between 0 and 30% of DWF.[5,7]
T_w = Average trade waste discharge, L/d
E_v = Rate of evaporation

The DWF can also be estimated from the data of water consumption as presented in Equation 6.2.

$$DWF = \chi*Q*POP \tag{6.2}$$

where:

DWF = Dry weather flow, m^3/s
χ = Percent water entering the sewerage system (usually ranging from 80 to 90% of water consumption)
Q = Water consumption, m^3/cs
POP = Population served, dimensionless

Maximum wastewater flow may be determined by using Equation 6.3.[9]

$$Q_{max} = a*DWF \tag{6.3}$$

where:

Q_{max} = Maximum daily wastewater flow, m^3/d
a = Factor (= 2 to 4)
DWF = Dry weather flow, m^3/s

Equation 6.4 gives an estimate of the minimum wastewater flow.[5,8]

$$Q_{min} = b'*Q_a \tag{6.4}$$

where:

Q_{min} = Minimum wastewater flow, m^3/d
b′ = Constant (with b′ = 30 to 50% for small communities, and b′ = 66 to 80% for large communities that are greater than 100,000 persons).
Q_a = Average wastewater flow, m^3/d

Equation 6.5 gives the relationship between maximum wastewater flows and the average flows called peaking factor.[9]

$$pf = Q_{max}/Q_a \tag{6.5}$$

where:

pf = Peaking factor given by Equation 6.6

$$pf = 5/(POP/1000)^{0.167} \tag{6.6}$$

where:

POP = Number of people served in thousands, dimensionless

Equation 6.6 is valid for a population size of 200 up to 1000.

Example 6.1

1. Write a computer program to determine the DWF for a sanitary sewer that is required to serve an area with a population number of POP, for an average daily sewage flow of Q (m^3/d). Take the daily infiltration in the area to be I (m^3/km*length of sewer) and total length of sewer as l (km).
2. An area with a population of 12,000 is to be served by a sanitary sewer. The average wastewater flow is 15 m^3/h per capita. The daily infiltration in the area is judged to be 55 m^3/km length of sewer. For a total sewer length of 7 km, compute the DWF.
3. Use the computer program developed in (1) to verify the manual solution obtained in (2).

Solution

1. For solution to Example 6.1 (1), see the following listing for Program 6.1.
2. Solution to Example 6.1 (2):
   - Given: POP = 12,000; Q = 15m^3/h*c, I_r = 55 m^3/length of sewer, length of sewer = 7 km.

- Use equation of DWF as: DWF = (POP*Q) + I_r + T_w – E_v.
- Find the average dry weather infiltration into the sewer as: I_r = 55 (m^3)*7 (km of sewer length) = 385 m^3/d.
- Assume both average trade waste discharge, T_w, and evaporation rate, E_v, to be negligible and find the DWF as: DWF = 12000*15/24 + 385 = 7885 m^3/d.

Listing of Program 6.1

```
DECLARE SUB box (r1%, c1%, r2%, c2%)      'Found in Program 2.1
'******************************************************************
'Program 6.1: Determination of The Dry Weather Flow
'******************************************************************
pi = 3.1415962#
g = 9.81

CALL box(1, 1, 20, 80)

LOCATE 2, 5
PRINT "Program 6.1: Determination of The Dry Weather Flow"
row = 3
LOCATE row, 5
INPUT "Enter the number of people to be served              ="; POP
LOCATE row + 1, 5
INPUT "Enter the average daily sewage flow Q (m3/h/c  ) ="; Q
LOCATE row + 2, 5
INPUT "Enter the daily infiltration rate  I (m3/Km)          ="; I
LOCATE row + 3, 5
INPUT "Enter the length of the sewer          L (km)          ="; L

Ir = I * L
Tw = 0
Ev = 0
DWF = POP * Q / 24 + Ir + Tw - Ev

LOCATE row + 5, 5
PRINT "The dry weather flow DWF        ="; DWF; "m3/d"

DO
LOOP WHILE INKEY$ = ""
END
```

Example 6.2

1. Write a computer program to determine the average, maximum, minimum, and peak wastewater flows, given population number and annual average water consumption (L/capita*d). Assume DWF amounting to any percent, a, of water consumption and minimum flow equal to any percent, b, of average flow.
2. Determine the different per capita wastewater flows (i.e., maximum daily, average daily, and minimum daily flow) for a population of 900 using the following data: annual average water consumption = 275 L/capita/d; DWF = 90% of water consumption; and minimum flow = 50% of average flow.
3. Use the computer program developed in (1) to verify the manual solution acquired in (2).

Solution

1. For solution to Example 6.2 (1), see the following listing of Program 6.2.
2. Solution to Example 6.2 (2):
   - Given POP = 900, Q = 275 L/c/d, DWF = 0.9*consumption, $Q_{min} = 0.5*Q_a$
   - Determine the average annual daily DWF, or wastewater flow = 0.9*water consumption = 0.9* 275 = 247.5 L/c.
   - Compute the peaking factor as: pf = $5/POP^{0.167} = 5/0.9^{0.167} = 5.09$.
   - Find maximum daily flow, or maximum 24-h flow = $pf*Q_a$ = 5.09*247.5 = 1259.5 L/c*d.

     Assuming a working time activity of 16 h, the design average daily flow will be = 1259.5*24/16 = 1889.7 L/c*d.
   - Determine the minimum daily flow as: $Q_{min} = 0.5*Q_a$ = 0.5*247.5 = 123.75 L/c*d.

Listing of Program 6.2

```
DECLARE SUB box (r1%, c1%, r2%, c2%)        'Found in Program 2.1
'*****************************************************************
'Program 6.2: Determination of Different Per Capita Wastewater Flows
'*****************************************************************
pi = 3.1415962#
g = 9.81

CALL box(1, 1, 20, 80)
LOCATE 2, 5
PRINT "Program 6.2: Determination of Different Per Capita Wastewater Flows"
row = 3
LOCATE row, 5
INPUT "Enter the number of people to be served        ="; POP
LOCATE row + 1, 5
INPUT "Enter the annual average water consumption  L/c.d="; Q
LOCATE row + 2, 5
INPUT "Enter the dry weather flow DWF as percentage of Q="; DWF
LOCATE row + 3, 5
INPUT "Enter the minimum flow as a percentage of Q      ="; Qmp

DWF = DWF * Q
Pf = 5 / (POP / 1000) ^ .167    'Peak factor
Qmax = Pf * DWF                 'maximum daily flow
Qd = Qmax * 24 / 16             'design average daily flow
Qmin = Qmp * DWF                  'minimum daily flow

LOCATE row + 5, 5
PRINT "Average annual daily DWF        ="; DWF; "L/c"
LOCATE row + 6, 5
PRINT "Maximum daily flow            ="; Qmax; " L/c.d"
LOCATE row + 7, 5
PRINT "Minimum daily flow            ="; Qmin; " L/c.d"

DO
LOOP WHILE INKEY$ = ""
END
```

6.3 Concept of Population Equivalent

Population equivalent (PE) of sewage or wastewater refers to a certain quality parameter — e.g., biochemical oxygen demand (BOD) or suspended solids (SS) — as a per capita contribution, compared to some quality parameter of the per capita contribution of a standard sewage. Domestic sewage is adopted as the standard. Thus, the PE of any sewage is the number of persons to the standard sewage. The relationship between PE and BOD_5 is presented in Equation 6.7.[7,50]

$$PE = BOD_5 * Q / BOD_s \tag{6.7}$$

where:

PE = Population equivalent

BOD_5 = 5-d BOD of wastewater, mg/L

Q = Wastewater flow rate, m^3/s

BOD_s = BOD of standard sewage (= 60 g of BOD_5 produced by an average person per day in the U.K. and 80 g of BOD_5 produced by an average person in the U.S. per day).

The main benefits of the population equivalent concept include:

- Estimation of the strength of industrial sewage for the purposes of waste treatment
- Measurement of changes of wastewater treatment for various industries
- Assessment of charges for factories and industrial firms

Example 6.3

1. Write a short computer program to find the population equivalent of an industrial establishment that produces wastewater of Q (m^3/d). The wastewater has a 5-d BOD of L (mg/L).
2. The daily wastewater production from a certain industry amounts to $6*10^6$L with a 5-d BOD of 370 mg/L. Determine the population equivalent for this industry.
3. Use the computer program developed in (1) to verify the manual solution in (2).

Solution

1. For solution to Example 6.3 (1), see the following listing of Program 6.3
2. Solution to Example 6.3 (2):
   - Given wastewater flow, Q = $6*10^6$ L/d, BOD = 370 mg/L.
   - Assume the average person exerts a 5-d BOD load of 0.06 kg/d.
   - Determine population equivalent as: PE = BOD_5 of waste*flow rate/BOD of standard sewage = ($370*10^{-3}$ g/L*$6*10^6$ L/d)/($0.06*10^3$ g/d) = 37,000.

Listing of Program 6.3

```
DECLARE SUB box (r1%, c1%, r2%, c2%)       'Found in Program 2.1
'****************************************************************
'Program 6.3: Population Equivalent of Industrial establishments
'****************************************************************
pi = 3.1415962#
g = 9.81

CALL box(1, 1, 20, 80)

LOCATE 2, 5
PRINT "Program 6.3: Population Equivalent of Industrial establishments"
row = 3
LOCATE row, 5
INPUT "Enter the daily wastewater production  Q (L/d)  ="; Q
LOCATE row + 1, 5
INPUT "Enter the 5 day BOD  in mg/L                    ="; BOD5

Ex = .06                    'Assumed per capita 5 day BOD load in kg/d
BOD5 = BOD5 / 1000            'convert to gm/L
PE = BOD5 * Q / (Ex * 10 ^ 3)   'Population equivalent

LOCATE row + 5, 5
PRINT "Population equivalent    PE      ="; PE

DO
LOOP WHILE INKEY$ = ""
END
```

6.4 Reasons for Treating Wastewater

The main reasons for treatment of wastewater include:

- Stabilize organic pollutants
- Reduce number of disease-causing agents found in sewage
- Prevent pollutants from entering water sources
- Reduce odors and other nuisances resulting from sewage
- Water reclamation and reuse
- By-product recovery and use

6.5 Wastewater Treatment Unit Operations and Processes

Wastewater treatment units can be classified simply:

1. *In situ* treatment works are small treatment plants (package plants) that handle the disposal of wastewater emerging from individual households or small rural communities.

2. Large treatment works are centrally located wastewater plants that treat the wastewater discharges from large metropolitan areas.

Table 6.1 gives a quick review of the most essential unit operation and processes used in water and wastewater treatment plants.

6.6 Preliminary Treatment: Grit Removal

Grit is composed of sand, gravel, cinders, ashes, metal fragments, glass, inert inorganic solids, pebbles, eggshells, bone chips, seeds, coffee and tea grounds, earth from vegetable washing, large organic particles such as food wastes, or other heavy solid material with specific gravities greater than those of the organic putrescent solids found in wastewater.

Problems in treating wastewater containing grit:

- Grit is an abrasive material.
- Grit will produce wear and tear on pumps and other treatment works.
- Grit absorbs oil and grease on its surface (grit becomes sticky and solidifies) in pipes, sumps, sedimentation tanks, or other areas of low hydraulic shear.
- Grit can occupy a valuable volume of the sludge digestor.

Grit chambers serve the following functions:

- Protect the moving mechanical equipment from abrasion and development of any abnormal wear and tear
- Reduce the formation of heavy deposits inside pipelines, channels, and conduits
- Reduce the frequency of digestor cleaning
- Facilitate the operation of treatment units

The quantity of grit in wastewater differs according to the status of the sewerage system, the quantity of storm water present, and the ratio of industrial to other sources of wastewater. Separation of grit particles in grit removal chambers relies on the difference in the specific gravity between the organic and inorganic solids. All particles are assumed to settle with a terminal velocity in agreement with Newton's law of motion as presented in Equation 6.8.[10]

$$v = (4*g[\rho_s - \rho]*d/3*C_D*\rho)^{0.5} = [4*g(s.g. - 1)*d/3*C_D]^{0.5} \quad (6.8)$$

where:

v = Terminal settling velocity, m/s

g = Gravitation acceleration, m/s^2

ρ_s = Mass density of particle, kg/m^3

ρ = Mass density of fluid, kg/m^3

d = Diameter of particle, m

s.g. = Specific gravity of settling particle, dimensionless.

TABLE 6.1
Water and Wastewater Treatment Units

Water treatment	
• Screening:	Bars, mesh, or strainer to remove large solids
• Flocculation	Mechanical mixing to favor agglomeration of solid particles
• Coagulation:	Addition of chemicals to coagulate suspended solids
• Sedimentation:	Settling out of flocculated substances
• Sand filtration:	Filtering out remaining suspended matter
• Disinfection:	Addition of disinfectant to kill harmful disease-causing agents
• Sludge treatment:	Thickening by gravity and then disposal
• Other:	Water softening, pH adjustment, fluoridation
Wastewater treatment	
Preliminary treatment:	
• Screening:	As above
• Grit removal:	Removing grit and inorganic matter (e.g., sand) but not organic matter
• Storm overflow:	Diverting sewage in excess of treatment plant capacity to storm water holding tanks
Primary treatment	
• Primary sedimentation:	Settling of suspended solids (only 40 to 60% removed), no chemicals added
Secondary treatment	
• Aerobic oxidation of organic matter:	Biodegrading organic matter through the action of microorganisms in a biological treatment unit such as activated sludge or trickling filter plant, etc.
• Secondary sedimentation:	Settling out of sludge containing microorganisms to produce a treated effluent
Tertiary treatment (effluent polishing, advanced treatment)	
• Finalizing treatment	Polishing of effluent by operations such as sand filters, microstrainers, etc.
Sludge Treatment	
• Anaerobic digestion:	Decomposing thickened sludge in absence of oxygen
• Gravity thickening:	Thickening of primary and secondary sludge
• Mechanical dewatering:	Removing water from sewage sludges by methods such as centrifuges, pressure or vacuum filters, etc.
• Drying beds:	Drying sewage sludge in open atmosphere
Sludge disposal	
• Composted to be used as a soil conditioner (digested only)	
• Dumped at sea (undigested)	
• Incinerated (normally undigested but thickened)	
• Landfilled (preferably digested and dewatered or dried)	

Source: Rowe, D.R. and Abdel-Magid, I.M., *Handbook of Wastewater Reclamation and Reuse*, CRC Press/Lewis Publishers, Boca Raton, FL, 1995. With permission.

$$C_D = \text{Drag coefficient (dimensionless)} = (24/Re) + [3/(Re)^{1/2}] + 0.34 \quad (6.9)$$

where:

Re = Reynolds number, dimensionless.

Particles within the grit chamber are scoured at a horizontal scouring velocity given by Equation 6.10.[10]

$$v_{sco} = [8*a(s.g. - 1)*g*d/f]^{1/2} \quad (6.10)$$

where:

v_{sco} = Horizontal velocity of flow (scour velocity), m/s
s.g. = Specific gravity of the particle, dimensionless
f = Darcy-Weisbach friction factor, dimensionless (usually, f = 0.02 to 0.003)[10]
a = A constant, dimensionless (usually a = 0.04 to 0.06)[10]

Generally, a grit removal system possesses the following properties:

- A long constant-velocity channel.
- A length-to-depth ratio of 10 (i.e., l/H = 10). In practice (owing to turbulence at inlet and outlet), values of l/h up to 25 are employed.
- A width-to-depth ratio of 2, i.e., B/h = 2.
- The width of the channel at any point above the invert has a settling velocity of 30 cm/s (to avoid removal of organic particles) and is given by Equation 6.11.[7,11]

$$B = 4.92*Q_{max}/d_{max} \quad (6.11)$$

where:

B = Width of the channel, m
Q_{max} = Maximum flow rate, m^3/s
d_{max} = Maximum depth of flow, m

Velocity control within the grit chamber may be made by a flow control device such as a standing wave flume, vertical throat, proportional flow weir, etc. For standing wave flumes, the flow rate is presented by Equation 6.12.[7,11]

$$Q = 1.71*y*d^{1.5} = 1.14*B*d^{1.5} \quad (6.12)$$

where:

Q = Flow rate, m^3/s
d = Depth of flow, m
y = Throat width, m

The throat width may be computed as given in Equation 6.13.

$$y = 2*B/3 \tag{6.13}$$

where:

B = Width of channel at any point above invert, m

Table 6.2 presents typical design parameters of grit chambers.

TABLE 6.2
Grit Chamber Typical Design Parameters[1,6,11,12]

Parameter	Value
Detention time (s)	60
Horizontal design velocity (cm/s)	30
Equivalent diameter of grit removed (mm)	0.2
Specific gravity of particles captured	2.65
Length of chamber (m)	$> 18*d_{max}$

Example 6.4

1. Write a computer program for designing a grit chamber that consists of N mechanically cleaned channels. Each channel of the chamber is required to carry a maximum flow of Q_{max} (m^3/s) with a maximum depth of d_{max} (m).
2. Design a grit chamber that consists of four mechanically cleaned channels. Each channel is to carry a maximum flow of 725 L/s with a maximum depth of 0.6 m (use $1 = 18*d_{max}$).
3. Use the computer program developed in (1) to verify the manual solution in (2).

Solution

1. For the solution to Example 6.4 (1), see the following listing of Program 6.4.
2. Solution to Example 6.4 (2):
   - Given Q_{max} = 725 L/s, d_{max} = 0.6 m, N = 4.
   - Determine the width of the channel as: $B = 4.92*Q_{max}/d_{max} = 4.92*0.725/0.6 = 5.95$ m.
   - Width of each channel = total width/number of channels = B/N = 5.95/4 = 1.5 m.
   - Find the throat width of the standing wave flume in channels that maintains the velocity as: y = (2/3)*B = (2/3)*1.5 = 0.99 m.
   - Determine length of the channel as: $1 = 18*d_{max} = 18*0.99 = 17.8$ m.

Listing of Program 6.4

```
DECLARE SUB box (r1%, c1%, r2%, c2%)      'Found in Program 2.1
'*****************************************************************
'Program 6.4: Designing a Grit Chamber
'*****************************************************************
```

```
pi = 3.1415962#
g = 9.81

CALL box(1, 1, 20, 80)

LOCATE 2, 25
PRINT "Program 6.4: Designing a Grit Chamber"
row = 3
LOCATE row, 5
INPUT "Enter the number of mechanically cleaned channels N="; N
LOCATE row + 1, 5
INPUT "Enter the flow per channel            Q (m3/s) ="; Q
LOCATE row + 2, 5
INPUT "Enter the required maximum depth of flow  dmax (m)  ="; dmax

B = 4.92 * Q / dmax          'width of channel
W = B / N                    'width of each channel
y = 2 / 3 * W                'throat width
L = 18 * y                   'length of channel

LOCATE row + 5, 5
PRINT "Width of each channel             ="; W; "  m"
LOCATE row + 6, 5
PRINT "Throat width                    ="; y; " m"
LOCATE row + 7, 5
PRINT "length of channel                ="; L; " m"

DO
LOOP WHILE INKEY$ = ""
END
```

6.7 Secondary Treatment (Aerobic and Biological)

6.7.1 Introduction

Secondary wastewater treatment serves the following functions:

- Stabilization of organic matter
- Coagulation and removal of nonsettleable colloidal solids
- Reduction of organic matter usually found in sewage sludge
- Reduction of growth nutrients in sewage, such as nitrogen and phosphorous

Secondary treatment facilities can be broadly classified as:

- *Attached growth processes*: In these processes, microorganisms are fixed or attached to a solid surface or media. Organic matter is brought in contact with the microorganisms by various mechanisms.[23]
- *Suspended growth processes*: In these processes, microorganisms (bacteria, fungi, and protozoa) and small organisms (rotifers and nematode worms) are free to move within the reactor and utilize the organic material for energy, growth, and reproduction.

6.7.2 Suspended Growth Systems (Aerobic Suspended Growth Process): Activated Sludge Process

Activated sludge (a.s.) is a form of a suspended growth system. This process is an aerobic, continuous (or semicontinuous, fill-and-draw) method used for biological wastewater treatment. This process includes carbonaceous oxidation and nitrification (see Figure 6.1). The activated sludge process utilizes all or part of the following:

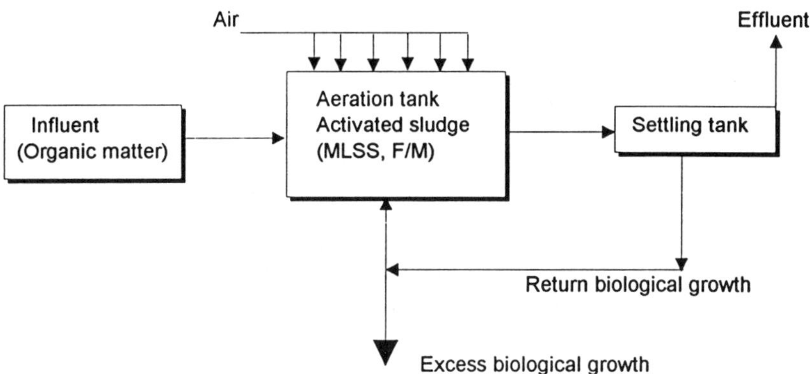

Figure 6.1
Activated sludge process; (MLSS = mixed liquor suspended solids. F/M = food-to-microorganism ratio.)

- Dissolved and colloidal biodegradable organic substances
- Unsettled suspended matter
- Mineral nutrients (e.g., phosphorous and nitrogen compounds)
- Volatile organic materials
- Other materials which can be sorbed on, or entrapped by, the activated sludge floc

Oxygen is provided to the activated sludge to:

- Supply needed oxygen for microbial oxidation and synthesis reactions[26]
- Properly mix contents of the aeration reactor
- Maintain microorganisms in suspension

The activated sludge floc adsorbs suspended and colloidal solids on its surface and, to some extent, soluble organic substances found in wastewater. Microbial activity transforms part of the available organic matter into a reserve food, thus rapidly reducing the BOD. Growth and multiplication of the aerobic microorganisms in the tank form an active biomass termed activated sludge. The activated sludge and wastewater in the aeration unit is called a mixed liquor.[8] The rate of removal of organic constituents depends upon the remaining BOD and the concentration of the activated sludge. The mixed culture of microbial species usually found in an activated sludge system includes viruses, bacteria, protozoa, and other organisms, singly or jointly grouped. Generally, the microbes are contained in a fabric of organic debris,

activity transforms part of the available organic matter into a reserve food, thus rapidly reducing the BOD. Growth and multiplication of the aerobic microorganisms in the tank form an active biomass termed activated sludge. The activated sludge and wastewater in the aeration unit is called a mixed liquor.[8] The rate of removal of organic constituents depends upon the remaining BOD and the concentration of the activated sludge. The mixed culture of microbial species usually found in an activated sludge system includes viruses, bacteria, protozoa, and other organisms, singly or jointly grouped. Generally, the microbes are contained in a fabric of organic debris, dead cells, or other waste products. The factors that influence operation of the process include sensitivity of organisms to nutrients, the wastewater composition, and other conditions (e.g., inorganic salt content, turbulence, pH, temperature, presence of other competing microorganisms, etc.).

6.7.3 Activated Sludge Process Kinetics

Kinetics of the activated sludge may be represented by the Monod equation for evaluation of the rate of limiting substrate biological growth. The equation relates the growth rate of organisms to the growth-limiting substrate concentration, as shown in Equation 6.14.

$$\mu_s = [(\mu_s)_{max}{*}s^*]/(K_s + s^*) \qquad (6.14)$$

where:

μ_s = Growth rate of microorganisms, per day = DR/R_u
DR = Dilution rate
R_u = Recycling ratio, dimensionless (6.15)
$(\mu_s)_{max}$ = Maximum growth rate of microorganisms, per day
K_s = Half-velocity constant (substrate concentration expressed in mg/L at half of the maximum growth rate)
s^* = Growth-limiting substrate concentration in solution, mg/L

When the substrate concentration is low, compared to the half velocity constant, Equation 6.14 may be simplified into a first-order equation, as presented in Equation 6.16.

$$\mu_s = (\mu_s)_{max}{*}s^*/K_s = K{*}s^* \qquad (6.16)$$

where:

μ_s = Growth rate of microorganisms
$(\mu_s)_{max}$ = Maximum growth rate of microorganisms
K_s = Half-velocity constant
K = Constant

2. A fully mixed continuous activated culture with partial feedback of cells is adopted for wastewater treatment. The recycling ratio of the process is 2. Values of the half-velocity constant and maximum growth rate of microorganisms are 1.5 mg/L and 1/h, respectively. Evaluate the concentration of the growth limiting substrate in the system for a dilution rate of 0.75/h.
3. Use the computer program developed in (1) to verify solution obtained in (2).

Solution

1. For the solution to Example 6.5 (1), see the following listing of Program 6.5.
2. Solution to Example 6.5 (2):
   - Given $R_u = 2$, DR = 0.75, $K_s = 1.5$, $\mu_{max} = 1$.
   - Find $\mu_s = DR/R_u = 0.75/2 = 0.375$.
   - Find the concentration of growth limiting substrate from the Monod equation as: $0.375 = 1*s^*/(1.5 + s^*)$: therefore, $s^* = 0.9$ mg/L.
3. Use the computer program written in (1) to verify computations attained in (2).

Listing of Program 6.5

```
DECLARE SUB box (r1%, c1%. r2%, c2%)      'Found in Program 2.1
'*******************************************************************
'Program 6.5: Determination of Growth Rate of Microorganisms
'*******************************************************************
pi = 3.1415962#
g = 9.81

CALL box(1, 1, 20, 80)

LOCATE 2, 10
PRINT "Program 6.5: Determination of Growth Rate of Microorganisms"
row = 5
LOCATE row, 5
INPUT "Enter the dilution rate  DR          (per hour)   ="; DR
LOCATE row + 1, 5
INPUT "Enter the recycling ratio Ru                      ="; Ru
LOCATE row + 2, 5
INPUT "Enter half-velocity constant   Kx                 ="; Kx
LOCATE row + 3, 5
INPUT "Enter the maximum growth rate  Umax   (/d) ="; Umax

Ux = DR / Ru
S = Ux * Kx / (Umax - Ux)

LOCATE row + 7, 3
PRINT "The growth limiting substrate concentration   ="; S; "  mg/L"

DO
LOCATE 22, 1
PRINT "Press any key to continue.........."
LOOP WHILE INKEY$ = ""
END
```

6.7.4 Factors Affecting the Activated Sludge Process

The main factors affecting the activated sludge process include volume and sludge loading, sludge age, sludge volume index, sludge density index, wastewater aeration time, wastewater flow and quality, mixed liquor suspended solids, dissolved oxygen in reactor, conditions of mixing and turbulence, wastewater temperature, and wastewater solids content.[1,4–6,10–14,19–27]

6.7.4.1 Volume and Sludge Loadings

The volume and sludge loadings are given by the ratio of food to microorganisms, or F/M (also referred to as sludge loading rate, SLR, or substrate loading). Equation 6.18 defines the F/M ratio.

$$\begin{aligned} F/M &= SLR \\ &= \text{mass of } BOD_5 \text{input to aeration basin/(MLVSS * tank volume)} \\ &= L_1 * Q / MLVSS * V \\ &= L / MLVSS * t \end{aligned} \tag{6.18}$$

where:

F/M	=	Food-to-microorganism ratio, per day
L_i	=	Influent BOD concentration, mg/L
Q	=	Wastewater flow, m^3/s
MLVSS	=	Mixed liquor volatile suspended solids concentration, mg/L
V	=	Tank volume, m^3
t	=	Tank retention time, d

Operation of an activated sludge unit at a high F/M ratio results in an incomplete metabolism of organic matter, inadequate BOD removal, and a lowering of the settling rate. On the contrary, operation of the process at a low F/M ratio increases the efficiency of organic matter removal, improves settleability of the activated sludge, and augments BOD removal. The volumetric organic loading, VOL rate, is defined by Equation 6.19.

$$VOL = Q*L_i/V = SLR*MLVSS \tag{6.19}$$

where:

VOL	=	Volumetric organic loading rate
SLR	=	Sludge loading rate, d

Table 6.3 outlines suitable values for sludge loading rate for particular treatment systems.

TABLE 6.3
SLR for Certain Treatment Units[1,11,12]

Unit	Reasonable value of SLR (d^{-1})
Conventional plants	0.3 – 0.35
Extended aeration	0.05 – 0.2
Step aeration	0.2 – 0.5

Note: SLR = sludge loading rate.

6.7.4.2 Sludge Age (SA), Mean Cell Residence Time (MCRT), Solids Retention Time (SRT), or Cell Age (CA)

Sludge age is defined as the total sludge in the biological treatment process divided by the daily waste sludge. This is presented in Equation 6.20.

$$\begin{aligned} SA &= [\text{mass sludge solids in aeration tank (kg)}]/ \\ &\quad [\text{mass sludge solids wasted (kg/d)}] \\ &= V*MLVSS/Q_w*SS \end{aligned} \quad (6.20)$$

where:

SA = Sludge age, d
V = Volume of aeration tank, m^3
MLVSS = Concentration of mixed liquor volatile suspended solids, mg/L
Q_w = Waste sludge flow, m^3/d
SS = Suspended solids in waste sludge, mg/L

The sludge age also is related to the reciprocal of the net microorganisms specific growth rate, as depicted in Equation 6.21.

$$1/SA = \{a(F/M)/Eff\} - k \quad (6.21)$$

where:

SA = Sludge age, d
a = Cell yield coefficient, mg cell per mg substrate
F/M = Food-to-microorganisms ratio, d^{-1}
Eff = Efficiency of BOD removal
k = Microorganisms endogenous decay coefficient, d^{-1}

For a completely mixed process operating under steady-state conditions, the sludge age usually is calculated by the relationship presented in Equation 6.22.

$$SA = V*SS/[Q_w*SS_R + (Q - Q_w)*SS_e] \quad (6.22)$$

where:

SA = Sludge age, d
V = Volume of aeration tank, m^3
Q_w = Waste sludge flow, m^3/d
SS_R = Suspended solids in recycle line, mg/L
Q = Influent volumetric flow rate, m^3/s
SS = Suspended solids in aeration tank, mg/L
SS_e = Suspended solids in final effluent, mg/L

Table 6.4 gives the value of the mean cell residence time for the various modifications of the activated sludge process.

TABLE 6.4
Mean Cell Residence Time for Various Modifications of the Activated Sludge Process[1,11]

Process	(SA) (d)
Contact stabilization	5–15
Conventional activated sludge	5–15
Extended aeration	20–30
High rate aeration	5–10
Modified aeration	0.2–0.5
Step aeration	5–15

Note: SA = sludge age.

Example 6.6

1. Write a computer program to determine the detention time (h), the volumetric organic loading rate (kg BOD/m^3/d), and the sludge loading ratio (kg BOD per kg MLVSS per day), given flow rate (m^3/d), BOD of incoming waste (mg/L), number of aeration tanks (activated sludge units), and their dimensions (m). Design the program to determine the sludge age for the activated sludge system given the daily sludge wasted (kg/d).
2. Settled sewage flowing at a daily rate of 5150 m^3 is introduced to an activated sludge plant. The sewage has a 5-day BOD of 245 mg/L. The aeration unit is composed of two aeration tanks each of dimensions: 3-m depth, 8-m width, 40-m length, with a MLVSS concentration of 2100 mg/L.
   - Determine the aeration unit detention time.
   - Find the volumetric organic loading rate.
   - Compute the sludge loading ratio of the aeration unit.

3. Determine the sludge age for the activated sludge system mentioned in (2) given that the daily sludge wasted was 495 kg.
4. Use the program developed in (1) to verify the manual computations of sections (2) and (3).

Solution

1. For the solution to Example 6.6 (1), see the following listing of Program 6.6.
2. Solution to Example 6.6 (2):
   - Given Q = 5150 m^3/d, L_1 = 245 mg/L, V for each tank = 3*8*40, MLVSS = 2100, number of tanks = 2.
   - Determine the volume of the aeration unit = volume of each tank*number of tanks = 3*8*40*2 = 1920 m^3.
   - Find the detention time: t = V/Q = 1920/5150 = 0.37 d = 8.9 h
   - Compute the volumetric organic loading rate as: VOL = Q*L_i/V = (5150 m^3/d*245*10^{-3} kg/m^3)/1920 = 0.66 kg BOD/m^3/d.
   - Determine the sludge loading ratio, SLR, as: F/M = SLR = L_i*Q/MLVSS*V = (245*10^{-3} kg/m^3*5150 m^3/d)/(2100*10^{-3} kg/m^3*1920 m^3) = 0.31 kg BOD/kg MLVSS/d.
3. Solution to Example 6.6 (3):
   - Given Q = 5150 m^3/d, L_i = 245 mg/L, V for the plant = 1920 m^3, MLVSS = 2100 mg/L, Q_w*SS = 495 kg/d.
   - Compute the SA as: SA = V*MLVSS/Q_w*SS = 1920*2100*10^{-3}/495 = 8.1 d.

Listing of Program 6.6

```
DECLARE SUB box (r1%, c1%, r2%, c2%)      'Found in Program 2.1
'*****************************************************************
'Program 6.6: Activated Sludge Processes
'*****************************************************************
pi = 3.1415962#
g = 9.81

CALL box(1, 1, 20, 80)
LOCATE 2, 30
PRINT "Program 6.6: Activated Sludge Processes"
row = 3
LOCATE row, 5
INPUT "Enter the flow rate                Q (m3/d)     ="; Q
LOCATE row + 1, 5
INPUT "Enter the number of aeration tanks              ="; N
row = 4
FOR I = 1 TO N
row = row + 1
LOCATE row, 5: PRINT "TANK NO."; STR$(I); ":"
LOCATE row, 20: INPUT "LENGTH ="; L(I)
LOCATE row, 35: INPUT "WIDTH ="; W(I)
LOCATE row, 50: INPUT "DEPTH ="; D(I)
NEXT I
LOCATE row + 1, 5
```

```
INPUT "Enter the 5-day BOD              (mg/L)        ="; Li
LOCATE row + 2, 5
INPUT "Enter the MLVSS concentration    (mg/L)        ="; MLVSS

'Determine total volume of units
V = 0!
FOR I = 1 TO N
V = V + L(I) * W(I) * D(I)
NEXT I

Li = Li / 1000                  'convert Kg/L
MLVSS = MLVSS / 1000            'convert to Kg/L
t = V / Q * 24                  'Detention time in hours
VOL = Q * Li / V                'Volumetric organic loading rate
SLR = Li * Q / (MLVSS * V)      'Sludge loading ratio

CALL box(1, 1, 20, 80)
LOCATE 2, 15: PRINT "Program 6.6: Activated Sludge Processes"
LOCATE 5, 5
PRINT "The detention time                 ="; t; " hr"
LOCATE 6, 5
PRINT "The volumetric organic loading rate  ="; VOL; " Kg BOD/m3/d"
LOCATE 7, 5
PRINT "The sludge loading ratio           ="; SLR; " Kg BOD/Kg MLVSS/d"
LOCATE 10, 5
INPUT "Do you want to compute the sludge age? Y/N :"; R$
IF UCASE$(R$) = "Y" THEN
     LOCATE 11, 5
     INPUT "Enter the daily sludge wasted     (Kg)="; DSL
     SA = V * MLVSS / DSL
     LOCATE 12, 5
     PRINT "The sludge age for this daily sludge ="; SA; " days"
END IF

DO
LOCATE 22, 1
PRINT "Press any key to continue.........."
LOOP WHILE INKEY$ = ""
END
```

6.7.4.3 Effects of Sludge Volume Index (Mohlman Sludge Volume Index)

The sludge volume index (SVI) estimates the settleability of the activated sludge, and it monitors the performance and operation of the aeration unit. The SVI is also defined as the volume in milliliters occupied by 1 g of activated sludge mixed liquor solids, dry weight, after settling for 30 min in a 1-L graduated cylinder. Equation 6.23 expresses mathematically the concept of the sludge volume index.

$$SVI = V_s * 1000 / MLVSS \tag{6.23}$$

where:

SVI = Sludge volume index, mL/g

V_s = Settled volume of sludge in a 1000-mL graduated cylinder in 30 min (mL/L or %)

1000 = mg/g

MLVSS = Mixed liquor volatile suspended solids, mg/L

Table 6.5 gives SVI values relating to the activated sludge.

TABLE 6.5
Activated Sludge Settleability According to SVI[1,11,18]

Value (mL/g)	Criteria
< 40	Excellent settling properties
40–75	Good settling properties
76–120	Fair settling properties
121–200	Poor settling properties
> 200	Bulking sludge

Note: SVI = sludge volume index.

6.7.4.4 Sludge Density Index or Donaldson Index

Sludge density index (SDI) is the reciprocal of the SVI multiplied by 100 as given by Equation 6.24.

$$SDI = 100/SVI \quad (6.24)$$

where:

SDI = Sludge density index, g/mL

SVI = Sludge volume index, mL/g

Sludge density index varies from around 2 for a good sludge to about 0.3 for a poor sludge, i.e., $0.3 < SDI < 2$.

Example 6.7

1. Write a computer program to compute the SVI (mL/g) and the sludge density index, SDI, (g/mL) of wastewater in a step aeration plant given the concentration of the mixed liquor suspended solids as MLSS (mg/L) and the volume of sludge settled after 1/2 h in a 1-L graduated cylinder, V_s (mL).
2. In a step aeration plant, the concentration of the mixed liquor suspended solids was 2400 mg/L. A sample of the wastewater was withdrawn for a sludge volume index, SVI, test. The volume of sludge settled after 1/2 h in a 1-L graduated cylinder was found to be 285 mL. Compute the SVI of the wastewater and the sludge density index, SDI.
3. Use the computer program of section (1) to verify the results obtained by manually solving section (2).

Solution

1. For solution to Example 6.7 (1), see the following listing of Program 6.7.
2. Solution to Example 6.7 (2):
   - Given V_s = 285 mL, MLVSS = 2400 mg/L.
   - Compute the sludge volume index of the sample as: SVI = V_s*1000/MLVSS = 285*1000/2400 = 119 mL/g.
   - From Table 6.5 it can be concluded that this SVI value signifies fair settling properties.
   - Determine the SDI as: SDI = 100/SVI = 100/119 = 0.84.

Listing of Program 6.7

```
DECLARE SUB box (r1%, c1%, r2%, c2%)      'Found in Program 2.1
'******************************************************************
'Program 6.7: Sludge Volume Index
'******************************************************************
pi = 3.1415962#
g = 9.81

CALL box(1, 1, 20, 80)
LOCATE 2, 35
PRINT "Program 6.7: Sludge Volume Index"
row = 3
LOCATE 3, 5
INPUT "Enter the concentration of mixed liquor suspended solids,  mg/L ="; MLVSS
LOCATE 4, 5
INPUT "Enter the Volume of sludge settling in half hour, mL               ="; Vx
SVI = Vx * 1000 / MLVSS
SDI = 100 / SVI

IF SVI <= 40 THEN c$ = "Excellent settling properties"
IF SVI > 40 AND SVI <= 75 THEN c$ = "Good settling properties"
IF SVI > 75 AND SVI <= 120 THEN c$ = "Fair settling properties"
IF SVI > 120 AND SVI <= 200 THEN c$ = "Poor settling properties"
IF SVI > 200 THEN c$ = "Bulking sludge"

LOCATE 7, 3
PRINT "The sludge volume  index ="; SVI, c$
LOCATE 8, 3
PRINT "The sludge density index ="; SDI

DO
LOCATE 22, 1
PRINT "Press any key to continue.........."
LOOP WHILE INKEY$ = ""
END
```

Table 6.6 gives general design criteria of the conventional activated sludge process.

TABLE 6.6
Design Criteria for Conventional Activated Sludge[1,2,11,18,26]

Item	Value
MLSS (mg/L)	1500–3000
Volumetric organic loading (g BOD per m^3 per d)	500–700
Aeration detention time (h) (based on average daily flow)	4–8
F/M ratio (g BOD per g MLSS per d)	0.1–0.6
Sludge retention time (d)	5–15
Sludge age (d)	3–4
Recycling ratio	0.25–0.5
Sludge yield index (kg solids per kg BOD removed)	0.7–0.9
BOS_5 removal (%)	85–95
Optimum pH for aerobic bacterial growth	6.5–7.5
Depth of aeration tanks (m)	3

Note: MLSS = mixed liquor suspended solids. BOD = biological oxygen demand. F/M = food-to-microorganism ratio.

6.8 Attached Growth Treatment Processes: Trickling Filter

6.8.1 Introduction

An attached growth culture is used to treat wastewater that comes in contact with microorganisms attached to surfaces of the media of the reactors. The trickling filtration process is a form of an attached growth system. In this process, settled wastewater is brought in contact with microorganisms attached to the filter bed media (see Figure 6.2). Wastewater is sprayed over a fixed media composed of a bed often packed with rocks or plastic structures. Thus, biological slimes develop and coat the filter surface area. These slimes provide a film consisting of heterotrophic organisms, facultative bacteria, protozoa, fungi, algae, rotifers, sludge worms, insect larvae, and snails. As the wastewater trickles over the media, microbes in the slime layer extract organic and inorganic matter from the liquid film. Suspended and colloidal particles are retained on the surfaces to be further converted to soluble products. Aerobic reactions are sustained by oxygen from the gas phase in the pores of the media through the liquid film to the slime layer. The digested waste products diffuse outward with the water flow or air currents through the voids in the filter medium. With time the biological film increases in thickness. This increase in film thickness is accompanied by a lowering of the content of oxygen and food. This condition eventually leads to anaerobic and endogenous metabolism at the slime–media interface, with increased detachment of organisms from the media. The shearing action of the flowing wastewater across the film then detaches and washes away the slime from the media. This process is referred to as filter sloughing off. Sloughing off develops because the microorganisms grow and multiply and the slime layer gets thicker until it is washed from the media by the incoming wastewater.

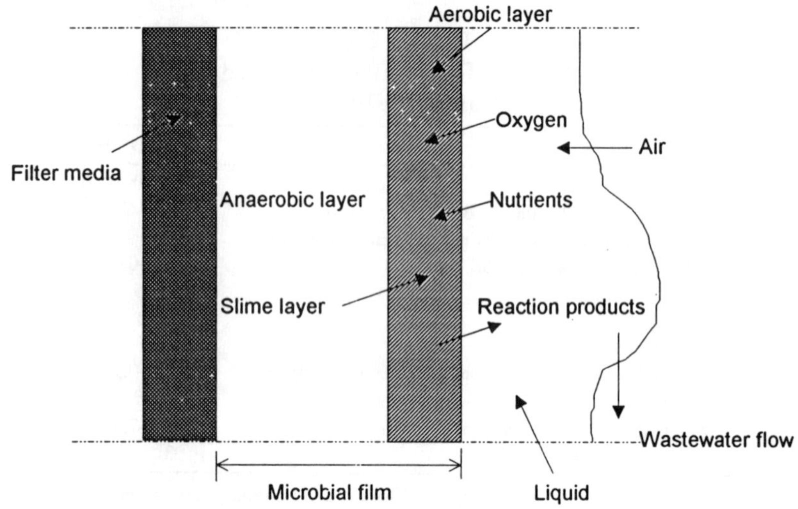

Figure 6.2
Schematic diagram of a trickling filter process.

Trickling filters are classified on the basis of hydraulic loading rate (shear velocities) and organic loading rate (metabolism in slime layer). As such, trickling filters are grouped into low rate and high rate filters (see Figure 6.3). Low rate filters are rather simple, having low hydraulic loading rates, and do not incorporate the recycling of wastewater effluent. A constant hydraulic loading is maintained in low rate filters by a dosing siphon or by a suction-level controlled pump. On the other hand, high rate filters have a high hydraulic loading maintained by recirculating part of the effluent from the filter or part of the final effluent.

6.8.2 Recirculation to the Trickling Filter

Recirculation of a portion of the treated effluent to the trickling filter is performed to:

- Increase biological solids in the system
- Provide continuous seeding with recirculated sloughed off solids
- Maintain uniform hydraulic load through filter
- Attain continuous rotation of the distributor arm
- Establish uniform organic load
- Dilute influent for production of better quality effluent
- Decrease thickness of biological slime layer
- Improve removal efficiency of pollutants

The recirculation ratio (R_u) lies between a value of 50 and 1000% of the raw wastewater flow. Usually, the recirculation ratio is between 50 and 300%.[10]

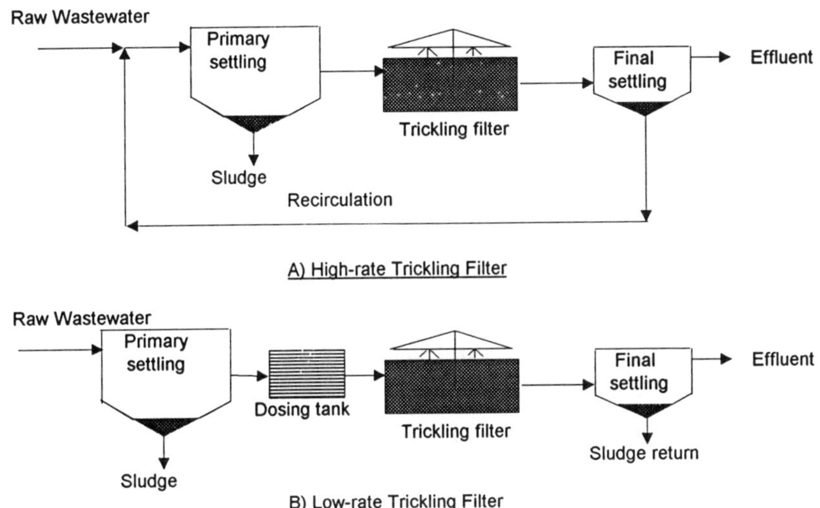

Figure 6.3
High (A) and low (B) rate trickling filters.

6.8.3 Efficiency of a Trickling Filter (BOD Removal Efficiency)

Computations of efficiency of a trickling filter are generally based on mathematical models and empirical formulae.

1. *Mathematical models*: Mathematical models usually assume the existence of uniform biological layer and even distribution of load within the trickling filter bed media. From an operational viewpoint, mathematical models are impractical for design purposes.
2. *Empirical formulas*: Empirical formulas are based on operational data collected from existing treatment plants. The collected data were analyzed to produce a design procedure.

Some of the most important formulas used in the design of trickling filters include the National Research Council, Velz, Rankin, and Rumpf formulas.

6.8.3.1 National Research Council Formula

The National Research Council (NRC) formula is based on wastewater treatment plants in U.S. military installations. Assumptions incorporated in the formula are

- Oxygen is the rate limiting factor with a soluble BOD in excess of 40 mg/L.
- Wastewater is not extremely dilute.
- The trickling filter is followed by a final settling tank or a secondary clarifier.
- Trickling filters treat settled domestic wastewater at a temperature of 20°C (temperatures different from this value have to be corrected for).

- The NRC formula includes the effect of secondary clarifiers.
- The maximum practical rate of recirculation predicted through the formula is 800%.

Equation 6.25 presents the NRC formula for a single-stage trickling filter.

$$Eff_1 = (L_i - L_e)/L_i = 100/\{1 + 0.44(W_1/V_1*F_1)^{1/2}\} \tag{6.25}$$

where:

Eff_1 = Efficiency of first-stage trickling filter, %
L_i = Influent BOD, mg/L
L_e = Effluent BOD from first stage, mg/L
W_1 = BOD load of first-stage filter, kg/d

$$W_1 = Q*L_i = BOD_{load} \tag{6.26}$$

where:

Q = Flow of wastewater, m^3/d
V_1 = Volume of single-stage filter media, m^3
F_1 = Single-stage recirculation factor

$$F_1 = (1 + R_u)/(1 + 0.1*R_u)^2 \tag{6.27}$$

$$R_u = \text{Recirculated flow } (Q_R)/\text{wastewater flow } (Q) = Q_R/Q \tag{6.28}$$

Equation 6.29 gives the NRC equation for a second stage trickling filter.

$$EFF_2 = [(L - L_e)/L_i]_2 = 100/\{1 + ((0.44[W_2/V_2*F_2]^{1/2})/[1 - Eff_1])\} \tag{6.29}$$

where

Eff_2 = Efficiency of second-stage trickling filter, %
W_2 = BOD load of second-stage filter = $Q*L_i = BOD_{load}$, kg/d
V_2 = Volume of second stage filter, m^3
F_2 = Second-stage recirculation factor
L_{i2} = Influent BOD to second-stage filter, mg/L
L_{e2} = Effluent BOD of second stage filter, mg/L

6.8.3.2 Velz Formula

The Velz formula is valid for BOD removals of 90% or less. The formula can be expressed as indicated in Equation 6.30.

$$L_e = \{(L_i + R_u*L_e)*e^{-k*h}\}/(1 + R_u) \tag{6.30}$$

where:

L_i = Influent BOD, mg/L
L_e = Effluent BOD, mg/L
k = Experimental coefficient
= 0.49 for high rate filters
= 0.57 for low rate filters
h = Filter depth, m
R_u = Recirculated flow/wastewater flow

6.8.3.3 Rankin Formula

The Rankin formula (Equation 6.31) may be used for design of a single-stage filter.

$$L_e = L_i/(3 + 2*R_u) \quad (6.31)$$

where:

L_e = BOD of the settled filter effluent, mg/L
L_i = BOD of the settled sewage, mg/L
R_u = Recirculation ratio

The Rankin formula applies to all plants treating sewage that have presedimentation, high rate filtration, and final settling tanks. This applies when the BOD does not exceed the value of 0.7 $kg/m^3/d$ and where recirculation, if applied, maintains a dosing rate between 93 and $244*10^3$/ha/d.[28]

6.8.3.4 Rumpf Formula

The Rumpf equation can be used to evaluate a trickling filter's efficiency, as presented in Equation 6.32.

$$Eff = 93 - (0.017*W/V) \quad (6.32)$$

where:

Eff = Trickling filter efficiency, %
W = Organic loading of the trickling filter, g BOD per day
V = Volume of the filter media, m^3

6.8.4 Overall Treatment Plant Efficiency

Many factors affect the operation and performance of a trickling filter, including organic loading, hydraulic flow rate, wastewater characteristics, rate of diffusability of food and air to the biological slime layer, temperature, etc. The overall treatment plant efficiency of a system comprised of a two-stage filter can be determined by using Equation 6.33.[14,32]

Eff_{over} = Overall treatment plant efficiency, %

Eff_{sed} = Percentage BOD removed in primary settling (usually is equal to 35%)

Eff_1 = BOD efficiency of first-stage filter and intermediate clarifier corrected for the operating temperature, %

Eff_2 = BOD efficiency of second-stage filter and final clarifier corrected for the operating temperature, %

Temperature correction can be carried out as shown by Equation 6.34.

$$Eff_T = Eff_{20}*(T_c)^{T-20} \tag{6.34}$$

where:

Eff_T = Efficiency at temperature, T %

Eff_{20} = Efficiency at a temperature of 20°C, %

T = Temperature, °C

T_c = Temperature correction factor, usually taken as equal to 1.035

Example 6.8

1. Write a computer program to compute the trickling filter efficiency using different formulas (NRC, Velz, Rumpf, and Rankin), given the population number, rate of population growth, clarifier efficiency, rate of wastewater flow to plant (m^3/s), 5-d BOD (mg/L), recirculation ratio, and filter diameter (m). Design the program so that the trickling filters determine volume, area, diameter, or depth and efficiency can be estimated at any temperature.
2. A treatment works includes primary sedimentation and trickling filtration to treat wastewater of 8,500 inhabitants. The sedimentation unit has an efficiency of 35%. The wastewater flows at a rate of 325 L/capita/d with a 5-d BOD of 310 mg/L. The trickling filter has the following characteristics: removal efficiency = 82%, recirculation ratio = 3:1, and filter depth = 2.5 m. Determine the trickling filter diameter using the NRC formula.
3. Use the computer program of section (1) to confirm computations in section (2).

Solution

1. For solution to Example 6.8 (1), see the following listing of Program 6.8.
2. Solution to Example 6.8 (2):
   - Given POP = 8500, efficiency of sedimentation = 35%, q = 325 L/capita/d, L_i = 310 mg/L, efficiency of filter = 82%, R_u = 3/1, h = 2.5 m.
   - Find the recirculation factor as: $F = (1 + R_u)/(1 + 0.1*R_u)^2 = (1 + 3)/(1 + 0.1*3)^2 = 2.37$.

- Find the wastewater flow: $Q = q*POP = 325*10^{-3}*8500 = 2762.5$ m^3/d.
- Find the BOD concentration emerging from the primary sedimentation and entering the trickling filter as: $L_i = BOD_i*(1 - Eff_{clarifier}) = 310*(1 - 0.35) = 201.5$ mg/L.
- Determine the influent BOD load to the filter as: $W = Q*L_i = 2762.5*201.5*10^{-3} = 556.6$ kg/d.
- Find the volume of the filter using the NRC efficiency equation as: $Eff = 100/\{1 + 0.44(W/V*F)^{0.5}\} = 82 = 100/\{1 + 0.44(556.6/V*2.37)^{0.5}\}$. This yields V = 944 m^3.
- Compute the surface area as: A = V/h = 944/2.5 = 377.6 m^2.
- Find the diameter of the filter as: $D = (4*A/\pi)^{0.5} = (4*377.6/\pi)^{0.5} = 22$ m.

Listing of Program 6.8

```
DECLARE FUNCTION log10! (x!)
DECLARE SUB box (r1%, c1%, r2%, c2%)        'Found in Program 2.1
'******************************************************************
'Program 6.8: Calculation of Trickling Filter Efficiency
'******************************************************************
pi = 3.1415962#
g = 9.81

m$(1) = "National Research Council formula"
m$(2) = "Velz formula"
m$(3) = "Rankine formula"
m$(4) = "Rumpf formula"
m$(5) = "Design a filter using NRC formula"
m$(6) = "Exit the program"
start:
CALL box(1, 1, 20, 80)
LOCATE 2, 15
PRINT "Program 6.8: Calculation of Trickling Filter Efficiency"
LOCATE 4, 3: PRINT "To compute Trickling Filter Efficiency using:"

row = 4
FOR I = 1 TO 6
LOCATE row + I, 10
PRINT I; m$(I)
NEXT I
DO
LOCATE row + 7, 3
INPUT "Select an option (1-6): "; opt
LOOP WHILE opt < 1 OR opt > 6
IF opt = 6 THEN END

CALL box(1, 1, 20, 80)
LOCATE 2, 15
PRINT "Program 6.8: Calculation of Trickling Filter Efficiency"
LOCATE 3, 3: PRINT m$(opt)

SELECT CASE opt
CASE 1
    'National Research Council formula
    LOCATE row + 1, 5
    INPUT "Enter influent BOD, kg/m3                    ="; Li
    LOCATE row + 2, 5
    INPUT "Enter the Effluent BOD,  mg/L(Enter 0 if unknown)="; Le
```

```
    IF Le = 0 THEN
        LOCATE row + 3, 5
        INPUT "Enter the wastewater flow, Q (m3/d)          ="; Q
        W = Q * Li
        LOCATE row + 4, 5
        INPUT "Enter the recirculation ratio, Ru            ="; Ru
        F1 = (1 + Ru) / ((1 + .1 * Ru) ^ 2)
        LOCATE row + 5, 5
        INPUT "Enter volume of filter, V (m3)               ="; V
        Eff = 100 / (1 + .44 * (W / (V * F1)) ^ .5)
    ELSE
        Eff = 100 * (Li - Le) / Li
    END IF

CASE 2
    'Velz formula
    LOCATE row + 1, 5
    INPUT "Enter influent BOD, mg/L                         ="; Li
    LOCATE row + 2, 5
    INPUT "Enter the flow rate of wastewater Q, m3/d        ="; Q
    LOCATE row + 3, 5
    INPUT "Enter the recirculated flow  Qr, m3/d            ="; Qr
    LOCATE row + 4, 5
    INPUT "Is it a high rate filter? Y/N                    ="; R$
    k = .49
    IF UCASE$(R$) = "Y" THEN k = .57
    LOCATE row + 5, 5
    INPUT "Enter filter depth, h  (m)                       ="; h
    Ru = Qr / Q
    A = (1 + Ru) * EXP(k * h)
    Le = Li / (A - Ru)
    Eff = 100 * (Li - Le) / Li


CASE 3
    'Rankine formula
    LOCATE row + 1, 5
    INPUT "Enter influent BOD, mg/L                         ="; Li
    LOCATE row + 2, 5
    INPUT "Enter the recirculation ration, Ru               ="; Ru
    Le = Li / (3 + 2 * Ru)
    Eff = 100 * (Li - Le) / Li

CASE 4
    'Rumpf formula
    LOCATE row + 1, 5
    INPUT "Enter the organic load of filter, W (g BOD/d)    ="; W
    LOCATE row + 2, 5
    INPUT "Enter the volume of the filter, V (m3)           ="; V
    Eff = 93 - (.017 * W / V)

CASE 5
    row = 4
    LOCATE row + 1, 5
    INPUT "Enter the population number                      ="; POP
    LOCATE row + 2, 5
    INPUT "Enter the rate of population growth, 0 if unknown ="; rate
    LOCATE row + 3, 5
```

```
        INPUT "Enter the 5-day BOD, mg/L                        ="; BOD5
        LOCATE row + 4, 5
        INPUT "Enter the recirculation ratio Ru                 ="; Ru
        LOCATE row + 5, 5
        INPUT "Enter the flow rate of wastewater Q, m3/c/d      ="; Qc
        LOCATE row + 6, 5
        INPUT "Enter the efficiency of sedimentation (%)        ="; Eff
        LOCATE row + 7, 5
        INPUT "Enter the removal efficiency      (%)            ="; Effr
        LOCATE row + 8, 5
        INPUT "Enter the filter depth, m                        ="; h

        IF Eff < 1 THEN Eff = Eff * 100
        IF Effr < 1 THEN Effr = Effr * 100
        F = (1 + Ru) / (1 + .1 * Ru) ^ 2            'recirculation factor
        Q = Qc * POP                                'wastewater flow
        Li = BOD5 * (1 - Eff / 100)                 'BOD concentration entering trickling filter
        W = Q * Li * 10 ^ -3                        'Influent BOD load to the filter
        V = W / (F * ((100 / Effr - 1) / .44) ^ 2)  'Volume of filter by NCR formula
        A = V / h                                   'Surface area
        D = (4 * A / pi) ^ .5                       'Diameter of filter
        LOCATE row + 10, 3
        PRINT "The diameter of the filter   ="; D; "  m"

CASE 6
        END
END SELECT

IF opt <> 5 THEN
LOCATE row + 10, 3
PRINT "EFFICIENCY    ="; Eff; " %"
END IF
DO
LOCATE 22, 1: PRINT "Press any key to continue............"
LOOP WHILE INKEY$ = ""

GOTO start
END

FUNCTION log10 (x)
   log10 = LOG(x) / LOG(10)
END FUNCTION
```

6.8.5 Trickling Filter Clarifier

Final or secondary clarifiers are required subsequent to trickling filtration in order to collect the relatively large particles of the sloughed-off biological slime solids or humus. In the design of secondary clarifiers, the following parameters need to be considered:[10]

- Neither thickening or hindered settling is assumed to develop in the secondary clarifier.
- Settling design criteria is based upon particle size and density.

- Surface overflow rate (v_s) lies between 25 and 33 m/d at average flow rates, not to exceed the value of 50 m/d at peak flow rates.
- Weir loading rates are in the range of 120 to 370 $m^3/d/m$ of the weir length at peak flow rate.

6.9 Combined Suspended and Attached Growth Systems Waste Stabilization Pond, Lagoon, or Oxidation Pond

6.9.1 Introduction

Waste stabilization ponds (WSP) are shallow earthen basins with a controlled shape or a natural depression that receives wastewater and retains it so that biological processes can proceed at acceptable levels. Waste stabilization ponds can be classified according to the type of biological activity taking place, such as anaerobic, facultative, or aerobic.

Anaerobic WSPs are deep ponds devoid of oxygen except for a relatively thin surface layer. This type of pond receives wastewater with a high organic load or wastewater with a high level of solids (i.e., not presettled in a settling unit operation such as a septic tank). The objectives of an anaerobic WSP include settling of solids, partial treatment of wastewater, and discharge of the effluent to a facultative pond. The main biological action here is achieved by anaerobic organisms. Usually, anaerobic ponds are used in series with facultative (or aerobic-anaerobic) ponds to provide complete treatment.

In facultative WSPs, aerobic conditions prevail in the upper surface mainly by using oxygen produced by algal organisms and to a lesser extent oxygen from the atmosphere. Anaerobic conditions develop at the bottom of the pond due to stagnant conditions. The depth of the aerobic and anaerobic zones is a function of mixing conditions, wind action, sunlight penetration, and the hours of daylight. Algae and bacteria are responsible for biodegradation of organic matter in the aerobic layers of the pond. Bacteria use oxygen for oxidation of waste organics to synthesize new cells and produce stable end products (such as carbon dioxide, nitrates, and phosphates). These inorganic compounds are used by algae, in the presence of sunlight, to yield new cells with end products such as oxygen. The oxygen produced is then used by the bacteria. This mutual benefit between bacteria and algae is termed a symbiotic relationship. The biological solids in the aerobic zone, together with heavy solids, settle to the anaerobic zone in the pond. These solids provide the needed food for the benthic anaerobic organisms. Solids in the anaerobic zone are then converted to organic acids and gases yet to be released in a soluble form to organisms in the aerobic zone of the pond. A facultative pond receives wastewater from a sewer system, or from an anaerobic pond. Effluent is then retained for some days, and the treated wastewater is discharged to a polishing pond. Figure 6.4 illustrates a schematic diagram of a facultative waste stabilization pond.

Tertiary (maturation) WSPs receive treated wastewater from facultative ponds and retain it for a period of time to improve its quality under aerobic conditions.

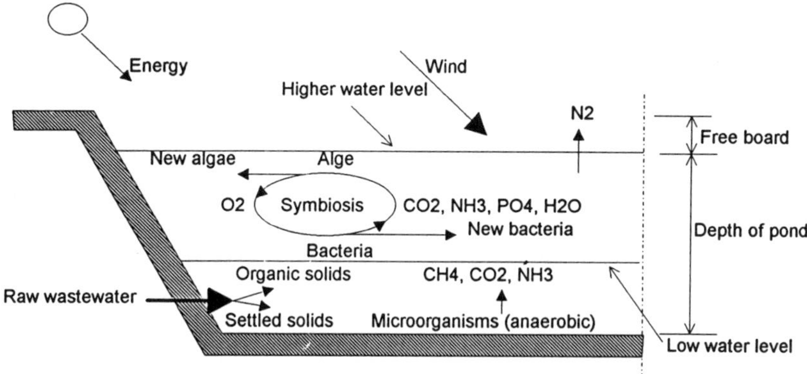

Figure 6.4
Schematic diagram of a facultative waste stabilization pond.

The effluent from a maturation pond can be used for landscape or crop irrigation. Maturation ponds can be used to grow fish and aquatic birds. Aerobic ponds are used primarily for the treatment of soluble organic wastes and effluents from wastewater treatment plants.

Waste stabilization ponds can be operated in series or in parallel or can function on an individual basis. In the series arrangement of ponds the wastewater or effluent is introduced from one pond to the next. Each pond in series discharges an effluent of a better quality than the one preceding it. Thus, this arrangement helps in treating unsettled raw sewage and improves the quality of the final effluent. In a parallel arrangement, ponds are placed adjacent to each other. The ponds simultaneously receive effluent from the same source and concurrently discharge the treated sewage to the same outlet or discharge point. The quality of the discharged effluent from parallel ponds is comparable. The advantage of such an arrangement is that during maintenance and repair programs one pond can remain in operation.

An individual WSP system of ponds can be used but a series of ponds is preferred. The series system of ponds can be selected for different reasons, such as obtaining higher treatment level, growing fish, or cleaning ponds without shutting off the collecting sewer system.

6.9.2 Waste Stabilization Pond Design

The design of a waste stabilization pond is based on such various interrelated parameters as:

1. *Site selection*: Gravity flow is practical in order to minimize pumping. Pumping of wastewater will increase the cost, use energy, and also require added operational and maintenance efforts.
2. *Soil*: The soil needs to be fairly impermeable, not sandy or gravely (to avoid seepage and groundwater contamination), and must be easy to excavate. Adequate soil provides material for constructing the embankments.

3. *Drainage*: Good drainage is required to provide for the discharge of the treated wastewater from the pond.
4. *Size*: The size provided for a waste stabilization pond must be large enough to accommodate the anticipated flows.
5. *Layout*: Ponds must be located at an appropriate distance from the neighboring houses and dwellings. The distance between the pond site and the nearest dwelling must be in excess of 200 m. The greater the distance between the ponds and the community, the better the situation.
6. *Wind direction*: Ponds must be located in a downwind direction from a community to avoid airborne pollution problems.

In the design of waste stabilization ponds the following parameters can be used:

6.9.2.1 General

- The expected daily flow of wastewater into the waste stabilization pond, is reported as Q, L/d.
- The average annual water temperature in the waste stabilization pond is reported as T, °C.
- The organic loading rate of the wastewater effluent is given as OL, g/L.
- The minimum required area of waste stabilization pond can then be calculated from the relationship given in Equation 6.35.

$$A = OL*Q/OL_{max} \tag{6.35}$$

where:

A = Minimum required area of waste stabilization pond, m^2

OL = Organic loading rate of the effluent, g/L

Q = Daily flow of sewage into the waste stabilization pond, L/d

OL_{max} = Maximum allowable organic load, $g/m^2/d = 2*T - 12$ (6.36)

T = Average annual water temperature, °C

- For a rectangular waste stabilization pond, the dimensional relationship of Equation 6.37 can be used.

$$l = [2 \text{ to } 3]*B \tag{6.37}$$

where

l = Length of the pond, m

B = Width of the pond, m

- The depth of a waste stabilization pond usually lies between 1 and 3 m, depending on type and volume of wastewater, climatic conditions, and waste stabilization pond classification.

6.9.2.2 Completely Mixed Conditions

- Assuming complete mixing conditions without solids recycling (i.e., a facultative pond), the relationship of the mass balance between the BOD entering and that leaving a waste stabilization pond is given by Equation 6.38.

$$\begin{aligned} BOD_{in} &= BOD_{out} + BOD_{consumed} \\ Q * L_i &= Q * L_e + V * k_p L_e \end{aligned} \tag{6.38}$$

where:

Q = Wastewater flow rate to the pond, m^3/s
L_i = Influent BOD, mg/L
L_e = Effluent BOD, mg/L
V = Volume, m^3
k_p = Removal rate constant for waste stabilization pond, d^{-1}

- The mathematical relationship for a single waste stabilization pond may be derived upon rearrangement of Equation 6.38, as presented in Equation 6.39.

$$L_e/L_i = 1/\{1 + (k_p * V/Q)\} = 1/\{1 + k_p * t\} \tag{6.39}$$

where:

L_e = Effluent BOD, mg/L
L_i = Influent BOD, mg/L
k_p = Removal rate constant for waste stabilization pond, d^{-1}
V = Volume, m^3
Q = Wastewater flow rate to the pond, m^3/s
t = Detention time, d

- For more than one waste stabilization pond arranged in series, the effluent of one pond becomes the influent to the next. Thus, a substrate balance written across a series of N ponds yields Equation 6.40.

$$L_e/L_i = 1/\{1 + [k_p * t/N]\}^N \tag{6.40}$$

where:

L_e = Effluent BOD, mg/L
L_i = Influent BOD, mg/L
t = Retention time, d
k_p = Removal rate constant for waste stabilization pond, d^{-1}
N = Number of ponds

The variation of the removal rate constant with the temperature can be calculated by Equation 6.41.

$$(k_p)_T = [k_p]_{20}*(T_c)^{[T-20]} \tag{6.41}$$

where:

$(k_p)_T$ = Removal rate constant for waste stabilization pond at temperature T, d^{-1}

$(k_p)_{20}$ = Removal rate constant for WSP at a temperature of 20°C, d^{-1}. (It relates to degradability of waste organics, temperature, and completeness of aeration mixing. It ranges between 0.3 to 1 per day to base e and at 20°C.)[14]

T_c = Temperature constant. This constant is a function of biodegradability and it can range from 1.03 to 1.12, usually taken equal to 1.035.

n = Number of waste stabilization ponds

- The value of the detention time can be found from the relationship shown in Equation 6.42.

$$t = V/Q \tag{6.42}$$

where:

V = Volume of the pond, m^3

Q = Rate of wastewater flow, m^3/s

- The depth of the waste stabilization pond may be assumed and the surface area determined from Equation 6.43.

$$A = V/H \tag{6.43}$$

where:

A = Area of waste stabilization pond, m^2

h = Depth of waste stabilization pond, m

Example 6.9

1. Write a computer program for the design of a waste stabilization pond given the 5-d BOD of wastewater (mg/L), flow rate (m^3/s), lowest temperature (°C), removal rate constant for the pond (d^{-1}), at any temperature (°C), temperature correction factor, effluent 5-d BOD from the pond (mg/L), and the depth of pond (m).
2. Wastewater with a 5-d BOD of 200 mg/L and flowing at a rate of 1.6 m^3/min is to be treated in a waste stabilization pond. Determine the surface area of the pond given the following date:
   - Lowest temperature = 23°C
   - Removal rate constant for the pond = 0.25/d at 20°C, temperature correction factor = 1.05
   - Effluent 5-d BOD from the pond = 25 mg/L

- Depth of pond = 1.4 m

3. Use the program established in (1) to verify computations obtained in (2).

Solution

1. For solution to Example 6.9 (1), see the following listing of Program 6.9.
2. Solution to example 6.9 (2):
   - Given Q = 1.6 m^3/min, L_i = 200 mg/L, L_e = 25 mg/L, $(k_p)_{20}$ = 0.25 /d, h = 1.4 m, T_{min} = 23°C, T_c = 1.05.
   - Find the removal rate constant for the pond at the lowest temperature of 25°C as: $[k_p]_{25} = [k_p]_{20}(1.05)^{[T-20]} = 0.25(1.05)^{[23-20]} = 0.29$ /d
   - Find the detention time of the pond as: $L_e/L_i = 1/(1 + k_p{*}t)$. Use equation to find detention time: 25/200 = 1/(1 + 0.29*t). This yields a detention time of t = 24.1875 d.
   - Compute the volume of the waste stabilization pond as: V = t*Q = 24.1875*1.6*60*24 = 55727.9 m^3.
   - Determine the surface area of the pond as: A = V/h = 55727.9/1.4 = 39805 m^2.

Listing of Program 6.9

```
DECLARE SUB box (r1%, c1%, r2%, c2%)        'Found in Program 2.1
'*****************************************************************
'Program 6.9: Design of a Waste Stabilization Pond
'*****************************************************************
pi = 3.1415962#
g = 9.81

CALL box(1, 1, 20, 80)
LOCATE 2, 25
PRINT "Program 6.9: Design of a Waste Stabilization Pond"
row = 5
LOCATE row, 5
INPUT "Enter the flow rate, Q (m3/min)              ="; Q
LOCATE row + 1, 5
INPUT "Enter the 5-day BOD (mg/L)                   ="; Li
LOCATE row + 2, 5
INPUT "Enter the lowest temperature, T (degree C)      ="; Tmin
LOCATE row + 3, 5
INPUT "Enter the removal rate constant for the pond(/d) ="; Kp
LOCATE row + 4, 5
INPUT "Enter the temperature correction factor       ="; Tc
LOCATE row + 5, 5
INPUT "Enter the effluent 5-day BOD, (mg/L)           ="; Lo
LOCATE row + 6, 5
INPUT "Enter the depth of the pond, d (m)            ="; h

Kpt = Kp * Tc ^ (Tmin - 20)       'removal rate constant at T
dt = (Li / Lo - 1) / Kpt          'Detention time
V = dt * Q * 60 * 24              'Volume of stabilization pond
A = V / h                         'Surface area of the pond

LOCATE row + 9, 3
PRINT "The required surface area of the pond   ="; A; " m2"
```

```
DO
LOCATE 22, 1
PRINT "Press any key to continue.........."
LOOP WHILE INKEY$ = ""
END
```

6.9.2.3 Maturation Waste Stabilization Pond

A maturation WSP pond is a polishing pond that enables the production of an effluent with improved bacteriological quality. The following points provide a summary of the most important relationships in connection with maturation ponds:

- The rate of die-off of fecal organisms in a single maturation pond can be estimated by using Equation 6.44.

$$N_e/N_i = 1/(1 + k'*t) \tag{6.44}$$

where:

N_e = Effluent bacterial number, number of bacteria per 100 mL
N_i = Influent bacterial number, number of bacteria per 100 mL
k' = Bacterial die-off rate, d^{-1}
t = Retention time, d

- The rate of die-off of fecal organisms in a multicelled maturation waste stabilization pond can be estimated by Equation 6.45.

$$N_e/N_i = 1/(1 + k'*t)^N \tag{6.45}$$

where:

N = number of waste stabilization ponds in series.

Waste stabilization ponds can be round, square, or rectangular in shape.

- The detention time of the maturation pond may be determined from the relationship between volume and flow as presented in Equation 6.46.

$$t = \dot{V}/Q \tag{6.46}$$

or

$$V = t*Q = A*h$$

from which the area, A, can be calculated.

Example 6.10

1. Write a computer program to find the retention time and volume of a pond required to reduce the bacterial number by any percent (a) with a bacterial die-off rate of k′ (per day), given the rate of wastewater flow to the pond as Q (m^3/d).

2. A polishing pond reduces the bacterial number by 99% with a bacterial die-off rate of 0.5 /d. Wastewater is introduced to the pond at the rate of 20 m^3/d. Find the necessary retention time and the volume of the pond.
3. Use the program established in (1) to verify computations obtained in (2).

Solution

1. For the solution to Example 6.10 (1), see the following listing of Program 6.10.
2. Solution to Example 6.10 (2):
   - Given $k' = 0.5$ /d, $N_e/N_i = (100 - 99)/100 = 0.01$, $Q = 20\ m^3/d$.
   - Find the detention time of the pond as: $N_e/N_i = 1/(1 + k'{*}t) = 0.01 = 1/(1 + 0.5{*}t)$. This yields t = 198 d.
   - Determine the volume of the pond as: $V = t{*}Q = 198{*}20 = 3960\ m^3$.

Listing of Program 6.10

```
DECLARE SUB box (r1%, c1%, r2%, c2%)        'Found in Program 2.1
'****************************************************************
'Program 6.10: Retention Time and Volume of a Pond
'****************************************************************
pi = 3.1415962#
g = 9.81

CALL box(1, 1, 20, 80)

LOCATE 2, 20
PRINT "Program 6.10: Retention Time and Volume of a Pond"
row = 5
LOCATE row, 5
INPUT "Enter the flow rate, Q (m3/day)                    ="; Q
LOCATE row + 1, 5
INPUT "Enter the bacterial die away rate, k (/d)          ="; k
LOCATE row + 2, 5
INPUT "Enter the percentage bacterial number reduction    ="; kil
rat = (100 - kil) / 100
t = (1 / rat - 1) / k        'detention time
V = t * Q                    'volume of pond
LOCATE row + 8, 3
PRINT "The detention time of the pond              ="; t; " days"
LOCATE row + 9, 3
PRINT "The volume of the pond                      ="; V; " m3"

DO
LOCATE 22, 1
PRINT "Press any key to continue.........."
LOOP WHILE INKEY$ = ""
END
```

Table 6.8 gives typical design information for waste stabilization ponds.

TABLE 6.8
Design Information for Waste Stabilization Ponds[1,11,26]

	Anaerobic	Facultative	Aerobic
Influent	Sewage with high organic load, high degree of solids	Sewage from a sewerage system or anaerobic pond	Sewage from facultative ponds
Treatment	Partial		
Disposal of effluent	to facultative pond	Maturation pond	Agricultural irrigation, fish farming, aquatic birds.
Depth (m)	2–4	1–1.5	1
Detention time (d)	8–20	20–180	5–10
Main biological action	Organisms that do not require DO for feeding and reproduction	Anaerobic and aerobic	Aerobic organisms.
Operation	Parallel or series connection	At least 3 ponds in series, parallel useful for large ponds	One or more, series or parallel.
Color	Grayish black	Green or brownish green	Green
Frequency of sludge removal (year)	2–12	8–20	Probably never
Optimum temperature (°C)	30	20	20
Oxygen requirement	—	—	0.7–1.4 times removed BOD
pH	6.8–7.2	6.5–9.0	6.5–8.0
Chemicals needed	Nutrients when there is a deficiency; no other chemicals	Nutrients when there is a deficiency; no other chemicals.	
Expected problems	Odors, large land requirements, groundwater pollution.	Odors when loading is high, groundwater pollution, reduction in biological activity in cold climates	Reduction in problems associated with biological activity under cold weather conditions.

6.10 Sludge Treatment and Disposal

6.10.1 Sludge Digestion

Sludge digestion is the controlled decomposition of organic matter present in the sludges. Sludge digestion facilitates volume reduction, conversion of solids to inert compounds, and removal of pathogenic microorganisms. The digestion process may be aerobic or anaerobic.

The aerobic process used to treat waste-activated humus or primary sludges, or a mixture thereof, in an open tank can be regarded as a modification of the activated sludge process.[26,50] The dominating reaction in aerobic digestion is endogenous respiration. The main advantages of aerobic digestion include reduction in odor and biodegradable material, improvement of the dewatering process, and reduction in the level of pathogenic microorganisms.

Anaerobic digestion is a fermentation process of sewage sludge in which facultative and anaerobic bacteria break down organic matter primarily into the gases carbon dioxide and methane. The process is carried down by acidic bacteria and methanogenic bacteria, respectively.[26,50] The anaerobic sludge digestion process is affected by many interrelated factors such as pH, temperature, nutrients present, toxic substances, volatile acids, ammonia, type and characteristics of decomposed materials, shock loads, and mixing conditions.[5] Acid formers convert solids to soluble organic acids and alcohols. Formation of acids results in a decrease in pH of the system and may end the action of acid formers. These species are then replaced by anaerobic bacteria (methane formers) which function within a narrow pH range of from 6.5 to 7.5. Methane formers convert the formed acids and alcohols, to carbon dioxide, and methane gas with traces of other gases such as hydrogen sulfide.[5]

The volumetric gas production, or specific yield, from an anaerobic sludge digester can be determined from the relationship presented in Equation 6.47.[29]

$$V_g = [Y_t{*}VS{*}(1 - \{k_n/(t{*}\mu_{smax} - 1 + k_n)\})]/t \tag{6.47}$$

where:

- V_g = Volumetric gas production rate, or the specific yield, m^3 gas per m^3 digester per d
- Y_t = Ultimate gas yield, m^3 gas per kg VS added
- VS = Concentration of influent volatile solids, kg/m^3
- k_n = Kinetic coefficient, dimensionless
- t = Hydraulic detention time, d
- μ_{smax} = Maximum specific growth rate of microorganisms, d^{-1}

Table 6.9 gives general design information for a conventional anaerobic digester.

TABLE 6.9
Design Information for the Conventional Anaerobic Digester[5,7,11,29]

Parameter	Value
Volatile solids loading ($kg/m^3/d$)	0.3–2
Volatile solids destruction (%)	40–50
Gas production (m^3 gas per kg VS)	0.2–1.5
Influent sludge solids ($kg/m^3/d$)	2–5
Total solids decomposition (%)	30–40
pH	6.5–7.4
Alkalinity concentration (mg/L)	2000–3500
Solids retention time (d)	30–90
Digester capacity, m^3/capita	0.1–0.17
Gas composition (%)	
Methane	65–70
Carbon dioxide	32–35
Hydrogen sulfide	Trace
Temperature (°C)	30–55

Example 6.11

1. Develop a computer program to find the volumetric gas production rate in an anaerobic digester, given specific yield (m^3 gas per m^3 digester per d), ultimate gas yield (m^3 gas per kg VS added), concentration of influent volatile solids (kg/m^3), kinetic coefficient, hydraulic detention time (d), and maximum specific growth rate of microorganisms (d^{-1}).
2. Compute the volumetric gas production rate from a beef waste, given that the influent volatile solids concentration to the digester amounts to 180 kg/m^3 for a hydraulic retention time of 15 d. The maximum specific growth rate of microorganisms is 0.15/d at a temperature of 20°C. Take the ultimate methane yield equal to 0.3 m^3 methane per kg VS added, kinetic coefficient equal to 1.5, and digester volume equal to 6 m^3.
3. Use the computer program of (1) to verify computations obtained in (2).

Solution

1. For the solution to Example 6.11 (1), see the following listing of Program 6.11.
2. Solution to Example 6.11 (2):
   - Given $Y_t = 0.3$, VS = 180, $k_n = 1.5$, t = 15, $\mu_{smax} = 0.15$.
   - Determine the volumetric gas production rate from the beef manure as: $V_g = [Y_t*VS*(1 - \{k_n/(t*\mu_{smax} - 1 + k_n)\})]/t = [0.3*180*(1 - \{1.5/(15*0.15 + 1.5)\})]/15 = 1.64$ m^3 methane gas per m^3 digester volume per day.
   - Determine daily amount of gas production = 1.64*6 = 9.84 m^3.

Listing of Program 6.11

```
DECLARE SUB box (r1%, c1%, r2%, c2%)      'Found in Program 2.1
'****************************************************************
'Program 6.11: Volumetric Gas Production Rate in Digesters
'****************************************************************
pi = 3.1415962#
g = 9.81

CALL box(1, 1, 20, 80)

LOCATE 2, 15
PRINT "Program 6.11: Volumetric Gas Production Rate in Digesters"
row = 5
LOCATE row, 5
INPUT "Enter the specific yield, Yt (m3 gas/Kg VS)           ="; yt
LOCATE row + 1, 5
INPUT "Enter the influent volatile solids concentration (Kg/m3) ="; VS
LOCATE row + 2, 5
INPUT "Enter the max specific growth rate of microorganisms(/d) ="; Uxmax
LOCATE row + 3, 5
INPUT "Enter the retention time  (days)                    ="; t
LOCATE row + 4, 5
INPUT "Enter the kinetic coefficient, Kn                   ="; Kn
LOCATE row + 5, 5
INPUT "Enter the volume of the digester (m3)                  ="; Vd


Vx = yt * VS * (1 - (Kn / (t * Uxmax - 1 + Kn))) / t
Px = Vx * Vd
LOCATE row + 8, 3
PRINT "The volumetric gas production rate      ="; Vx; " m3"
LOCATE row + 9, 3
PRINT "The daily amount of gas production      ="; Px; " m3"

DO
LOCATE 22, 1
PRINT "Press any key to continue.........."
LOOP WHILE INKEY$ = ""
END
```

6.10.2 Sludge Dewatering

6.10.2.1 Introduction

Sludge dewatering is a physical unit operation utilized in rendering the moisture content of a sludge with the following objectives:

- Reduction of sludge volume
- Promotion of the sludge handling process
- Retardation of biological decomposition
- Increase sludge calorific value by removal of excess moisture

- Removal of sludge odors
- Reduction of leachate production at sanitary landfills
- Safe use of sludge for landscape fertilizer or as a farmland fertilizer

The methods used for sludge dewatering include land disposal (drying beds, injection, ridge and furrow, spray irrigation, lagooning), vacuum filters, pressure filters, centrifuges, composting, elutration, and compaction. These methods are of a physical rather than chemical nature.

In the drying bed operation, sludge is spread in an open area to assist in the moisture reduction through evaporation or drainage according to prevailing climatic conditions, soil topography, hydrology of the area, meteorological factors, type and properties of sludge, nature and size of solid particles, bed dimensions especially the depth, shape of drying surface as related to air flow, land value, and management.[5,26,31,32] Collected filtrate is usually returned to the treatment plant, and the sludge is removed from the drying bed after it has drained and dried sufficiently to be stable.[50]

Vacuum filtration is a continuous operation carried out at pressures around 70% of the atmospheric pressure.[5] Chemical conditioning is used before vacuum filtration to agglomerate small particles. This method is used in larger facilities where space is limited, or when incineration is necessary for maximum volume reduction.[26] The factors that affect vacuum filter yield include type and properties of sludge, type and concentration of the conditioner and conditioning procedure, type and condition of the filter cloth, and the operational parameters of the filter.[5,26]

Pressure filtration is a batch process that dewaters a conditioned sludge. Application of pressure drains the liquor and retains the sludge cake. Factors influencing operation include the pressure applied, type and life of filter, sludge properties and conditioning, and collection and final disposal of the cake.

Centrifugation (solid bowl, disc, basket type) uses centrifugal and the Stoke's frictional forces to increase the settling rate of sludge solids. The method is used in large facilities where space is limited or when incineration is required.[26] Factors of importance include solids content, particle size and shape, electrostatic factors, viscosity effect, differential density between solid particles and liquid, disposal of centrate, availability of power, properties of sludge, rotational speed, hydraulic loading rate, depth of liquid pool in bowl, and use of polyelectrolyte to improve performance.[1,5,26]

In sludge composting, the sludge is mixed with a bulking material (e.g., wood chips, leaves, garbage, rags, paper, cardboard, etc.) to form an aerated, porous pile. Decomposition is achieved by aerobic and thermophilic bacteria. The method may be used to convert digested and undigested sludge cake.[26] Factors affecting the process include weather conditions, properties of sludge, and the aeration rate.

In elutriation, the sludge is washed with treated effluent to yield a porous sludge. During mixing of elutriant and sludge, substances such as carbonates and phosphates are removed from the sludge, together with products of decomposition and nonsettleable fine solids.[5,31]

In general, the main factors that affect dewatering practice include nature and content of the fine particles, sludge solids content, particle size and charge, shearing strength, protein content, pH, moisture content, anaerobic digestion, and the type and concentration of the filter aids and conditioners used.

6.10.2.2 Filtration of Sludge

Filtration may be defined as the separation of a solid from a suspension of a liquid. Filtration retains the solid on a screen or membrane and allows the filtrate to drain off. The methods used for determining how well a sludge dewaters include the cracking time test, capillary suction time (CST) test, filter leaf test, and the specific resistance test.

The cracking time is the time necessary for a formed cake to crack. The CTS is the time required by water from a sludge to travel a given distance along a filter paper.[50] The shorter the CST, the higher the dewaterability of the sludge. Long CST values indicate problems in the filtration of the sludge analyzed.[5] The filter leaf test stimulates the principle of operation of a vacuum filter, whereby the filter yield is estimated based on the drying cycle period of the sludge.[5,12] Coackley[37,39] advocated the use of the specific resistance concept whereby the ease of dewatering is given by the Carman and Coackley[30,36,37,43–47] equation for constant pressure, as presented in Equation 6.48.

$$t/V = (\mu * r_s * C/2 * P * A^2) * V + (\mu * R_m/P * A) \tag{6.48}$$

where:

- t = Time of filtration, s
- V = Volume of filtrate, m^3
- μ = Viscosity of filtrate, $N*s/m^2$
- r_s = Specific resistance of sludge cake, m/kg
- C = Solids content, kg/m^3
- P = Pressure applied, N/m^2
- A = Area of filtration, m^2
- R_m = Resistance of filter medium, m^{-1}

Equation 6.48 may be put in the form of Equation 6.49:

$$t/V = b * V + a \tag{6.49}$$

where:

$$b = (\mu * r_s * C)/(2 * P * A^2), \text{ and} \tag{6.50}$$

$$a = \mu * R_m/P * A \tag{6.51}$$

Plotting t/V vs. V yields a straight line of gradient b, and r_s is found as indicated in Equation 6.52.

$$r_s = (2*b*P*A^2)/(\mu*C) \quad (6.52)$$

where:

r_s = Specific resistance of sludge cake, m/kg
b = Slope of the straight line of t/V vs. V, s/m^6
P = Pressure applied, N/m^2
A = Area of filtration, m^2
μ = Viscosity of filtrate, $N*s/m^2$
C = Solids content, kg/m^3

The specific resistance to filtration may be defined as, "The resistance to filtrate flow caused by a cake of unit weight of dry solids per unit filter area."[33] Table 6.10 gives a general outline of the values of the specific resistance and gives an estimation of the degree of the sewage sludge dewaterability.

The value of r_s for most wastewater sludges changes with pressure as indicated in Equation 6.53.

$$r_s = r_s'*P^a \quad (6.53)$$

where:

r_s = Specific resistance to filtration at applied pressure P, m/kg
r_s' = A constant
a = Coefficient of compressibility (varies between 0 and 1).

TABLE 6.10
Sludge Characteristics as Related to Specific Resistance[5,11,33]

Specific resistance value	Sludge characteristics (m/kg)
10^{11}–10^{12}	Easily filtered sludge
10^{14}–10^{15}	Poorly filtered sludge

Source: From Rowe, D.R. and Abdel-Magid, I.M., *Handbook of Wastewater Reclamation and Reuse*, CRC Press/Lewis Publishers, Boca Raton, FL, 1995. With permission.

Equation 6.53 can be arranged in a straight line relationship form as presented in Equation 6.54.

$$\log(r_s) = a*\log(P) + \log(r_s') \quad (6.54)$$

The compressibility coefficient is taken to be the slope, a, of the straight line obtained by plotting log (r_s) vs. log (P). The greater the value of a, the more compressible is the sludge. Figure 6.5 shows a sketch of the apparatus used in determining the specific resistance of a sludge to filtration.

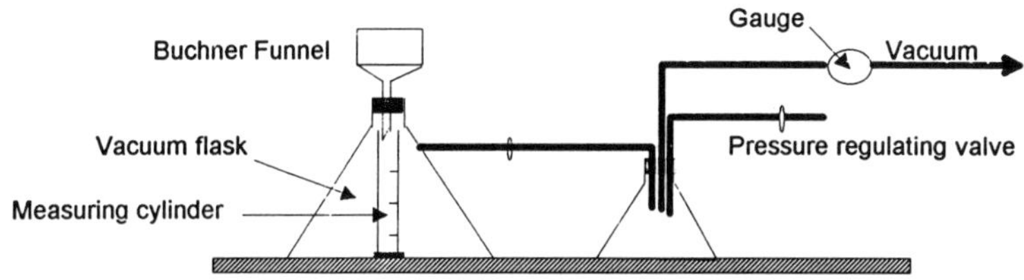

Figure 6.5
Specific resistance apparatus.

Example 6.12

1. Write a computer program to compute the specific resistance of a sample of a sludge given the variation of a collected volume of filtrate V (mL), with time t (s), vacuum applied P (N/m^2), filtrate viscosity μ ($N*s/m^2$), solids concentration C (kg/m^3), and area of filtration A (m^2). Design the computer program to show whether the sludge under consideration is amenable to dewatering by vacuum filtration.
2. The following data were obtained in an experiment in order to determine its specific resistance to filtration:

Volume of filtrate collected (mL)	Time taken to collect volume (min)
6.8	1
9.5	2
11.6	3
13.4	4
14.9	5

   - Find the specific resistance of the sludge, given vacuum pressure used = 60 kPa, dynamic viscosity of filtrate = $1.139*10^{-3}$ $N*s/m^2$, volume of sample used in filtration = 50 mL, solids concentration = 0.084 g/mL, diameter of filter paper used in the experiment = 7 cm.
   - Is this sludge amenable to dewatering by vacuum filtration? State your reasons.
3. Use the computer program of (1) to verify computations obtained in (2).

Solution

1. For the solution to Example 6.12 (1), see the following listing of Program 6.12.

2. Solution to Example 6.12 (2):
   - Given variation of volume of filtrate collected with time of filtration, $P = 60*10^3$ N/m^2, $\mu = 1.139*10^{-3}$ $N*s/m^2$, C = 0.084 g/mL, D = 7 cm.
   - Find the ratio of time to filtrate volume, t/V as:

	Time (s)				
	60	120	180	240	300
Time/Volume of filtrate collected (s/mL^2)	8.82	12.63	15.52	17.91	20.13
Volume of filtrate collected (mL)	6.8	9.5	11.6	13.4	14.9

   - Draw the straight line of the plot of t/V vs. V, and determine the slope of the straight line as: $b = 1.387*10^{12}$ s/m^6.
   - Find the area of filtration as: $A = \pi(0.07)^2/4 = 38.48*10^{-4}$ m^2.
   - Find the sludge solids concentration as: $C = 0.084*10^{-3}/10^{-6} = 84$ kg/m^3.
   - Compute the sludge specific resistance as: $r_s = 2*b*P*A^2/\mu*C = (2*1.387*10^{12}*60*10^3*(38.48*10^{-4})^2)/(1.139*10^{-3}*84) = 2.58*10^{13}$ m/kg.
   - Since the specific resistance of this sludge exceeds the value of $1*10^{12}$ m/kg (with $r_s = 2.58*10^{13}$ m/kg), then it will not filter well in a vacuum filter.

Listing of Program 6.12

```
DECLARE SUB box (r1%, c1%, r2%, c2%)      'Found in Program 2.1
'*****************************************************************
'Program 6.12: Specific Resistance of Sludge
'*****************************************************************
pi = 3.1415962#
g = 9.81

CALL box(1, 1, 20, 80)
LOCATE 2, 10: PRINT "Program 6.12: Specific Resistance of Sludge"
row = 5
LOCATE row, 5
INPUT "Enter the vacuum applied, P (N/m2)                        ="; P
LOCATE row + 1, 5
INPUT "Enter the filtrate viscosity, mu (Ns/m2)                  ="; mu
LOCATE row + 2, 5
INPUT "Enter the solids concentration, C (g/mL)                  ="; c
LOCATE row + 3, 5
INPUT "Enter the diameter of the filter paper used, d (m)        ="; d
LOCATE row + 4, 5
INPUT "Enter the number of measurements of data of volume & time ="; N
CALL box(1, 1, 20, 80)
LOCATE 2, 10: PRINT "Program 6.12: Specific Resistance of Sludge"
row = 3
LOCATE row, 3
PRINT "Enter the data collected for volume versus time:"
LOCATE row + 1, 3
PRINT "Volume of filtrate (mL)     Time (minutes)        T/V"
y1 = 0!
x1 = 0!
```

```
row = 4
FOR i = 1 TO N
row = row + 1
LOCATE row, 12: INPUT v(i)
LOCATE row, 40: INPUT t(i)
v(i) = v(i) / 10 ^ 6                'convert to m3
t(i) = t(i) * 60                    'convert to seconds
tov(i) = t(i) / v(i)
LOCATE row, 60: PRINT tov(i)
y1 = y1 + tov(i)
x1 = x1 + v(i)
NEXT i

'Compute averages for regression analysis
y1 = y1 / N          'average t/v
x1 = x1 / N          'average v
sum1 = 0!
sum2 = 0!
FOR i = 1 TO N
sum1 = sum1 + (tov(i) - y1) * (v(i) - x1)
sum2 = sum2 + (v(i) - x1) ^ 2
NEXT i

b = sum1 / sum2                     'the slope of the line
A = pi / 4 * d ^ 2                  'area of filtration
c = c * 10 ^ 3                      'convert to Kg/m3
rx = 2 * b * P * A ^ 2 / (mu * c)     'specific resistance

c$ = "case not covered by Table 6.10"
IF  rx < 10 ^ 12 THEN c$ = "Easily filtered sludge"
IF rx > 10 ^ 12 THEN c$ = "Poorly filtered sludge"

LOCATE row + 2, 3
PRINT "The specific resistance of the sludge ="; rx; " m/Kg : "
LOCATE row + 3, 20: PRINT c$

DO
LOCATE 22, 1
PRINT "Press any key to continue.........."
LOOP WHILE INKEY$ = ""
END
```

Example 6.13

1. Write a computer program to determine the compressibility of a sludge sample for a set of specific resistance values and corresponding pressures.
2. For a sample of sewage sludge the following specific resistance values were obtained for the corresponding applied vacuum pressures:

Pressure applied (kPa)	Specific resistance, $r*10^{-13}$ (m/kg)
293.04	57.85
586.075	104.45
1172.15	186.54
1758.225	256.84
2344.3	331.89

Determine the compressibility coefficient of the sludge.

3. Use program developed in (1) to verify solution obtained in (2).

Solution

1. For the solution to example 6.13 (1), see the following listing of Program 6.13.
2. Solution to Example 6.13 (2):
   - Given specific resistance values corresponding to applied pressures.
   - Determine log(P) and log(r) as presented in the table below:

Pressure applied (kPa)	Specific resistance $r*10^{-13}$ m/kg	log(P)	log(r)
293.04	57.85	5.466927	14.7623
586.075	104.45	5.767953	15.01891
1172.15	186.54	6.068983	15.27077
1758.225	256.84	6.245074	15.40966
2344.3	331.89	6.370013	15.52099

   - Plot log(r) as a function of log(P). Find the slope of the drawn line to correspond to compressibility coefficient of the sample as s = 0.84.
3. Use program developed in (1) to verify solution obtained in (2).

Listing of Program 6.13

```
DECLARE FUNCTION Log10! (x!)
DECLARE SUB box (r1%, c1%, r2%, c2%)      'Found in Program 2.1
'*****************************************************************
'Program 6.13: Compressibility of Sludges
'*****************************************************************
pi = 3.1415962#
g = 9.81

CALL box(1, 1, 20, 80)
LOCATE 2, 15: PRINT "Program 6.13: Compressibility of Sludges"
LOCATE 4, 3
INPUT "Enter the number of measurements of data of volume & time ="; N
LOCATE 5, 3
PRINT "Pressure Applied(kPa)   Spec.Resistance(/10^13),m/kg    Log(P)  Log(r)"
y1 = 0!
x1 = 0!
row = 5
FOR i = 1 TO N
row = row + 1
LOCATE row, 12: INPUT P
LOCATE row, 40: INPUT S
P = P * 1000              'convert to Pascal
S = S * 10 ^ 13
tov(i) = Log10(S)
v(i) = Log10(P)
LOCATE row, 60: PRINT v(i)
LOCATE row, 70: PRINT tov(i)
```

```
yl = yl + tov(i)
xl = xl + v(i)
NEXT i

'Compute averages for regression analysis
yl = yl / N          'average t/v
xl = xl / N          'average v
suml = 0!
sum2 = 0!
FOR i = 1 TO N
suml = suml + (tov(i) - yl) * (v(i) - xl)
sum2 = sum2 + (v(i) - xl) ^ 2
NEXT i

b = suml / sum2                    'the slope of the line

LOCATE row + 2, 3
PRINT "The compressibility coefficient of the sample ="; b

DO
LOCATE 22, 1
PRINT "Press any key to continue.........."
LOOP WHILE INKEY$ = ""
END

FUNCTION Log10 (x)
DECLARE FUNCTION Log10! (x!)
DECLARE SUB box (r1%, c1%, r2%, c2%)     'Found in Program 2.1
'*****************************************************************
'Program 6.13: Compressibility of Sludges
'*****************************************************************
pi = 3.1415962#
g = 9.81

CALL box(1, 1, 20, 80)
LOCATE 2, 15: PRINT "Program 6.13: Compressibility of Sludges"
LOCATE 4, 3
INPUT "Enter the number of measurements of data of volume & time ="; N
LOCATE 5, 3
PRINT "Pressure Applied(kPa)    Spec.Resistance(/10^13),m/kg       Log(P)  Log(r)"
yl = 0!
xl = 0!
row = 5
FOR i = 1 TO N
row = row + 1
LOCATE row, 12: INPUT P
LOCATE row, 40: INPUT S
P = P * 1000               'convert to Pascal
S = S * 10 ^ 13
tov(i) = Log10(S)
v(i) = Log10(P)
LOCATE row, 60: PRINT v(i)
LOCATE row, 70: PRINT tov(i)


yl = yl + tov(i)
xl = xl + v(i)
NEXT i
```

```
'Compute averages for regression analysis
y1 = y1 / N          'average t/v
x1 = x1 / N          'average v
sum1 = 0!
sum2 = 0!
FOR i = 1 TO N
sum1 = sum1 + (tov(i) - y1) * (v(i) - x1)
sum2 = sum2 + (v(i) - x1) ^ 2
NEXT i

b = sum1 / sum2                'the slope of the line

LOCATE row + 2, 3
PRINT "The compressibility coefficient of the sample ="; b

DO
LOCATE 22, 1
PRINT "Press any key to continue.........."
LOOP WHILE INKEY$ = ""
END


FUNCTION Log10 (x)
'Computes the logarithm of x to base 10
Log10 = LOG(x) / LOG(10)
END FUNCTION
```

6.10.2.3 Centrifugation

In centrifugation, settling and rejection of solids are parameters that are considered in the modeling and scale-up between two geometrically similar centrifuges. Settling of particles is estimated by the Sigma equation, while rejection of solids is measured by the Beta equation.[5,12]

The Sigma equation is illustrated in Equation 6.55 for two centrifuges having the same settling characteristics within the bowl.[5,12]

$$Q_1/\Sigma_1 = Q_2/\Sigma_2 \tag{6.55}$$

where:

Q_1 = Liquid flow rate into the first centrifuge, m^3/s
Σ_1 = A parameter related to properties of the first centrifuge
Q_2 = Liquid flow rate into the second centrifuge, m^3/s
Σ_2 = A parameter related to properties of the second centrifuge

The Sigma value for a solid bowl centrifuge can be calculated by using Equation 6.56.

$$\Sigma = ([v_r]^2 * V)/(g * Ln(r_2/r_1)) \tag{6.56}$$

where:

Σ = A parameter related to the characteristics of a centrifuge[5,12]
v_r = Rotational velocity of the bowl, rad/s
V = Liquid volume in the pool, m^3
g = Local gravitational acceleration, m/s^2
r_1 = Radius from centerline to the surface of the sludge, m
r_2 = Radius from centerline to inside bowl wall, m

Solids movement in a centrifugation system may be estimated by the Beta equation, Equation 6.57.[5,12]

$$W_1/\beta_1 = W_2/\beta_2 \quad (6.57)$$

where:

W_1 = Solids loading rate for the first centrifuge, kg/h
β_1 = Beta function for the first centrifuge
W_2 = Solids loading rate for the second centrifuge, kg/h
β_2 = Beta function for the second centrifuge

The Beta function may be computed by using Equation 6.58.

$$\beta = v_w{*}d_p{*}n{*}\pi{*}Z{*}D \quad (6.58)$$

where:

β = Beta function for a centrifuge
v_w = Difference in the rotational velocity between bowl and conveyor, rad/s
d_p = Distance between blades, or the scroll pitch, m
n = Number of leads
Z = Depth of sludge in the bowl, m
D = Bowl diameter, m

Example 6.14

1. Develop a computer program to find the flow rate at which a solid bowl centrifuge will perform similarly as an old one that is to be replaced in order to increase sludge production. Data given include solids concentration (%); rate of flow of sludge to the old centrifuge that needs dewatering (m^3/d); and characteristics of both centrifuges, in terms of bowl length (cm), bowl diameter (cm), bowl speed (rpm), bowl depth (cm), scroll pitch (cm), number of leads, and conveyor velocity (rpm). Assume that the new centrifuge is scaled up geometrically similar to larger.
2. The method of centrifugation is used to dewater a sludge of 3.5% solids content. The sludge enters the solid bowl centrifuge at an hourly rate of 0.5 m^3. Due to an increase

in quantity of sludge requiring dewatering, the centrifuge is scaled up to another geometrically similar and larger one. The properties of the two centrifuges are as tabulated below. Determine the flow rate at which the second centrifuge will operate similarly to the first one.

Characteristic	First centrifuge	Second centrifuge
Bowl length (cm)	24	60
Bowl diameter (cm)	20	40
Bowl speed (rpm)	4500	4000
Bowl depth (cm)	3	5
Scroll pitch (cm)	5	10
Number of leads	1	1
Conveyor velocity (rpm)	4450	3950

3. Use the computer program developed in (1) to verify computations obtained in (2).

Solution

1. For the solution to Example 6.14 (1), see the following listing of Program 6.14.
2. Solution to example 6.14 (2):

   Given C = 3.5%, Q = 0.5 m^3/hr = 12 m^3/d, properties of the two centrifuges.

   - Construct the following table using the given data:

Parameter	First centrifuge	Second centrifuge
1 (cm)	24	60
D (cm)	20	40
r_2 (cm)	20/2 = 10	40/2 = 20
v_r (rad/s)	$(2\pi/60)*4500 = 471$	$(2\pi/60)*4000 = 419$
Z (cm)	3	5.0
$r_1 = r_2 - Z$ (cm)	10 – 3 = 7	20 – 5 = 15
d_p (cm)	5	10
n (dimensionless)	1	1
v_w (rad/s)	4500 – 4450 = 50	4000 – 3950 = 50

   - Estimate settling of each centrifuge by using the Sigma equation as: $Q_1/\Sigma_1 = Q_2/\Sigma_2$.
     - Find volume as $V = 2\pi([r_1 + r_2]/2)*(r_2 - r_1)*1$.
     - Compute Sigma as $\Sigma = ([v_r]^2*V)/(g*Ln(r_2/r_1))$.

Parameter	First centrifuge	Second centrifuge
V (cm^3)	3847	33000
v_r (rad/s)	471	419
$Ln(r_2/r_1)$	Ln(10/7) = 0.35667	Ln(20/15) = 0.28768
g (cm^2/s)	981	981
Σ	2443409	20533306
Q (m^3/d)	12	?

- Compute the liquid flow rate into the second centrifuge as: $Q_2 = (Q_1/\Sigma_1)*\Sigma_2 = (12/2443409)*20533306 = 101$ m^3/d. With respect to settling of solids, the second centrifuge will achieve equal dewaterability when the flow rate is 101 m^3/d.
- Estimate movement of solids out of centrifuge by finding the Beta equation as: $W_1/\beta_1 = W_2/\beta_2$.
  - Determine the solids loading rate for the two centrifuges as: W = flow rate (m^3/d)*solids concentration*density. (Assume density of water = 1000 kg/m^3.)
  - Determine the Beta value for each centrifuge as: $\beta = v_w*d_p*n*\pi*Z*D$.

Parameter	First centrifuge	Second centrifuge
W (kg/d)	420	?
v_w (rad/s)	50	50
d_p (cm)	5	10
n (dimensionless)	1	1
Z (cm)	3.0	5.0
D (cm)	20	40
β	47,124	314,159

- Estimate solids movement of the second centrifuge as: $W_2 = (W_1/\beta_1)*\beta_2 = (420/47{,}124)*314{,}159 = 2800$ kg/d. This is equivalent to a value of 2800/(0.035*1000) = 80 m^3/d (assuming the same solids concentration of 3.5% and same density).
- In summary and to have similar dewaterability to the first centrifuge:
  - Sigma equation advocates settling properties for second centrifuge when liquid flow rate attains a value of 101 m^3/d.

  Beta equation reveals movement of solids out of the second centrifuge when the solids loading rate attains a value of 80 m^3/d.

  Since the scale-up procedure indicates that the lower value governs the centrifuge capacity, then in this case the solids loading rate governs. As such, the new centrifuge is not to be operated at a solids loading rate exceeding 80 m^3/d.

Listing of Program 6.14

```
DECLARE FUNCTION volume! (r1!, r2!, L!)
DECLARE FUNCTION sigma! (Vr!, V!, r1!, r2!)
DECLARE FUNCTION Beta! (Vw!, dp!, n!, z!, D!)
DECLARE SUB box (r1%, C1%, r2%, C2%)      'Found in Program 2.1
'*****************************************************************
'Program 6.14: Centrifugation
'*****************************************************************
pi = 3.1415962#
g = 9.81
CALL box(1, 1, 20, 80)

LOCATE 2, 30
PRINT "Program 6.14: Centrifugation"
LOCATE 3, 5
INPUT "Enter the solids concentration (%)                =";  C
LOCATE 4, 5
INPUT "Enter the flow rate of sludge to dewatered, Q (m3/hr)  =";  Q
```

```
LOCATE 6, 3
PRINT "Enter the characteristics of the two centrifuges:"
LOCATE 7, 3
PRINT "Characteristic        First Centrifuge        Second Centrifuge"
LOCATE 8, 3: PRINT "Bowl Length, cm"
LOCATE 8, 30: INPUT L1
LOCATE 8, 55: INPUT L2
LOCATE 9, 3: PRINT "Diameter, cm"
LOCATE 9, 30: INPUT D1
LOCATE 9, 55: INPUT D2
LOCATE 10, 3: PRINT "Speed, rpm"
LOCATE 10, 30: INPUT vs1
LOCATE 10, 55: INPUT vs2
LOCATE 11, 3: PRINT "Depth, cm"
LOCATE 11, 30: INPUT z1
LOCATE 11, 55: INPUT z2
LOCATE 12, 3: PRINT "Scroll Pitch, cm"
LOCATE 12, 30: INPUT dp1
LOCATE 12, 55: INPUT dp2
LOCATE 13, 3: PRINT "Number of leads"
LOCATE 13, 30: INPUT n1
LOCATE 13, 55: INPUT n2
LOCATE 14, 3: PRINT "Conveyor Velocity, rpm"
LOCATE 14, 30: INPUT Vw1
LOCATE 14, 55: INPUT Vw2

r11 = D1 / 2: r12 = r11 - z1
r21 = D2 / 2: r22 = r21 - z2
v1 = volume(r12, r11, L1)
v2 = volume(r22, r21, L2)

vr1 = 2 * pi / 60 * vs1
vr2 = 2 * pi / 60 * vs2
sig1 = sigma(vr1, v1, r12, r11)
sig2 = sigma(vr2, v2, r22, r21)

Q1 = Q * 24                   'flow in first centrifuge, m3/d
Q2 = Q1 * sig2 / sig1           'flow in the second centrifuge

Ro = 1000                     'Density of water (assumed), Kg/m3
Vw1 = vs1 - Vw1
Vw2 = vs2 - Vw2
b1 = Beta(Vw1, dp1, n1, z1, D1) 'beta for first centrifuge
b2 = Beta(Vw2, dp2, n2, z2, D2) 'beta for second centrifuge

C1 = C / 100                  'Concentration of solids in first centrifuge
W1 = Q1 * C1 * Ro               'solids loading rate in first centrifuge
W2 = W1 / b1 * b2               'solids loading rate in second centrifuge

LOCATE 16, 3: PRINT "Results:"
LOCATE 17, 3
PRINT "The flow rate in the second centrifuge  ="; Q2; " m3/d"
LOCATE 18, 3
PRINT "The solids movement in second centrifuge ="; W2; " m3/d"

DO
LOCATE 22, 1
PRINT "Press any key to continue.........."
```

```
LOOP WHILE INKEY$ = ""
END

FUNCTION Beta (Vw, dp, n, z, D)
pi = 22 / 7!
Beta = Vw * dp * n * pi * z * D
END FUNCTION

FUNCTION sigma (Vr, V, r1, r2)
g = 9.81
sigma = Vr ^ 2 * V / (g * LOG(r2 / r1))
END FUNCTION

FUNCTION volume (r1, r2, L)
pi = 22 / 7!
volume = 2 * pi * ((r1 + r2) / 2) * (r2 - r1) * L
END FUNCTION
```

6.11 Wastewater Disposal

6.11.1 Dilution

Discharges of small quantities of relatively dilute sludges and treated effluents find their way into water courses. Such wastewater discharges can affect the water quality in rivers, streams, lakes, estuaries, and oceans. To predict the effects imposed by these discharges on water courses, mathematical models have been formulated. Factors affecting natural purification processes include rate of flow and characteristics of body of water, reoxygenation processes, water use downstream from the disposal point, and the quantity of the wastewater discharged, etc.[5]

In Figure 6.6, application of the principle of mass balance between two points, upstream and downstream of the discharge point, gives an estimate of the degree of dilution in the river as presented in Equation 6.59 (dilution law).

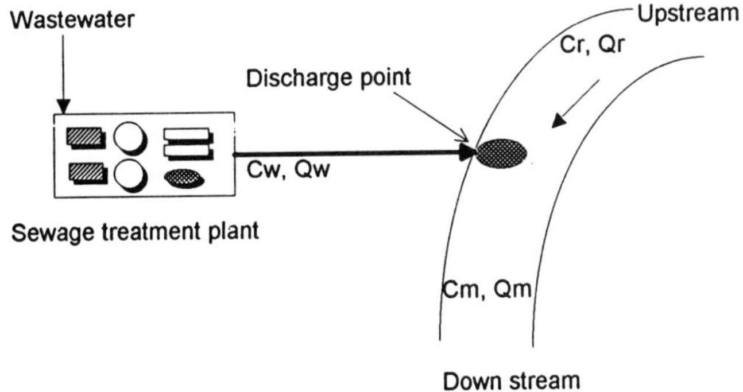

Figure 6.6
Waste water disposal into a stream (dilution).

$$C_w*Q_w + C_r*Q_r = C_m*Q_m \tag{6.59}$$

where:

C_w = Concentration of pollutant in effluent from wastewater treatment works, mg/L

Q_w = Wastewater flow rate to adjacent river, m^3/s

C_r = Concentration of same pollutant in river upstream of the discharge point, mg/L

Q_r = River flow rate, m^3/s

C_m = Concentration of the pollutant in mixture of river water and wastewater discharge, or concentration of pollutant in the river downstream of point of discharge, mg/L

Q_m = Rate of flow of mixture of river water and wastewater discharge, or rate of flow of river downstream of point of discharge, m^3/s

Example 6.15

1. Write a short computer program to determine the concentration of a pollutant in the mixture of effluent and river, C_m (mg/L), given concentration, C_w (mg/L), rate of flow, Q_w (m^3/s) of pollutant introduced to the river from the treatment facility, the river flow, Q_r (m^3/s), and the amount of pollutant initially found in it, C_r(mg/L).
2. A sewage treatment works discharges its wastewater at the rate of 8 m^3/s with a pollutant concentration of 15 mg/L to a receiving stream. The concentration of a particular pollutant upstream of the sewage treatment plant is 2 mg/L and the river flows at a rate of 720 m^3/min. Find the pollutant concentration downstream from the sewage treatment plant.
3. Use the program indicated in (1) to verify computations of (2) above.

Solution

1. For the solution to Example 6.15 (1), see the following listing of Program 6.15.
2. Solution to example 6.15 (2):
   - Given Q_w = 8 m^3/s, C_w = 15 mg/L, C_r = 2 mg/L, Q_r = 720 m^3/min = 12 m^3/s.
   - Compute the pollutant concentration of the mixture downstream of the treatment plant as: $C_m*Q_m = C_w*Q_w + C_{st}*Q_{st}$
   - Find the discharge of the mixture of river water flow and the treatment plant effluent as: $Q_m = Q_w + Q_{st}$ = 8 + 12 = 20 m^3/s.
   - Find the concentration of the pollutant of mixture as: C_m = (15*8 + 2*12)/20 = 7.2 mg/L.

Listing of Program 6.15

```
DECLARE SUB box (r1%, c1%, r2%, c2%)     'Found in Program 2.1
'*****************************************************************
'Program 6.15: Concentration of Pollutants in Rivers
'*****************************************************************
pi = 3.1415962#
g = 9.81
```

```
CALL box(1, 1, 20, 80)
LOCATE 2, 25: PRINT "Program 6.15: Concentration of Pollutants in Rivers"
row = 5
LOCATE row, 5
INPUT "Enter wastewater flow rate, Qw (m3/s)        ="; Qw
LOCATE row + 1, 5
INPUT "Enter the river flow, Qr (m3/s)               ="; Qr
LOCATE row + 2, 5
INPUT "Enter the effluent pollutant concentration, Cw(mg/L)         ="; Cw
LOCATE row + 3, 5
INPUT "Enter the concentration of pollutants in river, Cr(mg/L) ="; Cr

Qm = Qw + Qr
Cm = (Cw * Qw + Cr * Qr) / Qm
LOCATE row + 5, 3
PRINT "The concentration of pollutants in the river ="; Cm; " mg/L"

DO
LOCATE 22, 1
PRINT "Press any key to continue.........."
LOOP WHILE INKEY$ = ""
END
```

6.11.2 Disposal into Natural Waters

6.11.2.1 Introduction

Natural bodies of waters may be used for discharge of treated wastewater when discharge regulations and standards are met. The reduction of pollutant concentrations exceeding permit discharge regulations may be achieved by adoption of appropriate treatment unit operations and processes. Table 6.11 gives general classification patterns for rivers in terms of biochemical oxygen demand, suspended solids, and dissolved oxygen.

TABLE 6.11
River Classification Patterns[4,5,11]

Classification scheme	BOD_5^{20} (mg/L)	SS (mg/L)	DO, as a percent of saturation value
Very clean	≤1	≤4	
Clean	2	10	≥90
Fairly clean	3	15	75–90
Doubtful	5	21	50–75
Poor	7.5	30	<50
Bad	10	35	
Very Bad	≥20	≥40	

Source: From Rowe, D. R. and Abdel-Magid, I.M., *Handbook of Wastewater Reclamation and Reuse*, CRC Press/Lewis Publishers, Boca Raton, FL, 1995. With permission.

The adverse effects of improperly treated wastewater discharges to rivers or water courses include:[5]

- Prevention of usage of water course for beneficial purposes
- Deterioration in quality of a water course
- Endangering aquatic life
- Generation of odors, tastes, and nuisances
- Introduction of public epidemiological health hazards in form of classical diseases or chronic chemical poisoning
- Delaying the mechanisms of self-purification in a water course

6.11.2.2 Oxygen Renewal and Depletion in Rivers

The main sources of oxygen renewal or reoxygenation in a river are reaeration from the atmosphere and the photosynthesis of aquatic plants and algae.[1,5] Atmospheric reaeration may be evaluated by using Equation 6.60

$$r_r = k''(C_s - C) \tag{6.60}$$

where:

r_r = Rate of reaeration
C_s = Dissolved oxygen saturation concentration, mg/L
C = Dissolved oxygen concentration, mg/L
k'' = Reaeration constant, d^{-1} (to base e)

k'' can be estimated from Equation 6.61[1,5,28] or chosen from tables such as Table 6.12.

TABLE 6.12
Reaeration Constants[1,5]

Water body	Range of k″ at 20°C (to base e)
Small ponds and back waters	0.1–0.23
Sluggish streams and large lakes	0.23–0.35
Large streams of low velocity	0.35–0.46
Large streams of normal velocity	0.46–0.69
Swift streams	0.69–1.15
Rapids and waterfalls	>1.15

Source: From Rowe, D.R. and Abdel-Magid, I.M., *Handbook of Wastewater Reclamation and Reuse*, CRC Press/Lewis Publishers, Boca Raton, FL, 1995. With permission.

$$k'' = (294(Diff_T*v)^{0.5})/(h)^{1.5} \tag{6.61}$$

where:

v = Mean river velocity, m/s

h = Average depth of flow, m

$Diff_T$ = Molecular diffusion coefficient for oxygen, m^2/d (see Equation 6.62)

$$(Diff)_T = (Diff_c)*(T_c)^{(T-20)} \quad (6.62)$$

where:

$Diff_T$ = Molecular diffusion coefficient for oxygen at temperature of T °C, m^2/d

$Diff_c$ = Molecular diffusion coefficient for oxygen at temperature of 20°C, m^2/d; this may be taken to be equal to $1.76*10^{-4}$

T_c = Temperature correction coefficient that may be taken to be equal to 1.037.

T = Temperature, °C.

For temperature other than 20°C, k″ can be determined by using Equation 6.63.

$$(k'')_T = (k'')_{20}*(1.024)^{(T-20)} \quad (6.63)$$

where:

$(k'')_T$ = Reaeration constant at a temperature of T°C

$(k'')_{20}$ = Reaeration constant at a temperature of 20°C

Oxygen depletion (deoxygenation) in a river is due primarily to microbial metabolic activities and benthic deposits.[1,5] Deoxygenation by microbial decomposition of organic matter may be estimated as shown in Equation 6.64.

$$r_D = -\,k'*L_o*e^{-k'*t} \quad (6.64)$$

where:

r_D = Rate of deoxygenation.

k' = First order reaction rate constant, d^{-1}

L_o = Ultimate BOD at point of discharge, mg/L

t = Time, d

The effect of organic solids and sediments accumulated in the benthic layer may be estimated by the Fair et al. empirical equation, Equation 6.65.[5,48]

$$L_m = 3.14(10^{-2}*L_o)*T_c*VS*\{(5 + 160*VS)/(1 + 160*VS)\}((t)^{0.5} \quad (6.65)$$

where:

L_m = Maximum daily benthal oxygen demand, g/m^2
L_o = BOD_5^{20} of benthal deposit, g/kg volatile matter
T_c = Temperature correction factor
VS = Daily rate of volatile solids deposition, kg/m^2
t = Time during which settling takes place, d

6.11.2.3 Dissolved Oxygen Sag Curves in Rivers

Equation 6.66 gives the simple river oxygenation model of Streeter and Phelps.[1,5]

$$DO_t = \{k'*L_o[e^{-k'*t} - e^{-k''*t}]/(k'' - k')\} + DO_o*e^{-k'*t} \quad (6.66)$$

where:

DO_t = Oxygen deficit at time t, mg/L
k' = First order reaction rate constant, d^{-1}
L_o = Ultimate BOD at point of discharge, mg/L
k'' = Reaeration constant, d^{-1}
t = Time of travel in river from point of discharge, d
DO_0 = Initial oxygen deficit at point of waste discharge, at time t = 0

The limitations of the model are as follows:[1,4–6,10,32,49]

- It ignores effects of oxygen production by algae.
- It omits oxygen depletion by benthic deposits.
- It assumes one pollutional source or a point source.
- It neglects factors influencing organic load apart from BOD.
- It assumes steady-state conditions along each river reach.

In spite of these limitations, the model provides reasonable estimation and is used extensively.

The point of lowest dissolved oxygen concentration is an important point in the sag curve. This point represents the maximum impact on the dissolved oxygen deficit due to organic waste disposal. The critical oxygen deficit may be estimated as indicated in Equation 6.67.

$$DO_c = (k'*L_o*e^{-k'*tc})/k'' \quad (6.67)$$

where:

DO_c = Critical oxygen deficit, mg/L
k' = First order reaction rate constant, d^{-1}
L_o = Ultimate carbonaceous BOD at point of discharge, mg/L
t_c = Critical time required to reach the critical distance

The critical time can be found from d(DO)/dt = 0, as found in Equation 6.68.

$$t_c = (1/[k'' - k'])*Ln\{[k''/k'](1 - ([DO_o/L_o]*([k'' - k'))/k''\} \quad (6.68)$$

where:

t_c = Critical time, d
k'' = Reaeration constant, d^{-1}
k' = First order reaction rate constant, d^{-1}
DO_o = Initial oxygen deficit at the point of waste discharge, at time t = 0
L_o = Ultimate BOD at point of discharge, mg/L

The critical distance may be found from Equation 6.69:

$$x_c = t_c*v \quad (6.69)$$

where:

x_c = Critical distance, m
t_c = Critical time, d
v = Velocity of flow in the river, m/d

Example 6.16

1. Develop a computer program to find the temperature, DO, and 5-d BOD of a mixture of river water and wastewater, the initial dissolved oxygen deficit of the river just below wastewater treatment plant outfall, the distance downstream to critical DO, and the minimum DO in the river below the wastewater treatment work, Given the daily production of wastewater (m^3/s), treatment plant effluent BOD (mg/L), temperature of the wastewater (°C), dissolved oxygen in effluent of plant as percentage of saturation concentration (%), river flow (m^3/s), velocity of flow in river (m/s), temperature of river water before wastewater addition to river (°C), dissolved oxygen content in river as percentage of saturation concentration (%), river BOD (mg/L), first order reaction constant k′ (d^{-1}), reaeration constant k″ (d^{-1}), temperature correction for k′ and for k″.
2. A river flows at a rate of 12,600 m^3/h with a 5-d BOD of 1 mg/L and it is saturated with oxygen. Wastewater effluent is discharged to the river at the rate of 12 m^3/min with a 5-d BOD of 200 mg/L. The wastewater is 10% saturated with dissolved oxygen. The velocity of the river is 3 km/h. Assume the temperatures to be constant at 20°C, and k′ = 0.1, k″ = 0.4 /d. Compute for the mixture: the dissolved oxygen concentration, the 5-d BOD, the initial deficit in the river, the minimum dissolved oxygen content, and its location downstream from the wastewater treatment plant.

Solution

1. For the solution to Example 6.16 (1), see the following listing of Program 6.16.
2. Solution to Example 6.16 (2):

- Given Q_r = 12600/60*60 = 3.5 m³/s, BOD_r = 1 mg/L, $DO_r = C_s$, Q_w = 12/60 = 0.2 m³/s, BOD_w = 200 mg/L, $DO_w = 0.1*C_s$.
- Find the saturation value of oxygen, C_s, from Table A2 in Appendix as equal to 9.2 for a temperature T of 20°C and DO_w = 0.1*9.2 = 0.92 mg/L.
- Determine the BOD of the mixture of river water and wastewater effluent by using the dilution law as: $BOD_m = (C_r*Q_r + C_w*Q_w)/Q_m$ = (1*3.5 + 200*0.2)/(3.5 + 0.2) = 11.76 mg/L.
- Determine the dissolved oxygen content of the mixture: DO_m = (9.2*3.5 + 0.2*0.92)/(3.5 + 0.2) = 8.7524 mg/L.
- Determine the ultimate BOD of the mixture: $L_o = BOD_m/(1 - e^{-k'*5})$; $L_{om} = 11.76/(1 - e^{-0.1*5})$ = 29.88 mg/L.
- Compute the initial dissolved oxygen concentration as: $DO_o = C_s - DO_m$ = 9.2 – 8.7524 = 0.4476 mg/L.
- Find critical time: $t_c = (1/[k'' - k'])*Ln\{[k''/k'](1 - ([DO_mL_{om}]*([k'' - k']/k'')))\}$ = 4.468 d.
- Determine the critical oxygen as: $DO_c = (k'*L_o*e^{-k'*tc})/k''$; $DO_c = (0.1/0.4)*29.88*e^{-0.1*4.468}$ = 4.39 mg/L.
- Determine the position downstream where DO_c occurs: xc = v*tc = 3*(24h)*4.468 d = 321.7 km.

Listing of Program 6.16

```
DECLARE FUNCTION GetCs! (T!, Cs(), temp())
DECLARE SUB box (r1%, C1%, r2%, C2%)        'Found in Program 2.1
'****************************************************************
'Program 6.16:DO & BOD of River-Wastewater Mixtures
'****************************************************************
'Data from Table A2
DATA 14.62,14.23,13.84,13.48,13.13,12.8,12.48,12.17,11.87
DATA 11.59,11.33,11.08,10.83,10.6,10.37,10.15,9.95,9.74
DATA 9.54,9.35,9.17,8.99,8.83,8.68,8.53,8.38,8.22,8.07,9.72,7.77,7.63

DIM Cs(31), temp(31)

FOR I = 1 TO 31
temp(I) = I
READ Cs(I)
NEXT I

CALL box(1, 1, 20, 80)
LOCATE 2, 25
PRINT "Program 6.16: River-Wastewater Mixtures"
row = 1
LOCATE row + 2, 5
INPUT "Enter the daily production of wastewater, Qw (m3/s)       ="; Qw
LOCATE row + 3, 5
INPUT "Enter the temperature of wastewater, Tw(C)                ="; Tw
LOCATE row + 4, 5
INPUT "Enter the percentage of saturation dissolved oxygen in waste, % ="; PDOw
LOCATE row + 5, 5
INPUT "Enter the 20 C 5 day BOD of the waste, BOD205w. (mg/L)   ="; BODw
LOCATE row + 6, 5
```

```
INPUT "Enter the river flow, Qr (m3/s)                 ='  Qr
LOCATE row + 7, 5
INPUT "Enter the temperature of the river before the waste, Tr  ="; Tr
LOCATE row + 8, 5
INPUT "Enter the percentage of saturation dissolved oxygen in river, % ="; PDOr
LOCATE row + 9, 5
INPUT "Enter the 20 C 5-day BOD of the river, BOD205r(mg/L)    ="; BODr
LOCATE row + 10, 5
INPUT "Enter the velocity of the river, v (km/h)           ="; v
LOCATE row + 11, 5
INPUT "Enter the first order reaction constant(at 20 C), k' /d  ="; kd
LOCATE row + 12, 5
INPUT "Enter the reareation constant(at 20 C), k" /d        ="; kdd
LOCATE row + 13, 5
INPUT "Enter the temperature correction of k'              ="; ckd
LOCATE row + 14, 5
INPUT "Enter the temperature correction for k"             ="; ckdd

IF PDOw < 1 THEN PDOw = PDOw * 100
IF PDOr < 1 THEN PDOr = PDOr * 100
Csw = GetCs(Tw, Cs(), temp())        'DO concentration in wastewater at Tw
DOw = PDOw / 100 * Csw               'DO of wastewater at Tw
Csr = GetCs(Tr, Cs(), temp())        'DO concentration in river at Tr
DOr = PDOr / 100 * Csr               'DO of river at Tr
Qm = Qw + Qr                         'Mixture flow
Tm = (Qw * Tw + Qr * Tr) / Qm        'Mixture temperature
DOm = (Qw * DOw + Qr * DOr) / Qm     'Mixture DO
BOD5m = (Qw * BODw + Qr * BODr) / Qm   'Mixture BOD5
Lom = BOD5m / (1 - EXP(-kd * 5))     'Mixture Lo
Csm = GetCs(Tm, Cs(), temp())        'DO concentration in mixture at Tm
DOi = Csm - DOm                      'Initial oxygen deficit in mixture
kd20 = kd                            'Store value of kd at 20C
kd = kd * ckd ^ (Tm - 20)            'Corrected kd at temperature of mixture
kdd = kdd * ckdd ^ (Tm - 20)         'Corrected kdd at temperature of mixture
'determine critical time tc
tc = 1 / (kdd - kd) * LOG(kdd / kd * (1 - DOi / Lom * (kdd - kd) / kd))
v = v * 24                           'Convert to Km/d
xc = v * tc                          'position downstream where critical conditions exist
Dc = kd / kdd * Lom * EXP(-kd * tc)
DOc = Csm - Dc                       'DO at xc
BOD5 = Lom * EXP(-kd * tc)           'BOD5 at xc at temperature Tm
BOD205c = BOD5 * (1 - EXP(-kd20 * 5))  'BOD5 at xc at temperature 20 C

CALL box(1, 1, 20, 80)
LOCATE 2, 25
PRINT "Program 6.16: River-Wastewater Mixtures"
row = 2
LOCATE row + 3, 3: PRINT "RESULTS:"
LOCATE row + 4, 5: PRINT "Temperature of the mixture       ="; Tm; " C"
LOCATE row + 5, 5: PRINT "Dissolved oxygen of the mixture   ="; DOm; " mg/L"
LOCATE row + 6, 5: PRINT "5-day BOD of the mixture         ="; BOD5m; " mg/L"
LOCATE row + 7, 5: PRINT "Initial DO deficit in river      ="; DOi; " mg/L"
LOCATE row + 8, 5: PRINT "Critical time tc for minimum DO    ="; tc; " day"
LOCATE row + 9, 5: PRINT "Position where minimum DO occurs, xc="; xc; " Km"
LOCATE row + 10, 5: PRINT "Minimum DO in the river (critical) ="; DOc; " mg/L"
LOCATE row + 11, 5: PRINT "5-day BOD at xc at 20 C          ="; BOD205c; " mg/L"
```

```
DO
LOCATE 22, 1
PRINT "Press any key to continue.........."
LOOP WHILE INKEY$ = ""
END

FUNCTION GetCs (T, Cs(), temp())
'Get Dissolved oxygen concentration at zero chloride content, Cs
'Cs in mg/L and temperature in C

FOR I = 1 TO 30
IF T < temp(I) THEN EXIT FOR
NEXT I
j = I
IF T <= 0 THEN j = 1
IF T >= 30 THEN j = 31
GetCs = Cs(j)

END FUNCTION
```

6.11.3 Disposal into Lakes

Treated wastewater can find its way into lakes and reservoirs, especially in inland locations where nearby streams are not present. Lakes and reservoirs are often subject to significant mixing due to wind-induced currents.[1] The theoretical model for analysis assumes complete mixing conditions (for small lakes and reservoirs), constant flow rates, and biodegradation of the pollutant follows the first-order reaction ($r_c = K'*C$). Figure 6.7 is a schematic of a lake system receiving a contaminant. The material mass balance for this lake is given by Equation 6.70, which estimates the concentration of pollutant in the lake.[1]

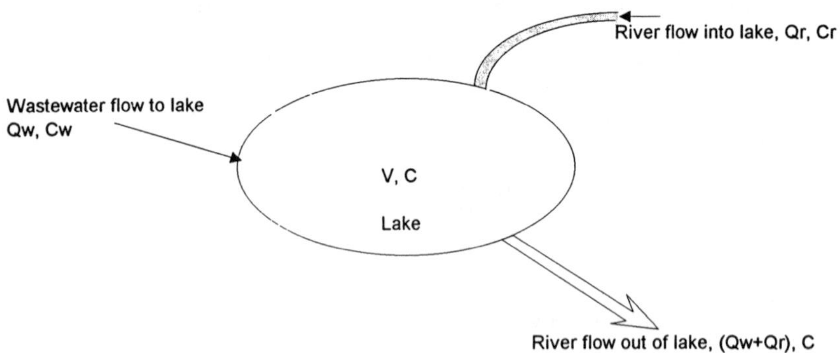

Figure 6.7
Disposal into a lake. (V = lake volume. C = concentration of pollutant.)

$$C = [W*(1 - e^{\beta*t})/\beta*V] + C_o*e^{-\beta*t} \qquad (6.70)$$

where:

C = Pollutant concentration in the lake and in effluent, kg/m^3

$$W = Q_r*C_r + Q_w*C_w \qquad (6.71)$$

Q_r = River flow rate into lake, m^3/s

C_r = Concentration of pollutant in river, kg/m^3

Q_w = Wastewater flow rate into lake, m^3/s

C_w = Concentration of pollutant in wastewater, kg/m^3

β = $(1/t) + k'$ (6.72)

t = Detention time = V/Q

k' = First-order decay constant (base e), d^{-1}

V = Lake volume

C_o = Concentration in the lake at time t = 0

The equilibrium concentration can be found by letting t equal ∞, as presented in Equation 6.73.

$$C_e = W/\beta*V \tag{6.73}$$

where:

C_e = Equilibrium waste concentration in the lake.

Example 6.17

1. Write a computer program to find the equilibrium concentration of a radioactive pollutant discharged into a small lake with average dimensions (m) L, B, d for length, breadth, depth, respectively, given concentration of pollutant entering the lake C_p (mg/L), half life of pollutant $t_{1/2}$, at a flow rate of q_p (m^3/s). Assume that wind currents enable complete mixing of the lake contents.
2. Use the program developed in (1) to determine the equilibrium concentration of a radioactive substance of a 12-h half life, discharged at a rate of 0.05 m^3/s with a concentration of 4 mg/L to a small lake of average dimensions 60*25*2 m for length, width, and depth, respectively.

Solution

1. For solution to Example 6.17 (1), see the following listing of Program 6.17.
2. Solution to Example 6.17 (2):
   - Given $t_{1/2}$ = 12/24 = 0.5 d, Q_r = 0 (no river flow out of lake), q_w = 0.05 m^3/s, C_w = 4, L = 60 m, B = 25 m, d = 2 m.
   - Determine W as: $W = Q_r*C_r + q_w*C_w$ = 0 + 0.05*4 = 0.2 g/s = 17280 g/d.
   - Determine detention time: $t = V/Q = L*B*d/(q_r + q_w)$ = 60*25*2/0.05*60*60*24 = 0.694 d.
   - Find first-order decay rate as: $L_t = L_o*e^{-k't}$, and when t = 0.5 d, $L_{1/2} = 1/2L_o$ or k' = – Ln1/2 = 0.693/d.
   - Determine the factor β as: $\beta = 1/t_o + k'$ = (1/0.694) + 0.693 = 2.133 /d.
   - Find the equilibrium concentration as: $C_e = W/\beta V$ = 17,280/(2.133*60*25*2) = 2.7 mg/L.

Listing of Program 6.17

```
DECLARE SUB box (r1%, c1%, r2%, c2%)      'Found in Program 2.1
'******************************************************************
'Program 6.17: Equilibrium Concentration of Pollutants in Lakes
'******************************************************************
pi = 3.1415962#
g = 9.81

CALL box(1, 1, 20, 80)

LOCATE 2, 10
PRINT "Program 6.17: Equilibrium Concentration of Pollutants in Lakes"
row = 5
LOCATE row, 5
INPUT "Enter the length of the lake, L (m)                =";  L
LOCATE row + 1, 5
INPUT "Enter the breadth of the lake, B (m)               =";  B
LOCATE row + 2, 5
INPUT "Enter the depth of the lake, d (m)                 =";  d
LOCATE row + 3, 5
INPUT "Enter the concentration of pollutants entering, Cp (mg/L)="; Cp
LOCATE row + 4, 5
INPUT "Enter the half life of pollutants, tw (hr)         =";  tw
LOCATE row + 5, 5
INPUT "Enter the flow rate of pollutants, Qp, (m3/s)      =";  Qp

tw = tw / 24            'convert to days
Qp = Qp * 3600 * 24     'convert to m3/d
Qw = Qp
Cw = Cp
Qr = 0                  'no flow out of the lake
Cr = 0
W = Qr * Cr + Qw * Cw
V = L * B * d           'volume
Q = Qr + Qw
t = V / Q               'detention time
k = -1 / tw * LOG(.5)
beta = 1 / t + k
Ceq = W / (beta * V)

LOCATE row + 8, 3
PRINT "The equilibrium concentration          ="; Ceq; " mg/L"
DO
LOCATE 22, 1
PRINT "Press any key to continue.........."
LOOP WHILE INKEY$ = ""
END
```

6.11.4 Disposal in Estuaries

An estuary signifies the zone where a river meets the sea. Tidal action is assumed to result in significant lateral mixing in reaches of the river adjacent to the estuary and increases the amount of mixing and dispersion of waste disposed along the length of channel. For continuous discharge of waste to an estuary, the concentration may be determined as presented in Equation 6.74.[1]

$$C = C_o{*}e^{j{*}x} \qquad (6.74)$$

where:

C = Concentration of waste discharged at time t

C_o = Initial concentration in estuary

$C_o = W/[Q(1 + 4k'{*}E/v^2)^{0.5}]$ (6.75)

$J = (v/2E){*}[(1 \pm (1 + 4{*}k'{*}E/v^2)^{0.5})]$ (6.76)

where:

E = Coefficient of eddy diffusion or turbulent mixing, m^2/s

v = Velocity of flow in river (= Q/A), m/s

A = Cross sectional area, m^2

k' = First-order decay constant, d^{-1}

W = Pollutant load as determined by Equation 6.71

The positive root for J refers to the upstream (–x) direction, and the negative root refers to the downstream (+x) direction.

Example 6.18

Write a computer program that enables determination of the concentration of a pollutant at any time t after its disposal to an neighboring estuary, given the relevant parameters.

Solution

See the following listing of Program 6.18.

Listing of Program 6.18

```
DECLARE SUB box (r1%, c1%, r2%, c2%)      'Found in Program 2.1
'*******************************************************************
'Program 6.18: Waste Concentration in Estuaries
'*******************************************************************
pi = 3.1415962#
g = 9.81

CALL box(1, 1, 20, 80)

LOCATE 2, 25
PRINT "Program 6.18: Waste Concentration in Estuaries"
row = 5
LOCATE row, 5
INPUT "Enter the cross sectional area of the river, A (m2)     ="; A
LOCATE row + 1, 5
INPUT "Enter the flow rate in river, Q (m3/s)                  ="; Q
LOCATE row + 2, 5
INPUT "Enter the value of k                                    ="; k
```

```
LOCATE row + 3, 5
INPUT "Enter the coefficient of turbulent mixing, E (m2/s)    ="; E
LOCATE row + 4, 5
INPUT "Enter the value of W                                   ="; W
LOCATE row + 5, 5
INPUT "Enter the distance where concentration is required, x (m)="; x

V = Q / A
j1 = V / (2 * E) * (1 + (1 + 4 * k * E / V ^ 2) ^ .5)
j2 = V / (2 * E) * (1 - (1 + 4 * k * E / V ^ 2) ^ .5)

IF x < 0 THEN j = j1 ELSE j = j2
C = W * EXP(-j * x) / (Q * (1 + 4 * k * E / V ^ 2) ^ .5)

LOCATE row + 8, 3
PRINT "The concentration of wastes  ="; C
DO
LOCATE 22, 1
PRINT "Press any key to continue.........."
LOOP WHILE INKEY$ = ""
END
```

6.12 Homework Problems

6.12.1 Discussion Problems

1. Define the following terms: wastewater, dry weather flow, population equivalent, and grit.

2. What are some of the problems associated with improper disposal of wastewater?

3. Outline the main sources of wastewater.

4. What are the reasons for the treatment of wastewaters?

5. Differentiate between wastewater treatment unit operations and processes. Give examples of each.

6. What are the problems associated with grit?

7. What are the reasons for installing secondary treatment units in a sewage treatment facility?

8. Differentiate between fixed-growth and suspended-growth systems. Give examples for each.

9. Define the following terms: activated sludge, food-to-microorganisms ratio, volumetric loading rate, hydraulic loading rate, cell age, sludge volume index, and sludge density index.

10. What are the factors that affect the activated sludge process?
11. Briefly describe the trickling filter process.
12. What are the merits of recirculating a portion of a treated effluent to a trickling filter unit?
13. Indicate how to estimate the efficiency of the performance of a trickling filter unit.
14. Indicate the assumptions made in the National Research Council formula for estimating efficiency of a trickling filter.
15. Define the following terms: waste stabilization pond and symbiotic relationship.
16. Differentiate between anaerobic, aerobic, facultative and maturation waste stabilization ponds. Indicate factors affecting design and operation of each.
17. Differentiate between aerobic and anaerobic digestion. Indicate the factors affecting each.
18. Outline the major methods that can be used to dewater sewage sludges. Indicate factors that govern the design and operation of each system. What demerits are to be expected in each system?
19. Outline the major methods that can be used to dewater sewage sludges. Indicate factors that govern the design and operation of each system. What demerits are to be expected in each system?
20. What are the major factors that influence filterability of sewage sludges?
21. Define the following terms: sludge filtration, specific resistance to filtration, compressibility of sludge, and Sigma equation.
22. What methods can be used to determine the filterability of a sludge sample? What are the limitations of each method?
23. What are the factors that affect the natural purification processes in a water course?
24. What are the limitations of the dilution law?
25. Show how to differentiate between a clean and a polluted river.
26. What are the effects of untreated wastewater discharges on surface water?
27. What are the main sources for reoxygenation of a river?
28. What are the limitations of the Streeter-Phelps model?

6.12.2 Specific Mathematical Problems

1a. Write a short computer program to determine the detention time for an activated sludge plant, given tank aeration volume, V (L/capita), MLVSS (mg/L), and sludge wasting rate, Q_w (kg/capita/d).

1b. Using this program, compute the solids content of the plant, given V = 48 L/capita, MLVSS = 2400 mg/L, and Q_w = 30 g/capita/d.

2a. Write a computer program that can be used to estimate the NRC-efficiency of a two-stage trickling filter unit, given the wastewater flow rate to the plant being Q (m^3/s); filter depth, h (m); influent BOD_5; L_i (mg/L); recirculation to the first filter, Ru_1; recirculation to the second filter, Ru_2; volume of the first filter, V_1 (m^3); and volume of the second filter, V_2 (m^3).

2b. Use this program to determine the efficiency of a two-stage trickling filter, given the following data: Q = 3 m^3/min, h = 2 m, L_i = 215 mg/L, Ru_1 = 150% of Q, Ru_2 = 125% of Q, V_1 = 250 m^3, and V_2 = 250 m^3. Use the NRC formula to compute the effluent BOD_5 of the two-stage trickling filter unit.

3a. Write a computer program to design a trickling filter to treat a wastewater flow of Q (m^3/s) with an average BOD of L_i (mg/L). The treatment works is assumed to have one single-stage, high-rate trickling filter with a BOD loading, BOD_1 (kg/m^3*s), and depth, h (m). The 5-d BOD^{20} removal ratio in the primary sedimentation tank and trickling filter are E_s, and E_f, respectively.

3b. Use this program to design a trickling filter given the following date: Q = 1600 m^3/d, L_i = 230 mg/L, h = 2 m, BOD_1 = 330 kg/m^3/d, E_s = 35%, and E_f = 62%.

4a. Use the NRC equation to write a short computer program to calculate the influent concentration of BOD of a wastewater, Y, in mg/L, introduced to a trickling filter, given filter diameter, D (m); filter depth, h (m); wastewater effluent, Q (m^3/s); recirculated flow of Q_r (m^3/s); and final effluent BOD_5, L_e (mg/L).

4b. Use this program to determine Y, given D = 10 m, h = 2.5 m, Q = 0.02 $m^3/_s$, Q_r = 1.6 m^3/min., L_e = 25 mg/L.

5a. Write a computer program to compare the dewaterability of two sludges, A and B, which have the same coefficient of compressibility "a". The value of the specific resistance of sludge A at a pressure of P_1 (Pa) is given as r_A m/kg. The specific resistance of sludge B is reported as r_B (s^2/g) at a pressure of P_2 (Pa).

5b. Use this program to compare the dewaterability of the two sludges A and B, given a = 0.75, $r_A = 2.3*10^{14}$ m/kg, P_1 = 60 kPa, $r_B = 3.7*10^8$ s^2/g, P_2 = 0.2 MPa.

6. Expand the computer program of Example 6.16 so that the dissolved oxygen sag curve can be plotted for a river receiving the treated waste discharges indicated in this example.

7a. Develop a computer program to determine the effluent 5-d BOD of a treated wastewater Y (mg/L). The river flows at a rate of q_r (m^3/h) with a 5-d BOD of BOD_r (mg/L) and is x% saturated with oxygen. The wastewater effluent is discharged to the river at a rate of Q_w (m^3/s) with a 5-d BOD of Y (mg/L) and it is z% saturated with dissolved oxygen. The critical oxygen deficit amounts to DO_c (mg/L). Temperatures are T_r and T_w degrees Celsius for the river water and wastewater effluent respectively.

7b. Use the program developed in 7a to determine the value of Y and the dissolved oxygen deficit after 3 d for the following data: $Q_r = 1800\ m^3/h$, $BOD_r = 1$ mg/L, x = 100%, $Q_w = 12\ m^3/min$, z = 10%, $DO_c = 2.5$ mg/L, $T_r = T_w = 20°C$, $k' = 0.1$ /d, and $k'' = 0.4$ /d.

References

1. Metcalf and Eddy, Inc., *Wastewater Engineering Treatment Disposal Reuse*, 3rd ed., McGraw-Hill, New York, 1991.
2. Barnes, D., Bliss, P.J., Gould, B.W., and Vallentine, H.R., *Water and Wastewater Engineering Systems*, Pitman Publishing, Marshfield, MA, 1981.
3. Callely, A.G., Forster, C.F.F., and Stafford, D.A., *Treatment of Industrial Effluents*, Hodder and Stoughton, London, 1977.
4. El-Hassan, B.M. and Abdel-Magid, I.M., *Environment and Industry: Treatment of Industrial Wastes*, Institute of Environmental Studies, Khartoum University Press, Khartoum, 1986.
5. Rowe, D.R. and Abdel-Magid, I.M., *Handbook of Wastewater Reclamation and Reuse*, CRC Press/Lewis Publishers, Boca Raton, FL, 1995.
6. Peavy, H.S., Rowe, D.R., and Tchobanoglous, G., *Environmental Engineering*, McGraw-Hill, New York, 1985.
7. Wilson, F., *Design Calculations in Wastewater Treatment*, E. and F.N. Spon, London, 1981.
8. American Society of Civil Engineers and the American Water Works Association, *Water Treatment Plant Design*, 2nd ed., McGraw-Hill, New York, 1990.
9. Mullick, M.A., *Wastewater Treatment Processes in the Middle East*, The Book Guild Ltd., Sussex, 1987.
10. Steel, E.W. and McGhee, T.J., *Water Supply and Sewerage*, 6th ed., McGraw-Hill, New York, 1991.
11. Abdel-Magid, I.M., *Selected Problems in Wastewater Engineering*, National Research Council, Khartoum University Press, Khartoum, 1986.
12. Vesilind, P.A. and Peirce, J.J., *Environmental Engineering*, Ann Arbor Science, Ann Arbor, MI, 1982.
13. Fair, G.M., Geyer, J.C., and Okun, D.A., *Water and Wastewater Engineering*, Vol. I and II, John Wiley & Sons, New York, 1966.
14. Hammer, M.J., *Water and Wastewater Technology*, John Wiley & Sons, New York, 1986.
15. American Water Works Association, *Water Quality and Treatment — A Handbook of Public Water Supplies*, McGraw-Hill, New York, 1971.
16. Black, J.A., *Water Pollution Technology*, Reston, VA, 1977.

17. Degremont, *Water Treatment Handbook*, 6th ed., Vol. 1 and 2, Degremont, Rueil-Malmaison, Cexex, France, 1991.
18. Ganczarczyk, J.J., *Activated Sludge Process: Theory and Practices*, Pollution Engineering and Technology Series No. 23, Marcel Dekker, New York, 1983.
19. Mara, D., *Sewage Treatment in Hot Climates*, John Wiley & Sons, New York, 1980.
20. Nathanson, J.A., *Basic Environmental Technology: Water Supply, Waste Disposal and Pollution Control*, Prentice-Hall, Englewood Cliffs, NJ, 1986.
21. Negulescu, M., *Municipal Wastewater Treatment*, Developments in Water Science Series No. 23, Elsevier, Amsterdam, 1986.
22. Salvato, J.A., *Environmental Engineering and Sanitation*, 4th ed., John Wiley & Sons, New York, 1992.
23. Berger, B.B., Ed., *Control of Organic Substances in Water and Wastewater*, Noyes Publishing, Park Ridge, NJ, 1987.
24. Frederick, S., Ed., *Standard Handbook for Civil Engineers*, McGraw-Hill, New York, 1976.
25. James, A., *An Introduction to Water Quality Modeling*, John Wiley & Sons, New York, 1984.
26. Vernick, A.S. and Walker, F.C., *Handbook of Wastewater Treatment Processes*, Pollution Engineering and Technology Series No. 19, Marcel Dekker, New York, 1981.
27. Tebbutt, T.H.Y., *Principles of Water Quality Control*, Pergamon Press, Oxford, 1992.
28. O'Connor, D. and Dobbins, W., The mechanism of reaeration in natural streams, *J. Sanitary Eng.*, SA6, 1956.
29. Gunnerson, C.G. and Stuckey, D.C., *Integrated Resources Recovery: Anaerobic Digestion Principles and Practice for Biogas Systems*, Technical Paper Number 49, World Bank, Washington, D.C., 1986.
30. Coackley, P., Sludge dewatering treatment, *Proc. Biochem.*, 2(3), 1967, 17.
31. Water Pollution Control Federation, *Sludge Dewatering*, Manual of Practice No. 20, Washington, D.C., 1969.
32. Viessman, W. and Hammer, M.J., *Water Supply and Pollution Control*, Harper & Row, New York, 1985.
33. Abdel-Magid, I.M., *The Role of Filter Aids in Sludge Dewatering*, Ph.D. thesis, University of Strathclyde, Glasgow, U.K., 1982.
34. Gale, R.S., *Recent Research on Sludge Dewatering, Water Pollution Research Laboratory*, Stevenage, Herts, 1971, 531–810.
35. Gale, R.S., Filtration theory with special reference to sewage sludges, *J. Water Pollut. Control*, 622–631, 1967.
36. Coackley, P., *Development in our Knowledge of Sludge Dewatering Behavior*, 8th Public Health Engineering Conference held in the Dept. of Civil Engineering, Loughborough University of Technology, 1975, 5.
37. Coackley, P., The theory and practice of sludge dewatering, *J. Instit. Public Health Eng.*, 64(1), 34, 1965.
38. Karr, P.R. and Keinath, T.M., Influence of particle size on sludge dewaterability, *J. Water Pollut. Control Fed.*, 1911–1930, 1978.
39. Coackley, P., Research on sewage sludge carried out in the civil Engineering department of the University College London, *J. Proc. Instit. Sewage Purification*, 1, 59–72, 1955.
40. Karr, P.R. and Keinath, T.M., Limitations of the specific resistance and CST tests for sludge dewatering, *Filtration Separation J.*, 543–544, 1978.

41. Swanwick, J.D. and Davidson, M.F., Determination of specific resistance to filtration, *Water Wastewater Treat. J.*, 8(8), 386–389, 1961.
42. Rudolfs, W. and Heukelekian, H., Relation between drainability of sewage and degree of digestion, *Sewage Works J.*, 6, 1073–1081, 1934.
43. Coackley, P. and Allos, R. The drying characteristics of some sewage sludges, *J. Proc. Inst. Sewage Purification*, 6, 557, 1962.
44. Coackley, P., *The Dewatering Treatment*, Ph.D. thesis, London University, U.K., 1953.
45. Nebiker, J.H., *Dewatering of Sewage Sludge on Granular Materials*, Environmental Engineering Report No. EVE-8-68-3, London, 1968.
46. Newitt, D.M., Oliver, T.R., and Pearse, J.F., The mechanism of the drying of solids, *Trans. Inst. Chem. Eng.*, 27, 1, 1949.
47. Carman, P.C., Fundamental principles of industrial filtration, *Trans. Inst. Chem. Eng.*, 16, 168–188, 1938.
48. Fair, G.M., Morris, F.C., Chang, S.L., Weil, I., and Burden, R.A., The behavior of chlorine as a water disinfectant, *J. Am. Water Works Assoc.*, 40, 1051, 1948.
49. Masters, G.M., *Introduction to Environmental Engineering and Science*, Prentice Hall, Englewood Cliffs, NJ, 1991.
50. Scott, J.S. and Smith, P.G., *Dictionary of Waste and Water Treatment*, Butterworths, London, 1980.

Section IV
Air Pollution and Control

Chapter 7

Air Pollution Control Technology

Contents

The introduction to this section presents a brief overview of the air pollution field. The next section deals with fundamental concepts that are needed to make calculations dealing with air pollution control. Also included here are computer programs based on mathematical equations and models relating to these fundamental concepts. The most commonly used air pollution control devices for gaseous and particulate pollutants are presented, such as absorption, adsorption, and combustion for gaseous contaminants and settling chambers; cyclones; electrostatic precipitators; venturi scrubbers; and baghouse filters for controlling particulate emissions. Each control device or technique is accompanied by a computer program that can aid in the design or evaluation of air pollution control equipment.

This section concludes with mathematical models that can be used for determining effective stack heights (plume rise) as well as dispersion models that can help estimate the concentrations of air pollutants dispersed in the atmosphere. Computer programs for both stack height calculations and dispersion models are included.

7.1 Introduction

The first step in directing compliance and enforcement of air pollution control laws and regulations is to have a credible and acceptable definition as to what air pollution is. In the U.S. the Code of Federal Regulations (40 CFR 52.741, July 1, 1994) indicates that air pollution means the presence of one or more contaminants in sufficient quantities and of such characteristics and duration as to be injurious to human, plant, or animal life, to health, or to property or to unreasonably interfere with the enjoyment of life or property.[1] All environmental regulations in the U.S. can be found in CFR Part 40.

Ambient (outdoor) unpolluted dry air consists by volume of about 78% nitrogen (N_2), 21% oxygen (O_2), 0.93% argon (Ar), 0.03% carbon dioxide (CO_2), and traces of other gases such as neon (Ne), helium (He), methane (CH_4), and krypton (Kr). The unpolluted dry air described above exists only in theory; all air contains natural contaminants such as pollen, fungi, spores, smoke, and dust particles from forest fires and volcanic eruptions. In contrast to natural air, pollutants are contaminants of an anthropogenic (manmade) origin.

Anthropogenic sources include transportation (mobile sources), fuel combustion from electric utilities, fuel combustion from other sources, industrial processes, waste disposal and recycling, and miscellaneous sources such as open burning, agricultural burning, recreation, etc. The two basic physical forms of air pollutants are particulate matter and gases. Particulate matter includes small solid or liquid particles such as dust, smoke, mists, and fly ash. Gases include substances such as carbon monoxide, sulfur dioxide, and volatile organic compounds.

In the U.S., the Environmental Protection Agency (EPA) has further classified air pollutants as criteria pollutants and noncriteria pollutants. Criteria pollutants are pollutants that have been identified as being both common and detrimental to human health and welfare. The EPA currently designates six pollutants as criteria pollutants:[2]

- Carbon monoxide (CO)
- Ozone (O_3)
- Sulfur Dioxide (SO_2)
- Particulate matter (PM) less than 10 μm in diameter
- Nitrogen dioxide (NO_2)
- Lead (Pb)

The U.S. Clean Air Act Amendments of 1990 established a new classification of noncriteria pollutants called hazardous air pollutants (HAPs). This Act listed 189 compounds as hazardous air pollutants and directed the EPA to investigate possible regulation of sources emitting these pollutants.

For each criteria pollutant, the EPA was required to set both a primary standard and a secondary standard. The purpose of the primary standard is to protect public health while secondary standards are set at a level to protect public welfare from any adverse effects. The primary and secondary standards for each of the criteria pollutants are shown in Table 7.1. Collectively, these standards are the National Ambient Air Quality Standards (NAAQS).[2,3]

For comparison, Table 7.2 presents air quality standards for other countries in the world, indicating the standard in micrograms per cubic meter (μg/m^3) and the averaging time. Table 7.3 presents the total air pollutant emissions (million tonnes per year) for the six major air pollutants in the U.S. for 1993.[4] Of these pollutants, transportation produced approximately 77% of the total carbon monoxide, 36% of the volatile organic compounds, 44% of the oxides of nitrogen, 22% of the PM_{10} particulates, 3% of the oxides of sulfur, and 32% of the lead. Transportation was responsible for 56% by weight of all the major contaminants emitted to the atmosphere in the U.S in 1993.[4]

As mentioned earlier, air pollutants can be divided into two classes, gaseous and particulate, and they generally require different prevention and control methods. The broad prevention methods to control air pollution emissions for both gaseous and particulates include:[2]

- Process change
- Changes in fuel
- Good operating practices
- Plant shutdowns

TABLE 7.1
National Ambient Air Quality Standards (NAAQS — 1993[2,3])

	Primary (health-related) standard level		Secondary (welfare-related) standard level	
Pollutant	Averaging time	Concentration	Averaging time	Concentration
Particulate matter<10 μm	Annual geometric mean — 24-h annual maximum	50 μg/m³	Same as primary	Same as primary
	24-h	150 μg/m³	Same as primary	Same as primary
SO_2	Annual arithmetic mean, 1 yr.	80 μg/m³ (0.03 ppm)		
	24-h	365 μg/m³ (0.14 ppm)	3-h	1300 μg/m³ (0.50 ppm)
CO	8-h	10,000 μg/m³ (9 ppm)	No secondary standard	No secondary standard
	1-h	40,000 μg/m³ (35 ppm)	No secondary standard	No secondary standard
NO_2	Annual arithmetic mean, 1 yr	100 μg/m³ (0.053 ppm)	Same as primary	Same as primary
O_3	Maximum daily 1-h average	235 μg/m³ (0.12 ppm)	Same as primary	Same as primary
Pb	Maximum quarterly arithmetic mean, 3-month	1.5 μg/m³	Same as primary	Same as primary

The basic principles involved in the control of gas emissions include:

- Combustion
- Adsorption
- Absorption
- Condensation

The most commonly used devices to control particulate emissions include:

- Settling chambers
- Cyclonics
- Electrostatic precipitation
- Venturi scrubbers
- Baghouse filters

TABLE 7.2
Air Quality Standards for Selected Pollutants in Several Countries (μg/m³)

Country	Suspended particulate	Sulfur oxides	Carbon monoxide	Nitrogen oxides
Canada	120/24 h 60–70/1-yr	450/1 h 150/24 h 30–60/ 1 yr	15,000/1 yr 6000/8 h	400/1 h 200/24 h 60–100/1 yr
Japan	200/1 h 100/24 h	300/1 h 120/24 h	11,100/8 h	100/24 h
Russia	150/24 h	157/24 h	5723/8 h 1145/24 h	113/24 h
Saudi Arabia	340/24 h 80/1 yr	800/1 h 400/24 h 85/1 yr	40,000/1 h 10,000/8 h	660/1 h 100/1 yr
West Germany	480/1/2 h	400/1/2 h	40,000/1 h 10,000/8 h	1,000/1/2 h

TABLE 7.3
Sources of Air Pollutants in the U.S., 1993[4a]

Source	CO	PM_{10}[b]	SO_3	VOCs[c]	NO_X	Pb	Total
Transportation	75.261	0.592	0.718	8.301	10.423	1.589	96.884
Fuel combustion (electric utilities)	0.322	0.270	15.836	0.036	7.782	0.062	24.308
Fuel combustion (industrial)	0.667	0.219	2.830	0.271	3.176	0.018	7.181
Fuel combustion (other)	4.444	0.723	0.600	0.341	0.732	0.417	7.257
Industrial processes	5.219	0.610	1.868	11.201	0.911	2.281	22.090
Waste disposal and recycling	1.732	0.248	0.037	2.271	0.084	0.518	4.890
Miscellaneous	9.506	0.0	0.011	0.893	0.296	n.a.[d]	10.706
Total	97.151	2.662	21.900	23.314	23.404	4.885	173.316

[a] In million short tons (2000). Reference 4.

[b] PM_{10} includes only those particles with aerodynamic diameter smaller than 10 μm. See Reference 4.

[c] VOCs = volatile organic compounds. See Reference 4.

[d] n.a. = not available.

Pollution control efforts utilizing these various principles and techniques have reduced the air pollution emissions of sulfur dioxide by 30% from 1970 to 1993; carbon monoxide, 18%, and volatile organic compounds (VOCs), 20%. The PM_{10} particulate emissions from 1985 to 1993 were reduced 10%; however, from 1970 to 1993, nitrogen emissions have increased about 11%. One of the greatest environmental success stories has been the reduction of lead in the ambient air. From 1984 to 1993, lead concentrations at 204 sampling sites in the U.S. have shown an 89% reduction.[4] Although there have been gains made in controlling air pollution, it must be remembered that some of these gains have been offset by an increase in all activities, especially motor vehicle registration and miles traveled.

In the past, identification and characterization of exposure of people to pollutants in the outdoor air has been emphasized. In recent years, attention has been directed at the effects of exposure to indoor air. Most people spend 80 to 90% of their time indoors and it has been shown that exposure to some pollutants can be two to five times higher indoors than outdoors.[2] Some of these indoor pollutants include carbon monoxide (CO), nitrogen oxides (NO_x), sulfur dioxides (SO_x), particulate matter (PM), asbestos, formaldehyde (HCHO), ozone, and radon gas (Ra-222). Some of the sources for these indoor air pollutants include emissions from combustion appliances, tobacco smoke, aerosol propellants, plastic furniture, rugs and curtains, refrigerants, paints, cleaners, and building materials. The broad measures that can be used to reduce exposure to indoor air pollution are careful selection and operation of household appliances, proper building design (selection of materials and construction), and, of course, proper ventilation.

While more attention is being paid to indoor air pollution on a regional, continental, and global basis, some of the major air pollution problems include:[5]

- Global warming (greenhouse effect due to carbon dioxide (CO_2) emissions)
- Acid rain (sulfur dioxide, SO_2, and nitrogen oxides, NO_x, transformed into acids in the atmosphere)
- Ozone layer depletion (increased ultraviolet radiation exposure caused by depletion of the ozone layer due to reaction with chlorine released by fluorocarbons emitted to the atmosphere)

7.2 Fundamental Concepts in Air Pollution

7.2.1 Units of Measurement

In the past, many confusing and conflicting units have been used in the air pollution field. At present the trend is to try to standardize the units by utilizing the metric system. For instance, on an international basis it is recommended that weights be reported in grams (g), milligrams (mg), or micrograms (μg) and volumes in cubic meters (m^3). The U.S. EPA has recommended using the units for particulates and gaseous pollutants as presented in Table 7.4.[6]

A common practice at present is to present the concentration of the gas contaminants first in μg/m^3, followed by parts per million (ppm), parts per hundred million (pphm) or parts per billion (ppb) by volume. For gases, ppm can be converted to μg/m^3 or vice versa by using the following equation:

$$\mu g / m^3 = \frac{\text{ppm} \times \text{molecular weight of gas} \times 10^3}{\text{liters / mole}} \tag{7.1}$$

TABLE 7.4
EPA Recommended Units of Measurement[6]

Parameter	Unit recommended
Particle fallout	milligrams/square centimeter/month (or year): mg/cm^2/month (or yr)
Outdoor airborne particulate matter	micrograms/cubic meter: $\mu g/m^3$
Gaseous materials	$\mu g/m^3$ or parts per million (ppm)
Standard conditions for reporting gas volumes	760 mmHg (1 atm) 25°C (STP)[a]
Particle counting	Number of particles/m^3 of gas
Temperature	°C
Pressure	mmHg (atm)
Sampling rate	m^3/min or L/min
Visibility	kilometers (km)

[a] STP = Standard temperature and pressure

For instance,

- At 1 atm and 0°C (273 K) L/mol = 22.41 L
- At 1 atm and 25°C (298 K), or standard temperature and pressure (STP), L/mol = 24.46 L

Example 7.1

1. Prepare a computer program relating the various elements in Equation 7.1 to convert ppm to $\mu g/m^3$ or vice versa for gaseous air contaminants.
2. Using the computer program developed in (1) determine the concentration of SO_2 in $\mu g/m^3$ when it has been reported to be 0.14 ppm by volume at STP.

Solution

1. For solution to Example 7.1 (1), see the following listing of Program 7.1.
2. Solution to Example 7.1 (2):

$$\mu g/m^3 = \frac{\text{ppm} \times \text{molecular wgt } \times 10^3}{24.46}$$

$$= \frac{0.14 \text{ ppm} \times 64.1 \text{ g/mol } \times 10^3}{24.46 \text{ L/mol}} = 367\ \mu g/m^3$$

Listing of Program 7.1

```
'***********************************************
'Example 7-1: Relating elements of equation 7-1
'Converts PPM to micro grams/m{3}
'***********************************************
COLOR 15, 1, 4
DIM form$(3)
form$(1) = "ug/m{3} = (ppm * Molecular Weight * 10{3}) / liter/mole "
form$(2) = "ppm = (ug/m{3} * liter/mole) / (Molecular Weight * 10{3})"
form$(3) = "EXIT"
title:
opt = 0
DO
  CLS
  PRINT "Exercise 7-1: Converts PPM (parts per million) to"
  PRINT "  micro grams/m{3} and vice versa"
  PRINT "Convert measure to:  "
  FOR i = 1 TO 3
    LOCATE i + 4, 13
   PRINT "(", i; ")  "; form$(i)
  NEXT i
  INPUT "Select the option 1 - 3:"; opt
LOOP WHILE opt < 1 OR opt > 3
IF opt < 3 THEN
    CLS
    PRINT "At 1 atm. and 0 degrees C (273K), liters / mole = 22.41 liters"
    PRINT "At 1 atm. and 25 degrees C (298K), liters/mole = 24.46"
    PRINT "1 atm. and 25 degrees C are known as Standard Temperature and Pressure"
    PRINT "u = micro"
    PRINT "a number in curly braces {} is a superscript number"
    PRINT "    m{3} = cubic meters"
    INPUT "Press ENTER to continue  "; ent$
    ELSE END
  END IF
SELECT CASE opt
  CASE 1
    CLS
    INPUT "Enter the measurement in ppm  "; ppm
    INPUT "Enter the molecular weight of the gas  "; MolWgt
    INPUT "Enter the liters/mole  "; lmole
      ugmol = (ppm * MolWgt) / lmole * 1000
    PRINT "micro grams/cubic meter = "; ugmol
    INPUT "Press ENTER to continue "; ent$
    GOTO title
  CASE 2
    CLS
    INPUT "Enter the measure in ug/m{3}  "; ugm
    INPUT "Enter the liters/mole  "; lmole
    INPUT "Enter the molecular weight of the gas  "; MolWgt
      ppm = (ugm * lmole / MolWgt) * .001
    PRINT "parts per million  = "; ppm
    INPUT "Press ENTER to continue "; ent$
    GOTO title
  CASE 3
    END
  END SELECT
END
```

Example 7.2

Using the computer program developed in Example 7.1 (1), convert 9.0 ppm CO to μg/m³ at 0°C (273 K), 1 atm, and at STP.

Solution

At 0°C (273 K) and 1 atm:

$$\mu g/m^3 = \frac{\text{ppm} \times \text{molecular wgt of gas} \times 10^3}{22.41 \text{ L/mol}} = \frac{9 \times 28 \times 10^3}{22.41} = 11{,}245\ \mu g/m^3$$

At STP, 25°C (298 K), and 1 atm:

$$\mu g/m^3 = \frac{\text{ppm} \times \text{molecular wgt of gas} \times 10^3}{22.46 \text{ L/mol}} = \frac{9 \times 28 \times 10^3}{24.46} = 10{,}302\ \mu g/m^3$$

7.2.2 Mole and Mole Fraction

The mole is a practical, simple unit that has helped make chemistry an exact and quantitative science. A mole of a substance is the molecular weight of the substance, expressed in mass units, where the molecular weight is the sum of the atomic weights of the atoms which compose the substance. For example, a molecule of hydrogen (2 atoms, H_2) has a molecular weight of 2; a molecule of oxygen (2 atoms, O_2) has a molecular weight of 32. Water (H_2O) has a molecular weight of 18. The atomic weight expresses the ratio of the weight of one atom to that of another. Since the atomic weight is a relative weight, the numerical value must be determined by reference to some standard. In 1961, the carbon-12 atom was adopted as the atomic weight standard with a value of exactly 12. Tables are readily available giving atomic weights (see Table A3 in the Appendix).

Avogadro in 1811 was the first to suggest that equal volumes of all gases at the same temperature and pressure have the same number of molecules. This was known as Avogadro's hypothesis and led to the discovery that one mole of gas, any gas, contained 6.02×10^{23} particles or entities; this value is now called Avogadro's number. By definition, one mole of a substance contains 6.02×10^{23} particles or constituents. It can be expressed as atoms per mole, molecules per mole, ions per mole, electrons per mole, or particles per mole:

6.02×10^{23} O atoms = 16.0 g O

6.02×10^{23} H atoms = 1.01 g H

6.02×10^{23} H_2O molecules = 18.0 g H_2O

6.02×10^{23} OH^- ions = 17.0 g of OH^-

Also, for example, 1 mole of carbon-12 contains 6.02×10^{23} carbon atoms. In mathematical terms the number of moles of a gas can be expressed as

$$n = \frac{m}{MW} \tag{7.2}$$

where:

n = Number of moles of a gas
m = Mass of the gas, g
MW = Molecular weight of the gas, g-mol

In a gaseous mixture the mole fraction for each gas is the ratio of the moles of the given gas divided by the total number of moles of all gases present in the mixture, or

$$X = \frac{n}{\Sigma n_i} \tag{7.3}$$

where:

X = Mole fraction for each gas present in the gaseous mixture
n = Moles of each gas present in the gaseous mixture
Σn_i = $n_1 + n_2 + ... n_i$ = Total number of moles present in the gaseous mixture

Example 7.3

A gaseous mixture contains 4.0 g O_2, 10 g N_2, 1.0 g CO, and 5.0 g CO_2. Determine the moles present for each gas, as well as the mole fraction for each gas. The total number of moles present and the mole fraction for each gas can be determined as follows.

Component	Weight (g)	Molecular weight	Moles	Mole fraction
O_2	4.0	32	0.125	0.20
N_2	10.0	28	0.357	0.56
CO	1.0	28	0.036	0.06
CO_2	5.0	44	0.114	0.18
TOTALS			0.632	1.00

The mole fraction has no dimensions and no units, and the sum of the mole fractions for a gas mixture equals unity.

Example 7.4

1. Write a computer program that can be used to calculate moles and mole fractions for each gas present in a gaseous mixture.
2. Using the computer program developed in (1) determine the moles and mole fraction for each of the gases present in the following mixture: 6.5 g methane (CH_4), 3.0 g carbon dioxide (CO_2), 0.1 g hydrogen sulfide (H_2S), 0.4 g nitrogen (N_2), and 0.1 g hydrogen (H_2). The total number of moles present and the mole fraction for each gas present can be determined as follows:

Component	Weight (g)	Molecular weight	Moles	Mole fraction
CH_4	6.5	16	0.406	0.750
CO_2	3.0	44	0.068	0.126
H_2S	0.1	34.1	0.003	0.005
N_2	0.4	28	0.014	0.026
H_2	0.1	2	0.050	0.093
TOTALS			0.541	1.000

Listing of Program 7.4

```
'*************************************************
'Example 7-4: Calculates moles and mole fractions
' for each gas present in a gaseous mixture
'*************************************************
COLOR 15, 1, 4
CLS
DIM gas$(10), Wgt(10), MolWgt(10), moles(10), MoleFrac(10)
opt = 0
DO
  PRINT "Exercise 7-4: Calculates moles and mole fractions"
  PRINT "             for each gas present in a gaseous mixture": PRINT
  PRINT "Formulas:  "
  PRINT "1.  moles = weight / molecular weight"
  PRINT "2.  mole fraction = moles / total moles": PRINT
  PRINT "Type 1 and ENTER to continue  "
  INPUT "Type 2 and ENTER to EXIT  "; opt
  LOOP WHILE opt < 1 OR opt > 2
  IF opt = 2 THEN
    END
   END IF
CLS
subsc = 1
totmoles = 0
totMoleFrac = 0
endsubsc = 0
DO
  CLS
  PRINT "Type in the name of the gas "
  INPUT "(Press ENTER without typing a gas name if you are done)"; gas$(subsc)
    IF (gas$(subsc) = "") THEN EXIT DO
  INPUT "What is the weight of the gas "; Wgt(subsc)
  INPUT "What is the molecular weight of the gas"; MolWgt(subsc)
     moles(subsc) = Wgt(subsc) / MolWgt(subsc)
```

```
        totmoles = totmoles + moles(subsc)
        subsc = subsc + 1
        endsubsc = subsc
LOOP UNTIL (subsc = 10)
subsc = 1
DO
   CLS
   MolFrac(subsc) = moles(subsc) / totmoles
   totMolFrac = totMolFrac + MolFrac(subsc)
   subsc = subsc + 1
LOOP UNTIL (subsc = endsubsc)
tabval = 6
subsc = 1
   PRINT TAB(25); "Molecular"; TAB(50); "Mole"
      tabval = tabval + 1
   PRINT TAB(7); "Gas"; TAB(15); "Weight"; TAB(25); "Weight"; TAB(35); "Moles"; TAB(50); "Fraction"
      tabval = tabval + 1
DO
   PRINT TAB(7); gas$(subsc);
   PRINT TAB(15); USING "###.####"; Wgt(subsc); TAB(25); MolWgt(subsc); TAB(35); moles(subsc); TAB(5
MolFrac(subsc);
      subsc = subsc + 1
      tabval = tabval + 1
LOOP UNTIL (subsc = endsubsc)
   PRINT TAB(35); "----------"; TAB(50); "-----------"
      tabval = tabval + 1
   PRINT TAB(35); USING "###.####"; totmoles; TAB(50); totMolFrac
END
```

7.2.3 Basic Gas Laws

7.2.3.1 Boyle's Law

Boyle's law states that the volume of a given quantity of an ideal gas varies inversely to its absolute pressure, with the temperature being held constant. A pracitcal and easy way to remember Boyle's law is provided by the following equation.

$$P_1V_1 = P_2V_2 \tag{7.4}$$

where:

P_1 = Initial gas pressure, atm
V_1 = Volume of gas at P_1, L
P_2 = Final gas pressure, atm
V_2 = Volume of gas at P_2, L

Example 7.5

1. Prepare a computer program relating the four elements in Boyle's law (Equation 7.4).
2. Using the computer program developed in (1), determine the volume of a gas held in a 100-L gas cylinder if the initial pressure was 35 atm and the final pressure was 10 atm, the temperature being held constant.

Solution

1. For solution to Example 7.5 (1), see the following lising of Program 7.5.
2. Solution to Example 7.5 (2):

$$P_1V_1 = P_2V_2$$

$$35 * 100 = 10 * V_2$$

$$V_2 = \frac{35 \times 100}{10} = 350 \text{ liters}$$

Listing of Program 7.5

```
DECLARE SUB box (r1%, c1%, r2%, c2%)
'************************************************
'Example 7-5: Relates the two elements in
' Boyle's Law
'************************************************
'
' Equation is "P1 * V1 = P2 * V2"
'
COLOR 15, 1, 4
CLS
DIM form$(2)
form$(1) = "P1 * V1 = P2 * V2"
form$(2) = "EXIT"
title:
opt = 0
DO
  CLS
  PRINT "Exercise 7-5: Relates the two elements in"
  PRINT "            Boyle's Law"
  FOR i = 1 TO 2
    LOCATE i + 4, 13
    PRINT "(", i; ")  "; form$(i)
  NEXT i
  INPUT "Select the option 1 or 2:"; opt
LOOP WHILE opt < 1 OR opt > 2
  IF opt < 2 THEN
    CLS
    PRINT "P1 = initial gas pressure, atm."
    PRINT "V1 = volume of gas at P1, liters"
    PRINT "P2 = final gas pressure, atm."
    PRINT "V2 = volume of gas at P2, liters": PRINT
    ELSE END
  END IF

SELECT CASE opt

  CASE 1
    PRINT " P1 * V1 = P2 * V2        Solving for V2"
    INPUT "Enter the initial gas pressure, (P1) (in atm.) "; ipres
    INPUT "Enter the volume of gas at P1, (V1) (in liters)  "; vol1
    INPUT "Enter the final gas pressure, (P2) (in atm.) "; fgas
```

```
      step1 = ipres * vol1
      step2 = step1 / fgas
    PRINT "volume of gas at P2, liters = "; step2; "liters"
    INPUT "Press ENTER to continue "; ent$
    GOTO title
  CASE 2
    END
  END SELECT
END
```

7.2.3.2 Charles's Law

Charles' law, also known as Gay-Lussac's law, states that the volume of a given mass of an ideal gas varies directly as the absolute temperature with the pressure being held constant.

$$V/T = \text{constant} \tag{7.5}$$

Combining the laws of Boyle and Charles into one expression gives Equation 7.6.

$$\frac{P_1V_1}{T_1} = \frac{P_2V_2}{T_2} \tag{7.6}$$

where:

P_1 = Initial gas pressure, atm
V_1 = Volume of gas at P_1, L
T_1 = Initial gas temperature, K
P_2 = Final gas pressure, atm
V_2 = Volume of gas at P_2, L
T_2 = Final gas temperature, K

Example 7.6

1. Prepare a computer program relating the various elements in Equation 7.6, which combines Boyle's and Charles' laws.
2. One very important use of the combined Boyle's and Charles' laws (Equation 7.6) is to compare an actual gas volume to its volume under a set of standard conditions. For most applications in regard to air pollution, standard conditions are 25°C (298 K), and 1 atm (760 mmHg). The term "standard conditions for temperature and pressure" is abbreviated STP. This principal can also be extended so as to compare actual volumetric flow rates to volumetric flow rates at standard conditions.

Listing of Program 7.6

```
'*************************************************
'Example 7-6: Relating elements of equation 7-4
'Boyle's Law
'*************************************************
```

```
' Equation is "P1 * V1 = P2 * V2"
COLOR 15, 1, 4
CLS
DIM form$(2)
form$(1) = "(P1 * V1) / T1 = (P2 * V2) / T2       Solving for V1"
form$(2) = "EXIT"
title:
opt = 0
DO
  PRINT "Exercise 7-6: Relates the combined elements in"
  PRINT "               Charle's and Boyle's Law"
  FOR i = 1 TO 2
    LOCATE i + 4, 13
    PRINT "(", i; ")  "; form$(i)
  NEXT i: PRINT
  INPUT "Select the option 1 or 2:"; opt: PRINT
LOOP WHILE opt < 1 OR opt > 2
  IF opt < 2 THEN
    PRINT "P1 = initial gas pressure, atm."
    PRINT "V1 = volume of gas at P1, liters"
    PRINT "T1 = initial gas temperature, C"
    PRINT "P2 = final gas pressure, atm."
    PRINT "V2 = volume of gas at P2, liters"
    PRINT "T2 = final gas temperature, C": PRINT
    ELSE END
  END IF
SELECT CASE opt
  CASE 1
    INPUT "Enter the final gas pressure, (P2) (in atm.)  "; fpres       'P2
    INPUT "Enter the volume of gas at P2, (V2) (in m{3}/min)  "; vol2     'V2
    INPUT "Enter the initial gas temperature, (T1) (in C)  "; itemp     'T1
    INPUT "Enter the initial gas pressure, (P1) (in atm.)  "; ipres     'P1
    INPUT "Enter the final gas temperature, (T2) (in C)  "; ftemp       'T2
      ktemp1 = 273 + itemp
      comp1 = fpres * vol2 * ktemp1
      ktemp2 = 273 + ftemp
      comp2 = ipres * ktemp2
      comp3 = comp1 / comp2
    PRINT : PRINT "volume of gas at P1, liters =  "; comp3; "m{3}/min"
    INPUT "Press ENTER to continue "; ent$
    CLS
    GOTO title
CASE 2
    END
  END SELECT
END
```

Example 7.7

Using the computer program developed in Example 7.6 section (1) determine the standard volumetric flow rate at STP for the following conditions:

Actual volumetric flow rate = 6 m^3/min, (V_2)

Actual operating temperature = 90°C (T_2)

Actual operating pressure = 1 atm (P_2)

STP = 25°C (T_1) and 1 atm (P_1)

Solution

Substituting in Equation 7.6:

$$\frac{P_1V_1}{T_1}=\frac{P_2V_2}{T_2}$$

$$V_1=\frac{P_2V_2T_1}{P_1T_2}=\frac{(1\text{ atm})(6\text{m}^3/\text{min})(273+25\text{ K})}{(1\text{ atm})(273+90\text{K})}=\frac{(1)(6)(298)}{(1)(363)}=4.92\text{m}^3/\text{min}$$

7.2.3.3 Ideal Gas Law

Ideal gases that obey Boyle's and Charles' laws, as well as Avogadro's hypothesis (see Section 7.2.2), comply with the ideal gas equation.

$$PV = nRT \tag{7.7}$$

where:

P = Absolute pressure, atm
V = Volume of a gas, L
T = Absolute temperature, K
n = Number of moles of a gas
R = Ideal or universal constant, (1) atm $(K)^{-1}$ $(g\text{-mol})^{-1}$

Typical values of R are as follows:

R = 0.08206 (1) atm $(K)^{-1}$ $(g\text{-mol})^{-1}$
R = 62.4 (1) (mmHg) $(K)^{-1}$ $(g\text{-mol})^{-1}$
R = 1.986 cal/g-mol-K

R is chosen in order to provide appropriate units consistent with the equation used.

As indicated earlier, the number of moles of a gas, n, can be determined if the mass and molecular weight of a gas are known.

$$n=\frac{m}{MW} \tag{7.2}$$

where:

n = Number of moles of gas
m = Mass of gas, g
MW = Molecular weight of gas, g-mol

Also, the density of a gas can be determined by rearranging Equation 7.7, where

$$\rho = \frac{m}{V} \tag{7.8}$$

where:

ρ = Gas density g/L
m = Mass of gas, g
V = Volume of gas, L

Substituting Equation 7.2 into Equation 7.7 gives

$$PV = \frac{mRT}{MW} \tag{7.9}$$

When Equation 7.9 is rearranged,

$$\frac{m}{V} = \frac{P(MW)}{RT} \tag{7.9}$$

Substituting Equation 7.8 into Equation 7.9 gives

$$\rho_a = \frac{P(MW)}{RT} \tag{7.10}$$

The units for this equation are the same as presented earlier.

No real gas obeys the ideal gas law exactly, although the "lighter" gases (hydrogen, oxygen, air, etc.) at ambient conditions approach ideal gas law behavior. The "heavier" gases, such as sulfur dioxide and hydrocarbons, particularly at high pressures and low temperatures, deviate considerably from the ideal gas law. Despite these deviations, the ideal gas law is routinely used in air pollution calculations.[9]

Example 7.8

1. Prepare computer programs for each of the equations dealing with the ideal gas law (Equations 7.7, 7,9, and 7.10).
2. Using the computer program developed in (1), determine the volume occupied by 1 mole of an ideal gas at 0°C (273 K) and 1 atm pressure.
3. Using the computer program developed for Equation 7.9, calculate the pressure exerted by 38.0 g of carbon monoxide in a volume of 25 L at 500 K; assume that the ideal gas law applies.
4. Using the computer program developed for Equation 7.9, determine the molecular weight of a gas if 0.6 g of the gas occupy 400 mL at 25°C and 760 mmHg (1 atm). (Standard conditions for temperature and pressure.)

5. Using the computer program developed for Equation 7.10, determine the density of air if it has a molecular weight of 29 and the air temperature is 93°C at 1 atm of pressure.

Solution

1. For solution to Example 7.8 (1), see following listing of Program 7.8.
2. Solution to Example 7.8 (2):

$$PV = nRT \tag{7.7}$$

where:

P = 1 atm
n = 1 g-mole
R = 0.08206 (l) atm $(K)^{-1}$ $(g\text{-mol})^{-1}$
T = 0°C (273 K)
V = Molar volume, (l)

$$(1\text{ atm})\ V = \frac{1\text{ g-mol }(0.08206\text{ l}*\text{atm})\ (273\text{ K})}{\text{g}-\text{mol K}}$$

$$V = 22.4\text{ L}$$

Thus, 1 mole of an ideal gas at 273 K and 1 atm occupies 22.4 L.

3. Solution to Example 7.8 (3):

$$PV = \frac{mRT}{MW} \tag{7.9}$$

$$P(25L) = \frac{38}{28}\left(\frac{(0.08206\ (l)atm)}{\text{g-mol K}}\right)(500\text{ K})$$

$$P = 2.22\text{ atm} = 1693\text{ mmHg }(1\text{ atm} = 760\text{ mmHg at }273\text{ K})$$

$$PV = \frac{mRT}{MW} \tag{7.9}$$

4. Solution to Example 7.8 (4):

$$1\text{ atm }(0.41) = \frac{0.6}{MW}\ \frac{(0.08206\ l\ (atm))(273+25\text{ K})}{\text{g-mol}}$$

which yields:

$$MW = 36\text{ g/g-mol}$$

5. Solution to Example 7.8 (5):

$$\rho = \frac{P(MW)}{RT}$$

$$= \frac{1 \text{ atm } (29 \text{ g})}{0.08206\ (1) \text{ atm } (K)^{-1}(g - mol)(273 + 93) \text{ K}} = 0.966 \text{ g/L or } 0.000966 \text{ g/cm}^3 \quad (7.10)$$

Listing of Program 7.8

```
'************************************************
'Example 7-8: Combining elements of Boyle's Law
'  and Charle's Law
'************************************************
'
' Equations 7.7, 7.9, and 7.10
'
COLOR 15, 1, 4
CLS
DIM form$(4)
form$(1) = "P * V = n * R * T"
form$(2) = "P * V = (m / MW) * R * T"
form$(3) = "r(rho) = (P * (mw)) / (R * T)"
form$(4) = "EXIT"
title:
opt = 0
DO
  PRINT "Exercise 7-8: Programs relating to the ideal gas law"
  PRINT " "
  FOR i = 1 TO 4
    LOCATE i + 2, 13
    PRINT "(", i; ")   "; form$(i)
  NEXT i: PRINT
  INPUT "Select the option 1 - 4:"; opt
LOOP WHILE opt < 1 OR opt > 4
  IF opt < 4 THEN
    CLS
    PRINT "P = 1 atmosphere"
    PRINT "V = molar volume, liters"
    PRINT "n = 1 g-mole  "
    PRINT "R = 0.08206 (liters) atm., (K){-1}, (g-mole){-1}  (a constant)"
    PRINT "T = 0 degrees C  (K = 273 + degrees C)"
    PRINT "m = weight of gas, grams"
    PRINT "MW = molecular weight of gas"
    PRINT "r(rho) = density of a gas (in g/liter or g/cm * 10{3})"
    PRINT
    ELSE END
  END IF
SELECT CASE opt
  CASE 1
    PRINT "Equation 7.7   "; form$(1)
    PRINT " (1 atm.) V = n * R * T            Solving for V"
    PRINT " 1 * V = 1 (g-mole) * 0.08206 (l.atm. / g-mole K) * 273 (K)"
    PRINT " V = 1 * 0.08206 * 273"
      step1 = 1 * .08206 * 273
```

```
  PRINT " V = 22.4 liters"
  PRINT "1 mole of an ideal gas at 273 K and 1 atm occupies 22.4 liters"
  INPUT "Press ENTER to continue "; ent$
  CLS
  GOTO title
 CASE 2
  PRINT "Equation 7.9   "; form$(2)
  PRINT " (1 atm.) V = (m / MW) * R * T          Solving for MW": PRINT
  INPUT "Enter the volume of the gas (V) (in liters) "; volgas   'V
  INPUT "Enter the weight of the gas (m) (in grams) "; vol2    'm
  INPUT "Enter the gas temperature, (T) (in C) "; gtemp        'T
  tottemp = 273 + gtemp
    step2 = vol2 * .08206 * tottemp
    step3 = step2 / volgas
  PRINT " MW = "; step3; " grams / gram mole"
  INPUT "Press ENTER to continue "; ent$
  CLS
  GOTO title
CASE 3
  PRINT "Equation 7.10   "; form$(3)
  PRINT " r(rho) = P * (MW) / R * T          Solving for r": PRINT
  INPUT "Enter the number of atmospheres (P) (in atm.) "; atm
  INPUT "Enter the molecular weight of the gas (MW) "; wgtgas   'MW
  INPUT "Enter the gas temperature, (T) (in C) "; gtemp        'T
  tottemp = 273 + gtemp
    step2 = .08206 * tottemp
    step3 = atm * wgtgas
    step4 = step3 / step2
  PRINT " r(rho) = "; step4; " grams/liter"
  INPUT "Press ENTER to continue "; ent$
  CLS
  GOTO title
 CASE 4
  END
 END SELECT
END
```

7.2.4 Van der Waal's Equation

As indicated, the ideal gas law is not strictly applicable to real gases. Van der Waal's equation contains two discreet constants which often provide a better representaion of the actual volumetric behavior of a gas:

$$\left(P + \frac{n^2 a}{V^2}\right)(V - nb) = nRT \tag{7.11}$$

where:

P = Absolute pressure, atm

V = Volume of a gas, L

n = Number of moles of gas

R = Ideal or universal gas constant, 0.08206 (1) atm $(K)^{-1}$ $(g\text{-mol})^{-1}$

T = Absolute temperature, K

a,b = Constants determined experimentally for each gas; these constants are available in most handbooks of chemistry (see Table 7.5)

a = L^2-atm/mol^2

b = L/mol

TABLE 7.5
Van der Waal's Constants for Some Gases[7]

Gas	a, in L^2 - atm/$(mol)^2$	b, in L/mol
O_2	1.382	0.03186
CO_2	3.658	0.04286
H_2	0.2453	0.02651
CH_4	2.300	0.04301
N_2	1.370	0.0387
NO	1.46	0.0289

Source: From Lide, D.R. and Frederikse, H.P.R., *CRC Handbook of Chemistry and Physics*, 76th ed., CRC Press, Boca Raton, FL, 1995, 6-48. With permission.

Example 7.9

1. Write a computer program that relates the various elements in Van der Waal's equation (Equation 7.11).
2. Using the computer program developed in section (1), determine the pressure developed by 660 g of CO_2 put into an 80.0-L container under the conditions stated in the solutions section. Also compare the results if the ideal gas law is used.

Solution

1. For solution to Example 7.9 (1), see the following listing of Program 7.9.
2. Solution to Example 7.9 (2):

$$V = 80.01$$

$$T = 130°C$$

$$a = 3.658 \frac{L^2 - atm}{(mol)^2}$$

$$b = 0.04286 \text{ L/mol}$$

Using Equation 7.2, $n = \frac{660}{44} = 15$ mol.

$$P = \frac{nRT}{V - nb} - \frac{n^2 a}{V^2}$$

$$= \frac{15(0.08206)(403)}{80 - 15(0.04286)} - \frac{(15^2(3.658))}{(80)^2} = \frac{496.0527}{79.3571} - 0.1286 = 6.122 \text{ atm} \qquad (7.11)$$

Using the ideal gas law

$$PV = nRT$$

$$P = \frac{nRT}{V} = \frac{15(0.08206)(403)}{80} = 6.20 \text{ atm} \qquad (7.7)$$

Van der Waal's equation gives 6.12 atm, and the ideal gas law gives 6.20 atm. The ideal gas law gives a pressure of 0.08 atm higher than Van der Waal's equation.

Listing of Program 7.9

```
'Example 7-9: Examining elements of Van Der
'  Waal's equation  (equation 7-11)
'*************************************************
'
' Equation 7.11
'
COLOR 15, 1, 4
CLS
DIM form$(2)
form$(1) = "P = ((n * R * T) / V - (n * b)) - (n{2} * a) / V{2}"
form$(2) = "EXIT"
title:
opt = 0
DO
  CLS
  PRINT "Exercise 7-9: Programs relating to VanderWaal's"
  PRINT "              equation"
  FOR i = 1 TO 2
    LOCATE i + 3, 13
    PRINT "(", i; ")   "; form$(i)
  NEXT i: PRINT
  PRINT "Type 1 and ENTER to continue  "
  INPUT "Type 2 and ENTER to EXIT  "; opt
LOOP WHILE opt < 1 OR opt > 2
  IF (opt = 2) THEN
    END
  END IF
  CLS
    PRINT "P = absolute pressure, atm."
    PRINT "V = volume of gas, liters"
    PRINT "n = number of moles of gas"
    PRINT "R = 0.08206 (liters) atm., (K){-1}, (g-mole){-1}  (a constant)"
```

```
    PRINT "T = absolute temperature   K"
    PRINT "a, b = constants determined for each gas"
    PRINT "a = (L{2} - atm.) / moles{2}"
    PRINT "b = L / mole": PRINT
    PRINT "Equation 7.11   ";
    PRINT "By Van Der Waal's Equation"
    PRINT "P = (n*R*T)/(V-(n*b) - (n{2}*a)/V{2}": PRINT
    INPUT "Enter the volume of gas, (V) (in liters)  "; volgas          'V
    INPUT "Enter the temperature of the gas, (T) (in C)  "; gastemp     'T
    INPUT "Enter the value of a, table 7-5  "; aval          'a
    INPUT "Enter the value of b, table 7-5  "; bval          'b
    INPUT "Enter the weight of the gas, grams  "; gaswgt
    INPUT "Enter the molecular weight of the gas  "; molwgt: PRINT
     n = gaswgt / molwgt
     tottemp = 273 + gastemp
     eq1n = n * .08206 * tottemp     'nominator "a"
     nb = n * bval
     eq1d = volgas - nb              'denominator "a"
     n2 = n * n
     eq2n = n2 * aval
     eq2d = volgas * volgas
     eq1 = eq1n / eq1d
     eq2 = eq2n / eq2d       '2nd half of formula
     pval = eq1 - eq2
    PRINT "P =  "; pval; " atm.": PRINT
    INPUT "Press ENTER to continue "; ent$
  CLS
    PRINT : PRINT : PRINT "By Ideal Gas Law"
    PRINT "P = n*R*T / V"
      eq3a = n * .08206 * tottemp
      pval2 = eq3a / volgas
    PRINT "   P  =  "; pval2: PRINT : PRINT
    PRINT "Van Der Waal's Equation gives "; pval
    PRINT "The Ideal Gas Law gives "; pval2: PRINT : PRINT
    INPUT "Press ENTER to continue "; ent$
    GOTO title
END
```

7.2.5 Dalton's Law of Partial Pressures

When a mixture of gases (having no chemical interaction) is present in a container, the total pressure exerted by the gas mixture is equal to the sum of the pressures which each gas would exert if it alone occupied the container.

$$P_{total} = p_1 + p_2 + p_3 + ... \tag{7.12}$$

The pressure exerted by each gas in the mixture is called its partial pressure.

$$\text{partial pressure} = \frac{p_i}{P_{total}} \tag{7.13}$$

There are two ways to express the fraction which one gaseous component contributes: either by mole fraction (Equation 7.3, n_i/n_{total}) or the pressure fraction (Equation 7.13, p_i/P_{total}).

Example 7.10

A gas mixture contains the following components. Using Equations 7.2 and 7.3, determine the number of moles of each gas present as well as the mole fraction for each gas.[9]

Gas	Weight (g)	Molecular weight	Moles	Mole fraction
O_2	20	32	0.625	0.885
SO_2	2	64	0.0313	0.044
SO_3	4.0	80	0.050	0.071
TOTAL			0.7063	1.000

The total number of moles in this gaseous mixture is 0.7063, using the moles present for each gas and the ideal gas law, Equation 7.7. Determine the partial pressure for each gas as well as the partial volumes when the gas temperature is 25°C and the gas is held in a 2-L container.

Solution

- For O_2,

$$P=\frac{(0.625)(0.08206)(298)}{2}=7.64 \text{ atm}$$

- For SO_2,

$$P=\frac{(0.0313)(0.08206)(298)}{2}=0.38 \text{ atm}$$

- For SO_3,

$$P=\frac{(0.050)(0.08206)(298)}{2}=0.61 \text{ atm}$$

Total pressure = 8.63 atm.

- Partial pressure for $O_2 = \dfrac{7.64}{8.63} = 0.885$

- Partial pressure for $SO_2 = \frac{0.38}{8.63} = 0.044$

- Partial pressure for $SO_3 = \frac{0.61}{8.63} = 0.071$ atm.

The partial volumes for each gas can also be calculated by using the ideal gas law, Equation 7.7.

- For O_2,

$$V = \frac{(0.625)(0.08206)(298)}{8.63} = 1770 \text{ L}$$

- For SO_2,

$$V = \frac{(0.0313)(0.08206)(298)}{8.63} = 0.089 \text{ L}$$

- SO_3,

$$V = \frac{(0.0375)(0.08206)(298)}{8.63} = 0.141 \text{ L}$$

Total volume = 2.000 L.

- Partial volume for $O_2 = \frac{1.770}{2} = 0.885$ L.

- Partial volume for $SO_2 = \frac{0.089}{2} = 0.044$ L.

- Partial volume for $SO_3 = \frac{0.141}{2} = 0.071$ L.

The partial pressure for each gas indicates the concentrations of that particular gas, and for most purposes the gas concentration governs the behavior of a gaseous substance regardless of the presence of other gases in the mixture.[7]

7.2.6 Henry's Law

Henry's law, simply stated, indicates that the concentration of a gas dissolved in a solvent under equilibrium conditions is proportional to the partial pressure of the gas above the solution at constant temperature. A mathematical equation describing this relationship is provided in Equation 7.14.

$$C_i = K_i p_i \tag{7.14}$$

where:

C_i = Concentration of gas i mol/L dissolved in solution, mol/L
K_i = Henry's law constant, mol/L
p_i = Partial pressure of the gas, atm

Table 7.6 presents Henry's law constants for some gases in water at 25°C.[8]

Example 7.11

1. Write a computer program relating the elements contained in Henry's law that will make it possible to determine the concenteration of a gas in water, provided Henry's gas constant is given.
2. Using the computer program prepared in (1), determine the solubility of oxygen in water at 25°C, knowing that air contains 20.95% oxygen by volume.

TABLE 7.6
Henry's Law Constants for Some Gases in Water at 25°C[8]

Gas	K, mol × L^{-1} × atm^{-1}
O_2	1.28×10^{-3}
CO_2	3.38×10^{-2}
H_2	7.90×10^{-4}
CH_4	1.34×10^{-3}
N_2	6.48×10^{-4}
NO	2.0×10^{-4}

Solution

1. For a solution to Example 7.11 (1), see the following listing of Program 7.11.
2. In calculating the solubility of a gas in water, a correction must be made for the partial pressure of water by subtracting it from the total pressure of the gas. At 25°C, the partial pressure of water is 0.0313 atm; values at other temperatures are readily obtained from standard handbooks. The concentration of oxygen in water saturated with air at 1 atm (760 mmHg) and 25°C (298 K) may be calculated as an example of a simple gas solubility calculation. Considering that dry air is 20.95% by volume oxygen and factoring in the partial pressure of water gives the following:

$$P_{O_2} = (1.00 \text{ atm} - 0.0313 \text{ atm}) \times 0.2095 = 0.2029 \text{ atm}$$

$$C_i = K_i p_i = 1.28 \times 10^{-3} \text{mol} \times L^{-1} \times \text{atm}^{-1} \times 0.2029 \text{ atm}$$

$$= 2.60 \times 10^{-4} \text{mol} \times L^{-1}$$

Since the molecular weight of oxygen is 32, the concentration of dissolved oxygen in water in equilibrium with air under the conditions given above is 8.32 mg/L, or 8.32 parts per million (ppm).[8]

Listing of Program 7.11

```
'***********************************************
'Example 7-11: Relates the components in Henery's
' Law to determine the concentration of a gas in
' water provided Henry's Gas Constant is given.
'***********************************************
'
'
COLOR 15, 1, 4
CLS
title:
DO
  opt = 0
  PRINT "Exercise 7-11:  Relates the components in Henry's"
  PRINT "Law to determine the concentration of a gas in "
  PRINT "water provided Henry's Gas Constant is given, and"
  PRINT "determine the solubility of a gas in water at 25 degrees."
  PRINT : PRINT "The formula is - C[i] = K[i] * P[i]"
  PRINT : PRINT "Type 1 and ENTER to continue  "
  INPUT "Type 2 and ENTER to EXIT  "; opt
LOOP WHILE opt < 1 OR opt > 2
  PRINT : PRINT
  INPUT "Type in the name of the gas  "; gas$
  PRINT "Enter the constant for this gas.  Enter the numbers "
  INPUT " as shown - X.XXE+/-x  (1.28E-3 for O[2]) from Table 7.6 "; constnt
  INPUT "Enter the partial pressure of gas (in atm.) "; pres
  INPUT "Enter the molecular weight of the gas (in g/mole/L) "; wgt
  PRINT : PRINT "C[i] = K[i] * p[i]  "
  PRINT "C[i] = "; constnt; " * "; pres; " * "; wgt
   comp = constnt * pres * wgt
   comp2 = comp * 1000
  PRINT "C[i] = "; comp; "g/liter or "; comp2; "mg/L": PRINT
  INPUT "Press ENTER to continue "; ent$
  CLS
  GOTO title
END
```

Example 7.12

Use the computer program developed in (1) of Example 7.11 to determine the weight of CO_2 that would be dissolved in one L of water at 25°C under a CO_2 pressure of 1 atm.

Solution

$$C_i = K_i p_i \tag{7.14}$$

From Table 7.6,

$K_i = 3.38 \times 10^{-2}$ mol $\times$ L^{-1} $\times$ atm^{-1}
$C_i = 3.38 \times 10^{-2}$ mol $\times$ L^{-1} $\times$ atm^{-1} $\times$ 1 atm
$C_i = 3.38 \times 10^{-2}$ mol $\times$ L^{-1}
1 mol CO_2 = 44 g
$C_i = 3.38 \times 10^{-2}$ mol $\times$ L^{-1} $\times$ 44 g/mol/L = 1.48 g/L or 1487 mg/L

7.2.7 Clausius-Clapeyron Equation

The solubilities of gases decrease with increasing temperature. This factor is taken into account with the following Clausius-Clapeyron equation:

$$\ln \frac{C_2}{C_1} = \frac{\Delta H}{R}\left(\frac{1}{T_1} - \frac{1}{T_2}\right) \qquad (7.15)$$

where:

C_1 = Gas concentration at absolute temperature T_1
C_2 = Gas concentration at absolute temperature T_2
ΔH = Heat of solution
R = Ideal or universal gas constant (1.987 cal/deg.mol)

Example 7.13

1. Prepare a computer program relating the various elements in Equation 7.15.
2. Using the computer program developed in (1), determine the concentration of oxygen in water at 40°C when the solubility of oxygen at 0°C is 14.60 mg/L and 7.54 mg/L at 30°C.

Solution

1. For solution to Examaple 7.13 (1), see the following listing of Program 7.13.
2. Using the Clausius-Clapeyron Equation 7.15, determine ΔH for 0°C and 30°C oxygen concentration:

$$\ln \frac{7.54}{14.60} = \frac{\Delta H}{1.986}\left(\frac{1}{273} - \frac{1}{303}\right)$$

$$-0.6607 = \frac{\Delta H(0.00366 - 0.00330)}{1.986}$$

$$\Delta H = \frac{-0.6607 \times 1.986}{0.000363} = -3618$$

The estimated solubility of oxygen in water at 40°C would then be

$$\ln\frac{C}{14.60}=\frac{-3618}{1.986}\left(\frac{1}{273}-\frac{1}{313}\right)$$

$$\ln\frac{C_2}{14.60}=-1820(0.00366-0.003195)$$

which yields C_2 = 6.22 mg/L estimated oxygen concentration at 40°C.

Listing of Program 7.13

```
'************************************************
'Example 7-13: Relates the components in the
' Clausis-Clapeyron Equation.
'************************************************
'
'
COLOR 15, 1, 4
CLS
opt = 0
title:
DO
  PRINT "Exercise 7-13:  Relates the components in the "
  PRINT "                Clausis-Clapeyron Equation ": PRINT
  PRINT "The formula is:  ": PRINT
  PRINT "ln (C2 / C1) = ({D}H / R) * ((1 / T1) - (1 / T2))": PRINT
  PRINT "Type 1 and ENTER to continue  "
  INPUT "Type 2 and ENTER to EXIT  "; opt
LOOP WHILE opt < 1 OR opt > 2
  IF opt = 1 THEN
    GOTO opening
  END IF
  IF opt = 2 THEN
    END
  END IF
opening:
CLS
  INPUT "Enter the gas concentration at absolute temp T1 (C1) (in mg/L) "; gasc1
  INPUT "Enter the temperature of the gas C1 (T1) (in C) (in mg/L) "; gast1
  INPUT "Enter the gas concentration at absolute temp T2 (C2) "; gasc2
  INPUT "Enter the temperature of the gas C2 (T2) (in C) "; gast2
    gasccomp = gasc2 / gasc1
    gasccompf = LOG(gasccomp)
    gastemp1 = 273 + gast1          'T1
    gastemp2 = 273 + gast2          'T2
    gastemp3 = 1 / gastemp1
    gastemp4 = 1 / gastemp2
    gastemp5 = gastemp3 - gastemp4
    delhen = (gasccompf * 1.986) / gastemp5
  PRINT
  PRINT "The Delta H (heat of solution) at "; T2; " degrees is "; delhen
  PRINT
  INPUT "Type in the new water temperature "; temp3: PRINT
    gastemp6 = 273 + temp3
```

```
    gastemp7 = gastemp3 - (1 / gastemp6)
    delhen2 = delhen / 1.986
    step1 = delhen2 * gastemp7
    step2 = LOG(gasc1)
    step3 = step1 + step2
    answ = EXP(step3)
  PRINT "C2 = "; answ; "mg/L   Estimated oxygen concentration at "; temp3; "degrees C"
  PRINT : INPUT "Press Enter to continue "; ent$
  CLS
  GOTO title
END
```

7.2.8 Reynold's Number

The Reynold's number (RE) relates inertial forces to viscous forces per unit volume for the medium involved (gas or liquid). Typical inertial force per unit volume of fluid:

$$\frac{\rho v^2}{g_c L} \tag{7.16}$$

Typical viscous force per unit volume of fluid:

$$\frac{\mu v}{g_c L^2} \tag{7.17}$$

The first expression divided by the second provides the dimensionless ratio known as Reynold's number:

$$Re = \frac{Lv\rho}{\mu} = \frac{\text{inertial force}}{\text{viscous force}} \tag{7.18}$$

where:

ρ = Density of the fluid (mass/volume), kg/m^3
v = Velocity of the fluid, m/s
g_c = Dimensional constant
L = Linear dimension, m
μ = Viscosity of the fluid, kg/m/s
Re = Reynold's number, dimensionless

The Reynold's number provides information on flow behavior. Laminar flow is normally encountered at at Reynold's number below 2100 in a tube, but it can persist up to Reynold's numbers of several thousand under special conditions. Under ordinary conditions of flow, the flow is turbulent at a Reynold's number above about 4000. Between 2100 and 4000, a transition region is found where the type of flow may be either laminar or turbulent. By definition, the Reynold's number is dimensionless.[9]

Example 7.14

1. Write a computer program that includes the elements contained in Equation 7.16 in order to calculate the Reynold's number.
2. Given the data in the Solutions section for the flow of an air stream through a circular duct, determine the Reynold's number for this air stream.

Solution

1. For solution to Example 7.14 (1), see the following listing of Program 7.14.
2. Solution to Example 7.14 (2):

ρ_a = 0.965 kg/m^3

v = 10.0 m/s

L = 0.50 m (diameter of duct)

μ_a = 2.15×10^{-5} kg/m/s

$$\text{Re} = \frac{\text{L}v\rho}{\mu}$$

$$= \frac{(0.50)(10.0\text{ m/s})(0.965\text{ kg/m}^3)}{2.15\times10^{-5}\text{ kg/m}} = 2.24\times10^5 \tag{7.18}$$

This would be turbulent flow. For most air pollution conditions the flow is usually in the turbulent range with high Reynold's numbers.

Listing of Program 7.14

```
'**************************************************
'Example 7-14: Calculates the Reynold's number.
'
'**************************************************
'
'
COLOR 15, 1, 4
CLS
title:
DO
  opt = 0
  PRINT "Exercise 7-14:  Calculates the Reynold's Number "
  PRINT "               (The number is usually high) ": PRINT
  PRINT "Formula 7.18 is:  ": PRINT
  PRINT "Re = (L * v * r (rho)) / u (viscosity)": PRINT
```

```
  PRINT "Type 1 and ENTER to continue  "
  INPUT "Type 2 and ENTER to EXIT  "; opt
LOOP WHILE opt < 1 OR opt > 2
  IF opt = 1 THEN
    GOTO opening
  END IF
  IF opt = 2 THEN
    END
  END IF
opening:
  PRINT : INPUT "Type in the diameter of the duct, (L) (in meters) "; dimen
  INPUT "Type in the velocity of the fluid, (v) (in m/sec) "; vel
  INPUT "Type in the density of the gas, (r (rho)) (in kg/m{3}) "; dens
  PRINT "Type in the viscosity of the fluid, (u) (in kg/m.sec) "
  INPUT "  use scientific notation (x.xxxE+/-x) "; visc: PRINT
    step1 = dimen * vel * dens
    step2 = step1 / visc
  PRINT "The Reynold's Number is ";
  PRINT USING "#.###^^^^"; step2: PRINT
  INPUT "Press ENTER to continue "; ent$
  CLS
  GOTO title
END
```

7.2.9 Stoke's Law

The terminal settling velocity of spherical particles larger than the free mean path in a fluid is described by Stokes' law, provided the particles are not so large that inertial effects arise from the fluid displaced by the falling particles. Terminal velocity is a constant value of velocity reached when all forces (gravity, drag, buoyancy, etc.) acting on a body balance. The sum of all the forces is then equal to zero (no acceleration). Particles smaller than about 50 μm reach a constant settling velocity within a fraction of a second. The terminal settling velocity is given by the following equation:

$$v_t = \frac{g d_p^2 (\rho_p - \rho_a)}{18\,\mu} \qquad (7.19)$$

where:

v_t = Terminal settling velocity, cm/s
g = Gravitational constant, cm/s^2
ρ_p = Density of the particle, g/cm^3
ρ_a = Density of air, g/cm^3
d_p = Diameter of particle, μm
μ = Viscosity of air, g/cm/s

For the determination of the terminal settling velocity of particles, three size regimes are considered here, based on a dimensionless constant k. Terminal velocity is a constant value of velocity reached when all forces (gravity, drag, buoyancy, etc.) acting on a body balance. The sum of all the forces is then equal to zero (no acceleration). A dimensionless constant k determines the appropriate range of the fluid-particle dynamic laws which apply.[9]

$$k = d_p(g\rho_p\rho/\mu^2)^{1/3} \quad (7.20)$$

where:

- k = Dimensionless constant which determines the range of the fluid-particle dynamic laws
- d_p = Particle diameter, cm
- g = Gravity force, cm/s^2
- ρ_p = Particle density, g/cm^3
- ρ = Fluid (gas) density, g/cm^3
- μ = Fluid (gas) viscosity, g/cm/s

The values of k that apply for the three regimes are as follows:

- The numerical value of k determines the appropriate law.
- $k < 3.3$, Stokes' law range
- $3.3 < k < 43.6$, intermediate law range
- $43.6 < k < 2360$, Newton's law range

The equations for each regime are

- For Stokes' law range:

$$v = gd_p^2\,(\rho_p - \rho_a)/\,18\,\mu \quad (7.19)$$

- For intermediate law range:

$$v = 0.153\,g^{0.71}\,d_p^{1.14}\rho_p^{0.71}\,/\mu^{0.43}\rho^{0.29} \quad (7.21)$$

- For Newton's law range:

$$v = 1.74(gd_p\rho_p/\rho)^{0.5} \quad (7.22)$$

When particles approach sizes comparable to the mean free path of the fluid molecules, the medium can no longer be regarded as continuous since particles can fall between the molecules at a faster rate than predicted by aerodynamic theory. To allow for this "slip", Cunningham's correction factor is introduced to Stokes' law.[9]

$$v = (gd_p^2\,\rho_p/\,18\mu)(C_f),\text{ where} \quad (7.23)$$

$$C_f = \text{Cunningham correction factor} = 1 + (2A\lambda/\,d_p) \quad (7.24)$$

where:

$A = 1.257 + 0.40e^{-1.10dp/2\lambda}$

λ = Mean free path of the fluid molecules (6.53×10^{-6} cm for ambient air)

The Cunningham correction factor is usually applied to particles equal to or smaller than 1 micron.[9]

Example 7.15

1. Write a computer program that provides for the calculation of the dimensionless constant k and thus provides information as to which appropriate terminal velocity equation to use.
2. Write a computer program that relates the various elements in Stokes' law (Equation 7.19) and will determine the terminal settling velocity for spherical particles.
3. Using the computer programs developed in (1) and (2), estimate the terminal settling velocity in cm/s for a 30-μm particle with a density of 2.31 g/cm³. The room air is 24°C with a room depth of 2.7 m. Estimate the time required for this particle to settle from the ceiling to the floor.

Solution

1. For a solution to Example 7.15 (1), see the following listing of Program 7.15.
2. For a solution to example 7.15 (2), first calculate the density of air at 24°C using Equation 7.10.

$$\rho_a = \frac{P(MW)}{RT} \tag{7.10}$$

where:

P = 1 atm

MW = 29 g $(\text{g-mol})^{-1}$ for air

R = 0.08206 (1) atm $(\text{g-mol})^{-1}$ $(\text{K})^{-1}$

T = K, (273 + °C)

$$\rho_a = \frac{1 \text{ atm} \times 29 \text{ g} \times (\text{g-mol})^{-1}}{0.08206\ (1) \text{ atm } (\text{g-mol})^{-1}\ (\text{K})^{-1} \times (273+24)\text{K}} = 1.19 \text{ g/L} = 0.00119 \text{ g/cm}^3$$

Using Equation 7.20, calculate the value of k in order to determine which equation to use to estimate the terminal settling velocity for the 30-μm particle.

$$k = d_p (g\rho_p\rho / \mu^2)^{1/3}$$

$$\mu_a \text{ at } 24°\text{C} = 0.00018 \text{ g/cm} * \text{s} \tag{7.20}$$

$$k = 30 \times 10^{-4} \text{cm} \left(\frac{981 \text{ cm/s}^2 \times 2.31 \text{ g/cm}^3 \times 0.00119 \text{ g/cm}^3}{(0.00018 \text{ g/cm} * \text{s})^2} \right)^{1/3}$$

If $k < 3.3$, the appropriate equation to use is Stokes' law:

$$v = \frac{gd_p^2(\rho_p - \rho_a)}{18\,\mu} \tag{7.19}$$

$$v = \frac{981 \text{ cm/s}^2 (30 \times 10^{-4} \text{ cm})^2 (2.31 \text{ g/cm}^3 - 0.00119 \text{ g/cm}^3)}{18(0.00018 \text{ g/cm} * \text{s})} = 6.29 \text{ cm/s}$$

The time required for this 30-μm particle to settle from the ceiling to the floor would be

$$\frac{270 \text{ cm}}{6.29 \text{ cm/sec}} = 43 \text{ seconds}$$

Listing of Program 7.15

```
'*************************************************
'Example 7-15: Provides for the calculation
' of the dimensionless constant k and provides
' information on which terminal velocity
' equation to use.
'Also relates the elements of Stoke's Law.
'*************************************************
'
'
COLOR 15, 1, 4
CLS
title:
DO
  opt = 0
  PRINT "Exercise 7-15:  Provides for the calculation of the dimensionless "
  PRINT " constant k and provides information on which terminal velocity "
  PRINT " formula to use.  Also relates the elements of Stoke's Law.": PRINT
  PRINT "The initial formulas are:  ": PRINT
  PRINT "r(rho) = (P * (MW)) / R * T                    Formula 7.10"
  PRINT "k = d[p] * (g * r[p] * (r / u{2})){1/3}        Formula 7.20"
  PRINT "v[t] = (g * d[p]{2} * (r[p] - r[a])) / 18 * u    Formula 7.19 ": PRINT
  PRINT "Type 1 and ENTER to continue  "
  INPUT "Type 2 and ENTER to EXIT  "; opt
LOOP WHILE opt < 1 OR opt > 2
  IF opt = 1 THEN
    GOTO opening
  END IF
  IF opt = 2 THEN
```

```
  END
 END IF
opening:
 PRINT : PRINT "To calculate the density of air:"
 INPUT "Type in the pressure, (P) (in atm.) "; pres
 INPUT "Type in the molecular weight of the air, (MW) (in g(g-mole){-1}) "; molwgt
 PRINT "The (R) is a constant (0.08206) "
 INPUT "Type in the temperature of the air, (T) (in C) "; temp
  comptemp = 273 + temp
  nom = pres * molwgt
  dem = .08206 * comptemp
  dens = nom / dem
  denscm = dens * .001
 PRINT "The density of the air at "; temp; "degrees C is  "; dens; " g/L"
 PRINT "                              and "; denscm; "g/cm * 10{3}"
 PRINT : INPUT "Press ENTER to continue "; ent$
 CLS
 PRINT "To calculate the value of k to determine which equation to use:"
 INPUT "Type in the particle diameter, in um.  (d[p]) "; diam
 PRINT "The gravity force is a constant (981 cm/sec{2})"
 INPUT "Type in the particle density, in g/cm{3} "; partden
 PRINT "The density of the gas is from the first equation  "
 INPUT "Type in the gas viscosity (from a table), in g/cm.sec "; visc
  nom2 = 981 * partden * denscm
  comp1 = nom2 / visc
  comp2 = comp1 / visc
  comp3 = comp2 ^ (.3333)
  diamcomp = diam * .0001
  comp4 = diamcomp * comp3
 PRINT
 PRINT "k =  "; comp4: PRINT
 PRINT "If k < 3.3 the appropriate equation to use is Stoke's Law range": PRINT
 PRINT "If k > 3.3 but < 43.6 the appropriate equation to use is"
 PRINT " the Intermediate law range.": PRINT
 PRINT "If k > 43.6 the appropriate equation to use is Newton's Law range": PRINT
 PRINT "This program will illustrate Stoke's Law ": PRINT
 INPUT "Press ENTER to continue "; ent$
 CLS
  compdiama = diam * diam
  compdiam = compdiama * .00000001#
  compdens = partden - denscm
  nom = 981 * compdiam * compdens
  denom = 18 * visc
  v = nom / denom
 PRINT "v = "; v; " cm/sec": PRINT
 INPUT "Type in the depth of the room  (in meters) "; depth: PRINT
 PRINT "The velocity from the preceding formula was "; v; " cm/sec"
  deptha = depth * 100
  time = deptha / v
 PRINT : PRINT "The room depth =  "; deptha; " cm"
 PRINT "                 ---------------- = "; time; " seconds "
 PRINT "  The velocity = "; v; " cm/sec"
 PRINT : INPUT "Press ENTER to continue "; ent$
 CLS
 GOTO title
END
```

7.2.10 Particle Size Distribution – Log Normal

The most important single parameter in evaluating the effects of and determining the best method for control of airborne particulates is the size of the particulate. Airborne particles are never all the same size; therefore, it is necessary to try to characterize and quantify the distribution of the various sizes involved. The most often used mathematical expresesion to do this evaluation is the log-normal distribution. There is, however, no law that says all particles follow the log-normal classification. In a number of industrial and ambient air studies, the size distribution of the particulate followed the "normal" or "Gaussian" distribution.[10]

The log-normal distribution of suspended particulate in the atmosphere is often used, as it is a relatively simple and convenient way to analyze the data. This is done by plotting on log-probability paper the particle size diameter on the ordinate (log scale) and the cumulative percentage less than or equal to a stated size on the abscissa (probability scale). If the data plots as a straight line on log-probability paper, this tends to indicate that the size distribution of the particulate is log-normal. The log-normal probability distribution for particulates can be described by the following mathematical expression.[11]

$$F(d) = \frac{\Sigma N}{\log \sigma_g \sqrt{2\pi}} \exp\left(-\frac{(\log d - \log d_g)^2}{2 \log^2 \sigma_g} \right) \tag{7.25}$$

where:

- $F(d)$ = Frequency of particle occurrence of diameter d
- ΣN = Total number of particles
- σ_g = Standard geometric deviation for the data
- d_g = Geometric mean diameter of the particles

If the size distribution data does plot a straight line on log-probability paper, then two parameters that characterize this data are the geometric mean diameter $d_g = d_{50}$ and the geometric standard deviation σ_g. With

$$\sigma_g = \frac{d_{84.13}}{d_{50}} = \frac{d_{50}}{d_{15.87}} \tag{7.26}$$

where:

- $d_{84.13}$ = Diameter such that particles constituting 84.13% of the total mass of particles are smaller than this size
- d_{50} = Geometric mean diameter
- $d_{15.87}$ = Diameter such that particles constituting 15.87% of the total mass of particles are smaller than this size
- σ_g = Geometric standard deviation

Example 7.16

Determine if the following particle size distribution data is log-normal. If the data indicates log-normal distribution, what then is the geometric mean diameter and the geometric standard deviation?

Data		Answer	
Particle size d (µm)	Weight of particles (µg)	Cumulative total	Cumulative percent
0.4	10	10	5.7
0.55	30	40	22.8
0.70	25	65	37.1
0.90	60	125	71.4
1.20	30	155	88.6
1.5	15	170	97.1
2.00	5	175	100

This data gives a straight line on log-probability paper and therefore can be considered to have a log-normal distribution. From the graph for Example 7.16:

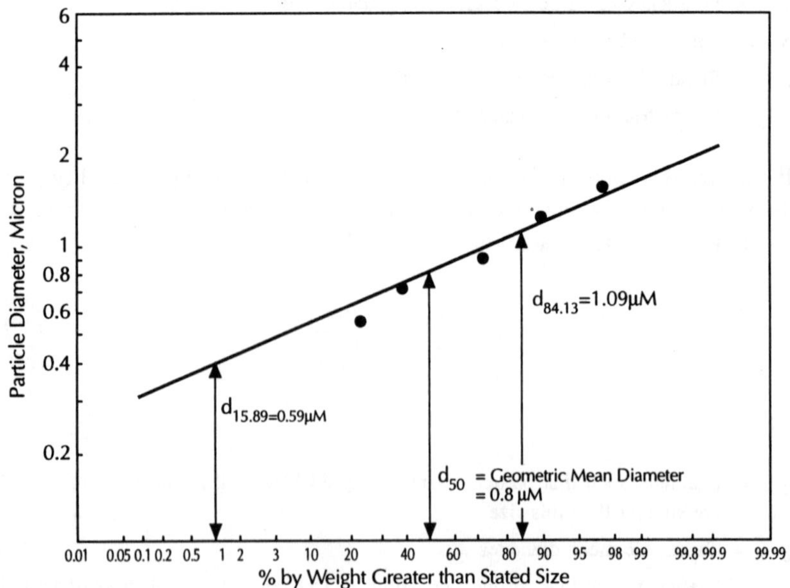

d_{50} = 0.8 μm or the geometric mean diameter

$d_{84.13}$ = 1.09 μm

$d_{15.87}$ = 0.59 μm

$$\sigma_g = \frac{0.8}{0.59} = \frac{1.09}{0.8} = 1.36, \text{ the geometric standard deviation}$$

Example 7.17

1. Develop a computer program that can be used to determine if the data for a particle size distribution sample is log normal. If the data gives a log-normal distribution, then include in the program a means by which the geometric mean diameter, d_{50}, $d_{15.87}$, and $d_{84.13}$ sizes can be determined, as well as the geometric standard deviation, σ_g.
2. Using the computer program developed in (1), determine if the following data for a particle size distribution is log normal. Also indicate the $d_{15.87}$, $d_{84.13}$, and d_{50} sizes as well as σ_g.

   Data:

Particle size d (μm)	Weight of particles (μg)
0.4	30
0.7	123
1.1	130
2.1	540
3.3	230
4.0	99
8.5	128

Solution

1. For a solution to Example 7.17 (1), see the following listing of Program 7.17.
2. A solution to Example 7.17 (2):

Particle size d (μm)	Weight of particles (μg)	Cumulative total	Cumulative percentage
0.4	30	30	2.3
0.7	123	153	12.0
1.1	130	283	22.1
2.1	540	823	64.3
3.3	230	1053	82.3
4.0	99	1152	90.0
8.5	128	1280	100.0

Plot the data on a log-probability paper as shown on graph for Example 7.17. Determine from the graph the following:

$d_{50} = 1.7\ \mu m$

$d_{15.87} = 0.8\ \mu m$

$d_{84.13} = 3.5\ \mu m$

$$\sigma_g = \frac{3.5}{1.7} = 2.1$$

$$\sigma_g = \frac{1.7}{0.8} = 2.1$$

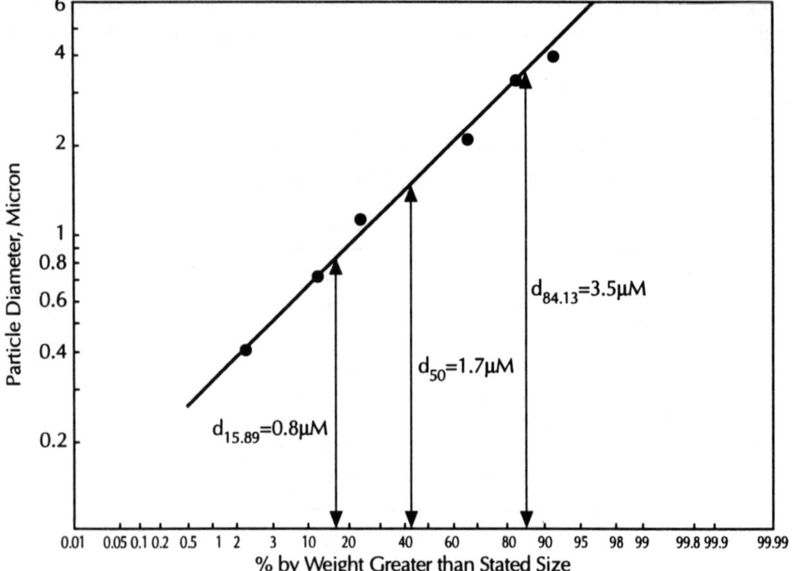

Listing of Program 7.17

```
'************************************************
'Example 7-17: Determine if distribution is
' log-normal and compute the geometric standard
' deviation.
'************************************************
'
'
COLOR 15, 1, 4
CLS
DIM Partsz(10), Wgt(10), cumtot(10), cumper(10)
title:
DO
```

```
   opt = 0
   PRINT "Exercise 7-17:  Determine if the particle size"
   PRINT "distribution data is log-normal and compute  "
   PRINT "the geometric standard deviation."
   PRINT : PRINT "Procedure:  "
   PRINT : PRINT "First determine the cumulative total for the "
   PRINT "particles and the cumulative percentages of each."
   PRINT : PRINT "Then the geometric standard deviation is computed."
   PRINT : PRINT "Type 1 and ENTER to continue  "
   INPUT "Type 2 and ENTER to EXIT  "; opt
LOOP WHILE opt < 1 OR opt > 2
   subsc = 1
   cumtot = 0
   cumtota = 0
   cumper = 0
   cumpera = 0
IF (opt = 2) THEN
   END
END IF
start:
 DO
   PRINT "Type in the particle size (in um) "
   INPUT "(Press ENTER without entering a size if you are done)"; partsize(subsc)
     IF (partsize(subsc) < .0001) THEN EXIT DO
   INPUT "What is the weight of the particles "; Wgt(subsc)
   cumtot(subsc) = cumtota + Wgt(subsc)
     cumtota = cumtota + Wgt(subsc)
     subsc = subsc + 1
     endsubsc = subsc
 LOOP UNTIL (subsc = 10)
subsc = 1
DO
     per(subsc) = (Wgt(subsc) / cumtota) * 100
     cumpera = cumpera + per(subsc)
     cumper(subsc) = cumpera
     subsc = subsc + 1
LOOP UNTIL (subsc = endsubsc)
tabval = 3
subsc = 1
   CLS
   PRINT TAB(5); "Particle size"; TAB(25); "Weight of "; TAB(40); "Cumulative "; TAB(55); "Cumulative "
   tabval = tabval + 1
   PRINT TAB(6); "d in um"; TAB(25); "particles"; TAB(41); "Total"; TAB(56); "Percent"
     tabval = tabval + 1
   PRINT TAB(25); "in ug"
     tabval = tabval + 1
DO
   PRINT TAB(8); USING "####.#"; partsize(subsc);
   PRINT TAB(25); USING "####.#"; Wgt(subsc);
   PRINT TAB(42); USING "####.#"; cumtot(subsc);
   PRINT TAB(58); USING "####.#"; cumper(subsc)
     subsc = subsc + 1
     tabval = tabval + 1
LOOP UNTIL (subsc = endsubsc)
   PRINT : PRINT "This data must now be graphed on log-probability paper."
   PRINT "d{50], d[84.13], and d[15.87] are read from the graph."
   PRINT : PRINT "The geometric standard deviation is computed as follows: "
   PRINT : PRINT "    d[84.13]         d[50]"
```

```
  PRINT "  ----------  =  ---------  = geometric standard deviation"
  PRINT "    d[50]          d[15.87] "
  PRINT : INPUT "Press ENTER to continue "; ent$
  CLS
  INPUT "What is d[84.13] "; d8
  INPUT "What is d[50] "; d5
  INPUT "What is d[15.87] "; d1
   gsd = d8 / d5
   gsd2 = d5 / d1
  PRINT "geometric standard deviation = "; d8; " / "; d5; " = "; gsd
  PRINT "geometric standard deviation = "; d5; " / "; d1; " = "; gsd2: PRINT
  INPUT "Press ENTER to continue "; ent$
  CLS
  GOTO title
END
```

7.3 Air Pollution Control Technology

7.3.1 Settling Chambers

The oldest and probably least effective air pollution control device is the gravity settling chamber. While simple to construct and having a low operating cost, settling chambers are still generally used only to collect large particles above 40 to 50 μm in size. However, if the particles have a high density, then particles as small as 10 to 15 μm can be collected. Settling chambers generally are simply an enlargement of the conduit carrying the contaminated airstream (see Figure 7.1).

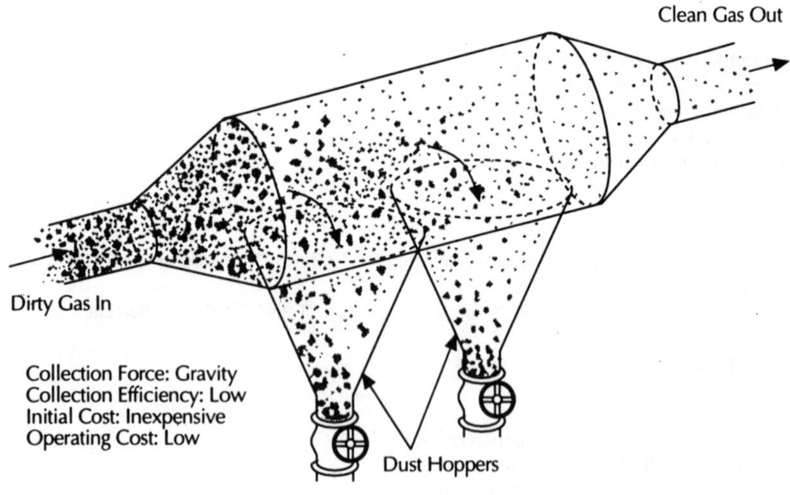

FIGURE 7.1
Settling chamber.

Settling chambers are often used as a precleaning stage for treating a contaminated gas stream in order to protect the other downstream collection equipment from hot, coarse, and abrasive materials.

If we assume that Stokes' law applies, we can derive a formula that can be used to calculate the minimum diameter of particles to be collected, at 100% theoretical efficiency in a chamber of given length L.

The following equation can be used to predict the largest-sized particle that can be removed with 100% efficiency in a settling chamber with given dimensions.

$$d_p = \left(\frac{18\,\mu\, v_h H}{gL\rho_p} \right)^{1/2} \tag{7.27}$$

where:

μ = Gas viscosity, g/cm*s
v_h = Horizontal gas velocity, cm/s
H = Height of chamber, cm
g = Gravitational constant 981 cm/s^2
L = Length of chamber, cm
ρ_p = Particle density, g/cm^3

Example 7.18

Estimate the minimum size of particle that will be removed with 100% efficiency from a settling chamber given the following data:

Dimensions of gravity settling chamber — 8 m wide, 6 m high, and 15 m long
Volumetric gas flow of 1.3 m^3/s
Particle density of 2.0 g/cm^3
Viscosity of the air in the gas stream of 1.85×10^{-4} g/cm*s
Gravitational constant of 981 cm/s^2

Solution

The velocity v_h can be determined from the continuity equation:

$$Q = VA$$

$$1.3\ m^3/s = V(6 \times 8\ cm)$$

$$V = 0.027\ m/s = 2.7\ cm/s$$

Substituting in Equation 7.27:

$$d_p = \left(\frac{18 \times (1.85 \times 10^{-4} g * cm/s) \times 2.7\ cm/s \times 600\ cm}{981\ cm/s^2 \times 2.0\ g/cm^3 \times 1500\ cm} \right)^{1/2}$$

$$= 0.000135\ cm - 1.35\ \mu m$$

Example 7.19

1. Prepare a computer program that can be used to predict the largest-sized particle that can be removed with 100% efficiency in a settling chamber (Equation 7.27).
2. Using the computer program developed in (1), estimate the largest-sized particle that can be removed with 100% efficiency in a settling chamber given the following data:

   Gravity settling chamber dimensions — 4 m wide, 1.2 m high, 8 m long

   Volumetric gas flow of 1.2 m^3/s

   Particle density of 2.5 g/cm^3

   Viscosity of the air at 27°C of 1.86×10^{-4} g*cm/s

Solution

1. For solution to Example 7.19 (1), see the following listing of Program 7.19.
2. A solution to Example 7.19 (2):
   - Calculate v_h from Q = VA:

$$1.2\ m^3/s = V(1.2 \times 4)$$

$$V = 0.25\ ms = 25\ cm/s$$

   - Substituting in Equation 7.27:

$$d_p = \left(\frac{18 \times 1.85 \times 10^{-4} \times 25 \times 120}{981 \times 2.5 \times 800} \right)^{1/2} = 0.0023\ cm = 23\ \mu m$$

Listing of Program 7.19

```
'**************************************************
'Example 7-19: Predict the largest sized particle
' that can be removed with 100 % efficiency in a
' settling chamber.
'**************************************************
'
'
COLOR 15, 1, 4
CLS
title:
DO
  opt = 0
  PRINT "Exercise 7-19: Predict the largest sized particle that"
  PRINT "can be removed with 100% efficiency in a settling chamber  "
  PRINT : PRINT "The formula:  "
  PRINT : PRINT "d[p] = (18 um v[h] H / g L r[p]){1/2} "
  PRINT : PRINT "The value for v[h] is computed first."
  PRINT : PRINT "This formula is Q = VA "
  PRINT : PRINT "Type 1 and ENTER to continue  "
  INPUT "Type 2 and ENTER to EXIT  "; opt
LOOP WHILE opt < 1 OR opt > 2
```

```
IF (opt = 2) THEN
  END
END IF
start:
  CLS
  PRINT "  Q = VA "
  PRINT "  V = Q / A "
  PRINT "Enter the value of the volumetric gas flow "
  INPUT "   measured in m{3}/sec "; measq
  INPUT "Enter the height of the chamber (in meters) "; meash1
  INPUT "Enter the width of the chamber (in meters) "; measw1
    comp1 = meash1 * measw1
    comp2 = measq / comp1
    comp3 = comp2 * 100
  PRINT : PRINT measq; " = V ("; meash1; " x "; measw1; ")"
  PRINT : PRINT "V = "; measq; " / "; comp2;
  PRINT : PRINT "V = "; comp2; " m/sec"
  PRINT : PRINT "V = "; comp3; " cm/sec"
  INPUT "Press ENTER to continue  "; ent$

  CLS
  PRINT "d[p] = (18 um v[h] h / g L r[p]){1/2}  "
  PRINT : PRINT "Enter the gas viscosity (in g/cm sec) "
  INPUT "   Enter as scientific - ex; 1.86E-5 "; visc
  INPUT "Enter the height of the chamber (in cm) "; hgt
  INPUT "Enter the length of the chamber (in cm) "; lnth
  INPUT "Enter the particle density (in g/cm{3}) "; dens
    comp6 = (18 * visc * comp3 * hgt) / (981 * dens * lnth)
    comp7 = comp6 ^ (.5)
    comp8 = comp7 * 10000
  PRINT : PRINT "d[p] = ((18 x "; visc; " x "; comp3; " x "; hgt; ") / (981 x "; dens; " x "; lnth; ")){1/2)"
  PRINT : PRINT "d[p] = "; comp7; " cm  =  "; comp8; " um"
  INPUT "Press Enter to continue "; ent$
  CLS
  GOTO title
END
```

7.3.2 Cyclones

Cyclones are the most commonly used air pollution devices for the removal of particulate from an air stream. Cyclones are inertial separators depending on centrifugal force as the separating mechanism. The separation of particulate in a cyclone is similar to that of gravitational settling, except that centrifugal instead of gravitational forces are used.

Cyclones are inexpensive to operate and have no moving parts. The cyclone operates with two vortexes, one spiraling downward at the outside of the cone and the other spiraling upward on the inside of the cone. The efficiency of a cyclone depends upon its size; a single, high-efficiency cyclone is most efficient at the removal of particles in the 40- to 50-μm size range, while small cyclones 25 cm in diameter are most efficient in the 15- to 20-μm size range. Figure 7.2 shows a conventional single, high-efficiency cyclone.

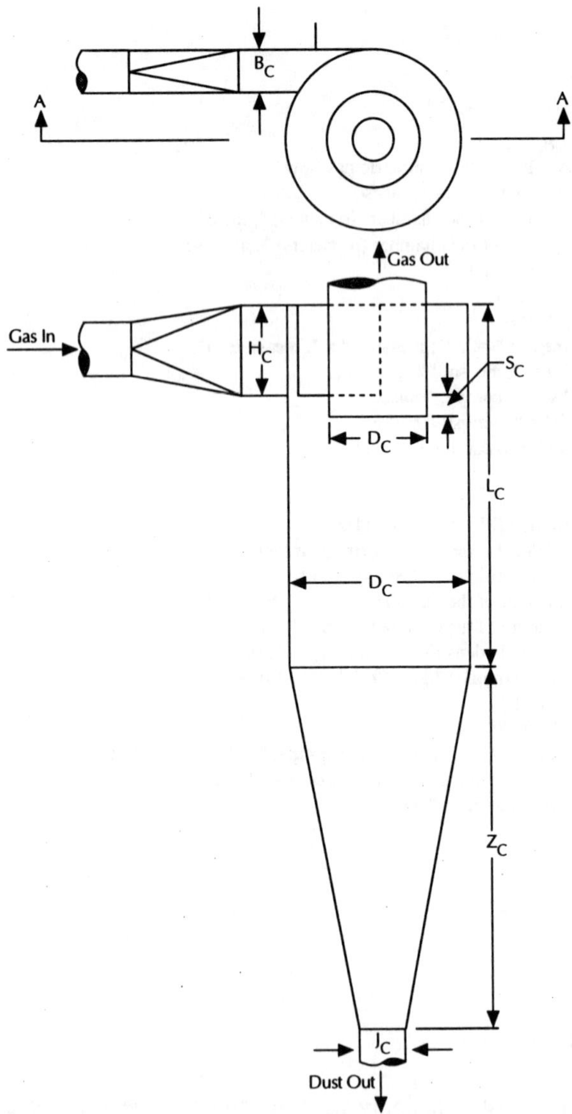

FIGURE 7.2
(A) typical dimension for single high-efficiency cyclone. (B) Nomenclature for a single high-efficiency cyclone.

An equation that can be used to calculate the size of particles that can be collected with a 50% efficiency (cut diameter) is as shown in Equation 7.28.

$$d_{50} = \left(\frac{9\,\mu B_c}{2\pi N_e v_i \rho_p} \right)^{1/2} \tag{7.28}$$

Major Cylinder Diameter	D_c
Major Cylinder Length	$L_c = 2\,D_c$
Cone Length	$Z_c = 2\,D_c$
Gas Outlet Diameter	$D_e = \frac{D_c}{2}$
Gas Outlet Length	$H_c + S_c = 5/8\,D_c$
Gas Inlet Height	$H_c = \frac{D_c}{2}$
Gas Inlet Width	$B_c = \frac{D_c}{4}$
Dust Outlet	$J_c = \frac{D_c}{4}$

Figure 7.2B

where:

d_{50} = Diameter of the particle that is collected with 50% efficiency (cut diameter)
μ = Gas viscosity, kg/m*s
B_c = Width of cyclone inlet, m
N_e = Number of effective turns within the cyclone
v_i = Inlet gas velocity, m/s
ρ_p = Density of the particulate matter, kg/m^3

The number of effective turns (N) within the cyclone can be approximated by Equation 7.29.[14]

$$N_e = \frac{1}{H_c}(L_c + Z_c/2) \qquad (7.29)$$

where:

H_c = Height of inlet duct, m
L_c = Length of vertical cylinder, m
Z_c = Length of cone section, m

The collection efficiency of particles both larger and smaller than the d_{50} (cut diameter) can be estimated by the use of Equation 7.30 or from Figure 7.3.[13,15]

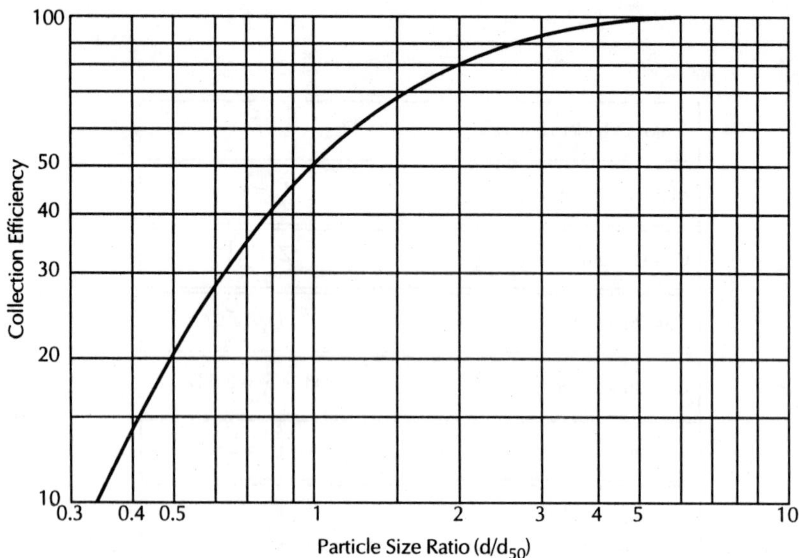

FIGURE 7.3
Cyclone efficiency vs. particle size ratio.

$$E = \frac{1.0}{\left[1 + \left(\frac{d_{50}}{d}\right)^2\right]} \times 100 \tag{7.30}$$

where:

E = Collection efficiency
d_{50} = Particle diameter with 50% collection efficiency (cut diameter), μm
d = Particle diameter of a specified size, μm

The pressure drop through a cyclone can be estimated by use of the following empirical equation:[16]

$$\Delta P = \frac{3950\, K\, Q^2\, P\rho}{T} \tag{7.31}$$

where:

ΔP = Pressure drop, m of water
Q = m^3/s of gas
P = Absolute pressure, atm
ρ = Gas density, kg/m^3
T = Temperature, K
K = Factor, function of cyclone diameter (approximate values are shown in Table 7.7)

TABLE 7.7
Values of K for Calculating Pressure Drop in Cyclones[16]

Cyclone diameter (cm)	K
74	10^{-4}
40	10^{-3}
20	10^{-2}
10	10^{-1}

Source: From Vesilind, P.A., Peirce, J.J., and Weiner, R.F., *Environmental Engineering*, 2nd ed., Butterworth-Heinemann, Stoneham, MA, 1988. With permission.

The following example demonstrates the use of Equations 7.28, 7.29, 7.30, and 7.31.

Example 7.20

Determine the size of the particles that can be collected with 50% efficiency (cut diameter) for a cyclone with the following dimensions and operating under the following conditions at a cement kiln plant. Also estimate the pressure drop through the cyclone as well as the removal efficiency for particles with a 20-μm diameter.

Flow rate, $Q = 1.6\ m^3/s$

Length of vertical cyclone cylinder, $L_c = 1$ m

Length of cyclone cone section, $Z_c = 1$ m

Cyclone diameter, $D_c = 0.50$ m

Cyclone inlet width, $B_c = 0.15$ m

Cyclone inlet height, $H_c = 0.25$ m

Specific gravity of particles, $\rho_p = 2000\ kg/m^3$

Gas viscosity (air) $\mu = 0.0756$ kg/m*h

Density of air, $\rho_a = 0.965\ kg/m^3$

Pressure P = 1 atm

Temperature = 93°C

Solution

From Equation 7.29, the estimated numbeer of effective turns (Ne) within the cyclone would be

$$N_e = \frac{1}{0.25}\left(1\text{ m} + \frac{1\text{ m}}{2}\right) = \frac{1.50}{0.25} = 6$$

The gas inlet velocity $v_i = Q/A$:

$$v_i = \frac{1.6\ m^3/s}{0.25\ m \times 0.15\ m} = 42.67\ m/s = 153,600\ m/h$$

From Equation 4.28, the d_{50} or cut diameter can be calculated.

$$d_{50} = \left(\frac{9 \times 0.0756\ kg/m*h \times 0.15\ m}{2 \times 3.14 \times 6 \times 153,600\ m/h \times 2000\ kg/m^3}\right)^{1/2} = 2.96 \times 10^{-6}\ m = 3\ \mu$$

Use Equation 7.31 to estimate the pressure drop through the cyclone:

Q = 1.6 m^3/s
P = 1 atm
ρ_a = 0.965 kg/m^3
T = 93°C
K = from Table 7.7, interpolation for a 50-cm diameter gives $K = 0.00074$, or 7.4×10^{-4}

$$\Delta P = \frac{3950(7.4 \times 10^{-4})(1.6\ m^3/s)^2(1\ atm)(0.965\ kg/m^3)}{(273 + 93\ K)} = 0.0197\ m = 1.97\ cm$$

For a 20-μm diameter particle, the removal efficiency can be estimated from Equation 7.30.

$$E = \frac{1}{\left[1 + \left(\frac{3}{20}\right)^2\right]} \times 100 = 98\%$$

From Figure 7.3, removal efficiency for a 20-μm diameter particle would be as follows:

$$\text{Relative size} = \frac{d_{20}}{d_{50}} = \frac{20}{2.6} = 7.7$$

From the graph, the efficiency would approach 100%.

Example 7.21

1. Prepare a computer program that can be used to estimate the size of particles that can be collected with 50% (cut diameter) efficiency with a cyclone of given dimensions. Include in the program a system for the determination of the number of effective turns

(Ne) within the cyclone, the collection efficiency for particles both smaller and larger than the d_{50} size, as well as an estimate of the pressure drop (ΔP) through the cyclone.

2. Using the computer program developed in (1), estimate for a single high-efficiency cyclone with the following dimensions and operating under the following conditions the size of the particles that can be collected with 50% efficiency, Ne, ΔP, and the collection efficiency for a 15-μm sized particle.

Flow rate $Q = 6\ m^3/s$

Length of vertical cyclone cylinder $L_c = 3$ m

Length of cyclone section $Z_c = 3$ m

Cyclone diameter $D_c = 1.5$ m

Cyclone inlet width $B_c = 0.375$ m

Cyclone inlet height $Hc = 0.75$ m

Specific gravity of particle $\rho_p = 1700\ kg/m^3$

Gas viscosity at 80°C $\mu = 0.0745$ kg/m*h

Density of air at 80°C $\rho_a = 1.001\ kg/m^3$

Pressure = 1 atm

K = factor for cyclone (D = 1.5 m); extrapolation from Table 4.7 gives 5×10^{-5}

Solution

1. For a solution to Example 7.21 (1), see the following listing of Program 7.21 for the determination of d_{50}, Ne, ΔP, and collection efficiency for particles smaller and larger than the d_{50} size.

2. A solution to Example 7.21 (2):
   - From Equation 7.29:

$$Ne = \frac{1}{0.75}\left(3m + \frac{3m}{2}\right) = 6$$

   - The gas inlet velocity $v_i = Q/A$:

$$v_i = \frac{6\ m^3/s}{0.375\ m \times 0.75\ m} = 21.33\ m/s = 76,800\ m/h$$

   - From Equation 7.28:

$$d_{50} = \left(\frac{9 \times 0.0745\ kg/m/h \times 0.375}{2 \times 3.14 \times 6 \times 76,800\ m/h \times 1700\ kg/cm^3}\right)^{1/2} = 0.000007\ m = 7\ \mu m$$

   - From Equation 7.31:

$$\Delta P = \frac{3950(5 \times 10^{-5})(6\ m^3/s)^2(1\ atm)(1.001\ kg/m^3)}{(273 + 80\ K)} = 0.020\ m = 2.0\ cm$$

- From Equation 7.30, efficiency for a 15-μm particle:

$$E = \frac{1.0}{\left[1+(7/15)^2\right]} \times 100 = \frac{1.0}{1.218} \times 100 = 82\%$$

- From Figure 7.3, d/d_{50} = 15/7 = 2.14. The collection efficiency = 82%.

Listing of Program 7.21

```
'**********************************************************
'Example 7-21: Estimate the size of particles that
' can be collected with a 50% (cut diameter) efficiency
' with a cyclone of given dimensions
'**********************************************************
'
'
COLOR 15, 1, 4
CLS
title:
DO
  opt = 0
  PRINT "Exercise 7-21:  Estimate the size of particles that can"
  PRINT "be collected with a 50% (cut diameter) efficiency with  "
  PRINT "a cyclone of given dimensions"
  PRINT : PRINT "The formulas:  "
  PRINT : PRINT "Ne = (1/h[c]) * (L[c] + (Z[c]/2)) "
  PRINT : PRINT "V[i] = Q/A"
  PRINT : PRINT "d[50] = (9uB[c] / 2piN[e]V[i]r[p])1/2"
  PRINT : PRINT "Delta P = 3950 K Q2 P r / T"
  PRINT : PRINT "E = 1 / (1 + (d[50]/d){2}) * 100"
  PRINT : PRINT "Type 1 and ENTER to continue  "
  INPUT "Type 2 and ENTER to EXIT  "; opt
LOOP WHILE opt < 1 OR opt > 2
IF (opt = 2) THEN
  END
END IF
formulaa:
  CLS
  PRINT "Ne = (1/h[c]) * (L[c] + (Z[c]/2)) "
  PRINT : INPUT "Enter the height of the inlet duct - H[c] - (in m) "; hgt1
  INPUT "Enter the length of vertical cylinder - L[c] - (in m) "; vert1
  INPUT "Enter the length of cone section - Z[c] - (in m) "; length1
  acomp1 = 1 / hgt1
   acomp2 = length1 / 2
   acomp3 = vert1 + acomp2
   aansw = acomp1 * acomp3
  PRINT : PRINT "Ne = "; aansw; ""
  PRINT : PRINT
formulab:
  PRINT "v[i] = Q / A  "
  PRINT : INPUT "Enter the Flow Rate - Q - (in m{3}/s) "; bq
  INPUT "Enter the cyclone inlet width - B[c] -(in m) "; bwidth
  INPUT "Enter the particle density - r[p] - (in kg/m{3}) "; dens
   bcomp1 = bwidth * hgt1
   bansw = bq / bcomp1
   banswb = bansw * 3600
```

```
  PRINT : PRINT "v[i] = "; bansw; " m/s  = "; banswb; "m/hr"
  INPUT "Press Enter to continue "; ent$
formulac:
  CLS
  PRINT "d[50] = (9uB[c] / 2piN[e]V[i]r[p])1/2"
  PRINT : INPUT "Enter the gas viscosity - u - (in kg/m*hr) "; visc
   ccomp1 = 9 * visc * bwidth
   ccomp2 = 2 * 3.14 * aansw * banswb * dens
   ccomp3 = ccomp1 / ccomp2
   cansw = ccomp3 ^ (.5)
   canswb = cansw * 1000000
  PRINT : PRINT "d[50] = "; cansw; "m  =  "; canswb; "um"
  PRINT : PRINT
formulad:
  PRINT : PRINT "Delta P = 3950 K Q{2} P r[a] / T"
  PRINT : PRINT "Enter the k factor from table 7.7 "
  INPUT "   use scientific notation - x.xxE-/+x "; factor
  INPUT "Enter the absolute pressure (in atm) "; pres
  INPUT "Enter the gas density - r[a] - (in kg/m{3}) "; gasdens
  INPUT "Enter the temperature - T - (in C) "; temp
   tcomp = 273 + temp
   qs = bq * bq
   dcomp1 = 3950 * factor * qs * pres * gasdens
   dansw = dcomp1 / tcomp
   danswb = dansw * 100
  PRINT : PRINT "Delta P = "; dansw; "m  =  "; danswb; "cm"
  INPUT "Press Enter to continue "; ent$
formulae:
  CLS
  PRINT "E = 1 / (1 + (d[50]/d){2}) * 100"
  PRINT : INPUT "Enter the particle diameter (in um) "; diam
   ecomp1 = canswb / diam
   ecomp2 = ecomp1 ^ 2
   ecomp3 = 1 + ecomp2
   ecomp4 = 1 / ecomp3
   eansw = ecomp4 * 100
  PRINT : PRINT "E = "; eansw; "percent "
  INPUT "Press Enter to continue "; ent$
  CLS
  GOTO title
END
```

7.3.3 Electrostatic Precipitators

Electrostatic precipitators (ESP) operate by ionizing air molecules into positive and negative ions. This ionization takes place when the air is passed through a very high direct current voltage, anywhere from 30,000 to 100,000 volts. As the negative ions move to the positive collecting electrode, they attach themselves to the particles in the gas stream. When the particles become charged by the negative ions they move to the positively charged collection surface (see Figure 7.4). The particles are removed from the collection surface either by rapping it or washing it.

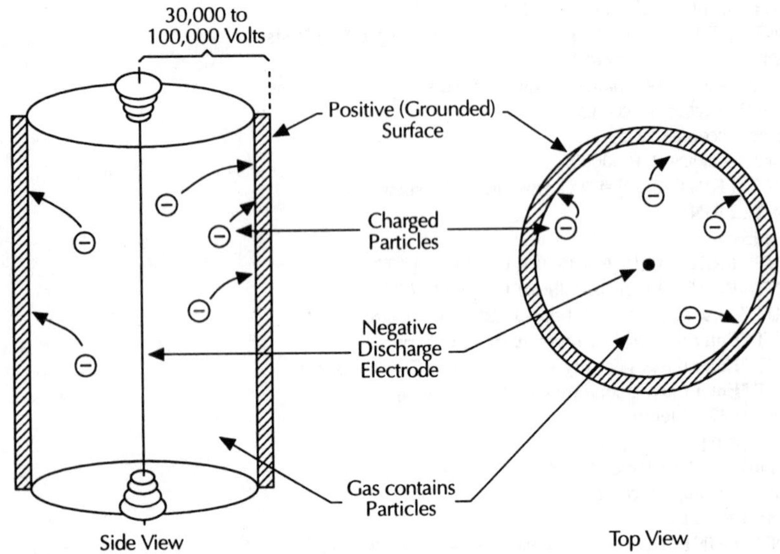

FIGURE 7.4
Schematic diagram of an electrostatic precipitator.

There are two types of precipitators, plate type and tube type (see Figure 7.5). The plate type tends to be used most often. The size of the collection plates used ranges from 1 to 3 m long and 15 m high. The distance between the plates ranges from 15 to 40 cm. Electrostatic precipitators can handle huge volumes of gas and are widely used at power plants where electrical power is readily available.

Electrostatic precipitators have many advantages, including:[11]

- High efficiency, especially for submicron particles
- Dry collection of dusts
- Low pressure drop
- Ability to collect mists and corrosive acids
- Low maintenance costs
- Low operating costs
- No moving parts
- Collection efficiency can be adjusted by unit size
- Ability to handle gases up to 815°C

Disadvantages include:

- High initial cost
- Frequently needs a precleaner

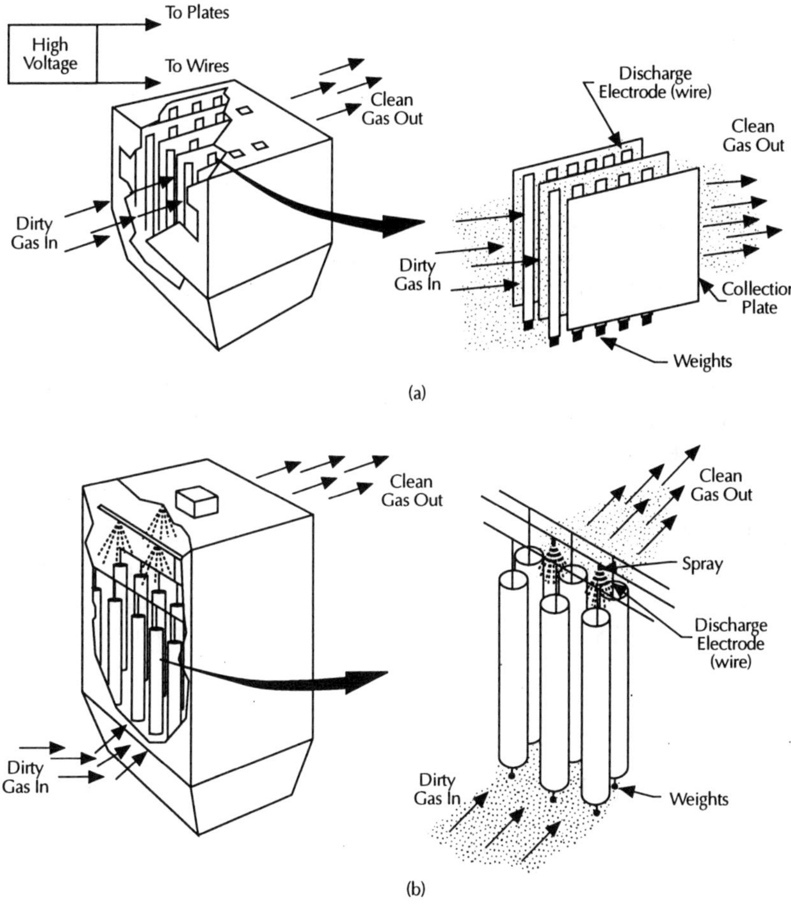

FIGURE 7.5
High-voltage electrostatic precipitators.

- Large space requirements
- Not suitable for collecting combustible particles such as grain or wood dust

The efficiency of an electrostatic precipitator can be estimated by the following empirical Equation 7.32 (Deutsch-Anderson equation).[9]

$$\eta = 1 - \exp\left(-{}^{Aw}/_{Q}\right) \tag{7.32}$$

where:

η = Fractional collection efficiency
A = Area of the collection plates, m^2
w = Drift velocity of the charged particles, m/s
Q = Flow rate of the gas stream, m^3/s

The drift velocity w is the velocity at which a particle approaches the collection surface and can be estimated by Equation 7.33.[9]

$$w = ad_p \tag{7.33}$$

where:

w = Drift velocity, m/s

a = Constant, a function of the charging field, the carrier gas properties, and the ability of the particle to accept an electrical charge, s^{-1}

d_p = Particle size, μm

Drift velocities commonly range from 0.02 to 0.2 m/s and can be determined experimentally. The following example demonstrates the calculations that can be used to evaluate the operation or design of an electrostatic precipitator.

Example 7.22

An electrostatic precipitator is to remove fly-ash particles of 0.5–μm diameter from a gas stream flowing at 0.7 m^3/s. Determine the plate area required to collect the 0.5-μm particles at 98% efficiency. The value for the constant a has been found to be 0.24×10^6/s.

Solution

Using Equation 7.33, the drift velocity will be

$$w = 0.24 \times 10^6 \ /s \times 0.5 \times 10^{-6} \ m = 0.12 \ m/s$$

Using Equation 7.32 and 98% removal efficiency:

$$0.98 = 1 - \exp - \left(\frac{0.12 \ m/s}{0.7 \ m^3/s} \times A \right)$$

which yields A = 23 m^2.

Considering that both sides of the plate are available for collecting particles, then the plate area required is 23/2 = 11.5 m^2. Use a square plate 3.4 by 3.4 m^3.

Example 7.23

1. Prepare a computer program that can be used to estimate the drift velocity for a plate-type electrostatic precipitator, given the constant a in Equation 7.33. Also include in the program the elements in Equation 7.32. Design the program so that any of the other elements can be calculated provided any of the other two elements are given.

2. Using the computer program developed in (1), determine the number of 6- by 6-m plates needed to collect 0.4-μm particles with a drift velocity constant a of 2.5×10^5/s. The total air flow is 30 m^3/s, and the collection efficiency is to be 90%.

Solution

1. For a computer program for Example 7.23 (1), see the following listing of program 7.23 which includes the elements in Equation 7.32 and 7.33.
2. A solution to Example 7.23 (2):
   - Using Equation 7.33:

$$w = 2.5 \times 10^5 \text{ /s} \times 0.4 \times 10^{-6} \text{ m} = 0.10 \text{ m/s}$$

   - Using Equation 7.32:

$$0.90 = 1 - \exp\left(-\frac{0.10 \text{ m/s}}{30 \text{ m}^3\text{/s}} \times A\right)$$

which yields A = 691 m^2. Each plate surface has $6 \times 6 \times 2 = 72$ m^2 surface area. Use 10 plates.

Listing of Program 7.23

```
'*******************************************************
'Example 7-23: Estimate the drift velocity for a plate-
' type electrostatic precipitator given the constant in
' equation 7.32.  The program can calculate any of the
' elements - given the other two
'*******************************************************
'
'
COLOR 15, 1, 4
CLS
title:
DO
  opt = 0
  PRINT "Exercise 7-23:  Estimate the drift velocity for a plate-type"
  PRINT "electrostatic precipitator given the constant a in equation "
  PRINT "7.32.  Given two elements, the program can calculate the "
  PRINT "other two. "
  PRINT : PRINT "The formulas:  "
  PRINT : PRINT "w = a*d[p]"
  PRINT "n = 1 -exp(-(A*w/Q)) "
  PRINT : PRINT "Type 1 and ENTER to continue  "
  INPUT "Type 2 and ENTER to EXIT  "; opt
LOOP WHILE opt < 1 OR opt > 2
IF (opt = 2) THEN
  END
END IF
velscreen:
  CLS
  PRINT "w = a*d[p]"
```

```
  PRINT : PRINT "Enter the value for the constant a "
  INPUT "  Use scientific notation (x.xxE+/-x) "; aconst
  INPUT "Enter the particle size - d[p] - (in um) "; partsize
    wcompa = aconst * partsize
    compvel = wcompa * .000001
  PRINT : PRINT "The Drift velocity - w  = "; compvel; " m/s"
  PRINT : PRINT
optscreen:
DO
  PRINT : PRINT "n = 1 -exp(-(A*w/Q)) "
  PRINT : PRINT "In the above formula, given w, this program solves for: ";
  PRINT "1.  n, the fractional collection efficiency"
  PRINT "2.  A, the area of the collection plates"
  PRINT "3.  Q, the flow rate of the gas stream"
  PRINT "4.  Start a new problem (recompute w)"
  PRINT "5.  Return to 1st screen"
  PRINT : INPUT "Select the option 1 - 5:   "; opt2
LOOP WHILE opt2 < 1 OR opt2 > 5
SELECT CASE opt2
   CASE 1
    CLS
    PRINT "Solving for n, the fractional collection efficiency"
    PRINT : PRINT "n = 1 -exp(-(A*w/Q)) "
    PRINT : INPUT "Enter the plate area - A - (in m{2}) "; area
    INPUT "Enter the flow rate of the stream - Q - (in m{3}/s) "; rate
      comp1 = (area * compvel) / rate
      comp2 = (EXP(-2.32))
      comp3 = 1 / (comp2)
      comp4 = 1 / comp3
      comp5 = 1 - comp4
      pcomp = comp5 * 100     'just trying it
    PRINT : PRINT "The fractional collection efficiency - n = "; pcomp; "%"
    INPUT "Press ENTER to continue  "; ent$
    GOTO optscreen
   CASE 2
    CLS
    PRINT "Solving for A, the area of the collection plates"
    PRINT "A = (Q (ln (1 / 1-n)) / w"
    INPUT "Enter the efficiency - n - (in %) "; perc
      IF (perc > 1) THEN
        compperc = perc * .01
        perc = compperc
      END IF
    INPUT "Enter the flow rate of the stream - Q - (in m{3}/s) "; rate
      comp1 = 1 - perc
      comp3 = LOG(comp1)
      comp4 = rate / compvel
      acomp = (comp4 * comp3) * -1
    PRINT "The area of the collection plates is "; acomp; "m{2}"
    PRINT : INPUT "Press ENTER to continue  "; ent$
    PRINT : PRINT "Determining the number of plates of a given size"
    INPUT "Enter the width of the collection plate (in m) "; platewidth
    INPUT "Enter the height of the collection plate (in m) "; plateheight
    platearea = platewidth * plateheight * 2
      numbplates = acomp / platearea
    PRINT : PRINT "The plate area is computed as width * height * 2 "
    PRINT "(the plates have two sides)"
    PRINT : PRINT USING "##"; numbplates;
```

```
    PRINT " plates will be required for this collection"
    PRINT : INPUT "Press ENTER to continue  "; ent$
    CLS
    GOTO optscreen
  CASE 3
    CLS
    PRINT "Solving for Q, the flow rate of the gas stream"
    PRINT "Q = (A w) / (ln (1 / 1-η))"
    INPUT "Enter the efficiency - n - (in %) "; perc
      IF (perc > 1) THEN
        compperc = perc * .01
        perc = compperc
      END IF
    INPUT "Enter the plate area - A - (in m{2}) "; area
      comp1 = 1 - perc
      comp2 = 1 / comp1
      comp3 = LOG(comp2)
      comp4 = area * compvel
      qcomp = comp4 / comp3
    PRINT "The flow rate (Q) is "; qcomp; "m{3}/s"
    INPUT "Press ENTER to continue  "; ent$
    GOTO optscreen
  CASE 4
    GOTO velscreen
  CASE 5
    CLS
    GOTO title
END SELECT
END
```

7.3.4 Venturi Scrubbers

A venturi scrubber passes a contaminated gas stream through a duct that has a converging venturi-shaped throat, followed by a diverging section (see Figure 7.6). The scrubbing liquid is injected at right angles to the incoming gas stream which breaks the liquid into small droplets that can then combine with the gaseous or particulate contaminants.

A venturi scrubber will remove from 92 to 99% of particulate in the size range from 0.02 to 0.5 μm, which makes them very effective for removal of submicron particles in smoke and fumes.[13] The advantages of using a venturi scrubber include:

- Collecting both particulate and gaseous contaminants
- Cooling high temperature gas streams
- Elimination of problems with fires or explosions
- Variable collection efficiency
- Recovery of a valuable by product is made possible

The disadvantages include:

- High power costs
- Collecting large volumes of liquid that require disposal

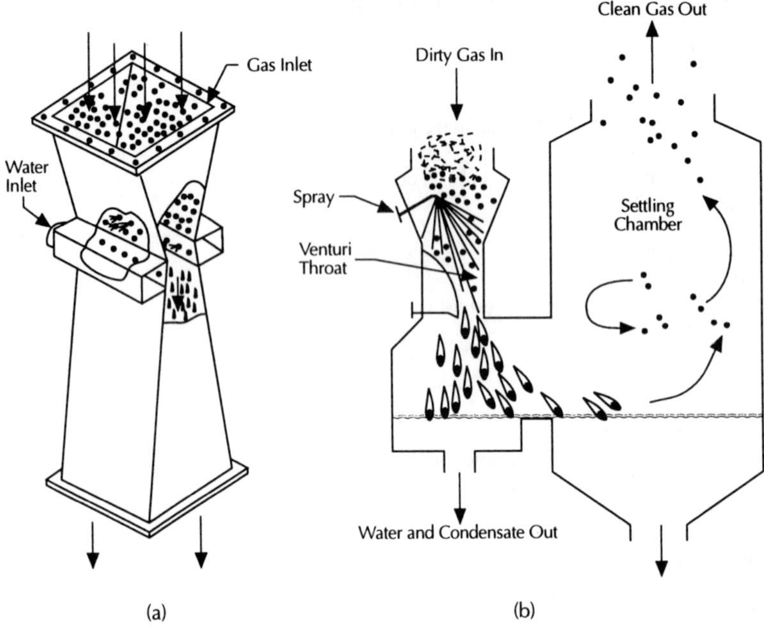

FIGURE 7.6
Venturi wet collector.

- Elimination of buoyancy of gas at stack exit by reduced gas temperature
- White vapor cloud at the stack exit that may concern surrounding community
- Problems with corrosion

Johnstone's Equation 7.34 can be used to evaluate the operation and design of venturi scrubbers:[9]

$$E = 1 - \exp\left[-k(Q_L / Q_G)\sqrt{\Psi}\right] \tag{7.34}$$

where:

E = Fractional collection efficiency

k = Correlation coefficient depending upon system geometry and operating conditions

Q_L/Q_6 = Liquid-to-gas ratio (Q_L, m^3 / s; Q_G, m^3/s)

Ψ = Interial impaction parameter

$$\Psi = C\rho_p v d_p^2 / 18 d_o \mu \tag{7.35}$$

where:

C = Cunningham correction factor, dimensionless
ρ_p = Particle density, kg/m^3
v = Gas velocity at venturi throat, m/s
d_p = Particle diameter, m
d_o = Average droplet diameter, m
μ = Gas viscosity, kg/m*s

The Cunningham correction factor may be approximated by using Equation 7.36.[17]

$$C = 1 + \frac{(6.21 \times 10^{-4})T}{d_p} \tag{7.36}$$

where:

T = Absolute gas temperature, K
d_p = Diameter of particle, μm

Under normal conditions, the Cunningham correction factor is taken as 1 for particles larger than 1 μm.

The average liquid droplet d_o may be determined by using Equation 7.37:[9]

$$d_0 = \frac{5000}{v} + 29.67\left(1000\frac{Q_L}{Q_G}\right)^{1.5} \tag{7.37}$$

where:

d_o = Average droplet diameter, μm

The other terms in Equation 7.37 are the same as presented earlier.

The pressure drop through a venturi scrubber can be estimated by the following empirical equation:[16]

$$\Delta P = v^2 L \times 10^{-6} \tag{7.38}$$

where:

ΔP = Pressure drop across the venturi scrubber, cm of water
v = Gas velocity through throat, cm/s
L = Water-to-gas volume ratio, L/m^3

The following example demonstrates the calculations that can be made to evaluate the operation and design of a venturi scrubber.

Example 7.24

Determine the collection efficiency of a venturi scrubber for particles 0.7 μm in diameter from a gas stream containing fly-ash. The liquid-to-gas ratio is 1.12×10^{-3} m^3 of liquid per m^3 of gas. Also determine the pressure drop across the venturi scrubber, given the scrubber described below:

Throat velocity = 90 m/s

Density of particles = 900 kg/m^3

Gas viscosity = 2.2×10^{-5} kg/m/s

Gas temperature = 100°C

Correlation coefficient K = 2000

Solution

- Using Equation 7.36, estimate the Cunningham correction factor:

$$C = 1 + \frac{(6.21 \times 10^{-4})T}{d_p} = 1.33$$

- Using Equation 7.37, estimate the average droplet diameter d_o:

$$d_0 = \frac{5000}{90} + 29.67(1000 \times 1.12 \times 10^{-3})^{1.5} = 90.7\ \mu m$$

- Using Equation 7.35, calculate the interial impaction parameter:

$$\Psi = \frac{(1.33 * 900\ kg/m^3)(90\ m/s)(0.7 \times 10^{-6}\ m)^2}{18(90.7 \times 10^{-6}\ m)(2.20 \times 10^{-5}\ kg/m/s)} = 1.47$$

$$\sqrt{\Psi} = 1.21$$

- Using Equation 7.34, determine the venturi scrubber's collection efficiency:

$$E = 1 - \exp[-2000 \times 1.12 \times 10^{-3} \times 1.21] = 1 - 0.07$$

$$= 93\% \text{ for removal of 0.7 sized particles}$$

- The pressure drop across the venturi scrubber can be calculated from Equation 7.38:

$$\Delta P = (9000\ cm/s)^2(1.12\ L/m^3)10^{-6} = 90.72\ cm = 35.7\ in\ H_2O$$

Example 7.25

1. Prepare a computer program that will estimate the efficiency of a venturi scrubber to remove particles of a specific size as well as the pressure drop through the venturi scrubber. Include in the program all the elements contained in Equations 7.34, 7.35, 7.36, and 7.38.
2. Using the program developed in (1), estimate the collection efficiency of a venturi scrubber at removing 3.2-μm sized particles. The particles have a density of 3000 kg/m^3. Also determine the pressure drop across the venturi scrubber. The venturi scrubber characteristics are as follows:

   Gas flow rate = 310 m^3/min = 5.2 m^3/s

   Liquid flow rate = 0.00139 m^3/s

   Gas flow velocity at throat = 100 m/s

   Correlation coefficient = 1500

   Temperature = 90°C

   Gas viscosity = 2.1×10^{-5} Kg/m/s

Solution

1. For a solution to Example 7.25 (1), see the following listing of Program 7.25, which includes elements in Equations 7.34, 7.35, 7.36, 7.37, and 7.38.
2. A solution to Example 7.25 (2):
   - Using Equation 7.36, estimate the Cunningham correction factor:

$$C = 1 + \frac{6.21 \times 10^{-4}(273+90)}{3.2} = 1.07$$

   - Using Equation 7.37, estimate the average droplet diameter:

$$d_0 = \frac{5000}{100\ \text{m/s}} + 29.67\left(\frac{1000 \times 0.00139\ \text{m}^3/\text{s}}{5.2\ \text{m}^3/\text{s}}\right)^{1.5} = 50 + 4 = 54\ \mu\text{m}$$

   - Using Equation 7.35, calculate the interial impaction parameter:

$$\Psi = \frac{1.07(3000\ \text{kg/m}^3)(100\ \text{m/s})(3.2 \times 10^{-6}\ \text{m})^2}{18(54 \times 10^{-6})(2.1 \times 10^{-5}\ \text{kg/m} * \text{s})} = 161$$

$$\sqrt{\Psi} = 12.69$$

   - Using Equation 7.34, determine the venturi scrubber efficiency for collecting 3.2-μm-sized particles:

$$E = 1 - \exp\left(-1500 \times \frac{0.00139}{5.2} \times 12.69\right) = .994 = 99.4\%$$

- Using Equation 7.38, estimate the pressure drop across the throat of the venturi scrubber:

$$\Delta P = (10,000 \text{ cm/s})^2 \left(\frac{1.39 \text{ L/s liquid}}{5.2 \text{ m}^3\text{/s gas}}\right) 10^{-6} = 26.73 \text{ cm} = 10.52 \text{ inches of } H_2O$$

Listing of Program 7.25

```
'**************************************************************
'Example 7-25: Estimate the efficiency of a Venturi scrubber
' to remove particles of a specific size as well as the
' pressure drop through the scrubber.
'**************************************************************
'
'
COLOR 15, 1, 4
CLS
title:
DO
  opt = 0

  PRINT "Exercise 4-25: Estimate the efficiency of a Venturi"
  PRINT "scrubber to remove particles of a specific size as "
  PRINT "well as the pressure drop through the scrubber."
  PRINT : PRINT "The formulas: "
  PRINT "E = 1 - exp [-k (Q[L] / Q[G]) SQR-IIP] "
  PRINT "(SQR-IIP = Square root of interial impaction parameter)"
  PRINT "IIP = C r[p] v d[p]{2} / 18 d[o] u"
  PRINT "C = 1 + ((6.21 x 10{-4})T) / d[p]"
  PRINT "d[o] = (5000 / v) + 29.67 (1000 (Q[L] / Q[G]){1.5}"
  PRINT "Delta P = v{2} L x 10{-6}"
  PRINT : PRINT "Type 1 and ENTER to continue "
  INPUT "Type 2 and ENTER to EXIT "; opt
LOOP WHILE opt < 1 OR opt > 2
IF (opt = 2) THEN
  END
END IF
screen1:
  CLS
  PRINT "C = 1 + ((6.21 x 10{-4})T) / d[p]"
  PRINT "d[o] = (5000 / v) + 29.67 (1000 (Q[L] / Q[G]){1.5}"
  PRINT : INPUT "Enter the gas temperature - T - (in C) "; temp
  INPUT "Enter the diameter of the particle - d[p] - (in um) "; dia
  INPUT "Enter the gas flow rate - (Q[G]) - (in m{3}/s) "; qg
  INPUT "Enter the liquid flow rate -(Q[L]) - (in m{3}/s) "; ql
  INPUT "Enter the throat velocity - v - (in m/s) "; vel
    tcomp = 273 + temp
    acomp1 = .000621 * tcomp
    acomp2 = acomp1 / dia
    c = 1 + acomp2
    acomp3 = 5000 / vel
    ratio = ql / qg
```

```
      acomp4 = 1000 * ratio
      acomp5 = acomp4 ^ (1.5)
      acomp6 = acomp5 * 29.67
      d = acomp3 + acomp6
    PRINT : PRINT "C = "; c
    PRINT : PRINT "d[o] = "; d
    PRINT : INPUT "Press ENTER to continue  "; ent$
  screen2:
    CLS
    PRINT "IIP = C r[p] v d[p]{2} / 18 d[o] u"
    PRINT : PRINT "Enter the gas viscosity - u -(in kg/m*s) "
    INPUT "    use scientific notation (x.xxE+/-x) "; visc
    INPUT "Enter the particle density - r[p] - (in kg/m{3}) "; dens
      dcomp = d * .000001
      diacomp = dia * .000001
      diasquar = diacomp ^ 2
      nom = c * dens * vel * diasquar
      denom = 18 * dcomp * visc
      IIP = nom / denom
      para = SQR(IIP)
    PRINT : PRINT "IIP = "; IIP; "    Square Root of IIP = "; para
    PRINT : INPUT "Press Enter to continue "; ent$
    PRINT : PRINT "E = 1 - exp [-k (Q[L] / Q[G]) SQR-IIP] "
    PRINT "(SQR-IIP = Square root of interial impaction parameter)"
    PRINT : INPUT "Enter the Correlation coefficient "; cc
      ratio = ql / qg
      ecomp1 = cc * ratio * para
      ecomp2 = EXP(-ecomp1)
      eff = (1 - ecomp2) * 100
    PRINT : PRINT "The efficiency is "; eff; "%"
    PRINT : INPUT "Press Enter to continue "; ent$
    PRINT : PRINT "Delta P = v{2} L x 10{-6}"
      lfr = ql * 1000
      ratio = lfr / qg
      pcomp1 = vel * 100
      pcomp2 = pcomp1 ^ 2
      pcomp3 = pcomp2 * ratio
      expon = 1 / 1000000
      drop = pcomp3 * expon
      dropb = drop * .3937
    PRINT : PRINT "The pressure drop is "; drop; "cm  or "; dropb; "inches of H[2]O"
    PRINT : INPUT "Press Enter to continue "; ent$
    CLS
    GOTO title
END
```

7.3.5 Baghouse or Fabric Filters

Baghouse or fabric filters are air pollution control devices that are used to separate solid particles from a gas stream. The particulate-laden gas stream is passed through woven or felted fabric; the fabric used for the filter varies from cotton to glass. The type of fabric used depends upon operational conditions such as temperature, pressure drop, chemical or physical degradation, cleaning methods, and, of course, cost and the life of the fabric. Filter bags usually are tubular or envelope shaped and are capable of removing over 99% of particles down to 0.3 μm, as well as substantial quantities of particles as small as 0.1 μm. Filter bags range from 1.8 to 12 m in length and 0.1 to 0.4 m in diameter. A typical baghouse filter is shown in Figure 7.7.

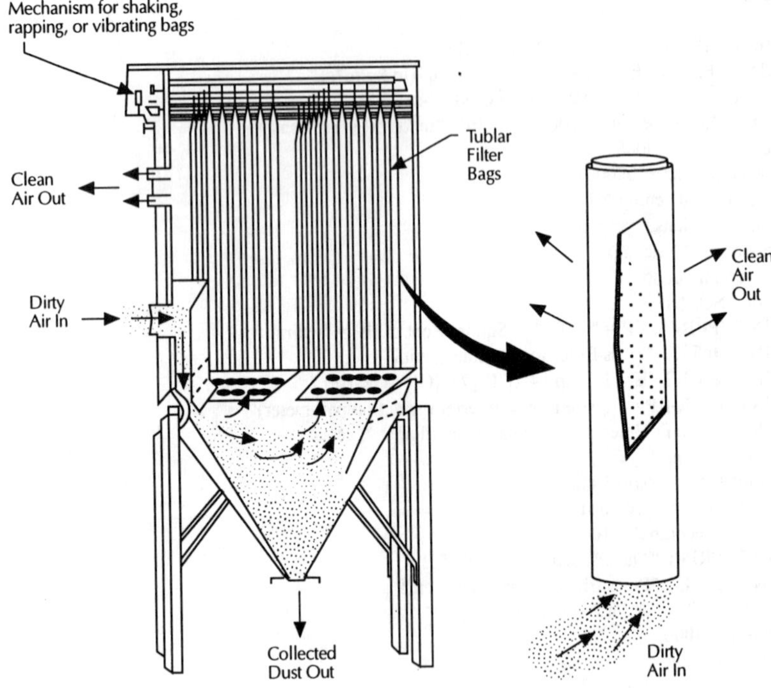

FIGURE 7.7
Baghouse filter.

The particles are captured and retained on the fibers of the fabric by means of interception, inertial impaction, Brownian diffusion, thermal precipitation, gravitational settling and electrostatic attraction. Once a mat or cake of particulate has collected on the fabric surface, further collection of particles is accomplished by sieving. The fabric then serves mainly as a supporting structure for the dust mat responsible for the high collection efficiency. Periodically the accumulated particulates are removed by mechanical shaking, reverse air cleaning, or pulse jet cleaning.[18] The advantages of fabric filters include:[11]

- High efficiency for collection of particles as small as 0.1 μm
- Moderate power requirements
- Reasonably low pressure drops
- Dry disposal of recovered particulate
- Operation over a wide range of particulates
- Modular design for simple add-ons

Disadvantages include:

- High capital cost
- Large space required
- High maintenance and replacement costs
- Cooling required for high temperature gas streams
- Fabric can be harmed by corrosive chemicals
- Control of moisture in particulate required
- Potential for fire or explosions

Equations that can be used to evaluate and aid in the design of fabric filters are

$$A_f = Q/v_f \tag{7.39}$$

where:

A_f = Total filtering area, m^2
Q = Volume of gas stream, m^3/min
v_f = Filtering velocity (air-to-cloth ratio), m/min

For cylindrical bags:

$$A_b = \pi dh \tag{7.40}$$

where:

A_b = Filtering area for each bag, m^2
π = 3.1416
d = Diameter of bag, m
h = Length of bag, m

The number of bags required then, is

$$N = \frac{A_f}{A_b} \tag{7.41}$$

where:

N = number of bags required

The cleaning frequency for a baghouse filter can be estimated from Equation 7.42, provided the desirable pressure drop is known, or, if the cleaning frequency is known, then the pressure drop can be estimated.[17,19]

$$\Delta P = S_e v_f + k_2 c v_f^2 t \tag{7.42}$$

where:

ΔP = Pressure drop
S_e = Effective residual drag
v_f = Gas velocity at filter surface
k_2 = Specific resistance coefficient of collected particulate
c = Concentration of particulate in the gas stream
t = Filtration time

Values for S_e and k_2 can be found in publications or handbooks such as Reference 17. Following is an equation which includes values for S_e and k_2. This equation can be used under specified conditions to estimate the pressure drop through a fabric filter, provided the required time between cleanings is known or vice versa.[9]

$$\Delta P = 1.6 v_f + 8.52 c v_f^2 t \tag{7.43}$$

where:

ΔP = Pressure drop, cm
v_f = Filtering velocity, m/min
c = Particulate concentration in the air stream, kg/m^3
t = Time required between cleanings, min

The following example demonstrates the calculations that can be made in order to evaluate the operation or design of a fabric filter.

Example 7.26

Determine the filter cloth area and the required frequency of cleaning given the following data:

Volume of contaminated air stream $Q = 400\ m^3/min$
Filtering velocity $v_f = 0.8$ m/min
Particulate concentration in the air stream $c = 0.03\ kg/m^3$
Diameter of the filter bag = 0.2 m
Length of the filter bag = 5.0 m
System designed for cleaning of filter when pressure drop reaches 9 cm of water

Solution

- The total filter area required according to Equation 7.39:

$$A_f = \frac{400\ m^3/min}{0.8\ m/min} = 500\ m^2$$

- Using Equation 7.40, the filtering bag area, which is 0.2 m in diameter and 5.0 m long, can be calculated:

$$A_b = 3.14(0.2\ m)(5.0\ m) = 3.14\ m^2/bag$$

- The number of bags required can be determined from Equation 7.41:

$$N = \frac{50\ m^2}{3.14\ m^2} = 159,\ \text{or}\ 160\ \text{bags}$$

- The frequency for cleaning these bags can be estimated from Equation 7.43:

$$9\ cm = 1.6(0.8\ m/min) + 8.52(0.03\ kg/m^3)(0.80\ m/min)^2 t$$

which yields t = 47 min.

Example 7.27

1. Prepare a computer program that will aid in the design and operation of a baghouse (fabric filter unit). Include in the program the elements in Equations 7.39, 7.40, 7.41, and 7.43.
2. Using the program developed in (1), determine the number of filtering bags required and the cleaning frequency of a plant employing a baghouse filter, given the following design data:

Volume of contaminated gas stream $Q = 12\ m^3/s$

Filtering velocity $v_f = 2$ m/min

Particulate concentration in the air stream $c = 0.025\ kg/m^3$

Diameter of filter bag = 0.3 m

Length of filter bag = 8 m

The baghouse filter system to be cleaned when pressure drop reaches 15 cm.

Solution

1. For a solution to Example 7.27 (1), see the following listing of Program 7.27, which includes the elements in Equations 7.39, 7.40, 7.41, and 7.43.
2. A solution to Example 7.27 (2):
   - From Equation 7.39, the total filter area required is

$$A_f = \frac{12\ m^3/s \times 60\ s/min}{2\ m/min} = 360\ m^2$$

- From Equation 7.40, the surface area for each bag is

$$A_b = 3.14(0.3\ m)(8\ m) = 7.54\ m^2$$

- From Equation 7.41, the number of bags required is

$$N = \frac{360\ m^2}{7.54\ m^2} = 47.7, \text{ or } 48 \text{ bags}$$

- From Equation 7.43, the estimated frequency of cleaning is

$$15\ cm = 1.6(2\ m/min) + 8.52(0.025\ kg/m^3)(2\ m/min)^2 t$$

which yields t = 13.8 min.

Listing of Program 7.27

```
'************************************************************
'Example 7-27: A program to aid in the design and operation
' of a baghouse (Fabric Filter unit).
'************************************************************
'
'
COLOR 15, 1, 4
CLS
title:
DO
  opt = 0
  PRINT "Exercise 7-27:  A program to aid in the design and operation"
  PRINT "of a baghouse (Fabric Filter Unit)  "
  PRINT : PRINT "The formulas:  "
  PRINT "A[f] = Q / v[f] "
  PRINT "A[b] = pi d h"
  PRINT "N = A[f] / A[b]"
  PRINT "Delta P = 1.6 v[f] + 8.52 c v[f]{2}t"
  PRINT : PRINT "Type 1 and ENTER to continue  "
  INPUT "Type 2 and ENTER to EXIT  "; opt
LOOP WHILE opt < 1 OR opt > 2
IF (opt = 2) THEN
  END
END IF
screen1:
  CLS
  PRINT "A[f] = Q / v[f] "
  PRINT : INPUT "Enter the volume of contaminated gas stream (Q) (in m{3}/s) "; vol
  INPUT "Enter the filtering velocity (v[f]) (in m/min) "; vel
   volcomp = vol * 60
   af = volcomp / vel
  PRINT : PRINT "A[f] = "; af; " m{2}"
  INPUT "Press ENTER to continue  "; ent$
```

```
screen2:
  PRINT : PRINT "A[b] = pi d h"
  PRINT : INPUT "Enter the diameter of the bag - d - (in m) "; dia
  INPUT "Enter the length of the bag - h - (in m) "; length
    ab = 3.1416 * dia * length
  PRINT : PRINT "A[b] = "; ab; " m{2}"
  INPUT "Press Enter to continue "; ent$
screen3:
  PRINT : PRINT "N = A[f] / A[b]"
  PRINT "All the factors for this equation are"
  PRINT "from the previous equations"
    n = af / ab
  PRINT : PRINT "N = ";
  PRINT USING "#####"; n;
  PRINT " bags"
  PRINT "(Rounded to the nearest whole number)"
  INPUT "Press Enter to continue "; ent$
formulad:
  CLS
  PRINT "Delta P = 1.6 v[f] + 8.52 c v[f]{2}t"
  PRINT : INPUT "Enter the pressure drop  - Delta P - (in cm) "; drop
  PRINT "Enter the particulate concentration in"
  INPUT "the air stream (in kg/m{3}) "; conc
   comp1 = 1.6 * vel
   compvel = vel ^ 2
   comp2 = 8.52 * conc * compvel
   comp4 = drop - comp1
   compt = comp4 / comp2
  PRINT : PRINT "t = "; compt; " minutes"
  PRINT : INPUT "Press Enter to continue "; ent$
  CLS
  GOTO title
END
```

7.3.6 Combustion Processes

Control of gaseous air contaminates can be categorized as adsorption, absorption, condensation, or combustion. Gaseous pollutants can be controlled by a wide variety of devices applying one or more of these basic principles. While combustion is one of the major sources of air pollution, it is also an important air pollution control method.

Combustion is defined as the burning or rapid oxidation of organic (fuel) compounds accompanied by the release of energy in the form of heat and light. Scientifically, the terms combustion and incineration have the same definition and often are used interchangeably.[20] Combustion or incineration can be further categorized as direct-flame combustion, thermal combustion, or catalytic combustion.

Only one catalytic combustion system will be considered here. In 1971, a homogeneous catalytic system for the oxidation of carbon monoxide was developed and was later patented in 1974.[21,22] This catalytic system was further developed, and by 1985, 10 patents were issued in the U.S., Canada, England, and Japan.[23] The first homogeneous and heterogeneous catalytic system (1974) contained palladium (II) salts and copper (II) salts, along with a specified balance of copper (II) halide and a nonhalide copper (II) salt. This catalyst was effective at oxidizing CO to CO_2 and SO_2 and SO_3.

After modifications and further development, this oxidizing catalyst in 1985 was composed of palladium, copper, and nickel on an alumina substrate. The catalyst was produced by impregnating the alumina substrate with a halide salt solution of palladium chloride, nickel chloride, copper chloride, and copper sulfate. The catalyst could remove by oxidation, adsorption, or decomposition such gases as carbon monoxide, hydrogen sulfide, hydrogen cyanide, sulfur dioxide, and ozone present in dilute concentrations in air.[23] See Figure 7.8 for the gaseous contaminants that can be removed by this catalytic system at both room temperatures and elevated temperatures.[2]

This catalytic system has many potential applications, some of which are listed here:[24,25]

- Home appliances (kerosene heaters, stove hoods, etc.)
- Indoor air filtration (home and office)
- Industrial air filtration
- Automotive emissions and passenger compartment (air conditioning unit)
- Home or public parking garages
- Warehouse and dock areas
- Airline industry
- Mine safety equipment
- Fire safety equipment
- Carbon monoxide analytical equipment and monitors
- Cigarette filters

One application of this catalyst was marketed in 1985 by Teledyne Water Pik. One stage of the four-stage "Teledyne Water Pik Instapure Filtration System" used this catalyst for the removal of indoor gaseous contaminants.[26]

Not only is the percent reduction of the air contaminant important but also parameters such as contact time and the reaction rate coefficient. To calculate the percent reduction of the air contaminant by the catalyst, Equation 7.44 can be used.

$$\% \text{ removal} = \frac{C_o - C_e}{C_0} \times 100 \tag{7.44}$$

where:

C_o = Initial air contaminant concentration, μg/m^3
C_e = Effluent air contaminant concentration, μg/m^3

The contact time can be determined by using Equation 7.45.

$$t = \frac{S_c}{f_r} \tag{7.45}$$

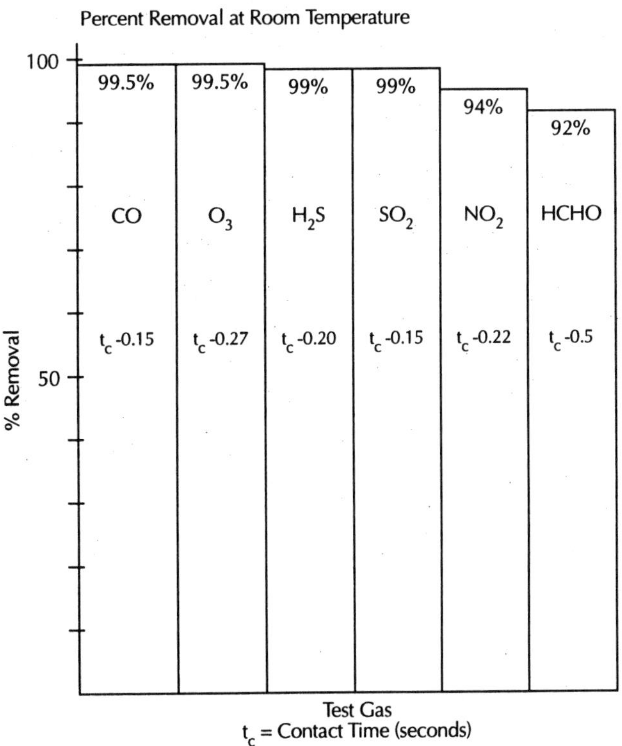

FIGURE 7.8
(A) Teledyne's room-temperature catalyst at room temperature effectively removes carbon monoxide (CO), ozone (O_3), hydrogen sulfide (H_2S), sulfur dioxide (SO_2), formaldehyde (HCHO), and nitrogen dioxide (NO_2). (B) Teledyne's room-temperature catalyst at elevated temperature effectively removes hydrogen cyanide (HCN), propane (C_3H_8), methane (CH_4), and isobutane (C_4H_{10}). (From *Teledyne Technology Introduces a New Room Temperature Catalyst,* Form 1007-F-33, Teledyne Water Pik Research & Development Laboratory, Ft. Collins, CO, 1983. With permission.)

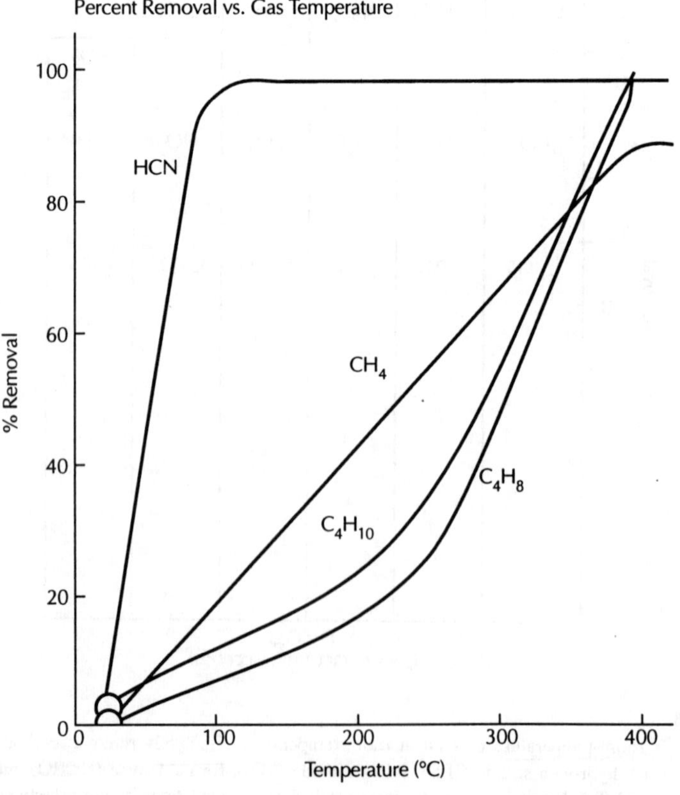

Percent Removal vs. Gas Temperature
100
80
60
40
20
0
0
100
200
300
400
Temperature (°C)
% Removal
HCN
CH4
C4H10
C4H8

FIGURE 7.8B

where:

t = Contact time, s
s_c = Space around the catalyst, mL
f_r = Contaminated gas flow rate, mL/s

A mathematical model (first-order reaction) proved valuable in evaluating the various catalysts. This mathematical model was verified by testing various catalytic systems using a fixed quantity of catalyst and a constant initial air contaminant concentration (C_0), but the gas flow was varied, which in turn varied the contact time (t). The data from these tests yielded a straight line when C_e/C_0 vs. t was plotted on semi-log paper. This indicated the air contaminant removal by the catalyst was a first-order reaction.

Equation 7.46 is the mathematical model describing this process.

$$\frac{C_e}{C_o} = e^{-kt} \tag{7.46}$$

where:

e = Base of natural logarithm, 2.71828
k = Reaction rate coefficient, s^{-1}
t = Catalytic contact time, s

C_o and C_e are the same as previously indicated.

In general, the longer the contact time with the catalyst the greater the reduction in the air contaminant levels. For most catalytic systems the contact time needs to be on the order of a few hundredths of a second.

The reaction rate coefficient k is a single number that measures the overall activity of a catalyst. With equal contact time, the higher the k value the greater the ability of the catalyst to remove air contaminants. It was found that for this catalytic system, k values greater than 10 s^{-1} indicated a catalytic system worthy of further development. The following example demonstrates how Equations 7.44, 7.45, and 7.46 can be used to evaluate a catalyst.

Example 7.28

1. A 2.5-g catalytic sample was tested for its removal efficiency of carbon monoxide (CO) from an air stream. The initial CO concentration was 23,300 µg/m³ (C_o). The CO concentration in the air stream after passing through the catalyst was 1770 µg/m³ (C_e). The airstream flow rate was 700 mL/min. The space around the catalytic support system was 0.6 mL. Determine the percent CO removal, the contact time in seconds, and the reaction rate coefficient.
2. Prepare a short computer program to solve this problem.

Solution

1. A solution to Example 7.28 (1):
   - Percent CO removal can be determined by using Equation 7.44:

$$\% \text{ removal} = \frac{23,330\ \mu g/m^3 - 1,770\ \mu g/m^3}{23,300\ \mu g/m^3} \times 100 = 92.4\%$$

   - Contact time can be calculated by using Equation 4.45:

$$\text{contact time} = \frac{0.6\ \text{mL}}{700\ \text{mL/min}/\ 60\ \text{s/min}} = 0.051\ \text{s}$$

   - Reaction rate coefficient can be determined by using Equation 7.46:

$$\frac{C_e}{C_o} = e^{-kt}$$

$$\frac{1,770\ \mu g/m^3}{23,300\ \mu g/m^3} = e^{-k(0.051\,\text{sec})}$$

   which yields K = 50.6 s^{-1}.

   - This would indicate that with 92.4% removal of CO at a contact time (t) of 0.051 s and a reaction rate coefficient (k) of 50.6 s^{-1}, further development of this catalyst should be considered.

2. For a solution to example 7.28 (2), see following listing of Program 7.28.

Listing of Program 7.28

```
'****************************************************************
'Example 7-28: A program that can be used to evaluate a
' catalytic system.
'****************************************************************
'
'
COLOR 15, 1, 4
CLS
title:
DO
  opt = 0
  PRINT "Exercise 7-28:  A program that can be used to evaluate"
  PRINT "a catalytic system."
  PRINT : PRINT "The formulas:  "
  PRINT "% Removal = ((C[o] - C[e]) / C[o]) * 100 "
  PRINT "t = s[c] / f[r]"
  PRINT "C[e] / C[o] = e{-kt}"
  PRINT : PRINT "Type 1 and ENTER to continue  "
  INPUT "Type 2 and ENTER to EXIT  "; opt
LOOP WHILE opt < 1 OR opt > 2
IF (opt = 2) THEN
```

```
  END
END IF
screen1:
  CLS
  PRINT "% Removal = ((C[o] - C[e]) / C[o]) * 100 "
  PRINT : INPUT "Enter the initial concentration (C[o]) (in um/m{3} "; co
  INPUT "Enter the concentration after the catalyst (C[e]) (in um/m{3}) "; ce
    perc = ((co - ce) / co) * 100
  PRINT : PRINT "The percentage of removal = "; perc
  PRINT : INPUT "Press ENTER to continue  "; ent$
screen2:
  PRINT : PRINT "t = s[c] / f[r]"
  PRINT : INPUT "Enter the space around the catalyst (s[c]) (in ml) "; space
  INPUT "Enter the air stream flow rate (f[r]) (in ml/min) "; rate
    time = space / (rate / 60)
  PRINT : PRINT "The contact time is = "; time; " sec"
  PRINT : INPUT "Press Enter to continue "; ent$
screen3:
  CLS
  PRINT "C[e] / C[o] = e{-kt}"
  PRINT : PRINT "All the factors for this equation are"
  PRINT "from the previous equations"
   acomp = LOG(ce / co)
   k = acomp / time * -1
  PRINT : PRINT "k = "; k; "sec{-1}"
  PRINT : INPUT "Press Enter to continue "; ent$
  GOTO title
  CLS
END
```

Example 7.29

1. Prepare a computer program that can be used to evaluate a catalytic system including the percentage reduction of an air contaminant, contact time, and reaction rate coefficient. This program should incorporate all the elements in Equation 7.44, 7.45, and 7.46.
2. Using the computer program developed in (1), indicate which of the following catalysts should be considered for further development. The air space around the catalyst was found to be 0.24 mL per g of catalyst.

Catalyst	Weight of catalyst	Flow rate (mL/s)	CO initial (CO_o) (μm/m³)	CO final (C_e) (μm/m³)
1	0.5	20	33,700	26.600
2	1.0	13	30,490	14,670
3	1.5	4	13,210	3100
4	2.0	12	19,720	1610

Solution

1. For solution to Example 7.29 (1), see the following listing of Program 7.29, which includes all the elements in Equations 7.44, 7.45, and 7.46.

2. Solution to Example 7.29 (2):

Catalyst	Removal (%)	Contact time (s)	Reaction rate constant k (s^{-1})
1	21	0.006	39.4
2	52	0.018	39.6
3	77	0.09	16.1
4	92	0.04	62.6

Of these four catalysts, number 4 shows the greatest promise for further development.

Listing of Program 7.29

```
'*************************************************************
'Example 74-29: A program that can be used to evaluate a
' catalytic system. Evaluates multiple catalysts.
'*************************************************************
'
COLOR 15, 1, 4
CLS
DIM cat(15), wgt(15), flow(15), initial(15), final(15)
title:
DO
  opt = 0
  PRINT "Exercise 7-29:  A program that can be used to evaluate"
  PRINT "a catalytic system.  Evaluates multiple catalysts"
  PRINT : PRINT "The formulas:  "
  PRINT "% Removal = ((C[o] - C[e]) / C[o]) * 100 "
  PRINT "t = s[c] / f[r]"
  PRINT "C[e] / C[o] = e{-kt}"
  PRINT : PRINT "Type 1 and ENTER to continue  "
  INPUT "Type 2 and ENTER to EXIT  "; opt
LOOP WHILE opt < 1 OR opt > 2
IF (opt = 2) THEN
  END
END IF
  subsc = 1
  endsubsc = 0
screen1:
DO
  CLS
  PRINT "Enter the number of the catalyst "
  INPUT "(Press ENTER without a number entered if you are done)"; cat(subsc)
    IF (cat(subsc) = 0) THEN EXIT DO
  INPUT "Enter the weight of the catalyst "; wgt(subsc)
  INPUT "Enter the flow rate (in ml/sec) "; flow(subsc)
  INPUT "Enter the initial concentration of the gas (in um/m{3}) "; initial(subsc)
  INPUT "Enter the final concentration of the gas (in um/m{3}) "; final(subsc)
    subsc = subsc + 1
    endsubsc = subsc
LOOP UNTIL (subsc = 15)
  CLS
```

```
  INPUT "Enter the air space around the catalyst (in ml per gram) "; space
subsc = 1
DO
    aspace(subsc) = space * wgt(subsc)
    perc(subsc) = ((initial(subsc) - final(subsc)) / initial(subsc)) * 100
    ctime(subsc) = aspace(subsc) / flow(subsc)
    comp1(subsc) = LOG(final(subsc) / initial(subsc))
    rrate(subsc) = (comp1(subsc) / ctime(subsc)) * -1
    subsc = subsc + 1
LOOP UNTIL (subsc = endsubsc)
  INPUT "Press ENTER to continue  "; ent$

  CLS
  PRINT TAB(27); "Space"; TAB(48); "Initial"; TAB(60); "Final"
  PRINT TAB(16); "Wt of"; TAB(27); "around"; TAB(40); "Flow"; TAB(48); "conc."; TAB(60); "conc."
  PRINT TAB(6); "Catalyst"; TAB(16); "Catalyst"; TAB(27); "Catalyst"; TAB(40); "Rate"; TAB(48); "um/m{3}";
TAB(60); "um/m{3}"
   tabval = 10
   subsc = 1
DO
  PRINT TAB(8); cat(subsc); TAB(18); wgt(subsc); TAB(28); aspace(subsc); TAB(41); flow(subsc); TAB(48);
initial(subsc); TAB(60); final(subsc)
    subsc = subsc + 1
    tabval = tabval + 1
LOOP UNTIL (subsc = endsubsc)
  INPUT "Press ENTER to continue  "; ent$
  subsc = 1
  CLS
  PRINT TAB(20); "Removal"; TAB(30); "Contact Time"; TAB(45); "Reaction Rate"
  PRINT TAB(8); "Catalyst"; TAB(20); "%"; TAB(30); "(sec)"; TAB(45); "Constant k(sec{-1})"
   tabval = 9
DO
  PRINT TAB(10); cat(subsc);
  PRINT TAB(22); USING "###."; perc(subsc);
  PRINT TAB(32); USING "###.###"; ctime(subsc);
  PRINT TAB(47); USING "###.##"; rrate(subsc)
    subsc = subsc + 1
    tabval = tabval + 1
LOOP UNTIL (subsc = endsubsc)
  PRINT : INPUT "Press ENTER to continue  "; ent$
  CLS
  GOTO title
END
```

7.3.7 Packed Column Absorption Towers

Absorption is a process by which a liquid (absorbent) is used to remove one or more soluble gases (absorbate) from a contaminated air stream. The liquid absorbent may react chemically or physically with the absorbate to remove it from the gas stream. Typical absorbents are water or dilute basic or acidic solutions.

Chemical absorption involves a liquid absorbent reacting with a pollutant to yield a nonvolatile product. A typical process here is the reaction of SO_2 and aqueous H_2O_2 to produce sulfuric acid. Physical absorption involves the physical dissolving of the pollutant in a liquid. This process is generally reversible. The ideal absorbent would be relatively nonvolatile, inexpensive, noncorrosive, stable, nonviscous, non-flammable, and nontoxic. In many cases, distilled water fulfills many of these characteristics.[6] The types of absorption equipment used for air pollution control include:

- Spray towers
- Spray chambers
- Venturi scrubbers
- Packed column (towers)

The amount of absorbate (gas) that dissolves in the absorbent (liquid) depends upon the properties of the absorbent and the absorbate, the absorbate (gas) concentration, the partial pressure of the pollutant, the temperature of the system, the turbulence, the flow rate, and the type of air pollution control equipment used. For dilute solutions, as is the usual case in air pollution control, the relationship between partial pressure and the concenteration of the absorbate (gas) in the absorbent (liquid) is given by Henry's law (see Section 7.2.6).

Only the packed column absorption type of air pollution control equipment will be considered here. One of the most efficient ways to contact the absorbent with the absorbate is the countercurrent flow system shown in Figure 7.9A.[27] In this system, the gas and liquid run in opposite directions. The absorbate (gas) enters at the bottom and the absorbent (liquid) at the top of the column. With this arrangement, the high concentration pollutant (gas) is absorbed into the absorbent (liquid) with a high pollutant concentration, and the lower concentration pollutant (gas) is absorbed into the absorbent (liquid) which has no contaminants present. As indicated by Henry's law, the concentration of a gas dissolved in a solvent under equilibrium conditions is proportional to the partial pressure of the gas above the solution at constant temperature.

Applying this principle from the bottom to the top of the packed tower makes it possible to determine the mole fraction of the pollutant (absorbate) in the gas phase in equilibrium with the mole fraction of the pollutant (absorbate) in the liquid phase. When this data is plotted with the gas mole fraction on the ordinate (y-axis) and the liquid mole fraction on the abscissa (x-axis), an equilibrium line is produced (see Figure 7.9B).

For absorption to take place, the operating line must be above the equilibrium line. The vertical distance between the equilibrium line and the operating line indicates the degree of saturation of the liquid with the pollutant (see Figure 7.9B). The operating line in this case is considered to be a straight line, and an equation depicting this line follows:

$$L_1\left(\frac{X_A}{1-X_A}-\frac{X_B}{1-X_B}\right)=G_1\left(\frac{Y_A}{1-Y_A}-\frac{Y_B}{1-Y_B}\right) \tag{7.47}$$

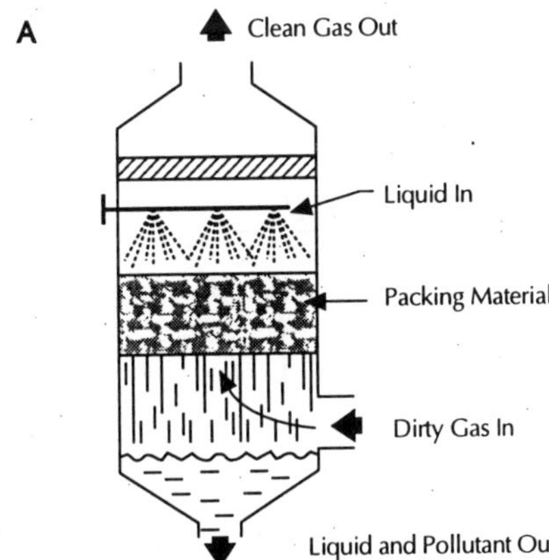

FIGURE 7.9
(A) Counter-current packed column: typical counter-current packed-column absorber system. (B) Counter-current packed column. (C) Counter-current packed column: graphical representation of the packed column.

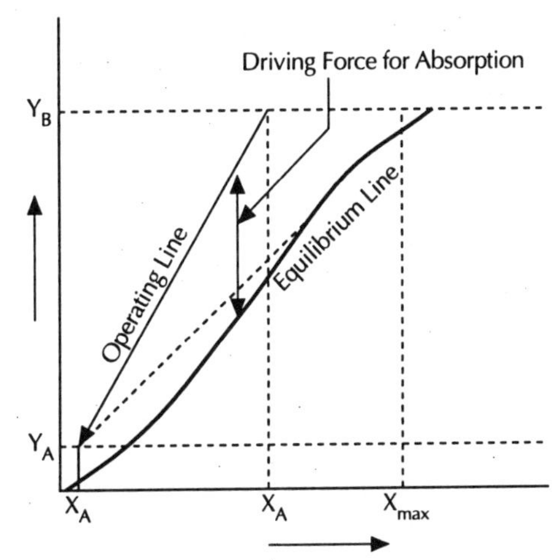

where:

L_1 = Liquid flow, kg*mol/h
G_t = Gas flow, kg*mol/h
X_A = Mole fraction of the pollutant in the liquid phase into the column
X_B = Mole fraction of the pollutant in the liquid phase out of the column
Y_A = Mole fraction of the pollutant in the gas phase out of the column
Y_B = Mole fraction of the pollutant in the gas phase into the column

See Figure 7.9C for a graphical representation of these terms.

The following example demonstrates how Equation 7.47 can be used to evaluate and aid in the design of packed columns used for air pollution control. While the

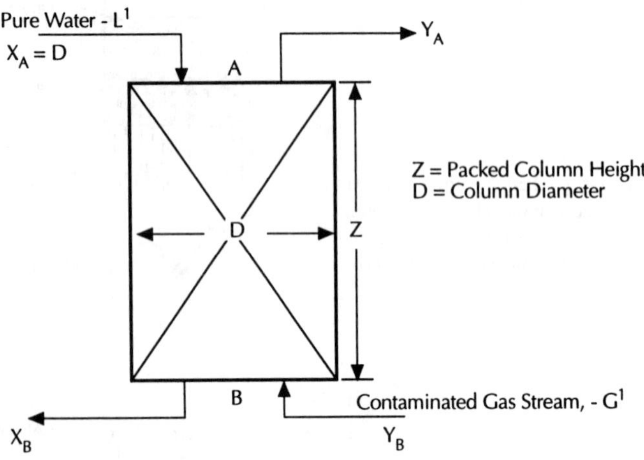

FIGURE 7.9C

following example illustrates some of the principles used for packed column evaluation and design, there are many other factors that must be taken into account, such as the number of transfer units, the height of an overall gas transfer unit, the type of packing, and the required packed column height and diameter. However, it is not possible to go into all these functions and formula here.

Example 7.30

A 500-kg/h air stream contains a mixture of acetone in air (CH_3–CO–CH_3; molecular weight of acetone = 58.08, molecular weight of air at 25°C = 28.970) to be treated in a countercurrent packed column absorber which has a diameter of 0.45 m and is packed with Raschig rings. The equilibrium equation was found to be $Y_B = 2.53X_B$, where Y_B = mole fraction of acetone in air and X_B = mole fraction of acetone in water.

The operating temperature is 25°C at 1 atm pressure. The acetone in the air stream entering at the bottom of the column is 50 mg/L, and it is required that the allowable level of acetone leaving the column be reduced to 2.5 mg/L. What is the required flow rate of water in kg/h (acetone free, X_A = 0) to lower the acetone concentration to the desired level?

Solution

- Inlet acetone concentration = 50 mg/L = 0.050 kg/m^3.
- Air at 25°C has a density of 1.185 kg/m^3:

$$Y_B = \frac{\dfrac{0.050\ \text{kg}/\text{m}^3}{58\ \text{MW}-\text{acetone}}}{\dfrac{1.185\ \text{kg}/\text{m}^3}{28.970\ \text{MW of air}}} = \frac{0.000862}{0.0409} = 0.021$$

- Outlet acetone concentration = 2.5 mg/L = 0.0025 kg/m:

$$Y_A = \frac{\dfrac{0.025\ \text{kg}/\text{m}^3}{58\ \text{MW}-\text{acetone}}}{\dfrac{1.185\ \text{kg}/\text{m}^3}{28.970\ \text{MW of air}}} = \frac{0.0000431}{0.0409} = 0.001$$

- The incoming distilled water has no acetone present, so:

$$X_A = 0$$

$$Y_B = 2.53X_B$$

$$X_B = \frac{0.021}{2.53} = 0.0083$$

- Substituting these values in Equation 7.47 gives:

$$L_1\left(\frac{0}{1-0} - \frac{0.0083}{1-0.0083}\right) = G_1\left(\frac{0.001}{1-0.001} - \frac{0.021}{1-0.021}\right)$$

which yields:

$$\frac{L_1}{G_1} = \frac{-0.0204}{-0.00837} = 2.437$$

- The number of moles of inlet air based on the following air-acetone mixture is calculated as follows:

Acetone (0.021×58.08) = 1.22
Air (0.979×28.970) = <u>28.36</u>
Combined mixtures MW = 29.58

- The kg/mol of air = $0.979 \times \dfrac{500\ \text{kg/h}}{29.58}$ = 16.55 kg*mol/h:

$$L_1 = 2.437G$$

$$G_1 = 16.55\ \text{kg} * \text{mol}/\text{h}$$

$$L_1 = 2.437(16.55)\ \text{kg} * \text{mol}/\text{h of } H_2O = 40.33\ \text{kg} * \text{mol}/\text{h of } H_2O$$

$$1\ \text{kg} * \text{mol of } H_2O = 18\ \text{kg}$$

- Water flow rate = 40.33 × 18 = 726 kg/h

Example 7.31

1. Prepare a computer program that incorporates all the elements in Equation 7.47 and thus provides a means to evaluate or aid in the design of packed columns used for air pollution control.
2. Using the computer program prepared in (1) and given the following data, determine the liquid flow required to remove 97% of SO_2 from the stack gas of a coal fired furnace.

   Column diameter = 3.00 m

   Operating temperature = 25°C

   Operating pressure = 1 atm

   Gas flow rate = 30,000 m³/h

   SO_2 inlet concentration = 20,000 ppm

   Incoming liquid is pure water, $X_A = 0$

Solution

1. For solution to Example 7.31 (1), see the following listing of Program 7.31, which includes all the elements in Equation 7.47.
2. - Convert 20,000 ppm SO_2 to μg/m³ using Equation 7.1 at STP:

$$\mu g/m^3 = \frac{20,000\ \text{ppm} \times 64 \times 10^3}{24.46\ \text{L/mol}} = 52,330,335\ \mu g/m^3$$

$$= 52.3\ g/m^3 = 0.0523\ kg/m^3$$

Inlet concentration SO_2 = 0.0523 kg/m³.

- Air at 25°C has a density of 1.185 kg/m³:

$$Y_B = \frac{\dfrac{0.0523\ kg/m^3}{64\ \text{MW of } SO_2}}{\dfrac{1.185\ kg/m^3}{28.970\ \text{MW of air}}} = \frac{0.0008171}{0.04090} = 0.02$$

- Outlet SO_2 concentration = 0.03 × 0.0523 kg/m³ = 0.001569.

$$Y_A = \frac{\dfrac{0.001569\ kg/m^3}{64\ \text{MW of } SO_2}}{\dfrac{1.185\ kg/m^3}{28.970\ \text{MW of air}}} = \frac{0.0000245}{0.04090} = 0.0006$$

- $X_A = 0$ for the incoming distilled water.

- Given $Y_B = 30X_B$,

$$X_B = \frac{0.02}{30} = 0.00067$$

- Substituting these values in Equation 7.47 gives:

$$L_1\left(\frac{0}{1-0} - \frac{0.00067}{1-0.00067}\right) = G_1\left(\frac{0.0006}{1-0.0006} - \frac{0.02}{1-0.02}\right)$$

which yields:

$$\frac{L_1}{G_1} = \frac{-0.0198}{-0.00067} = 29.69$$

- The kg*mol of air is 30,000 m³/h × 1.185 kg/m³ = 35,550 kg/h.
- Inlet gas composition:
  0.98 = percentage air
  0.02 = percentage SO_2
- kg*mol of air is 0.98 × $\frac{35,550 \text{ kg/h}}{28.970 \text{ MW of air}}$ = 1202.6 kg*mol/h.
- $L_1 = 29.69G_1$.
- $L_1 = 29.69(1202.6) = 35,707$ kg*mol/h of H_2O.
- 1 kg*mol of water = 18 kg.
- Water flow rate = 35,707*18 = 642,729 kg/h, or 178 kg/s.

Listing of Program 7.31

```
'*****************************************************************
'Example 7-31: A program that can be used to evaluate or aid
' in the design of packed columns absorption towers used for air
' pollution control.
'*****************************************************************
'
'

COLOR 15, 1, 4
CLS
title:
DO
  opt = 0
  PRINT "Exercise 7-31: A program that can be used to evaluate"
  PRINT "or aid in the design of packed columns absorption towers"
  PRINT "used for air pollution control."
  PRINT : PRINT "The formula: "
  PRINT "L[l] * ((X[a] / (1 - X[a]) - (X[b] / 1 - X[b])) = "
  PRINT "G[l] * ((Y[a] / (1 - Y[a]) - (Y[b] / 1 - Y[b]))  "
  PRINT : PRINT : PRINT "Type 1 and ENTER to continue  "
  INPUT "Type 2 and ENTER to EXIT  "; opt
LOOP WHILE opt < 1 OR opt > 2
  IF (opt = 2) THEN
  END
```

```
END IF
screen1:
  CLS
  INPUT "Enter the inlet concentration of the gas (in ppm) "; inconc
  INPUT "Enter the molecular weight of the gas "; gasmolwgt
  INPUT "Enter the air density (in kg/m{3}) "; airdens
  INPUT "Enter the molecular weight of the air "; airmolwgt
  INPUT "Enter the percent removal of the gas (in %) "; outconc
  PRINT "The incoming distilled water has none of the gas,"
  PRINT "   therefore   X[A] = 0"
screen2:
    inconck = inconc *  gasmolwgt*10^-6/24.46
    outconck = inconc * (1 - (inconc/100))
    yb1 = inconck / gasmolwgt
    yb2 = airdens / airmolwgt
    yb = yb1 / yb2
    ya1 = outconck / gasmolwgt
    ya = ya1 / yb2
    xa = 0
  PRINT : PRINT "Enter the value for the # in the following given"
  INPUT "   equilibrium equation - Y[B]  = # X[B] "; given
    xb = yb / given
    compl1 = (xa / (1 - xa)) - (xb / (1 - xb))
    compg1 = (ya / (1 - ya)) - (yb / (1 - yb))
    compdl1 = compg1 / compl1
    gas = yb * gasmolwgt
    compair = 1 - yb
    air = compair * airmolwgt
    combmolwgt = gas + air
  INPUT "Enter the gas flow rate (in m3/hr) "; strea
     stream = strea*airdens
    g1 = compair * (stream / airmolwgt)
  PRINT : PRINT "A kg*mole of water = 18 kg "
  PRINT : INPUT "Press Enter to continue "; ent$
  PRINT : PRINT "L[1] = "; compdl1; "G"
  PRINT "G[1] = "; g1; "kg.mole/hr of H[2]O"
    l1 = compdl1 * g1
    flowrate = l1 * 18
  PRINT "L[1] = "; l1; " kg.mole of water "
  PRINT "Water flow rate = "; flowrate; " kg/hr"
  PRINT : INPUT "Press Enter to continue "; ent$
  CLS
  GOTO title
END
```

7.3.8 Adsorption

Adsorption is a phenomenon by which gases, liquids, and solutes within liquids are attracted, concentrated, and retained at a boundary surface.[6] In air pollution control, this process involves the retention of molecules from the gas phase onto a solid surface. A typical process consists of passing a contaminated gas stream through a container filled with an absorbent such as activated carbon, activated alumina, silica gel, acid-treated clay, molecular sieve, Fuller's earth, or magnesia.

In the air pollution field, gas adsorption can be used to remove volatile organic compounds (VOCs), H_2S, SO_2, and NO_2 from contaminated air streams. In industry, adsorption is used to recover valuable volatile solvents such as benzene, ethanol, trichloroethylene, and freon.

Adsorption involves either physical adsorption (physiosorption) or chemical adsorption (chemisorption). With physical adsorption, the attractive forces consist of Van der Waal's interactions, dipole-dipole interactions, and/or electrostatic interactions.[6] An example of physical adsorption is when a contaminant in a gas stream is adsorbed onto activated carbon. In physical adsorption, the adsorbed layer generally is considered to be several molecules thick (multilayer adsorption).

In chemical adsorption the contaminant gas molecules form a chemical bond with the adsorbent, and the gas (adsorbate) is held strongly to the solid surface (adsorbent) by valence forces. An example of chemical adsorption is the adsorption of oxygen by activated charcoal to form carbon monoxide (CO) and carbon dioxide (CO_2). The CO and CO_2 are not easily removed from the activated carbon and this can only be accomplished at elevated temperatures. Chemical adsorption is usually limited to the formation of a single layer of molecules on the surface of the adsorbent (monolayer adsorption).[6] The variables affecting gas adsorption are

- Concentration of the contaminant in the gas stream
- Surface area of the adsorbent
- Temperature of the system
- Presence of other competing contaminant gases
- Characteristics of the adsorbent, such as weight, electrical polarity, and chemical reactivity

The relationship between the amount of pollutant adsorbed by the adsorbent under equilibrium conditions and at a constant temperature is called an adsorption isotherm. An isotherm is a graphical representation of the adsorbent's capacity vs. the partial pressure of the absorbate (contaminant) at a particular temperature.[20] Adsorption isotherms are useful in that they provide a means of evaluating:

- Quantity of gas adsorbed at various concentrations
- Adsorption capacity of the adsorbent at various gas concentrations
- Adsorption capacity of the adsorbent for various gases
- Surface areas for a given amount of adsorbent

A graphical plot of the amount of the gas absorbed per gram of adsorbent at various gas concentrations at equilibrium and under conditions of constant temperature produces isotherms with many shapes. Specific isotherms have been developed by Freundlich, Langmuir and Brunauer, Emmett, and Teller (BET).

Only the Freundlich isotherm will be considered here. The Freundlich isotherm is a purely empirical equation valid for monomolecular physical and chemical adsorption. The equation is

$$\frac{X}{M} = KC^{\frac{1}{n}} \tag{7.48}$$

where:

X/M = Mass (X) of the element or contaminant adsorbed from solution per unit mass of adsorbent (M)

K, n = Constants fitted from the experimental data

C = Concentration of the contaminant in the gas phase at equilibrium

By taking the logarithm of both sides, this equation is converted to a linear form.

$$\log \frac{X}{M} = \log K + \frac{1}{n} \log C \tag{7.49}$$

If the experimental data fits the Freundlich adsorption isotherm, a plot of log X/M vs. log C gives a straight line as shown in Figure 7.10. If a vertical line is erected from a point on the horizontal scale corresponding to the initial contaminant concentration (C_0) and the isotherm extrapolated to intersect that line, the X/M value at this point of intersection can then be read from the vertical scale. The value of $(X/M)_{C0}$ represents the amount of contaminant adsorbed per unit weight of adsorbent when that adsorbent is in equilibrium with the initial contaminant concentration. This represents the ultimate sorption capacity of the adsorbent for that contaminant.[28]

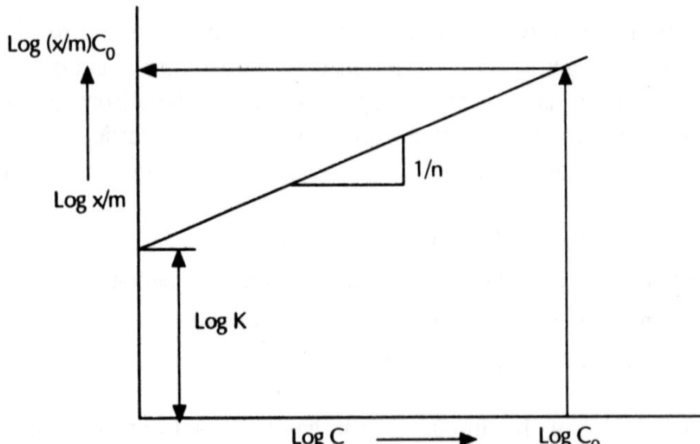

FIGURE 7.10
Straight-line form of the Freundich adsorption isotherm.

The Freundlich equation is most useful for dilute solutions over small concentration ranges. The 1/n value represents the slope or change in rate of effectiveness in uptake with varying amounts of adsorbent, and K, the ordinate intercept, the fundamental effectiveness of the adsorbent. High K and high n values indicate high adsorption capacities; low K and high n values indicate low adsorption; low n values or a steep slope indicate high adsorption at high contaminant levels and low adsorption at low contaminant levels.[29] Both graphical and computer analyses can be made for the Freundlich adsorption isotherm.[30]

In the air pollution control field, adsorption systems can be either nonregenerative or regenerative. They also can be fixed, moving, or fluidized beds. Nonregenerative adsorptions systems have thin beds and are economical only when the contaminant level in the gas stream is in the $\mu g/m^3$ or ppm range. Regenerative beds are generally thick and are designed to handle considerable contaminant loadings with the additional advantage of the recovery of a valuable solvent. (See Figures 7.11A and B.)

Adsorption of a contaminant from an air stream continues until the bed capacity has been reached; at this point the adsorbent is saturated with the adsorbate. The concentration of the contaminant in the exit gas stream begins to rise rapidly, and the adsorber must be regenerated or disposed of. This point in the adsorption system is called the breakthrough capacity of the bed.

Following is an equation that can be used to calculate the breakthrough capacity of an adsorption bed:[9]

$$C_B = \frac{0.5C_s(MTZ) + C_s(D - MTZ)}{D} \tag{7.50}$$

where:

C_B	=	Breakthrough capacity, fractional
C_s	=	Saturation capacity, fractional
MTZ	=	Mass transfer zone, cm
D	=	Adsorption bed depth, cm

The degree of saturation of an adsorption bed is defined as follows:[31]

$$C_s = \frac{WAE}{WAT} \tag{7.51}$$

where:

C_s	=	Saturation capacity, fractional
WAE	=	Weight of adsorbate, g or kg
WAT	=	Weight of adsorbent, g or kg

When an adsorbent bed is regenerated, it is not economical to remove absolutely all the adsorbate (contaminant). The residual contaminant in the adsorbent after regeneration is called the "heel". The practical capacity or working capacity of an adsorbent can be computed as follows:[9]

$$W_c = C_B - H - PF \tag{7.52}$$

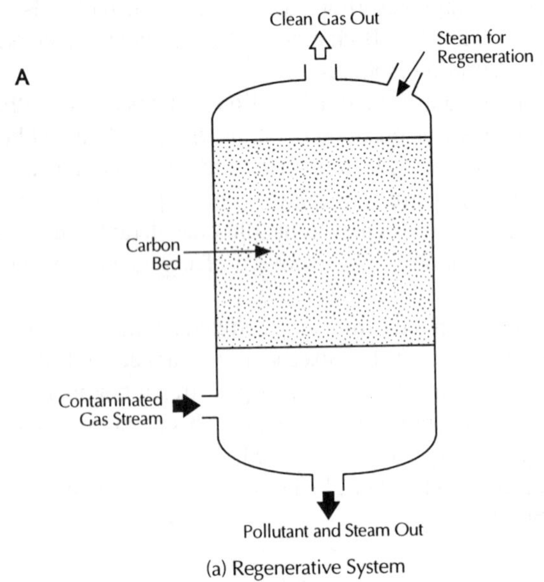

FIGURE 7.11
(A) Schematic of a regenerative adsorption type system. (From Lloyd, W.G. and Rowe, D.R., *Environ. Sci. Technol.*, 5, 11, 1971. With permission.) (B) Schematic of a nonregenerative adsorption type system (From Corman, R.D. and Black, D., *Controlling Air Pollution*, American Lung Association, Washington, D.C., 1974.With permission.)

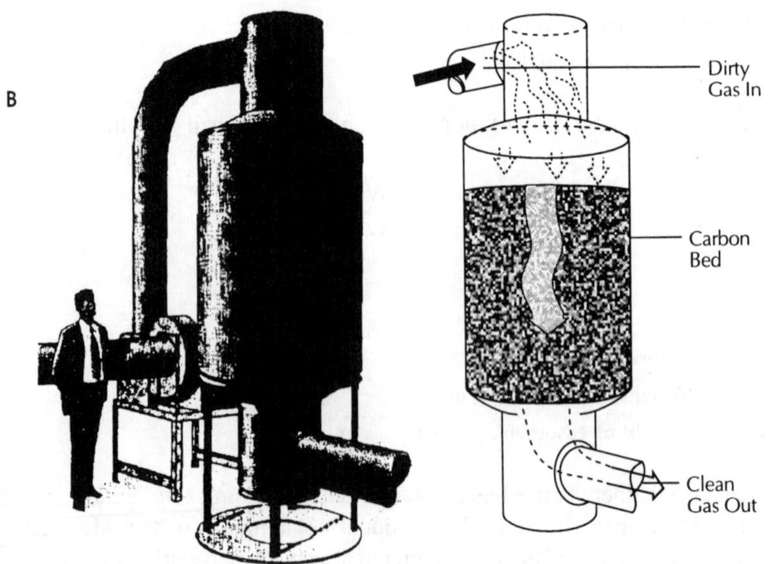

where:

W_c = Working capacity of adsorbent, fractional
H = Heel, fractional
PF = Packing factor, fractional

The working capacity is lower than the breakthrough or saturation capacity. The packing factor is determined experimentally for each adsorbent. The time required before an adsorbent needs to be regenerated is an important parameter in the operation of an adsorption system. An estimate of the time an adsorbent can be left in operation can be computed by using the following equation:[31]

$$t = \frac{2.41 \times 10^7 SW}{EQMC} \tag{7.53}$$

where:

t = Duration of adsorbent service before saturation, h
S = Proportionate saturation of sorbent, fractional
W = Weight of adsorbent, kg
E = Sorption efficiency, fractional
Q = Airflow rate through sorbent bed, m^3/h
M = Average molecular weight of sorbed vapor, g/mol
C = Entering vapor concentration, ppm by volume

Following are examples using Equations 7.48 to 7.53.

Example 7.32

Given the following data for the uptake or adsorption of benzene by activated carbon from a gas stream, determine if this data fits the Freundlich isotherm; if so, then determine the K and n values as well as the ultimate adsorption capacity of the activated carbon $(X/M)_{C_0}$ for benzene at its initial concentration.

Activated carbon (g) (M)	Residual benzene concentration (g/m^3) (C_0 = 30)	Benzene adsorbed (g) (X)	(X/M)
0	—	—	—
200	20	11	0.055
300	15	15	0.050
500	11	19	0.038
650	8	22	0.034
900	5	25	0.028
1600	2	28	0.018

Solution

The data plots a straight line on log-log paper, indicating the data fits the Freundlich isotherm (see the following graph). The K and 1/n constants are 0.012 and 0.52, respectively. The ultimate capacity of the activated carbon for benzene adsorption at 30 g/m^3 is 0.07 g/g, or 7 g of benzene can be adsorbed by 100 g of this activated carbon.

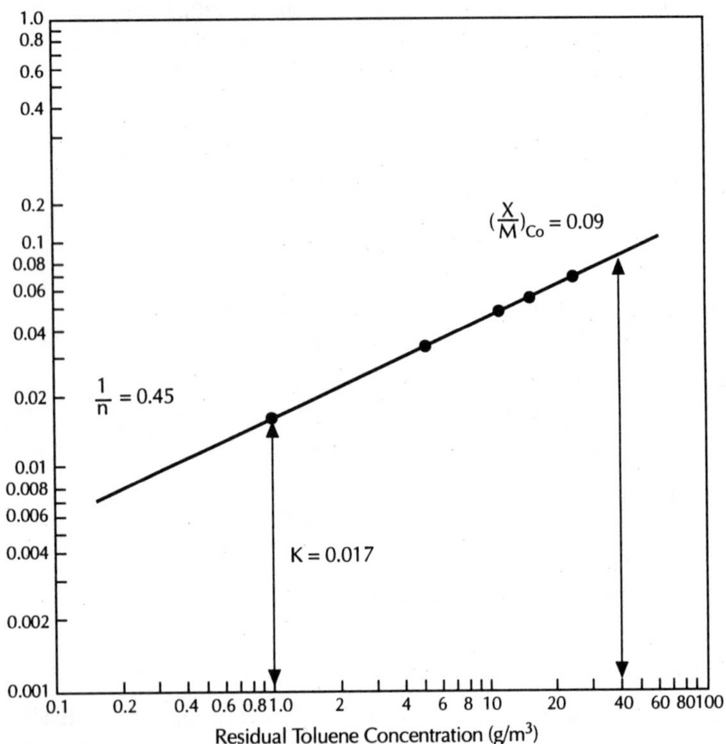

Example 7.32

Example 7.33

1. Prepare a computer program incorporating all the elements in Equation 7.48 (Freundlich isotherm) that makes it possible to determine the constants K and n as well as the adsorbent's ultimate capacity $(X/M)_{C0}$.
2. Using the computer program developed in (1), determine the K and n values as well as the ultimate capacity of this activated carbon $(X/M)_{C0}$ for toluene adsorption.

Solution

1. For solution to Example 7.33 (1), see following listing of Program 7.33.
2. This data plots a straight line on log-log paper indicating it follows the Freundlich isotherm (see the following graph). The K and 1/n constants are 0.017 and 0.45, respectively. The ultimate capacity of this activated carbon for toluene adsorption at 40 g/m^3 was 0.09 g/g, or 9 g of toluene can be adsorbed on 100 g of activated carbon.

Activated carbon (g) (M)	Residual toulene concentration (g/m³) (C_0 = 40 g/m³)	Toulene adsorbed (g) (X)	(X/M)
100	32	8	0.08
240	24	16	0.067
430	15	25	0.058
600	11	29	0.048
1000	5	35	0.035
2500	1	39	0.016

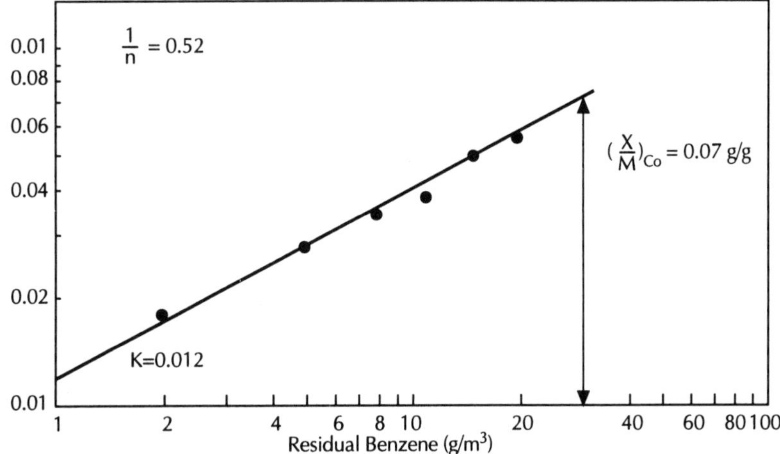

Example 7.33

Listing of Program 7.33

```
'************************************************
'Example 7.33
'Freundlich Adsorption Isotherm:
' A retyped version from the older program.
'************************************************
'
COLOR 15, 1, 4
CLS
PRINT "Exercise 7-33:  A program to compute using the Freundlich"
PRINT "              Adsorption Isotherm"
PRINT : PRINT :
DIM M(15), C(15), X(15), Q(15): Z = 2.302585093#
  INPUT "Name of Adsorption material:  "; N$
  INPUT "Name of Contaminant material:  "; C$
  INPUT "Number of measurements:  "; T
  INPUT "Initial Concentration (g/m{3}):  "; CO
```

```
  FOR I = 1 TO T: PRINT
  PRINT "Measurement #"; I; " grams of "; N$; " "; : INPUT M(I)
  INPUT "g/m{3} of Contaminant remaining:  "; C(I)
    X(I) = CO - C(I)
    Q(I) = X(I) / M(I)
    NEXT I
  PRINT
  INPUT "Press ENTER to continue "; ent$
  CLS
' Print data
  PRINT TAB(1); " [M]"; TAB(16); " [C]"; TAB(29); " [X]"; TAB(43); "[X/M]"
  PRINT TAB(11); "Concentration"; TAB(28); "Amount"
  PRINT TAB(13); "remaining"; TAB(27); "adsorbed"
  PRINT "grams"; TAB(15); "g/m{3}"; TAB(29); "g/m{3}"; TAB(43); "mg/g"
  PRINT "-------"; TAB(11); "----------------"; TAB(27); "----------"; TAB(43); "-----"
  FOR I = 1 TO T
  PRINT M(I), C(I), X(I), Q(I)
  NEXT I
  PRINT : PRINT "Initial Concentration was "; CO; " g/m{3}"
  PRINT : PRINT
  INPUT "Press ENTER to continue "; ent$
  CLS
  PRINT TAB(1); "C"; TAB(10); "log C"; TAB(20); "log X"; TAB(30); "log M"; TAB(40); "log(X/M)"; TAB(
"log(C^2)"; "logC*(log[X/M])": PRINT
  FOR I = 1 TO T
  C1 = LOG(C(I)) / Z: REM  ln (C(I))
  C2 = C1 ^ 2
    IF X(I) = 0 THEN C3 = 0: GOTO step2
  C3 = LOG(X(I)) / Z: REM ln (X(I))
step2:
  C4 = LOG(M(I)) / Z
  C5 = C3 - C4
  C6 = C1 * C5
  A1 = A1 + C1: A2 = A2 + C5: A3 = A3 + C2: A4 = A4 + C6
  PRINT C(I); TAB(8); C1; TAB(20); C3; TAB(30); C4; TAB(40); C5; TAB(50); C2; TAB(65); C6
  NEXT I: PRINT : PRINT : PRINT
  PRINT "sum of log(C)'s:  "; A1
  PRINT "sum of log(X/M)'s:  "; A2
  PRINT "sum of (log C)^2's:  "; A3
  PRINT "sum of (log C * log[X/M]):  "; A4: PRINT : PRINT
  D = (T * A3) - (A1 ^ 2)
  D1 = (A2 * A3) - (A4 * A1)
  D2 = (T * A4) - (A2 * A1)
  S = D2 / D: I = D1 / D
  S = ABS(S)
  PRINT "Slope    ="; S
  I = 10 ^ I
  PRINT "Intercept = "; I
  V = I * (CO ^ S)
  PRINT "Ultimate sorption capacity of absorbent = "; V; "g/g"
  PRINT : PRINT "*****************************************************************"
END
```

Example 7.34

Calculate the breakthrough capacity and the working capacity of an adsorption bed using Equations 7.50 and 7.52, given the following data (neglect the packing factor, PF).

Depth of adsorption bed = 1 m (100 cm)
Saturation capacity = 42%
MTZ (mass transfer zone) = 10 cm
Heel = 2%

Solution

- C_B = breakthrough capacity:

$$C_B = \frac{0.5(.42)(10) + .42(100 - 10)}{100} = \frac{2.10 + 37.8}{100} = .399 = 39.9\% \qquad (7.50)$$

- W_c = working capacity (neglect PF):

$$W_c = C_B - H - PF = 39.9 - 2.0 = 37.9\% \qquad (7.52)$$

Example 7.35

1. Prepare a computer program that incorporates all the elements in Equations 7.50 and 7.52 so that the breakthrough and working capacity of an absorbent can be determined if given the depth of bed, the saturation capacity, mass transfer zone, and heel.
2. Using the computer program developed in (1) and given the following data, calculate the breakthrough capacity and the working capacity for this adsorbent bed.

Depth of adsorption bed = 50 cm
Saturation capacity = 0.75
MTZ = 3 cm
Heel = 2.5%

Solution

1. For a solution to Example 7.35 (1), see the following listing of Program 7.35.
2. A solution to Example 7.35 (2):

$$\text{Breakthrough capacity, } C_B = \frac{0.5(.75)(3) + .75(50 - 3)}{50} = 0.7275 = 72.8\%$$

$$\text{Working capacity, } W_c = 72.8 - 2.5 = 70.3\%$$

Listing of Program 7.35

```
'*************************************************************
'Example 7-35: A program to calculate the breakthrough capacity
' and the working capacity of an adsorption bed
'*************************************************************
'
COLOR 15, 1, 4
CLS
title:
DO
  opt = 0
  PRINT "Exercise 7-35:  A program to calculate the breakthrough"
  PRINT "capacity and the working capacity of an adsorption bed"
  PRINT : PRINT "The formulas:  "
  PRINT "C[B] = (0.5 * C[s] * (MTZ) + C[s] * (D - MTZ)) / D "
  PRINT "W[c] = C[B] - H - PF"
  PRINT : PRINT "Type 1 and ENTER to continue  "
  INPUT "Type 2 and ENTER to EXIT  "; opt
LOOP WHILE opt < 1 OR opt > 2
  IF (opt = 2) THEN
  END
END IF
screen1:
  CLS
  INPUT "Enter the saturation capacity (C[s]) "; cap
  INPUT "Enter the value of MTZ (MTZ) (in cm)"; mtz
  INPUT "Enter the adsorption bed depth (D) in cm) "; depth
  INPUT "Enter the heel value (H) in %"; heel
    comp1 = .5 * cap * mtz
    comp2 = depth - mtz
    comp3 = cap * comp2
    comp4 = comp1 + comp3
    cb1 = comp4 / depth
    cb = cb1 * 100
    wc = cb - heel
  PRINT : PRINT "Breakthrough capacity (C[B]) = "; cb; "%"
  PRINT : PRINT "Working Capacity (W[c]) = "; wc; "%"
  PRINT : INPUT "Press ENTER to continue  "; ent$
  CLS
  GOTO title
END
```

Example 7.36

An activated carbon adsorption bed is used to remove trichloroethylene (TCE) from an air stream at 21°C and 1 atm. The actual flow rate is 12,730 m^3/h, and the inlet TCE concentration is 2000 ppm by volume (MW of TCE = 131.5). Estimate the service time of the adsorbent before saturation. The proportionate saturation of the adsorbent is taken to be 0.2. The activated carbon bed is in a cylinder type configuration, 1.36 m in radius and 1 m deep ($V = \pi r^2 h = 3.14 \times 1.85 \times 1\ m = 5.8\ m^3$). The density of the adsorbent is 480 kg/m^3. The sorption efficiency is 99.5%.

Solution

Using Equation 7.53, the estimated service time of the adsorbent would be

$$t = \frac{2.41 \times 10^7 \times 0.2 \times 5.8\ m^3 \times 480\ kg/m^3}{0.995 \times 12,730\ m^3/h \times 2000\ ppm \times 131.5} = 4.03\ h$$

Example 7.37

1. Prepare a computer program that incorporates all the elements in Equation 7.53 which makes it possible to estimate the adsorbent's service time before regeneration.
2. An adsorption bed of activated carbon is to be used to remove acetone (MW = 58.08) from an air stream. The bed is 4 m thick and has a surface area of 6 m^2. The gas flow rate is 5,000 m^3/h. The density of the activated carbon is 400 kg/m^3. The inlet acetone concentration is 110,000 ppm. The proportionate saturation of the adsorbent is taken to be 0.2. The sorption efficiency is 99.5%. Using Equation 7.53, estimate the service time of the adsorbent before regeneration is required.

Solution

1. For a solution to Example 7.37 (1), see the following listing of Program 7.37, which incorporates all the elements in Equations 7.48, 7.50, and 7.53.
2. A solution to Example 7.37 (2):

$$t = \frac{2.41 \times 10^7 \times 0.2 \times 4\ m \times 6\ m^2 \times 400\ kg/m^3}{0.995 \times 5000\ m^3/h \times 58.08 \times 110,000\ ppm} = 1.46\ h$$

Listing of Program 7.37

```
'*************************************************************
'Example 7-37: A program to estimate the adsorbent's service
' time before regeneration.
'*************************************************************
'
'
COLOR 15, 1, 4
CLS
title:
DO
  opt = 0
  PRINT "Exercise 7-37:  A program to estimate the adsorbent's"
  PRINT "service time before regeneration."
  PRINT : PRINT "The formula:  "
  PRINT "t = (2.41E7 * S * W) / (E * Q * M * C) "
  PRINT : PRINT "Type 1 and ENTER to continue  "
  INPUT "Type 2 and ENTER to EXIT  "; opt
LOOP WHILE opt < 1 OR opt > 2
  IF (opt = 2) THEN
  END
```

```
END IF
screen1:
  CLS
  PRINT "t = (2.41E7 * S * W) / (E * Q * M * C) "
  INPUT "Is the adsorption bed square (1) or cylindrical (2)? "; opt
    IF (opt = 2) THEN GOTO cshape
    IF (opt = 1) THEN GOTO sshape
    IF (opt > 2) THEN GOTO screen1
      thick = 0
      surface = 0
      v = 0
cshape:
  CLS
  INPUT "Enter the depth of the bed of adsorbent (in m) "; thick
  INPUT "Enter the diameter of the cylinder (in m) "; surface
    v = 3.14 * thick * surface
  GOTO screen2
sshape:
  CLS
  INPUT "Enter the thickness of the bed of adsorbent (in m) "; thick
  INPUT "Enter the surface area of the adsorbent (in m{2}) "; surface
    v = thick * surface
  GOTO screen2
screen2:
  PRINT : PRINT "t = (2.41E7 * S * W) / (E * Q * M * C) "
  PRINT "Enter the proportionate saturation of"
  INPUT "the adsorbent (fractional) (x.x) "; S
  INPUT "Enter the density of the adsorbent (in kg/m{3}) "; dens
  INPUT "Enter the adsorbent efficiency (fractional .xxx) "; e
  INPUT "Enter the gas flow rate through the bed (in m{3}/hr) "; q
  INPUT "Enter the molecular weight of the sorbed vapor (in gm/mole) "; m
  INPUT "Enter the entering vapor concentration (in ppm by volume) "; C
    comp1 = 2.41E+07 * S * v * dens
    comp2 = e * q * m * C
    t = comp1 / comp2
  PRINT : PRINT "The time - t = "; t; " hours"
  INPUT "Press ENTER to continue  "; ent$
  CLS
  GOTO title
END
```

7.4 Air Quality Modeling

7.4.1 Introduction

Dispersion modeling is a procedure used to estimate the ambient air pollutant concentrations at various locations (receptors) downwind of a source, or any array of sources, based on emission rates, release specifications, and meteorological factors such as wind speed, wind direction, atmospheric stability, mixing height, and ambient temperature.[32] Air quality models can be categorized into four generic classes: Gaussian, numerical, statistical, or empirical and physical. Gaussian models generally are considered to be state-of-the-art techniques used for estimating the environmental impact of nonreactive pollutants. Numerical models are more appropriate than Gaussian models for multi-source applications that involve reactive pollutants.

Statistical or empirical models frequently are used in situations where incomplete scientific understanding of the physical and chemical processes make use of a Gaussian or numerical model impractical. Physical modeling involves the use of wind tunnels or other fluid modeling facilities. Physical modeling is a complex process and applicable to a limited geographic area of only a few square kilometers.[23]

In the U.S., the 1977 Clean Air Act amendments required that air quality models be used to identify potential violations of the National Ambient Air Quality Standards (NAAQS) and to determine emission limits.[33]

A wide variety of air quality models are available to evaluate and simulate atmospheric dispersion processes. Two modeling systems used by the EPA are the Industrial Source Complex (ISC2) model and the SCREEN2 model. The SCREEN2 model is a scoping model that can be used to evaluate the air quality and estimate whether or not a given source is likely to pose a threat and cause the NAAQS to be exceeded. This type of dispersion model is used first, before going to a regulatory model in order to evaluate air quality conditions. The Industrial Source Complex model is more precise than a screening model and uses local data to predict levels of pollutants at a specific place.[2] The EPA ISC2 model is a Gaussian dispersion model and has both a long-term module (ISCLT) and a short-term module (ISCST).[20]

The Office of Air Quality Planning and Standards (OAQPS) of the U.S. EPA, through the Technology Transfer Network (TTN), has established an electronic bulletin board system (BBS) which allows remote users with either terminals or microcomputers to dial up and have access to numerous air quality models.[34] Only two air quality models will be considered here: the basic Gaussian model and the SCREEN2 model. First, however, the equations, formulas, models, or procedures used to estimate the effective stack height will be presented.

7.4.2 Effective Stack Height

To apply the various Gaussian dispersion models, one important element in the models is the height at which the pollutant is emitted. Most stack emissions have an initial upward momentum and buoyancy. The buoyancy is due to the hot gases being ejected. The temperature differential between the hot stack gases (less dense) and the surrounding cooler ambient air cause the plume to rise into the atmosphere. The distance above the stack that a plume will rise into the atmosphere before leveling off is called the plume rise, Δh. The plume rise is affected not only by temperature but also by wind speed, molecular weight of the gases being emitted, the physical stack height, its inside diameter, and the ambient pressure.

The emission height used in the dispersion models is the effective stack height which is not only the physical stack height (h) but also includes the plume rise (see Figure 7.12A).[35]

$$H = h + \Delta h \tag{7.54}$$

where:

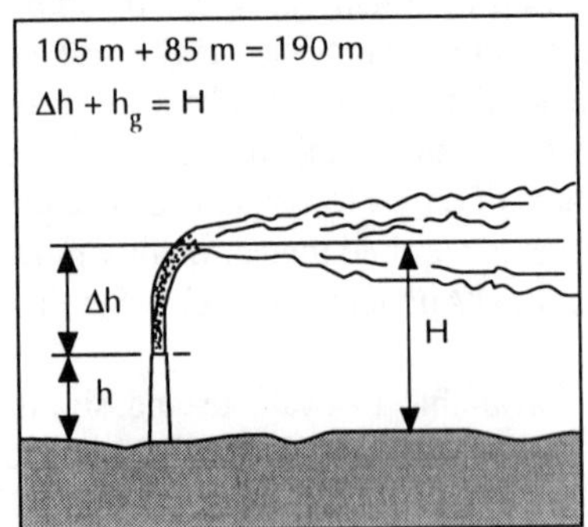

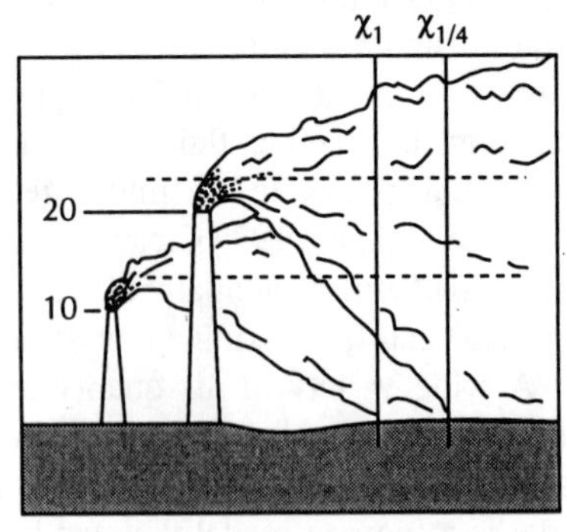

FIGURE 7.12
(A) An example of plume rise calculations. (B) An example of the reduced ground level contaminant concentration due to doubling the effective stack height.

H = Effective stack height, m
h = Physical stack height, m
Δh = Rise of the plume above the stack, m

The plume rise which is part of the effective stack height is an important element in estimating the maximum downwind groundlevel concentration of a pollutant emitted from a stack. The maximum downwind groundlevel concentration of a pollutant is reduced approximately by the inverse square of the effective stack height. For example, if the effective stack height is doubled, the maximum downwind groundlevel concentration on the center line of the plume will be reduced by a factor of four (see Figure 7.12B).

Numerous systems are available to estimate the effective stack height. Following are the names of a few of these equations, models, or procedures.

- Bryant-Davidson
- Bosanquet
- Concawe
- Briggs
- Moses and Carson
- Holland

Only the well known Holland equation will be presented here:[36]

$$\Delta h = \frac{v_s d}{u}\left[1.5 + 2.68(10)^{-3} p\left(\frac{T_s - T_a}{T_s}\right)d\right] \tag{7.55}$$

where:

v_s = Stack gas velocity, m/s

u = Mean wind speed at stack height, m/s

d = Stack inner diameter, m

p = Atmospheric pressure, mb

T_s = Stack gas temperature, K

T_a = Atmospheric temperature, K

Equation 7.55 is valid for neutral stability conditions; however, Holland suggests that the plume rise be adjusted by a factor of from 1.1 to 1.2 for unstable conditions such as stability types A and B and from 0.8 to 0.9 for stability conditions E and F (see Table 7.8).[37,38]

TABLE 7.8
Pasquill Stability Types[37,38]

Surface wind speed (m/s)	Day: Incoming solar radiation (sunshine)			Night	
	Strong	Moderate	Slight	Thinly overcast or ≥ 4/8 low cloud	≤ 3/8 low cloud
< 2	A	A–B	B		
2	A–B	B	C	E	F
4	B	B–C	C	D	E
6	C	C–D	D	D	D
> 6	C	D	D	D	D

Note: A = extremely unstable. B = moderately unstable. C = slightly unstable. D = neutral. E = slightly stable. F = moderately stable. Neutral class D, should be assumed for overcast conditions during day or night.

To estimate the mean wind speed (u) at the top of the stack, an empirical formula can be used. This formula generally is considered appropriate for estimating wind speeds at various heights up to 700 to 1000 m. This simple formula, or power law, is as follows:

$$\frac{v}{v_o} = \left(\frac{z}{z_o}\right)^k \tag{7.56}$$

where:

v = Wind speed at height z, m/s

v_0 = Wind speed at anemometer level z_0, m/s

k = Exponent or coefficient, dimensionless

The exponent k in the past has been taken generally as 1/7; however, recent research has provided values for k depending upon the stability class. Table 7.9 presents k values for each stability class.[39] The following examples demonstrate the use of Equations 7.54, 7.55, and 7.56.

TABLE 7.9
Average Values of Wind Profile Power Law Exponents (k) by Stability Class[39]

Pasquill stability class	Average value of exponent
A	0.141
B	0.176
C	0.174
D	0.209
E	0.277
F	0.414
G[a]	0.435

[a] Pasquill stability class G is considered to be very stable.

Source: From Touma, J.S., *J. Air Pollut. Control Assoc.*, 27, 863, 1977. With permission.

Example 7.38

Determine the effective stack height given the following data.

Physical stack height, h = 183 m inside diameter, with d = 6 m
Wind velocity at anemometer level (2 m above ground) = 5 m/s
Air temperature = 10°C
Atmospheric pressure = 1000 mbar
Stack gas velocity = 16 m/s
Stack gas temperature = 135°C
Class B Pasquill stability type

Solution

- Determine mean wind speed at the top of the stack; use Equation 7.56.

$$u = (5\,m/s)\left(\frac{183\,m}{2\,m}\right)^{0.176} = 11\,m/s$$

For Class B stability, k = 0.176 (see Table 7.9).

- Convert temperature to K:

$$Ta = 273 + 10 = 283 \text{ K}$$

$$Ts = 273 + 135 = 408 \text{ K}$$

- Substitute the given and calculated values in Holland's equation (7.55) to determine the plume rise, Δh:

$$\Delta h = \frac{(16 \text{ m/s})(6 \text{ m})}{11 \text{ m/s}}\left[1.5 + 2.68(10)^{-3}(1000 \text{ mbar})\left(\frac{408 \text{ K} - 283 \text{ K}}{408 \text{ K}}\right)6 \text{ m}\right] = 56 \text{ m}$$

- Use Equation 7.54 to determine the effective stack height.

$$H = 183 \text{ m} + 56 \text{ M} = 239 \text{ m}$$

Example 7.39

1. Prepare a computer program that will determine the effective stack height, given all the required elements contained in Equations 7.54, 7.55, and 7.56.
2. Using the computer program developed in (1), compute the effective stack height, given the following data.

Physical stack height h = 50 m with inside diameter d = 3.5 m

Wind velocity at anemometer level (4 m above ground level) = 10 m/s

Air temperature = 0°C

Atmospheric pressure = 1000 mbar

Stack gas velocity = 20 m/s

Stack gas temperature = 150°C

Class C Pasquill stability type

Solution

1. For a solution to Example 7.39 (1), see the following listing of Program 7.39 which includes the elements in Equations 7.54, 7.55, and 7.56. This computer program can be used to estimate the effective stack height.
2. A solution to Example 7.39 (2):
   - Determine the mean wind speed at the top of the stack (50 m); use Equation 7.56:

$$u = 10 \text{ m/s}\left(\frac{50 \text{ m}}{4 \text{ m}}\right)^{0.174} = 15.5 \text{ m/s}$$

For class C stability, k = 0.174 (see Table 7.9).

- Convert temperature to K:

$$Ta = 0 + 273 = 273\ K$$

$$Ts = 150 + 273 = 423\ K$$

- Substitute values in Holland's equation (7.55) to determine the plume rise, Δh:

$$\Delta h = \frac{20\ m/s(3.5\ m)}{15.5\ m/s}\left[1.5 + 2.68(10^{-3})(1000\ mb)\left(\frac{423\ K - 273\ K}{423\ K}\right)3.5\ m\right] = 21.8\ m$$

- Use Equation 7.54 to determine the effective stack height, which would be

$$H = 50\ m + 21.8\ m = 71.8\ m$$

Listing of Program 7.39

```
'**************************************************************
'Example 7-39: A program to determine the effective stack
' height.
'**************************************************************
'
'
COLOR 15, 1, 4
CLS
title:
DO
  opt = 0
  PRINT "Exercise 7-39:  A program to determine effective stack height"
  PRINT : PRINT "The formulas:  "
  PRINT "H = h + (delta)h "
  PRINT "(delta)h = ((v[s]*d)/u)*(1.5+2.68(10){-3}r((T[s]-T[a])/T[s])d)"
  PRINT "v/v[o] = (z/z[o]){k}"
  PRINT : PRINT "Type 1 and ENTER to continue  "
  INPUT "Type 2 and ENTER to EXIT  "; opt
LOOP WHILE opt < 1 OR opt > 2
  IF (opt = 2) THEN
  END
END IF
screen1:
  CLS
  PRINT "v/v[o] = (z/z[o]){k}"
  INPUT "Enter the wind velocity at anemometer level (u) (in m/s) "; wind
  INPUT "Enter the physical stack height (in m) "; sheight
  INPUT "Enter the anemometer height above ground level (in m) "; aheight
  INPUT "Enter stability type value from Table 7.9 "; stabil
    comp1 = sheight / aheight
    comp2 = comp1 ^ stabil
    u = wind * comp2
  PRINT "Wind speed at the top of the stack = "; u; "m/s"
screen2:
  PRINT : PRINT "(delta)h = ((v[s]*d)/u)*(1.5+2.68(10){-3}r((T[s]-T[a])/T[s])d)"
  INPUT "Enter the stack gas velocity (in m/s) "; vel
  INPUT "Enter the inside diameter of the stack (in m) "; dia
```

```
INPUT "Enter the atmospheric pressure (in mb) "; pres
INPUT "Enter the air temperature (in C) "; airtemp
INPUT "Enter the stack gas temperature (in C) "; stemp
  ta = atemp + 273
  ts = stemp + 273
  comp3 = (vel * dia) / u
  compt = (ts - ta) / ts
  comp4 = pres * compt * .001
  comp5 = 2.68 * comp4 * dia
  comp6 = 1.5 + comp5
  dh = comp3 * comp6
PRINT : PRINT "The plume rise, (delta)h = "; dh; "m"
  h = sheight + dh
PRINT : PRINT "The effective stack height = "; h; "m"
PRINT : INPUT "Press ENTER to continue  "; ent$
CLS
GOTO title
END
```

7.4.3 Dispersion Models

7.4.3.1 Basic Gaussian Dispersion Model

Many dispersion models have been developed, most of the models in use today are based on the work of Pasquill and modified by Gifford. The following binormal Gaussian plume equation relates disperson in the x (downwind) direction as a function of variables in all directions of a three-dimensional space.

The concentration (χ) of a gas or aerosol (< 20 μm) calculated at ground level for a distance downwind (x) is expressed as follows:

$$\chi(x, y) = \frac{Q}{\pi \sigma_y \sigma_z \bar{u}} \exp\left[-\frac{1}{2}\left(\frac{y}{\sigma_y}\right)^2\right] \exp\left[-\frac{1}{2}\left(\frac{H}{\sigma_z}\right)^2\right] \tag{7.57}$$

where:

(x,y) = Receptor coordinates, m
χ = Ground level concentration, g/m^3
Q = Emission rate, g/s
H = Effective stack height, m
$\bar{u}$ = Mean wind speed, m/s
σ_y, σ_z = Dispersion coefficients, m
π = 3.14159
exp = Base of natural logs, 2.7182818

Figure 7.13A depicts a graphical presentation of this equation.[37]

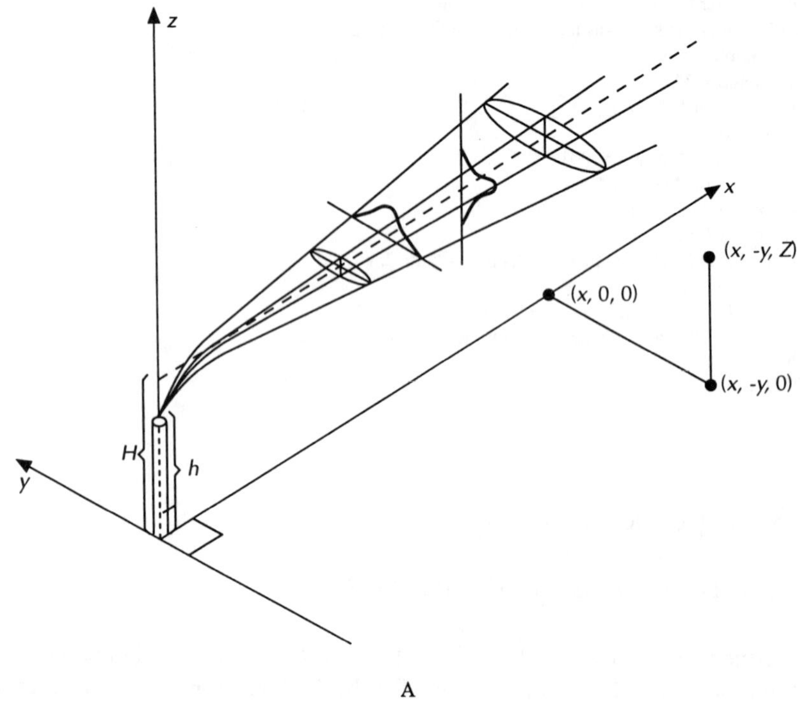

A

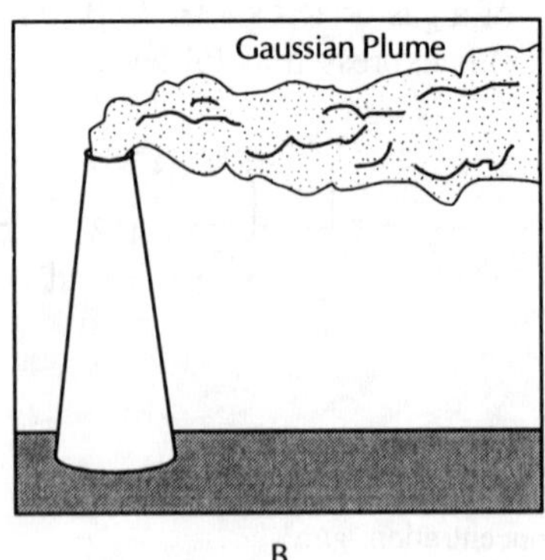

B

FIGURE 7.13
(A) Coordinate system showing Gaussian distribution in the horizontal and vertical. (B) Flat terrain for Gaussian distribution.

The Gaussian distribution equation expresses downwind groundlevel contaminant concentrationw when the terrain is approximately flat (see Figure 7.13B). Uneven terrain such as valleys, hills, and mountains makes it necessary to modify the Gaussian plume distribution. These modifications are exponential and vary depending upon the specific air quality model used.[38]

Values for the dispersion coefficients σ_y and σ_z not only depend on the downwind distances but also on the atmospheric stability (lapse rates). Values for σ_y and σ_z for various distances downwind (x) with various stability categories can be determined by using Figures 7.14A and 7.14B.[37] The Pasquill stability types or categories were previously presented in Table 7.8.

Equation 7.57 can be simplified if only the ground level downwind concentrations along the center line of the plume are needed. In this case, y = 0, and the equation then becomes

$$\chi_{x,0} = \frac{Q}{\pi \bar{u} \sigma_y \sigma_z} \exp\left[-\frac{1}{2}\left(\frac{H}{\sigma_z}\right)^2\right] \tag{7.58}$$

The units for this equation are the same as for Equation 7.57. See Figure 7.15A for a schematic presentation of this equation.[37,38]

Equation 7.58 may be further simplified if the effective stack height is H = 0. In this case, the source, such as a burning dump, is at groundlevel. The following equation can then be used to calculate the groundlevel downwind concentrations:[37,38]

$$\chi = \frac{Q}{\pi \bar{u} \sigma_y \sigma_z} \tag{7.59}$$

The units for this equation are the same as for Equation 7.57 and 7.58. Figure 7.15B is a graphical schematic presentation of this equation.[37,38]

The maximum downwind groundlevel concentration of a contaminant occurs on the center line of the plume. The maximum downwind groundlevel concentration can be estimated by taking the differential of a modified version of Equation 7.57, resulting in the following expression:[18]

$$\sigma_z = 0.707H \tag{7.60}$$

This equation holds true provided σ_z/σ_y are constant with downwind distance x. This gives a rough approximation of the maximum groundlevel concentration and is a much better approximation for unstable conditions than for stable conditions.[37]

The distance at which the maximum groundlevel concentration occurs can then be estimated by using the vertical dispersion coefficient σ_z (Figure 7.14B) and the appropriate stability type. The following examples demonstrate the use of the dispersion modeling equations.

Example 7.40

A modern 700-MW coal-fired power plant operates under the following conditions:

Stack height = 203 m

Inside stack diameter = 7 m

Stack gas exit velocity = 16 m/s

Stack gas temperature = 160°C

Ambient air temperature = 24°C

The average wind speed at anemometer level (2 m above ground level) = 4 m/s

Atmospheric presesure = 1000 mb

Atmospheric conditions are neutral stability (type D)

Emission rate for SO_2 from the stack = 4000 g/s

Determine the effective stack height using Holland's equation (7.55). What is the maximum downwind ground-level concentration and how far is this from the plant? What would the expected groundlevel concentration be at 300 m crosswind from the point on the plume center line where the maximum downwind groundlevel concentration occurred? Do the emissions from this plant cause the U.S. NAAQS, 24-h standards for SO_2 at groundlevel downwind from the plant to be exceeded?

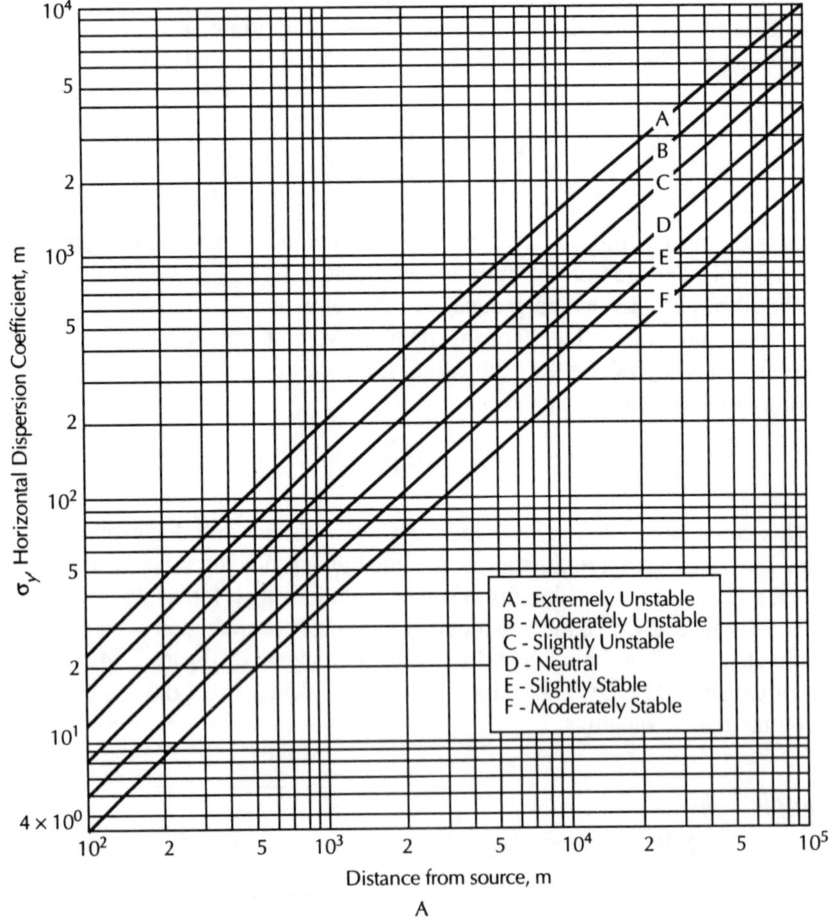

FIGURE 7.14
(A) Horizontal dispersion coefficient, σ_y, as a function of downwind distance from the source. (B) Vertical dispersion coefficient, σ_z, as a function of downwind distance from the source.

Solution

- Determine the mean wind speed at the top of the stack; use Equation 7.56:

$$u = 4\ m/s\left(\frac{203\ m}{2\ m}\right)^{0.209} = 10.5\ m/s$$

k = 0.209 for neutral (type D) stability conditions; see Table 7.9.

- Convert temperatures to K:

$$Ta = 24 + 273 = 297\ K$$

$$Ts = 160 + 273 = 433\ K$$

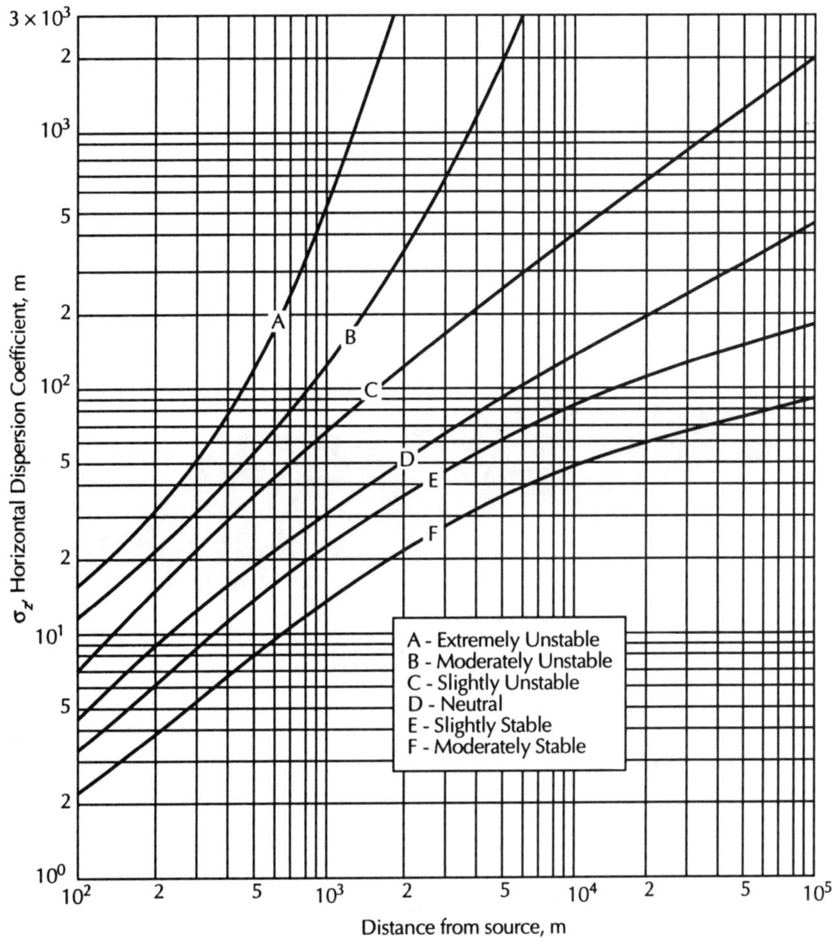

FIGURE 7.14B

- Use Holland's equation (7.55) to determine the plume rise, Δh:

$$\Delta h = \frac{16\ / s(7\ m)}{10.5\ m/s}\left[1.5 + 2.68(10)^{-3}(1000\ mbar)\left(\frac{433\ K - 297\ K}{433\ K}\right)7\ m\right] = 78.9\ m$$

- Use Equation 7.54 to determine the effective stack height, which would be

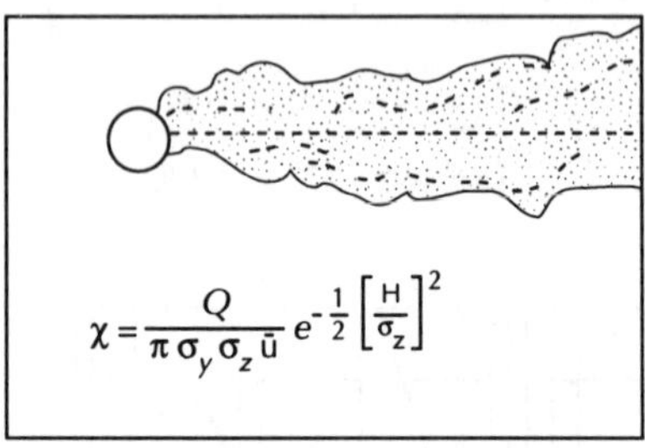

A

B

FIGURE 7.15
(A) Schematic of groundlevel concentration underneath the center line of the plume. (B) Schematic of contaminants generated under groundlevel conditions.

$$203\ m + 78.9\ m = 281.9\ m\ (use\ 282\ m)$$

- Determine the location for the maximum groundlevel concentration; use Equation 7.60:

$$\sigma_z = 0.707H = 0.707(282\ m) = 199\ m$$

From Figure 7.14B and under neutral type D stability, σ_z reaches a value of 199 m at a distance of 20 km from the stack.

- Thus, χ-max occurs approximately 20 km downwind from the plant. Determine the maximum ground-level concentration at 20 km; use Equation 7.58. From Figure 7.14A and 7.14B:

$$\sigma_y = 1000 \text{ m}$$

$$\sigma_z = 199 \text{ m}$$

$$\chi \max = \frac{4000 \text{ g/s}}{\pi \times 10.5 \text{ m/s} \times 199 \text{ m} \times 1000 \text{ m}} \exp -\frac{1}{2}\left(\frac{282 \text{ m}}{199 \text{ m}}\right)^2$$

$$= 0.0002233 \text{ g/m}^3 = 223 \,\mu\text{g/m}^3$$

- Determine the expected groundlevel concentration at 300 m crosswind from the point of maximum downwind groundlevel concentration; use Equation 7.57.

$$\chi = \frac{4000}{\pi \times 10.5 \times 199 \times 1000} \exp -\frac{1}{2}\left(\frac{300}{1000}\right)^2 \exp -\frac{1}{2}\left(\frac{282}{199}\right)^2 = 213 \,\mu\text{g/m}^3$$

- From Table 7.1, the primary U.S. NAAQS standard for the 24-h SO_2 level is 365 μg/m^3, while the 3-h secondary SO_2 standard is 1300 μg/m^3. In this case, the emissions of SO_2 from this plant do not cause the downwind SO_2 ambient air concentrations to exceed the standards. However, field testing would be necessary to confirm this.

Example 7.41

1. Prepare a computer program that incorporates all the elements in Equations 4.57, 4.58, 4.59, and 4.60.
2. Using the computer program developed in (1), compute the maximum downwind groundlevel concentration given the following data:
   - A manufacturing plant emits nitrogen dioxide (NO_2) on a continuous 24-h/d basis. What is the concentration of NO_2 at a receptor 8 km downwind of the plant? Does this plant cause concentrations at the receptor to be in excess of the primary and secondary U.S. NAAQS?
   - The effective stack height for this plant is 60 m.
   - The NO_2 emission rate is 100 g/s.
   - The wind speed at the top of the stack is 15.5 m/s.
   - Pasquill stability type is moderately stable, type F.

Solution

1. For a solution to Example 7.41 (1), see the following listing of Program 7.41.
2. A solution to Example 7.41 (2):

- Use Equation 7.58. From Figures 7.14A and 7.14B at 8 km and F stability:

$$\sigma_y = 220\,\text{m}$$

$$\sigma_z = 42\,\text{m}$$

$$\chi = \frac{100\,\text{g/s}}{3.14 \times 15.5\,\text{m/s} \times 220\,\text{m} \times 42\,\text{m}} \exp -\frac{1}{2}\left(\frac{60\,\text{m}}{42\,\text{m}}\right)^2$$

$$= 0.000080\,\text{g/m}^3 = 80\,\mu\text{g/m}^3$$

- The U.S. NAAQS for NO_2 (Table 7.1) for the primary and secondary standard is 100 $\mu g/m^3$, with an averaging time of 1 year. The receptor in this case would not be exposed to NO_2 concentrations above the standard.

Listing of Program 7.41

```
'*****************************************************************
'Example 7-41: A program to compute the maximum downwind
' ground-level concentration.
'*****************************************************************
'
'
COLOR 15, 1, 4
CLS
title:
DO
  opt = 0
  PRINT "Exercise 7-41:  A program to compute the maximum"
  PRINT "downwind ground-level concentration."
  PRINT : PRINT "The formulas:  "
  PRINT "The symbols used in these formulas are not available"
  PRINT "in Basic.  Refer to the text for formulas 7.57, "
  PRINT "7.58, 7.59, and 7.60"
  PRINT : PRINT "Type 1 and ENTER to continue  "
  INPUT "Type 2 and ENTER to EXIT  "; opt
LOOP WHILE opt < 1 OR opt > 2
  IF (opt = 2) THEN
  END
END IF
screen1:
DO
  opta = 2
  CLS
  PRINT "The dispersion cooefficients (sigma)[y] and (sigma)[z]"
  PRINT "will be represented by the letter S."
  PRINT "If the mean wind speed at the top of the stack is"
  PRINT "provided, type 1 and ENTER, "
  INPUT "if not, type 2 and ENTER "; opta
LOOP WHILE opta < 1 OR opta > 2
  IF (opta = 1) THEN
  GOTO windp
END IF
wind:
  CLS
```

```
  INPUT "Enter the wind velocity at anemometer level (u) (in m/s) "; wind
  INPUT "Enter the stack height (in m) "; sheight
  INPUT "Enter the anemometer height above ground level (in m) "; aheight
  INPUT "Enter stability type value "; stabil
    comp1 = sheight / aheight
    comp2 = comp1 ^ stabil
    u = wind * comp2
  PRINT : PRINT "Wind speed at the top of the stack = "; u; "m/s"
  PRINT : INPUT "Press ENTER to continue  "; ent$
  GOTO screen2
windp:
  CLS
  INPUT "Enter the wind speed at the top of the stack (in m/s) "; u
screen2:
DO
  optb = 2
  CLS
  PRINT "If the effective stack height for this plant is"
  PRINT "provided, type 1 and ENTER,"
  INPUT "if not, type 2 and ENTER "; optb
LOOP WHILE optb < 1 OR optb > 2
  IF (optb = 1) THEN
  GOTO heightp
END IF
  CLS
  PRINT "(delta)h = ((v[s]*d)/u)*(1.5+2.68(10){-3}r((T[s]-T[a])/T[s])d)"
  PRINT : INPUT "Enter the stack gas velocity (in m/s) "; vel
  INPUT "Enter the inside diameter of the stack (in m) "; dia
  INPUT "Enter the atmospheric pressure (in mb) "; pres
  INPUT "Enter the air temperature (in C) "; airtemp
  INPUT "Enter the stack gas temperature (in C) "; stemp
    ta = atemp + 273
    ts = stemp + 273
    comp3 = (vel * dia) / u
    compt = (ts - ta) / ts
    comp4 = pres * compt * .001
    comp5 = 2.68 * comp4 * dia
    comp6 = 1.5 + comp5
    dh = comp3 * comp6
  CLS
  PRINT : PRINT "The plume rise, (delta)h = "; dh; "m"
    h = sheight + dh
  PRINT : PRINT "The effective stack height = "; h; "m"
  PRINT : INPUT "Press ENTER to continue  "; ent$
  GOTO screen3
heightp:
  CLS
  INPUT "Enter the effective stack height for this plant (in m) "; h
  INPUT "Enter the emission rate (in g/s) "; q
  PRINT "The Pasquill stability type data is provided "
  PRINT "S[y] is read from the tables"
  INPUT "Enter the value of S[y] from the tables "; sy
  PRINT "The value of S[z] is = .707 * effective stack height"
  PRINT : INPUT "Press ENTER to continue  "; ent$
screen3:
  CLS
    sz = h * .707
    den1 = 3.14 * u * sy * sz
```

```
    half1 = q / den1
    den2 = -.5 * ((h / sz) * (h / sz))
    half2 = EXP(den2)
    xg = half1 * half2
    xu = xg * 1000000
  PRINT : PRINT "The maximum downwind ground-level concentration at 8 km"
  PRINT "and F stability is "; xu; " ug/m{3}"
  PRINT : INPUT "Press ENTER to continue  "; ent$
  CLS
  GOTO title
END
```

7.4.3.2 U.S. Environmental Protection Agency Dispersion Models

As indicated at the beginning of this section the U.S. EPA Office of Air Quality Planning and Standards, through the Technology Transfer Network, has established an electronic bulletin board system which allows anyone in the world to dial up and exchange information about air pollution.[40] Environmental Protection Agency personnel, state and local agencies, the private sector, and foreign countries all have access to this network.[40]

To access the 18 bulletin boards currently available on the network, a computer, modem, and communications software set with the following parameters are needed:

Data bits: 8

Parity: N

Stop bits: 1

Terminal emulation: VT100 or VT/ANSI

Duplex: Full

Following is a short explantation of the communications software parameters. Each 0 or 1 in a binary system is a bit, 0 and 1 are binary numbers. A byte is a string of binary digits and usually contains eight. "Data bits" indicates that a packet of informaiton is being processed, while "Stop bits" indicates that no information is being transmitted. "Parity" indicates that an extra bit has been added to a computer word to detect errors. "Terminal emulation" designates a communications program that makes a personal computer act like a terminal for the purpose of interacting with a remote computer. The terms VT100 or VT/ANSI are options that are used in order to access particular computer programs. ANSI stands for American National Standards Institute. "Full duplex" indicates a method of transmitting data that allows for the simultaneous sending and receiving of data. Using the communications software, the number to call is (919) 541-5742 for modems up to 14,400 (bps), or a voice help line is available at (919) 541-5384. Of the bulletin boards currently available, the SCRAM (Support Center for Regulatory Air Models) provides regulatory air quality model computer codes, meteorological data, and documentation, as well as modeling guidance.[40]

Table 7.10 presents the dispersion models presently available on the SCRAM BBS.[41] These models can be downloaded, and some files on the BBS can be uploaded. The dispersion models are in the FORTRAN computer language. The regulatory dispersion models are used in the U.S. in three programs. The first program, Prevention of Significant Deterioration (PSD), was designed to prevent deterioration of air quality in areas where it is already better than that required by the National Ambient Air Quality Standards. Modeling is used to verify that the air quality does not exceed these standards.[35]

TABLE 7.10
Models on the SCRAM BBS[41]

Regulatory models	Screening models	Other models
BLP (S)	COMPLEX I (S)	CMB7 (S,U)
CALINE3 (S,U)	CTSCREEN (S,DU)	COMPDEP (S,U)
CAL3QHC (S,U)	LONGZ (S)	DEGADIS21 (S)
CDM-2 (S,U)	RTDM 3.2 (S)	FDM (S,U)
CRSTER (S)	RVD2(S)	MESOPUFF II (S)
CTDMPLUS (S,DU)	SCREEN2 (S,U)	PLUVUE2 (S)
EKMA (S)	SHORTZ (S,U)	RPM-IV (S,U)
ISC2 (S,U,DU)	TSCREEN (S,U)	SDM (S)
MPTER (S,U)	VALLEY (S)	TOXLT (S,U)
OCD (S)	VISCREEN (S,DU)	TOXST (S,U)
RAM (S,U)		
UAM (S,DU)		

Note: S indicates source code. U indicates user's guide. DU indicates documentation updates.

The New Source Review is a second program for which dispersion modeling is used. The New Source Review program concerns the effects on air quality of either building new pollution sources or making certain modifications to existing sources.[35] The third programs where the regulatory dispersion models are used is called the Control Strategy Evaluation Program. The EPA uses the regulatory models to evaluate the air pollution control programs of the individual states and they are called State Implementation Plans (SIPs).[35] The regulatory dispersion models generally are used to evaluate air contaminant concentrations over at least a 1-year period or longer.

The screening or scoping models are used first to determine if a given source is likely to pose a threat to air quality. These models may be run with limited meteorological data and receptor sites. Screening methods allow for a fast estimate of whether a source may cause the National Ambient Air Quality Standards or PSD increments to be exceeded. The screening models used for the air quality estimates are not expensive to run, and the mathematical equations are simpler to solve.

The EPA recommends using a screening model first in the modeling process. If the screening model indicates that the source may cause the National Ambient

Air Quality Standards or the PSD increments to be exceeded, then a refined analysis must be made.[35] Most urban air toxic assessments have been conducted using the screening or scoping models.

The other models category is dispersion models, which have not as yet gained regulatory or screening status and have not gone through public review and comment. These dispersion models are available for use, but with the above qualifications. Following is a brief description of just one of the screening dispersion models.[42]

7.4.3.3 SCREEN2 Dispersion Model

The SCREEN2 dispersion model uses the Gaussian plume model that incorporates source-related factors and meteorological factors to estimate pollutant concentration from continuous sources. It is assumed that the pollutant does not undergo any chemical reactions and that no other removal processes, such as wet or dry deposition, act on the plume during its transport from the source.[42] The Gaussian model equations and the interactions of the source-related and meteorological factors were presented in Section 7.4.3.1 and more details are presented in Reference 37.

The basic equation used in the SCREEN2 model for determining groundlevel concentrations under the plume center line is[42]

$$
\begin{aligned}
\chi = Q/(2\pi u_s \sigma_y \sigma_z)\{&\exp\left[-\frac{1}{2}((z_r - h_e)/\sigma_z)^2\right] \\
&+\exp\left[-\frac{1}{2}(z_r + h_e)/\sigma_z)^2\right] \\
&+\sum_{N=1}^{k}[\exp\left[-\frac{1}{2}((z_r - h_e - 2Nz_i)/\sigma_z)^2\right] \\
&+\exp\left[-\frac{1}{2}((z_r + h_e - 2Nz_i)/\sigma_z)^2\right] \\
&+\exp\left[-\frac{1}{2}((z_r - h_e + 2Nz_i)/\sigma_z)^2\right] \\
&+\exp\left[-\frac{1}{2}((z_r + h_e + 2Nz_i)/\sigma_z)^2\right]]\}
\end{aligned}
\tag{7.61}
$$

where:

χ = Concentration, g/m^3
Q = Emission rate, g/s
π = 3.141593
u_s = Stack height wind speed, m/s
σ_y = Lateral dispersion parameter, m
σ_z = Vertical dispersion parameter, m

z_r = Receptor height above ground, m

h_e = Plume center line height, m

z_i = Mixing height, m

k = Summation limit for multiple reflections of plume off of ground and elevated inversion, usually ≤ 4

This SCREEN2 model can be used for point sources or area sources.[42]

For point sources, the inputs needed are as follows:

- Emission rate, g/s
- Stack height, m
- Stack inside diameter, m
- Stack gas exit velocity, m/s, or flow rate (ACFM or m^3/s)
- Stack gas temperature, K
- Ambient temperature, K (use default of 293 K if not known)
- Receptor height above ground (may be used to define flagpole receptors)
- Urban/rural option (U = urban, R = rural)

For area sources, the inputs needed are as follows:

- Emission rate, $g/s/m^2$
- Source release height, m
- Length of side of the square area, m
- Receptor height above ground, m
- Urban/rural option (U = urban, R = rural)

Other information that must be considered in the SCREEN2 model follows:

- Mixing height
- Worst-case meteorological conditions
- Plume rise for point source
- Dispersion parameters
- Buoyancy-induced dispersion
- Building downwash
- Fumigation
- Shoreline fumigation
- Complex terrain 24-h screen

For stable conditions and/or mixing heights greater than or equal to 10,000 m, unlimited mixing is assumed and the summation term in Equation 7.61 is assumed to be zero. The mixing height used in Equation 7.61 is based on Randerson calculations (see Reference 44). The worst-case meteorological conditions use the various stability classes (see Table 7.8) and wind speeds to identify the "worst-case" scenario.

The plume rise equations developed by Briggs are used in the SCREEN2 model to estimate the plume rise. The plume rise equations come from Reference 45, 46, and 47. The dispersion coefficients σ_z, vertical, and σ_y, lateral, were designed to be used in the Gaussian plume models and were first published by Pasquill in 1959 (see Figures 7.14A and B). The dispersion coefficients used for rural and urban sites in the SCREEN2 model come from Reference 48. The dispersion coefficients (σ_y and σ_z) are adjusted in most cases to account for the effects of buoyancy-induced dispersion using the following equations:

$$\sigma_{ye} = \left[\sigma_y^2 + \left(\frac{\Delta h}{3.5}\right)^2\right]^{0.5}$$

$$\sigma_{ze} = \left[\sigma_z^2 + \left(\frac{\Delta h}{3.5}\right)^2\right]^{0.5}$$

where:

σ_{ye} = Adjusted lateral dispersion coefficient, m
σ_{ze} = Adjusted vertical dispersion coefficient, m
Δh = Distance-dependent plume rise, m

The building downwash, cavity, and wake screening problems encountered in dispersion modeling are dealt with in References 48 and 49.

When an inversion layer occurs a short distance above a plume source and superadiabatic conditions prevail below the stack, the plume is said to be fumigating. In this case, the air contaminants are suddently brought to groundlevel. In the SCREEN2 model, the inversion breakup calculations are based on procedures described by Turner in Reference 37.

A special-purpose model has been developed to consider the worst-case impact for complex terrain, 24-h screen. This dispersion model is called VALLEY and is a modified Gaussian model.[50] This model assumes F stability conditions (E for urban) and a stack height wind speed of 2.5 m/s. The width of the sector of uniform air contaminant concentration is assumed to be 22.5 degrees. The following information is taken directly from Reference 42.

> If the plume height is at or below the terrain height for the distance entered, then SCREEN2 will make a 24-hour average concentration estimate using the VALLEY screening technique. If the terrain is above stack height but below plume center line height, then SCREEN2 will make a VALLEY 24-hour estimate (assuming F or E and 2.5 m/s), and also estimate the maximum concentration across a full range of meteorological conditions using simple terrain procedures with terrain "chopped off" at physical stack height, and select the higher estimate. Calculations continue until a terrain

height of zero is entered. For the VALLEY model concentration SCREEN2 will calculate a sector-averaged ground-level concentration with the plume center line height (h_e) as the larger of 10.0 m or the difference between plume height and terrain height.

7.5 Homework Problems

7.5.1 Discussion Problems

1. Define:

   - Charles' law
   - Boyle's law
   - Ideal Gas law
   - Henry's law
   - Dalton's law
   - Reynold's law
   - Stoke's law
   - Air pollution

2. What are the four major components which make up the troposphere of the earth and what is the percentage of each?

3. What are the two basic physical forms of air pollution?

4. What are the six criteria pollutants designated by the U.S. EPA?

5. In the U.S., what are the four principal sources (categories) of air pollution, and what are the major pollutants emitted by these sources?

6. If you were to read an air pollution control report, in what units would you expect to see the following concentrations reported?

   - SO_2
   - CO
   - Pb
   - PM_{10}

7. Indicate the basic principles involved in the control of gaseous air pollution emissions.

8. Indicate the commonly used devices for control of particulate air pollution emissions.

9. In the following equations, what does each term represent? Give the units for each, as well as the name associated with each equation.

$$\left(P+\frac{n^2a}{V^2}\right)(V-nb)=nRT$$ Name: ____________________

$$\ln\frac{C_2}{C_1}=\frac{\Delta H}{R}\left(\frac{1}{T_1}-\frac{1}{T_2}\right)$$ Name: ____________________

$$\frac{X}{M}=KC^{\frac{1}{n}}$$ Name: ____________________

$$v=\frac{gd_p^2(\rho_p-\rho_a)}{18\,\mu}$$ Name: ____________________

10. What industrial plant operating conditions would dictate the use of an electrostatic air cleaning device instead of a fabric filter?

11. What control technology do we have available for control of particulate airborne contaminants? Indicate the size range for which each type of unit is most efficient for removal.

12. What is meant by effective stack height and what are the controlling factors?

13. What are the four general categories of air quality models?

14. Multiple choice and matching questions:

    a. Dispersion models estimate the ambient air concentrations as a function of:
        (1) Source location
        (2) Emission strengths
        (3) Terrain features
        (4) Meteorological conditions
        (5) All of the above

    b. The primary mobile source of air pollution in the U.S. is
        (1) Diesel trucks
        (2) Automobiles
        (3) Airplanes
        (4) Lawn mowers

    c. Which air pollution control device generates a large volume of wastewater?
        (1) Venturi scrubbers
        (2) Cyclone
        (3) Settling chamber
        (4) Electrostatic precipitator

    d. A substance that alters the rate of a chemical reaction without the substance itself being changed or consumed by the reaction is a(n):

(1) Additive
(2) Catalyst
(3) Add-on
(4) Stimulus

e. Match the type of atmospheric stability to the appropriate Pasquill-Gifford stability categories:

__ Unstable (1) E-F
__ Neutral (2) A-B-C
__ Stable (3) D

f. The σ_y and σ_z used in the Gaussian dispersion formulas are defined as:

(1) Atmospheric pressure at points y and z
(2) Standard deviations of pollutant concentration in the horizontal and vertical directions
(3) Temperature variations in the y and z directions
(4) None of the above

g. Match the category of air quality model with its definition.

__ Empirical 1. Investigate pollutant dispersion for complicated situations
__ Numerical 2. Derived from an analysis of source data, meteorological data, and air quality
__ Physical 3. Use complex equations to simulate the effects of turbulence, chemical transformations, deposition, etc. on pollutant transport and dispersion.
__ Gaussian 4. Techniques for estimating the impact of nonreactive pollutants; use simple algebraic expressions for reactive pollutants

h. Plume rise from a stack is due to:

(1) Heat and type of pollutant
(2) Momentum and buoyancy
(3) Composition of the stack
(4) None of the above

i. Settling chambers use _______ to remove solid particles from a gas stream.

(1) Electrostatic pressure
(2) Gravity
(3) Hydraulic simulation
(4) Filtration

15. Draw a schematic diagram for each of the following air pollution control devices and briefly describe the mechanisms, principles, or processes involved in their operation.

a. Settling chambers
b. Electrostatic precipitators
c. Cyclone cleaner
d. Venturi scrubber
e. Packed tower

16. Indicate the equipment needed to access the EPA, OAQPS, BBS, and SCRAM regulatory dispersion models on the Internet; also indicate the parameters required on the communications software. Define or explain each of these parameters.

17. What are the three categories for the SCRAM BBS models?

18. How are the regulatory dispersion models and the screening models used by the EPA?

19. What data or information is needed in order to use the SCREEN2 model?

7.5.2 Specific Mathematical Problems

1a. Write a computer program to determine the partial pressure exerted by a gas, given the temperature of the gas mixture, the prevailing pressure, and the gas concentration.

1b. Use the program developed in (a) to find the partial pressure exerted by carbon monoxide in a gas mixture. The gas mixture which is at 25°C and 1 atm pressure contains 150 mg/L of carbon monoxide (CO) gas.

2a. Write a computer program to find the percentage of the margin of safety lost if the temperature of a gas in a tank rises to T_2°C from a temperature of T_1, given the gas pressure and the pressure the tank is able to withstand.

2b. A tank is tested to be able to withstand 17 atm of pressure. The tank is filled with gas at 25°C and 10 atm of pressure. What percentage of the margin of safety is lost if the temperature goes to 100°C? Use the program developed in (a) to check your answer.

3a. Write a short computer program that allows the determination of the volume of a certain gas, given its weight, temperature, and pressure.

3b. Use the computer program developed in (a) to calculate the volume of 36 g of SO_2 at 25°C and an absolute pressure of 1 atm.

4a. A cylinder containing 100 g of carbon monoxide (CO) fell off a laboratory table and broke the release valve, permitting the gas to excape into the closed laboratory room that was 20 m long, 10 m wide, and 3 m high. What would the average CO concentration be in this room in μg/m³ and ppm?

4b. Write a short computer program to solve the problem outlined in (a).

5a. Write a computer program to find the new volume of a gas, the temperature of which has been increased from a certain value to another, given its initial pressure.

5b. Use the program developed in (a) to solve the following problem: When the temperature of 25 mL of dry carbon dioxide (CO_2) is increased from 10°C to 35°C at a pressure of 1 atm, what then is the volume of the gas?

6a. Write a short computer program to compute the number of moles present of each gas in a gas mixture composed of three different gaseous substances enclosed in a container of a known volume, given the weight of each gas. Let the program determine the total pressure exerted by the gas mixture. The program should also estimate the percentage of each volume of gas present in the container.

6b. Use the program developed in (a) to solve the following problem: A 25-L volume cylinder at 25°C contains 1 g of methane (CH_4), 2 g of nitrogen (N_2), and 12 g of oxygen (O_2). What is the number of moles of each gas present? What is the partial pressure exerted by each gas? What is the total pressure exerted by the gas mixture, and what is the percentage by volume of each gas present in the cylinder?

7. Write a simple computer program to change the concentration of a certain pollutant from μg/m³ to ppm by volume given the relevant atomic weights. Use the program to fill in the blanks in the following table:

Pollutant	Concentration (μg/m³)	Concentration (ppm by volume)
SO_2	60	
NO_2		0.70
CO	55	

Note: S = 32; O = 16; N = 14; C = 12

8. Use the computer program developed for Example 7.15 to solve the following problem: According to Stoke's law, what would be the terminal settling velocity of a 1.0-μm-radius particle of quartz that has a specific gravity of 2.75, and the temperature is 27°C?

9a. It is desired that all particles with a size of 50 μm and large be removed with 100% efficiency in a settling chamber which has a height of 0.8 m. The air moves at 1.6 ft/s and its temperature is 80°F. The particle density is 2.2 g/cm³. What minimum length of chamber should be provided?

9b. Write a short computer program to solve the problem of section (a).

10a. Write a computer program to determine the particle size collected in a cyclone cleaner, given the outer diameter of the cyclone, the number of effective turns for the cyclone, air flow, temperature and pressure of gas, density of particles, the gas velocity at the entrance to the cyclone, and the efficiency of the system.

10b. Use the program developed in (a) to solve the following problem: A cyclone cleaner has an outer circumference of 251 cm and handles 3.7 m^3/s of contaminated air at a temperature of 30°C and 1 atm of pressure. The density of the particles is 2.0 g/m^3. The gas velocity at the entrance to the cyclone is 15 m/s. Determine the particle size in microns collected with 50% efficiency if the number of effective turns for the cycone is 4.6.

11a. Calculate the fabric area, in square meters, required in a baghouse to treat 6500 m^3/min of particulate-laden gas stream at a removal efficiency of 99.82%. The baghouse unit operates at an air-to-cloth ratio of 0.72 m/min.

(1) 6000 (3) 9000
(2) 5500 (4) 15,000

11b. Calculate the area A, in square meters, for an electrostatic precipitator required to process 4000 m^3/min of a particulate-laden gas stream with a removal efficiency of 99.5%. The drift velocity **w** has been determined to be 0.07 m/s.

(1) 6500 (3) 55,000
(2) 10,000 (4) 60,500

11c. A plant has in inlet loading to a baghouse of 23,000 mg/m^3. The average filtration velocity is 3 m/min and the gas flow rate is 700 m^3/min. What is the air-to-cloth ratio for this operation?

(1) 76 m/min (3) 233 m^2/min
(2) 3 $m^3/min/m^2$ (4) 2 m/min

11d. At the point of maximum concentration downwind from a source with an effective stack eight of 70 m, σ_z is closest to:

(1) 15 m (3) 50 m
(2) 58 m (4) 80 m

11e. The concentration at 1000 meters downwind of a 60 meter (effective stack height) source emitting 100 grams/s under type B stability conditions with a 6 meter/sec mean wind speed is?

(1) 7.0×10^{-5} g/m^3 (3) 2.6×10^{-4} g/m^3
(2) 1.5×10^{-3} g/m^3 (4) 5.3×10^{-3} g/m^3

11f. Write short computer programs to solve each of the aforementioned multiple choice questions.

12a. Write a short computer program to estimate the collection efficiency of a venturi scrubber at removing particles of given densities. Let the program determine the pressure drop across the venturi scrubber, given the relevant characteristics of the venturi scrubber.

12b. Use the program developed in (a) to solve the following problem: Estimate the collection efficiency of a venturi scrubber at removing 4-μm-sized particles. The particles have a density of 2750 kg/m³. Also, determine the pressure drop across the venturi scrubber. The venturi scrubber has the following characteristics:

Gas flow rate = 7 m³/s

Liquid flow rate = 0.002 m³/s

Gas flow rate at the throat = 80 m/s

Correlation coefficient = 1700

Temperature = 100°C

13a. Write a computer program to find the geometric mean diameter and the geometric standard deviation for a certain aerosol given the results for the mass distribution of the aerosol.

13b. Use the program developed in (a) to solve the following problem: Following are the results for the mass distribution of an aerosol. From this determine the geometric mean diameter and the geometric standard deviation for this aerosol.

Average size (μm)	% by weight
0.25	0.1
1.0	0.4
2.0	9.5
3.0	20.0
4.0	20.0
5.0	15.0
6.0	11.0
7.0	8.0
8.0	5.5
10.0	5.5
14.0	4.0
20.0	0.8
>20.0	0.2

14a. Write a computer program to compute the collection efficiency of an electrostatic precipitator for a certain particle size having a certain drift velocity. Let the program determine the efficiency for other particle diameters given the ESP specifications.

14b. Use the program developed in (a) to determine the collection efficiency of the electrostatic precipitator described below for a particle size of 0.70-μm diameter having a drift velocity of 0.20 m/s. What would be the efficiency for 1.0-μm particles? ESP specifications:

Plate height = 8 m

Plate length = 6 m

Number of passages = 6

Plate spacing = 0.3 m

Gas flow rate = 20 m^3/s

15a. A 2.5-g sample of catalyst was tested for its efficiency for removal of CO from an air stream. The initial CO concentration entering the catalyst was 160 ppm; the CO concentration after passing through the catalyst was found to be 2.5 ppm. Determine the percent removal of CO for this catalyst, the contact time or space velocity, and the reaction rate coefficient for the catalyst. The space around the catalyst was found to be 1.3 cm^3 and the air stream in which the CO is present had a flow rate of 800 cm^3/min.

15b. Write a computer program to solve the problem indicated in (a).

16a. Develop a computer program to find the effective stack height for a power plant given the physical stack height, stack inside diameter, stack gas effluent velocity, effluent gas temperature, ambient air temperature, and mean wind speed.

16b. Use the computer program developed in (a) to determine the effective stack height for a power plant operating under the following conditions:

Physical stack height = 45 m

Stack inside diameter = 3 m

Stack gas effluent velocity = 16 m/s

Effluent gas temperature = 160°C

Ambient air temperature = 22°C

Mean wind speed = 5 m/s

17a. A manufacturing plant emits nitrogen oxides under the conditions set forth below on a continuous 24-h/d basis. What is the concentration of NO_2 at a receptor 760 m downwind of the plant? Does this plant by itself cause concentrations at the receptor to exceed U.S. primary and secondary ambient annual air quality standards for NO_2?

Physical stack height = 55 m (situated on level terrain)

Stack inside diamter = 1 m

Stack exit velocity = 4.6 m/s

Stack temperature = 120°C

NO_2 emission rate = 450 kg/h

Mean wind speed u = 3.8 m/s

t_{air} = 10°C

p = 1000 mbar

Pasquill stability category D

17b. Develop a computer program to solve the problem mentioned in (a).

18a. Improve the computer program for Example 7.33 to solve the following problem given in 18b.

18b. Given the following data for the adsorption of vinyl chloride by activated carbon,

Activated carbon (g)	Concentration of vinyl chloride remaining (g/m³)
2.0	80
3.5	50
5.0	31
16.0	2.7
25.0	1.0

Note: Control or initial concentration at the start of the test = 155 g/m³.

Determine if the data follows the Freundlich isotherm. If the data does follow the Freundlich isotherm:

(1) Determine the k and n values.

(2) Determine the ultimate capacity of this activated carbon $(X/M)_{C_0}$ for vinyl chloride adsorption at the control or initial concentration.

19a. An effluent air stream contains 5.0% SO_2 by volume. It is desired to reduce this concentration to 2500 mg/m³ by passing the gas stream through a packed column adsorption tower before releasing the air stream to the atmosphere. The inlet-gas flow rate is 30 m³/min measured at 1 atm and 22°C, and the absorber is the counter-flow type. The equilibrium of SO_2 in water is assumed to be reasonably represented by the following relationship: $Y_B = 30X_B$ (Y_B = mole fraction of SO_2 in the air and X_B = mole fraction of SO_2 in the water). Estimate the minimum amount of water (SO_2-free) required as a solvent, in cubic meters per second, to meet the desired requirements.

19b. Develop a suitable computer program to solve the problem outlined in (a).

References

1. Definitions, *Fed. Regist.*, 52.741 (40), July 1, 1994.
2. Air Pollution Training Institute, *Air Pollution Control Orientation Course*, (S1:422), U.S. Environmental Protection Agency, Research Triangle Park, NC, 1992.
3. National primary and secondary ambient air quality standards, *Fed. Regist.*, 50.1 (40), July 1, 1993.
4. Council on Environmental Quality, *24th Annual Report of the Council on Environmental Quality*, Washington, D.D., 1993.
5. Peavy, H.S., Rowe, D.R., and Tchobanoglous, G., *Environmental Engineering*, McGraw-Hill, New York, 1985.
6. Air Pollution Training Institute, *Atmosphere Sampling*, (S1:435), U.S. Environmental Protection Agency, Research Triangle Park, NC, 1983.
7. Lide, D.R. and Frederiske, H.P.R., Eds., *CRC Handbook of Chemistry and Physics*, 76th ed., CRC Press, Boca Raton, FL, 1995–1996, 6–48.
8. Manahan, S.E., *Environmental Chemistry*, 6th ed., CRC Press, Boca Raton, FL, 1994, 117.
9. Air Pollution Training Institute, *Control of Gaseous and Particulate Emissions*, (S1:412D), U.S. Environmental Protection Agency, Research Triangle Park, NC, 1984.
10. Rowe, D.R. et al., Indoor-outdoor relationship of suspended particulate matter in Riyadh, Saudi Arabia, *APCA J.*, 35 (1), 1985.
11. U.S. Department of Health and Human Services, *The Industrial Environment — Its Evaluation and Control*, National Institute for Occupational Safety and Health, Cincinnati, OH, 1973, 159, 641.
12. Air Pollution Training Institute, *Control Techniques for Gaseous and Particulate Pollutants*, (SI:422), U.S. Environmental Protection Agency, Research Triangle Park, NC, 1981.
13. Danielson, J.A., Ed., *Air Pollution Engineering Manual*, Office of Air and Water Programs, U.S. Environmental Protection Agency, Research Triangle Park, NC, 1973, 92, 107.
14. Cooper, C.D. and Alley, F.C., *Air Pollution Control: A Design Approach*, PWS Engineering, MA, 1986, 110.
15. Theodore, L. and DePaola, V., Predicting cyclone efficiency, *J. Air Pollut. Control Assoc.*, 80, 1132, 1980.
16. Vesilind, P.A., Peirce, J.J., and Weiner, R.F., *Environmental Engineering*, 2nd ed., Butterworth-Heinemann, Stoneham, MA, 1988, 431, 434.
17. Calvert, S., Englund, H.M., *Handbook of Air Pollution Technology*, John Wiley & Sons, New York, 1984, 104, 252.
18. Wark, K. and Warner, C.F., *Air Pollution*, 2nd ed., Harper & Row, New York, 1981, 210.
19. Williams, C.E., Hatch, T., and Greenburg, L., Determination of cloth area for industrial air filters, *Heating Piping Air Cond.*, 12, 259, 1940.
20. Lee, C.C., *Environmental Engineering Dictionary*, 2nd ed., Government Institutes, Rockville, MD, 1992, 76.
21. Lloyd, W.G. and Rowe, D.R., Homogeneous catalytic oxidation of carbon monoxide, *Environ. Sci. Technol.*, 5, 11, 1971.
22. Lloyd, W.G. and Rowe, D.R., Palladium Compositions Suitable as Oxidation Catalysts, U.S. Patent 3,849,336, 1974.

23. Zackay, V.F. and Rowe, D.R., Catalyst of Palladium, Copper and Nickel on a Substrate, U.S. Patent 4,459,269, July 10, 1984; U.S. Patent 4,521,530, June 4, 1985.
24. *Teledyne Technology Introduces a New Room Temperature Catalyst*, Form 1007-F-33, 1730 E. Prospect St., Fort Collins, CO, 1983.
25. Rowe, D.R. and Lloyd, W.G., Catalytic cigarette filter for carbon monoxide reduction, *J. Air Pollut. Control Assoc.*, 28 (3), 1978, 253.
26. *Teledyne Instapure Filtration System*, Form 1035-F-35, 1730 E. Prospect St., Fort Collins, CO, 1985.
27. Corman, R.D. and Black, D., *Controlling Air Pollution*, American Lung Association, Washington, D.C., 1974.
28. Culp, R.L., Wesner, G.M., and Culp, G.L., *Handbook of Advanced Wastewater Treatment*, 2nd ed., Van Nostrand Reinhold, New York, 1978, 180.
29. Ford, D.L., *Process Design in Water Quality Engineering, New Concepts and Developments*. XI: *Coagulation and Precipitation and Carbon Adsorption*, sponsored by the Department of Environmental and Water Resources Engineering, Vanderbilt University, Nashville, TN, November, 1975.
30. Chansler, J.M., Lloyd, W.G., and Rowe, D.R., Soil sorption of zinc according to the Freundlich isotherm, *Florida Water Resources Journal*, 47, 9, 1995, 32–34.
31. Turk, A., Adsorption, in *Air Pollution*, Vol. IV, 3rd ed., Engineering Control of Air Pollution, Stern, A., Ed., 1977, 329.
32. Office of Air Quality Planning And Standards, *Assessing Multiple Pollutant Multiple Source Cancer Risks From Urban Air Toxics*, 450/2-89-010, U.S. Environmental Protection Agency, Research Triangle Park, NC, 1989.
33. Office of Air Quality Planning And Standards, *Guideline on Air Quality Models*, 450/2-78-027, U.S. Environmental Protection Agency, Research Triangle Park, NC, 1978.
34. Office of Air Quality Planning And Standards, *Technology Transfer Network*, U.S. Environmental Protection Agency, Research Triangle Park, NC, 1991.
35. Air Pollution Training Institute, *Introduction to Dispersion Modeling*, (SI:410), U.S. Environmental Protection Agency, Research Triangle Park, NC, 1983, 4–19.
36. Holland J.Z., *A Meteorological Survey of the Oak Ridge Area*, Atomic Energy Commission, Rep. ORO-99, Washington D.C., 1953, 540.
37. Turner, D.B., *Workbook of Atmospheric Dispersion Estimates*, National Air Pollution Control Administration, Cincinnati, OH, 1969, 31.
38. Air Pollution Training Institute, *Basic Air Pollution Meteorology*, (SI:409), U.S. Environmental Protection Agency, Research Triangle Park, NC, 1982.
39. Touma, J.S., Dependence of the wind profile power law on stability for various locations, *J. Air Pollut. Control Assoc.*, 27, 863, 1977.
40. Office of Air Quality Planning And Standards, U.S. Environmental Protection Agency, Technology Transfer Network, Research Triangle Park, NC, 27711, January, 1994.
41. Atkinson, D.G., Meteorologist (NOAA), OAQPS, U.S. Environmental Protection Agency, Research Triangle Park, NC, personal communication, May, 1995.
42. Office of Air Quality Planning And Standards, *SCREEN2 Model User's Guide*, 450/4-92-006, U.S. Environmental Protection Agency, Research Triangle Park, NC, 1992.
43. Office of Air Quality Planning And Standards, *Assessing Multiple Pollutant, Multiple Source Cancer Risks From Urban Air Toxics, 450/2-89-010,* U.S. Environmental Protection Agency, Research Triangle Park, NC, 1989, A-xii.

44. Randerson, D., Atmospheric boundary layer, in *Atmospheric Science and Power Production*, Randerson, D., Ed., DOE/TIC-27601, U.S. Department of Energy, Washington, D.C., 1984.
45. Briggs, G.A., *Plume Rise*, USAEC Critical Review Series, TID-25075, National Technical Information Service, Springfield, VA, 1969.
46. Briggs, G.A., *Diffusion Estimation for Small Emissions*, NOAA ATDL, Contribution File No. 79 (draft), Oak Ridge, TN, 1973.
47. Briggs, G.A., Plume Rise Predictions, in *Lectures on Air Pollution and Environmental Impact Analysis*, Haugen, D.A., Ed., American Meteorological Society, Boston, MA, 1975, 59.
48. U.S. EPA, *Industrial Source Complex (ISC2) Dispersion Model User's Guide*, 450/4-92-008, U.S. Environmental Protection Agency, Research Triangle Park, NC, 1992.
49. Hosker, R.P., Flow and Diffusion Near Obstacles, in *Atmospheric Science and Power Production*, Randerson, D., Ed., DOE/TIC-27601, U.S. Department of Energy, Washington, D.C., 1984.
50. Burt, E.W., *VALLEY Model User's Guide*, 450/2-77-018, U.S. Environmental Protection Agency, Research Triangle Park, NC, 1977.

Appendix

Appendix

Contents

TABLE A1
Physical Properties of Water

Temperature (°C)	Temperature (°F)	Density at 1 atm (g/cm³)	Dynamic viscosity, in centipoises (10^{-2} dyn-s cm^{-2})	Kinematic viscosity, in centistokes (10^{-2} cm^2 s^{-1})	Surface tension against air dyn cm^{-1})	Vapor pressure (mmHg)
5	41.0	0.999965	1.5188	1.5189	74.92	6.543
6	42.8	.999941	1.4726	1.4727	74.78	7.013
7	44.6	.999902	1.4288	1.4289	74.64	7.513
8	46.4	.999849	1.3872	1.3874	74.50	8.045
9	48.2	.999781	1.3476	1.3479	74.36	8.609
10	50.0	.999700	1.3097	1.3101	74.22	9.209
11	51.8	.999605	1.2735	1.2740	74.07	9.844
12	53.6	.999498	1.2390	1.2396	73.93	10.518
13	55.4	.999377	1.2061	1.2069	73.78	11.231
14	57.2	.999244	1.1748	1.1757	73.64	11.987
15	59.0	.999099	1.1447	1.1457	73.49	12.788
16	60.8	.998943	1.1156	1.1168	73.34	13.634
17	62.6	.998774	1.0875	1.0889	73.19	14.530
18	64.4	.998595	1.0603	1.0618	73.05	15.477
19	66.2	.998405	1.0340	1.0357	72.90	16.477
20	68.0	.998203	1.0087	1.0105	72.75	17.535
21	69.8	.997992	0.9843	0.9863	72.59	18.650
22	71.6	.997770	.9608	.9629	72.44	19.827
23	73.4	.997538	.9380	.9403	72.28	21.068
24	75.2	.997296	.9161	.9186	72.13	22.377
25	77.0	.997044	.8949	.8976	71.97	23.756
26	78.8	.996783	.8746	.8774	71.82	25.209
27	80.6	.996512	.8551	.8581	71.66	26.739
28	82.4	.996232	.8363	.8395	71.50	28.349
29	84.2	.995944	.8181	.8214	71.35	30.043
30	86.0	.995646	.8004	.8039	71.18	31.824
31	87.8	.995340	.7834	.7871	*71.02	33.695
32	89.6	.995025	.7670	.7708	*70.86	35.663
33	91.4	.994702	.7511	.7551	*70.70	37.729
34	93.2	.994371	.7357	.7399	*70.53	39.898
35	95.0	.99403	.7208	.7251	70.38	42.175
36	96.8	.99368	.7064	.7109	*70.21	44.563
37	98.6	.99333	.6925	.6971	*70.05	47.067
38	100.4	.99296	.6791	.6839	*69.88	49.692

Source: From Van del Leeden, F., *The Water Encyclopedia*, 2nd ed., Lewis Publishers, Chelsea, MI, 1990. With permission.

TABLE A2

Saturation Values of Dissolved Oxygen in Water Exposed to Water-Saturated Air Containing 20.9% Oxygen Under a Pressure of 760 mmHg[a]

Temperature (°C)	Dissolved oxygen (mg/L) — Chloride concentration in water (mg/L): 0	5000	10,000	Difference per 100 mg chloride	Temperature (°C)	Vapor pressure (mm)
0	14.6	13.8	13.0	0.017	0	5
1	14.2	13.4	12.6	0.016	1	5
2	13.8	13.1	12.3	0.015	2	5
3	13.5	12.7	12.0	0.015	3	6
4	13.1	12.4	11.7	0.014	4	6
5	12.8	12.1	11.4	0.014	5	7
6	12.5	11.8	11.1	0.014	6	7
7	12.2	11.5	10.9	0.013	7	8
8	11.9	11.2	10.6	0.013	8	8
9	11.6	11.0	10.4	0.012	9	9
10	11.3	10.7	10.1	0.012	10	9
11	11.1	10.5	9.9	0.011	11	10
12	10.8	10.3	9.7	0.011	12	11
13	10.6	10.1	9.5	0.011	13	11
14	10.4	9.9	9.3	0.010	14	12
15	10.2	9.7	9.1	0.010	15	13
16	10.0	9.5	9.0	0.010	16	14
17	9.7	9.3	8.8	0.010	17	15
18	9.5	9.1	8.6	0.009	18	16
19	9.4	8.9	8.5	0.009	19	17
20	9.2	8.7	8.3	0.009	20	18
21	9.0	8.6	8.1	0.009	21	19
22	8.8	8.4	8.0	0.008	22	20
23	8.7	8.3	7.9	0.008	23	21
24	8.5	8.1	7.7	0.008	24	22
25	8.4	8.0	7.6	0.008	25	24
26	8.2	7.8	7.4	0.008	26	25
27	8.1	7.7	7.3	0.008	27	27
28	7.9	7.5	7.1	0.008	28	28
29	7.8	7.4	7.0	0.008	29	30
30	7.6	7.3	6.9	0.008	30	32

[a] Saturation at barometric pressures other than 760 mm (29.92 in.), C_s' is related to the corresponding tabulated values, C, by the equation:

Source: From Hammer, M.J., *Water and Wastewater Technology*, 2nd ed., Prentice Hall, Englewood Cliffs, NJ, 1986. With permission.

$$C_s' = C_s \frac{P - p}{760 - p}$$

where:

C_s' = Solubility at barometric pressure P and given temperature, mg/L
C_s = Saturation at given temperature from table, mg/L
P = Barometric pressure, mm
p = Pressure of saturated water vapor at temperature of the water selected from table, mm

New Notation / Previous IUPAC Form / CAS Version

Key to Chart: Atomic Number → 50; Symbol → Sn; 1993 Atomic Weight → 118.71; Oxidation States → +2 +4; Electron Configuration → 18 18 4

1 Group IA	2 IIA	3 IIIA IIIB	4 IVA IVB	5 VA VB	6 VIA VIB	7 VIIA VIIB	8 VIIIA VIII	9 VIIIA VIII	10 VIIIA VIII	11 IB	12 IIB	13 IIIB IIIA	14 IVB IVA	15 VB VA	16 VIB VIA	17 VIIB VIIA	18 VIIA	Shell
1 H +1 -1 1.00794 1																	2 He 0 4.0020602 2	K
3 Li +1 6.941 2-1	4 Be +2 9.012182 2-2											5 B +3 10.811 2-3	6 C +2 +4 -4 12.011 2-4	7 N +1 +2 +3 +4 +5 -1 -2 -3 14.00674 2-5	8 O -2 15.9994 2-6	9 F -1 18.9984032 2-7	10 Ne 0 20.1797 2-8	K-L
11 Na +1 22.989768 2-8-1	12 Mg +2 24.3050 2-8-2											13 Al +3 26.981539 2-8-3	14 Si +2 +4 -4 28.0855 2-8-4	15 P +3 +5 -3 30.973762 2-8-5	16 S +4 +6 -2 32.066 2-8-6	17 Cl +1 +5 +7 -1 35.4527 2-8-7	18 Ar 0 39.948 2-8-8	K-L-M
19 K +1 39.0983 -8-8-1	20 Ca +2 40.078 -8-8-2	21 Se +3 44.955910 -8-9-2	22 Ti +2 +3 +4 47.867 -8-10-2	23 V +2 +3 +4 +5 50.9415 -8-11-2	24 Cr +2 +3 +6 51.9961 -8-13-1	25 Mn +2 +3 +4 +7 54.93805 -8-13-2	26 Fe +2 +3 55.845 -8-13-2	27 Co +2 +3 58.93320 -8-15-2	28 Ni +2 +3 58.6934 -8-16-2	29 Cu +1 +2 63.546 -8-18-1	30 Zn +2 65.39 -8-18-2	31 Ga +3 69.723 -8-18-3	32 Ge +2 +4 72.61 -8-18-4	33 As +3 +5 -3 74.92159 -8-18-5	34 Se +4 +6 -2 78.96 -8-18-6	35 Br +1 +5 -1 79.904 -8-18-7	36 Kr 0 83.80 -8-18-8	-L-M-N
37 Rb +1 85.4678 -18-8-1	38 Sr +2 87.62 -18-8-2	39 Y +3 88.90585 -18-9-2	40 Zr +4 91.224 -18-10-2	41 Nb +3 +5 92.90638 -18-12-1	42 Mo +6 95.94 -18-13-1	43 Tc +4 +6 +7 (98) -18-13-2	44 Ru +3 101.07 -18-15-1	45 Rh +3 102.90550 -18-16-1	46 Pr +2 +3 106.42 -18-18-0	47 Ag +1 107.8682 -18-18-1	48 Cd +2 112.411 -18-18-2	49 In +3 114.818 -18-18-3	50 Sn +2 +4 118.710 -18-18 -4	51 Sb +3 +5 -3 121.760 -18-18-5	52 Te +4 +6 -2 121.760 -18-18-6	53 I +1 +5 +7 -1 126.90447 -18-18-7	54 Xe 0 131.29 -18-18-8	-M-N-O
55 Cs +1 132.90543 -18-8-1	56 Ba +2 137.327 -18-8-2	57* La +3 138.9055 -18-9-2	72 Hf +4 178.49 -32-10-2	73 Ta +5 180.9479 -32-11-2	74 W +6 183.84 -32-12-2	75 Re +4 +6 +7 186.207 -32-13-2	76 Os +3 +4 190.23 -32-14-2	77 Ir +3 +4 192.217 -32-15-2	78 Pt +2 +4 195.08 -32-18-1	79 Au +1 +3 196.96654 -32-18-1	80 Hg +1 +2 200.59 -32-18-2	81 Tl +1 +3 204.3833 -32-18-3	82 Pb +2 +4 207.2 -32-18-4	83 Bi +3 +5 208.98037 -32-18-5	84 Po +2 +4 (209) -32-18-6	85 At (210) -32-18-7	86 Rn 0 (222) -32-18-8	-N-O-P
87 Fr +1 (223) -18-8-1	88 Ra +2 226.025 -18-8-2	89** Ac +3 227.028 -18-9-2	104 Unq +4 (261) -32-10-2	105 Unp (262) -32-11-2	106 Unh (263) -32-12-2	107 Uns (262) -32-13-2	108 Uno (265) -32-14-2	109 Une (266) -32-15-2	110 Uun (269) -32-16-2									O P Q

															Shell
* Lanthanides	58 Ce +3 +4 140.115 -19-9-2	59 Pr +3 140.90765 -21-8-2	60 Nd +3 144.24 -22-8-2	61 Pm +3 (145) -23-8-2	62 Sm +2 +3 150.36 -24-8-2	63 Eu +2 +3 151.965 -25-8-2	64 Gd +3 157 .25 -25-9-2	65 Tb +3 158.92534 -27-8-2	66 Dy +3 162.50 -28-8-2	67 Ho +3 164.93032 -29-8-2	68 Er +3 167.26 -30-8-2	69 Tm +3 168.93421 -31-8-2	70 Yb +3 +5 173.04 -32-8-2	71 Lu +3 174.967 -32-9-2	N O P
** Actinides	90 Th +4 232.0381 -18-10-2	91 Pa +5 +4 231.03588 -20-9-2	92 U +3 +4 +5 +6 238.0289 -21-9-2	93 Np +3 +4 +5 +6 237.048 -22-9-2	94 Pu +3 +4 +5 +6 (244) -24-8-2	95 Am +3 +4 +5 +6 (243) -25-8-2	96 Cm +3 (247) -25-9-2	97 Bk +3 +4 (247) -27-8-2	98 Cf +3 (251) -28-8-2	99 Es +3 (252) -29-8-2	100 Fm +3 (257) -30-8-2	101 Md +2 +3 (258) -31-8-2	102 No +2 +3 (259) -32-8-2	103 Lr +3 (206) -32-9-2	O P Q

The new IUPAC format numbers the groups from 1 to 18. The previous IUPAC numbering system and the system used by Chemical Abstracts Service (CAS) are also shown. For radioactive elements that do not occur in nature, the mass number of the most stable isotape is given in parentheses.

References

1. G. J. Leigh, Editor, *Nomenclature of Inorganic Chemistry*, Blackwell Scientific Publications, Oxford, 1990.
2. *Chemical and Engineering News*, 63(5), 27, 1985.
3. Atomic Weights of the Elements, 1993, *Pure & Appl. Chem.*, 66, 2423, 1994.

Note: The larger and smaller labels reflect two different numbering schemes in common usage.

TABLE A3
The Periodic Table of the Elements

Source: From Manahan, S.E., *Fundamentals of Environmental Chemistry*, Lewis Publishers, Chelsea, MI, 1993. With permission.

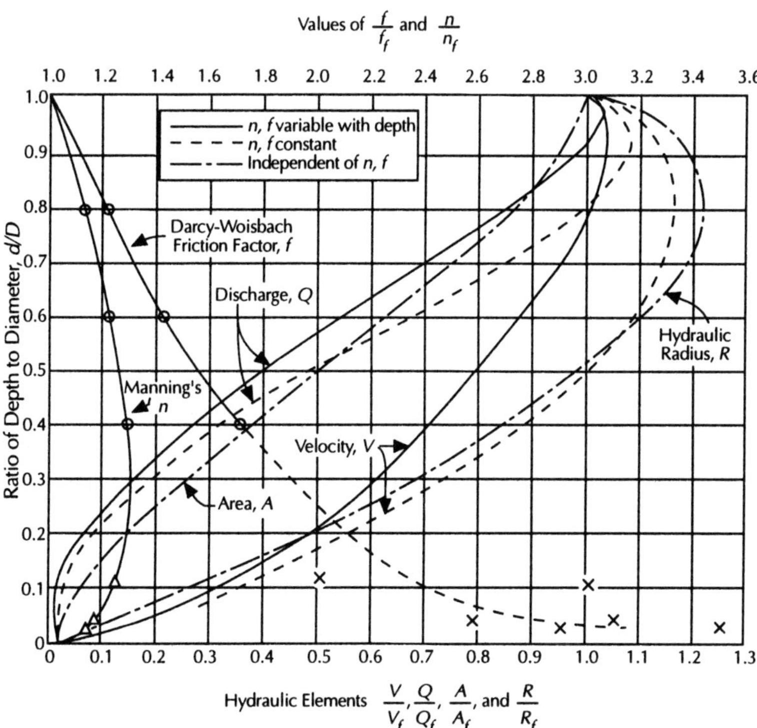

FIGURE A.4
Hydraulic elements graph for circular sewers. (From the Joint Task Force of the American Society of Civil Engineers and the Water Pollution Control Federation, *Gravity Sanitary Sewer Design and Construction*, ASCE Manuals and Reports on Engineereing Practice No. 60, ASCE, WPCF, New York, 1982. Reprinted by permission of ASCE)

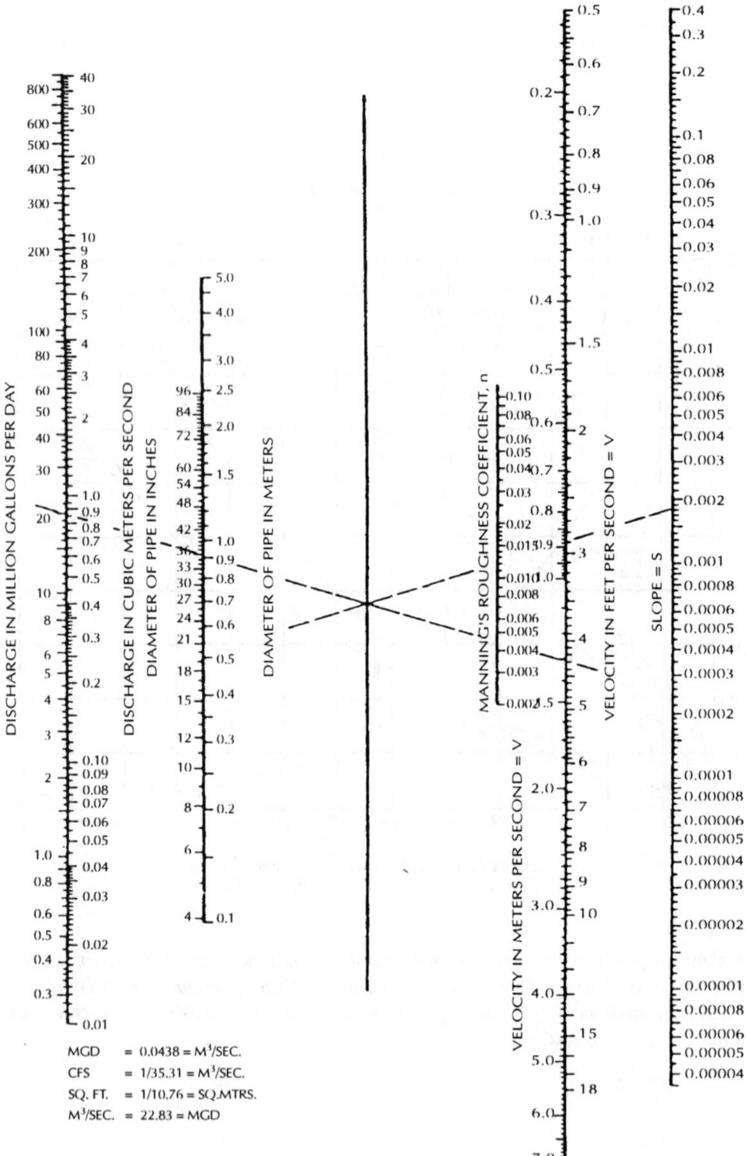

FIGURE A.5
Alignment chart for flow in pipes. (From *Gravity Sanitary Sewer Design and Construction*, ACSE Manuals and Reports on Engineering Practice, no. 60, American Society of Chemical Engineers, Water Pollution Control Federation, New York 1982. Reproduced by permission of the publisher American Society of Civil Engineers).

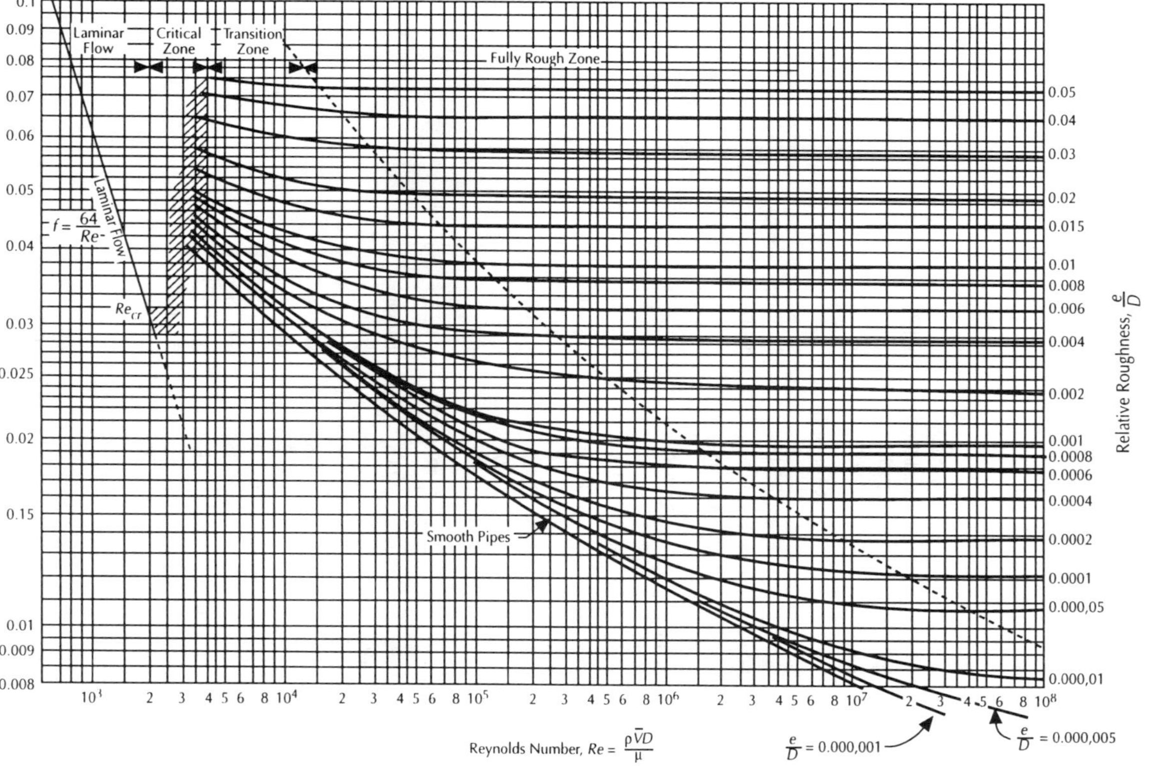

FIGURE A.6
Friction factor to fully developed flow in circular pipes. (From Fox, R.W. and McDonald, A.T., *Introduction to Fluid Mechanics*, John Wiley & Sons, New York, 1985. With permission. Moody, L.F., *Friction Factors for Pipe Flow*, Trans. *ASME*, 66 (8), 671, 1944. With permission.)

TABLE A7
Conversion Table

Multiply	by	to obtain
	Area	
acre	0.4047	ha
acre	43560	ft^2
acre	4047	ft^2
cm^2	0.155	in^2
ft^2	0.0929	m^2
hectare, ha	2.471	acre
ha	10^4	m^2
in^2	6.452cm^2	cm^2
km^2	0.3861	$mile^2$
m^2	10.76	ft^2
mm^2	0.00155	in^2
	Concentration	
mg/L	8.345	lb/million U.S. gal
ppm	1	mg/L
	Density	
g/cm^3	1000	kg/m^3
g/cm^3	1	kg/L
g/cm^3	62.43	lb/ft^3
g/cm^3	10.022	lb/gal (Br.)
g/cm^3	8.345	lb/gal (U.S.)
kg/m^3	0.001	g/cm^3
kg/m^3	0.001	kg/L
kg/m^3	0.6242	lb/ft^3
	Flow rate	
ft^3/s	448.8	gal/min
ft^3/s	28.32	L/s
ft^3/s	0.02832	m^3/s
ft^3/s	0.6462	M gal/d
gal/min	0.00223	ft^3/s
gal/min	0.0631	L/s
L/s	15.85	gal/min
M gal/d	1.547	ft^3/s
m^3/h	4.4	gal/min
m^3/s	35.31	ft^3/s
	Length	
ft	30.48	cm
in.	2.54	cm
km	0.6214	mile
km	3280.8	ft
m	3.281	ft
m	39.37	in
m	1.094	yd
mile	5280	ft
mile	1.6093	km
mm	0.03937	in.

TABLE A7 (CONTINUED)
Conversion Table

Multiply	by	to obtain
yard	0.914	m
	Mass	
g	$2.205*10^{-3}$	lb
kg	2.205	lb
lb	0.4536	kg
lb	16	oz
ton	2240	lb
tonne, t	1.102	ton (2000 lb)
	Power	
Btu	252	cal
Btu	778.2	ft-lb
Btu	$3.93*10^{4}$	Hp-h
Btu	1055	J
Btu	$2.93*10^{-4}$	kW-h
HP	0.7457	kW
	Pressure	
atm	33.93	ftH_2O
atm	29.92	inHg
atm	$1.033*10^{4}$	kg/m^2
atm	760	mmHg
atm	10.33	m water
atm	$1.013*10^{5}$	N/m^2
bar	10^5	N/m^2
cm water	98.06	N/m^2
inH_2O	1.8665	mmHg
inHg	0.49116	lb/in^2
inHg	25.4	mmHg
inHg	3386	N/m^2
kPa	0.145	$lb/in.^2$, psi
lb/in^2	0.0703	kg/cm^2
lb/in^2	6895	N/m^2
mmHg	13.595	kg/m^2
mmHg	0.01934	$lb/in.^2$
mmHg	133.3	N/m^2
mmHg	1	torr
torr	133.3	N/m^2
	Temperature	
Celsius	(9C/5) + 32	F
Fahrenheit (F)	5(F – 32)/9	Celsius (C)
Kelvin	C + 237.16	K
Rankine	F + 459.67	F
	Velocity	
cm/s	0.03281	ft/s
cm/s	0.6	m/min
m/s	196.8	ft/min
m/s	3.281	ft/s

TABLE A7 (CONTINUED)
Conversion Table

Multiply	by	to obtain
ft/min	0.508	cm/s
ft/s	30.48	cm/s
ft/s	1.097	km/h
mile/hr	1.609	km/h
poise (g/cm.sec)	0.1	N.s/m^2
	Viscosity	
centipoise	0.01	g/cm.s
centistoke	0.01	cm^2/s
stoke	10^{-4}	m^2/s
	Volume	
ft^3	6.229	gal (Br.)
ft^3	7.481	gal (U.S.)
ft^3	28.316	L
ft^3	0.02832	m^3
gal (Br.)	0.1605	ft^3
gal (U.S.)	0.1337	ft^3
gal (U.S.)	0.833	gal (Br.)
gal	3.785	L
in.^{3}	16.39	cm^3
L	0.03532	ft^3
L	0.22	gal (Br.)
L	0.2642	gal (U.S.)
L	0.001	m^3
m^3	35.314	ft^3
m^3	1000	L

References

1. Van del Leeden, F., *The Water Encyclopedia*, 2nd ed., Lewis Publishers, Chelsea, MI, 1990.
2. Hammer, M.J., *Water and Wastewataer Technology*, 2nd ed., Prentice Hall, Englewood Cliffs, NJ, 1986.
3. Manahan, S.E., *Fundamentals of Environmental Chemistry*, Lewis Publishers, Chelsea, MI, 1993.
4. Joint Task Force of the American Society of Civil Engineers and the Water Pollution Control Federation, *Gravity Sanitary Sewer Design and Construction*, ACSE Manuals and Reports on Engineering Practice, no. 60, ASCE, WPCF, New York, 1982.
5. Fox, R.W. and McDonald, A.T., *Introduction to Fluid Mechanics*, 3rd ed., John Wiley & Sons, New York, 1985.
6. Moody, L.F., Friction factors for pipe flow, *Trans. ASME*, 66 (8) 671, 1944.

References

Index

Index

B

C

D

E

F

J

K

L

M

N

O

P

T

Z